Heribert Kahmen · Wolfgang Faig

Surveying

W DE G

Walter de Gruyter · Berlin · New York · 1988

Authors
Professor Dr.-Ing. Heribert Kahmen
Institut für Landesvermessung
und Ingenieurgeodäsie
Technische Universität Wien
Vienna, Austria

Professor Dr.-Ing. Wolfgang Faig
Faculty of Engineering
University of New Brunswick
P.O. Box 4400
Fredericton, N.B., Canada
E3B 5A3

This book contains 474 illustrations and 23 tables

Printed on acid free paper
(ageing resistant-ph 7, neutral)

Deutsche Bibliothek Cataloging in Publication Data

Kahmen, Heribert:
Surveying / Heribert Kahmen ; Wolfgang Faig. – Berlin ; New York : de Gruyter, 1988
Einheitssacht.: Vermessungskunde 〈engl.〉
ISBN 3-11-008303-5
NE: Faig, Wolfgang:

Library of Congress Cataloging in Publication Data

Kahmen, Heribert, 1940–
Surveying.

Bibliography: p.
Includes index.
1. Surveying. I. Faig, Wolfgang. II. Title.
TA545.K24 1988 526.9 88-3899
ISBN 0-89925-022-X (U.S.)

 – Typesetting and Printing: Tutte GmbH, Salzweg-Passau – Binding: Lüderitz & Bauer GmbH, Berlin – Cover design: Rudolf Hübler, Berlin – Printed in Germany.

Preface

This book, entitled "Surveying," represents a translation and thorough revision of the German three-volume series "Vermessungskunde I, II, III," which are presently in the 16th, 14th, and 12th editions, respectively. This series is very popular and has the largest distribution of books on surveying within the German speaking population. The three volumes enjoy a long and illustrious tradition. Professor Dr. Ing. Paul Werkmeister, from the Technical University of Hannover, W.-Germany, was the initial author responsible for the early editions that appeared between 1910 and 1949. His successor at the Technical University of Hannover was Prof. Dr. Ing. Walter Grossmann who continued the work and authored the editions between 1950 and 1980. Afterwards, Prof. Dr. Ing. Heribert Kahmen, who succeded Prof. Grossmann in Hannover and subsequently, in 1986, moved on to the Technical University in Vienna, Austria, took over the responsibility for the series, rewriting and completely modernizing the books. The present English edition was written in cooperation with Prof. Dr. Ing. Wolfgang Faig from the University of New Brunswick in Fredericton, N.B., Canada.

The book has been written in such a way that it can serve as an introductory text for the uninitiated while having enough depth to serve the seasoned professional as well. The rapid developments in surveying instrumentation and methodology have led to significant changes and new approaches. This book should prove to be valuable for the continuing education of professionals in a wide spectrum of the geosciences and related fields, such as surveying, civil engineering, cartography, architecture, geography, planning, etc.

The authors acknowledge with thanks the helpful support of their colleagues at the universities of Hannover, Vienna, and New Brunswick, especially the contributions of Mr. H.J. Kramer, who provided the graphic design, and of Ms. W. Wells, whose editing skills in English proved to be invaluable for the completion of this volume.

November 1987 Heribert Kahmen Wolfgang Faig

Contents

1. Fundamentals

1.1 Introduction

Surveying deals primarily with geometric measurements on the earth's surface, the computation of derived quantities, e.g. coordinates, areas etc., and the representation of these numerical data in graphical form, such as in plans or maps. This requires a solid knowledge of instrumentation and measuring procedures, as well as of computation methods and accuracy evaluations.

A surveyor is not only charged with providing results derived from his measurements, but also has to give an indication of the quality and reliability of these. This requires a clear understanding of the functional and stochastic relationships between measured quantities and derived results, as well as a solid understanding of the external factors that influence the measurements. These may be global (e.g. curvature of the earth's surface, in fact the very definition of it, gravity acceleration, changes in earth magnetism, atmospheric refraction), regional (e.g. reference surfaces), or local (e.g. propagation characteristics of electromagnetic waves in air, weather influences, instrumental and personal errors). He has to be able to select equipment and procedures to meet certain standards and specifications, and be flexible enough to design his own when circumstances require it (e.g. on construction sites or in difficult terrain).

Whether maps or plans have to be prepared, boundaries defined and marked, engineering structures laid out and staked, or deformations monitored, it is a surveyor who is called upon to efficiently take measurements and to come up with results that meet the given specifications, thereby serving the society and supporting many professionals who use this information every day.

1.2 Reference Surfaces

In order to be able to position points in planimetry or height, a reference surface is necessary. For this, a level surface is most suitable, because the vertical axes of most surveying instruments are placed in the direction of gravity with the aid of level bubbles. Such a level surface would be normal to the gravity at each point. On the earth, a natural surface of this type would be the ocean surface, if it were in a mean position, free of other external influences, such as tides, currents, wind, etc. Such a surface, which can be defined uniquely for the whole earth, is considered a mathematical figure of the earth, and denoted as the *"geoid"* based on the Greek word for earth (Fig. 1.2.1).

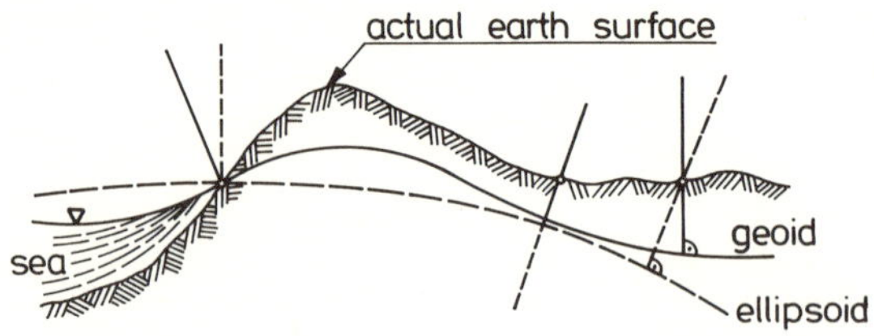

Fig. 1.2.1 Geoid, ellipsoid, actual earth surface

The surface of the sea is formed according to the gravity. The gravity has irregularities due to mass distribution inside the earth, which causes the geoid to be an irregular surface. The geoid does, however, resemble a *rotational ellipsoid* and hardly deviates by more than 80 m from it. By international agreement in 1967, its equatorial semi-diameter has been set at 6 378 160 m with a shortening factor of 1 : 298.25 for the rotational axis in respect to the equatorial axis. This means that the rotational axis is only 3% shorter than the equatorial axis.

If a section of the *earth's actual surface* is to be surveyed, all surface points are considered as if they were projected along the gravity vector onto the geoid. Then the area of the survey is defined as the area of the projection of the section onto the geoid, and the horizontal distance between two points is the shortest distance between the projected points as measured on the geoid. The elevation (above mean sea level) of a point is its separation from the geoid as measured along the plumb line, while the elevation difference between two points equals the difference in mean sea level elevations.

Due to the rather small discrepancies between the geoid and rotational ellipsoid, the latter can be used as a reference surface for planimetric surveys, especially for medium sized countries. This has the advantage that the computations can be referred to a mathematically defined surface. For smaller countries, provinces, or states, a sphere is chosen with its radius equalling the radius of the earth's curvature at the centre of the area. If the surveying area does not exceed 10 km in diameter, the tangent plane to the geoid is sufficient as a reference surface.

These simplifications are not permissible for elevation measurements for two reasons: The curvature of the earth's surface is such that a tangent plane to the earth, considered as either ellipsoid or sphere, would deviate from it by about 100 m at 35 km distance from the point of tangency. Secondly, differences between the ellipsoid and the geoid are significant for elevation measurements. Elevations are therefore always referred to the geoid – or mean sea level, as it is called by the practitioner.

1.3 Measuring Systems and Units

1.3.1 From the Archive Metre to the SI System

Based on a suggestion by the Paris Academy of Science, the French National Assembly decided in 1791 to introduce a uniform standard of length, called the *"metre"*, which represents one ten millionth of an earth meridian. The length of the metre was derived in subsequent years using several arc measurements. In order to permit reproduction at any time, a prototype was produced of platinum and stored at the French National Archive. This *"Archive Metre"* forms the basis of the metric system and is the reference to units of area, volume and weight as well as length.

In the following decades, several countries adopted the system. In 1875 these nations formed the *"International Metric Convention"* to spread international recognition of the metric system. All independent nations of the earth were invited to join the convention. The nations also agreed on the formation of an *International Bureau for Measure and Weight* in Breteuil near Paris. However, decisions on changes were left to meetings of delegates from participating nations, which were called *"General Conferences on Measure and Weight"* from then on.

As the first major project, the Bureau produced a new *metre prototype of platinum-iridium with an x-shaped cross section* culminating 10 years of experimentations. This prototype was designed to define the metre more exactly than the Archive Metre, and it was accepted by the first general conference in 1889 as the new international metre prototype. The metre was then defined as the separation of two line marks on the prototype of Breteuil at 0 °C. All nations who had joined the Convention received a copy of this prototype. In view of the continuing progress in physical science as well as the increasing accuracy requirements, in time this definition became insufficient.

Without altering the length of the metre, this definition was changed by decision of the 11th General Conference for Measure and Weight on 14 October 1960. Since then, the metre has been defined as 1 650 763.73 times the wavelength of the radiation emitted by the atoms of ^{86}Kr – an isotope of the inert gas crypton with mass 86 –when changing from state $5d$ to state $2p_{10}$. Under certain conditions, this radiation can be realized with a so-called "Engelhard-lamp", when placed inside a cold bath at 63 Kelvin.

Independent of the metric convention, which only included the unit metre, square metre, cubic metre and kilogram, the electromagnetic units, Volt, Ampere, Ohm and Watt, were introduced in the decades after 1875. The Italian physisist Giovanni Giorgi recognized, in 1901, that it is possible to form a coherent system of units using these electromagnetic units together with the units metre, kilogram and second. Such a system would require only four basic units if the definitions of the electromagnetic units were formulated somewhat differently. Subsequently, another 15 units were defined, including the Kelvin for the thermodynamic temperature and the Candela for light intensity. All these units have been reduced to (at present) 7 *base units* according to Giorgi's procedure.

This excellent system was accepted in 1954 by the 10th General Conference for Measure and Weight. The 11th General Conference (1960) named it *"Système Internationale"*, or in short SI.

In the following sections, rules are presented which affect surveying in one way or another.

1.3.2 Basic Rules of SI

The 7 basic units and their abbreviations are as follows:

for length: metre = m
for mass: kilogram = kg
for time: second = s
for electric current: Ampere = A
for light intensity: Candela = cd
for quantity of substance: Moll = mol

Other units can be derived from the basic ones by multiplication. If the factor 1 is used, coherent SI units are obtained, such as

for area: $1\ \mathrm{m}^2$
for velocity: $1\ \mathrm{m\ s}^{-1}$
for acceleration: $1\ \mathrm{m\ s}^{-2}$
for force: $1\ \mathrm{m\ kg\ s}^{-2}$, called 1 Newton (N)
for pressure: $1\ \mathrm{m}^{-1}\ \mathrm{kg\ s}^{-1} = 1\ \mathrm{N/m}^2$

Non-coherent units are obtained by multiplication with different factors as well as with powers of 10, for example:

area $10^2\ \mathrm{m}^2 = 1\ \mathrm{a}$
acceleration $10^{-2}\ \mathrm{s}^{-2} = 1\ \mathrm{Gal}$
force $10^{-5}\ \mathrm{m\ kg\ s}^{-2} = 1\ \mathrm{dyn}$
pressure $10^5\ \mathrm{m}^{-1}\ \mathrm{kg\ s}^{-2} = 10^5\ \mathrm{N/m}^2 = 1\ \mathrm{bar}$

and

force $9.80665\ \mathrm{m\ kg\ s}^{-2} = 9.80665\ \mathrm{N} = 1\ \mathrm{kp}$
pressure $101\,325\ \mathrm{m}^{-1}\ \mathrm{kg\ s}^{-2} = 101\,325\ \mathrm{N/m}^2 = 1\ \mathrm{atm}$.

The utilization of prefixes leads to decimal multiples and fractions of these units. These are denoted as:

	term	prefix		term	prefix
10^1	deca	da	10^{-1}	deci	d
10^2	hecto	h	10^{-2}	centi	c
10^3	kilo	k	10^{-3}	milli	m
10^6	mega	M	10^{-6}	micro	μ
10^9	giga	G	10^{-9}	nano	n
10^{12}	tera	T	10^{-12}	pico	p

1.3.3 Basic SI Units

1.3.3.1 The Units for Length, Area and Volume

These are based on the form accepted by the 1875 metric convention, and have since been expanded only upwards and downwards by several powers of ten.

a) The SI unit for length is the base unit metre (m). Using the prefixes from section 1.3.2, the following units are obtained:

1 decametre = 10^1 m = 1 dam	1 decimetre = 10^{-1} m = 1 dm
1 hectometre = 10^2 m = 1 hm	1 centimetre = 10^{-2} m = 1 cm
1 kilometre = 10^3 m = 1 km	1 millimetre = 10^{-3} m = 1 mm
1 megametre = 10^6 m = 1 Mm	1 micrometre = 10^{-6} m = 1 μm
1 gigametre = 10^9 m = 1 Gm	1 nanometre = 10^{-9} m = 1 nm
1 terametre = 10^{12} m = 1 Tm	1 picometre = 10^{-12} m = 1 pm

b) The SI unit for area is the derived unit square metre (m^2). Using the mentioned prefixes one obtains:

1 are = 10^2 m^2 = 1 a	1 square decimetre = 10^{-2} m^2 = 1 dm^2
1 hectare = 10^4 m^2 = 1 ha	1 square centimetre = 10^{-4} m^2 = 1 cm^2
1 square kilometre = 10^6 m^2 = 1 km^2	1 square millimetre = 10^{-6} m^2 = 1 mm^2
etc.	etc.

c) The SI unit for volume is the derived unit cubic metre (m^3). With the respective prefixes, dm^3, cm^3, mm^3 have been derived from it.

One litre has been retained as a special denotation for 1 cubic decimetre. The formal connection between the litre and the unit for mass has been discontinued (1 L equals the volume of 1 kg pure water at maximum density under 1 atm pressure).

1.3.3.2 SI Units for Plane Angles

There are three systems in use for angular units, namely: sexagesimal graduation, centesimal graduation, and radiants (arc definition). The first two are as follows:

– *sexagesimal graduation:*

$$1 \text{ full circle} = 360^\circ \text{ (degrees)}$$
$$1^\circ = 60' \quad \text{(minutes)}$$
$$1' = 60'' \quad \text{(seconds)}$$

– *centesimal graduation:*

1 full circle = 400 gon
1 gon = 100 cgon (centigon)
1 cgon = 10 mgon (milligon)

In SI, the sexagesimal graduation has been retained with its units degree, minute and second because of its close relationship to astronomy and to the world grid.

The arc of an angle is given by the ratio between the arc b as bisected by the sides of the angle α with vertex in its centre, and the radius r of the circle (Fig. 1.3.1). The unit of the arc is the angle for which this ratio equals 1, i.e. for $b = r$. This angle is called the *"radian"*, because it is obtained by the length of the radius on the circumference of the circle. The arc of the full circle is therefore 2π, while it is $\pi/2$ for a right angle.

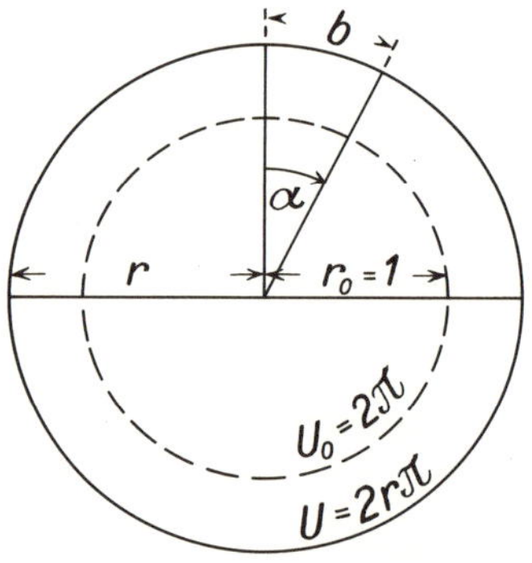

Fig. 1.3.1 Definition of the arc of an angle

The arc is the quotient of two lengths, which renders the radian as a derived SI unit. Specifically, one radian equals the plane angle which, as centre-angle of a circle with radius 1 m, cuts an arc of 1 m length from the circumference. This is illustrated in Fig. 1.3.2 (unit circle), where the centre-angle α, the corresponding arc b, and the radius r have been indexed with zero (0).

To simplify procedures for the practitioners, the following units have been derived using radians:

$$1 \text{ full circle} = 2\,\pi \,\text{rad} = 360^\circ = 400 \text{ gon}$$
$$1 \text{ right angle} = \pi/2 \text{ rad} = 90^\circ = 100 \text{ gon}$$

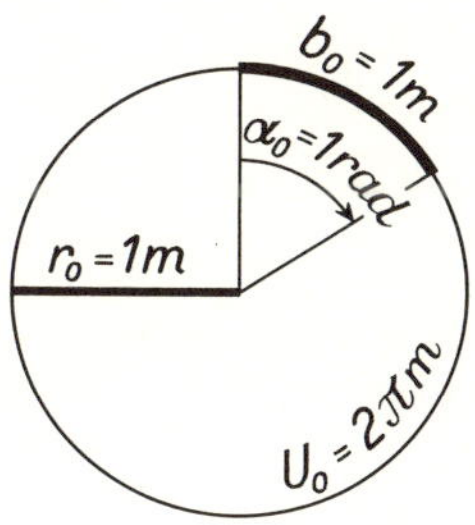

Fig. 1.3.2 The arc for a unit circle

$$1 \text{ degree} = \frac{\pi}{180} \text{rad} = 1°$$

$$1 \text{ minute} = \frac{\pi}{180 \cdot 60} \text{rad} = 1'$$

$$1 \text{ sec} = \frac{\pi}{180 \cdot 60^2} \text{rad} = 1''$$

$$1 \text{ gon} = \frac{\pi}{200} \text{rad} = 1 \text{ gon}$$

$$1 \text{ centigon} = \frac{\pi}{200 \cdot 10^2} \text{rad} = 1 \text{ cgon}$$

$$1 \text{ milligon} = \frac{\pi}{200 \cdot 10^3} \text{rad} = 1 \text{ mgon} .$$

1.3.3.3 Special Surveying Units

In surveying, the reciprocals of $\pi/180°$, or $\pi/200$, gon are used so often that the symbol ϱ has been introduced as:

$$180/\pi = \varrho^{(0)} \quad \text{or} \quad 200/\pi = \varrho^{(\text{gon})} .$$

A common application follows from the ratio (see Fig. 1.3.1):

$$\alpha : b = 360° : 2\pi .$$

For the unit circle (Fig. 1.3.2), b_0 and r_0 are equal to 1, and α equals 1 rad, then

$$\alpha^{(0)} = \frac{180°}{\pi} \cdot \varrho^{(0)} \quad \text{or} \quad \alpha(\text{gon}) = \frac{200 \text{ gon}}{\pi} \varrho(\text{gon}) .$$

The angular values of ϱ in the sexagesimal and centesimal systems are therefore equal to the radian in the respective system, namely

$$\varrho^{(0)} = 57.295779\ldots \qquad \varrho^{(\mathrm{gon})} = 63.661977$$
$$\varrho^{(\prime)} = 3437.7467\ldots \qquad \varrho^{(\mathrm{cgon})} = 6366.1977$$
$$\varrho^{(\prime\prime)} = 206264.8\ldots \qquad \varrho^{(\mathrm{mgon})} = 63661.977$$

A second application based on Fig. 1.3.1 uses

$$b : 2\pi r = \alpha : 360^\circ .$$

If both sides are multiplied by 2π, the common formula

$$b : r = \alpha : \varrho$$

is obtained, where b and r, or α and ϱ, have to be in corresponding units.

Example: A 150 m long line has to be rotated by an angle of 10′. What is the linear distance b by which the free end of the line is shifted sideways?

$$b[\mathrm{m}] = r[\mathrm{m}] \cdot \frac{\alpha'}{\varrho'} = \frac{150 \cdot 10}{3438} = 0.436\ \mathrm{m}\,.$$

A third application lies in the transfer from one angular system to the other. Although there are numerous tables, the following identities can easily be used with an electronic calculator:

$$1^\circ = 10/9\ \mathrm{gon} = 1.111\ldots\ \mathrm{gon} \qquad 1\ \mathrm{gon} = 0.9^\circ$$
$$1' = 1.851\,851\,85\ldots\ \mathrm{cgon} \qquad 1\ \mathrm{cgon} = 0.54'$$
$$1'' = 0.308\,641\,975\,308\ldots\ \mathrm{mgon} \qquad 1\ \mathrm{mgon} = 3.24''$$
$$1' \approx 2\ \mathrm{cgon}; \quad 1'' \approx 0.3\ \mathrm{mgon} = 3 \cdot 10^{-4}\ \mathrm{gon}.$$

In civil engineering construction, the elevation differences are often given in percent of the distance or the ratio $1 : n$, rather than with the slope angle α. In round figures, the following correspondence can be established:

%	1	2	3	5	10	25	50	100
$\alpha^{(\mathrm{gon})}$	0.6	1.3	1.9	3.2	6.3	16.1	29.5	50
$\alpha^{(0)}$	0.5	1.2	1.7	2.9	5.7	14.5	26.5	45
$1 : n$	1:100	1:50	$1:33^1/_3$	1:20	1:10	1:4	1:2	1:1

1.3.4 Less Frequently Used SI Units

1.3.4.1 The (Derived) SI Unit for Pressure or Mechanical Tension

This SI unit which is used in barometric surveys, is the Pascal (Pa). 1 Pascal equals the pressure uniformly acting on an area of 1 m^2 when the force of 1 Newton is applied normal to it. 10^5 Pa are equal to 1 bar (section 1.3.2), and 10^2 Pa are equal to 1 millibar (mbar).

1.3.4.2 Temperature

The base unit for (thermodynamic) temperature (T) is the Kelvin. It is defined as the 273.15th part of the thermodynamic temperature of the triple point of water (freezing point). This unit also holds for temperature differences. The specific difference of a thermodynamic temperature T from the temperature $T_0 = 273.15$ K is denoted as Celsius temperature (t), which means:

$$t = T - T_0 = T - 273.15\,\mathrm{K}\,.$$

Celsius temperatures are denoted as degrees Celsius (°C). The difference Δt of two Celsius temperatures $t_1 = T_1 - T_0$ and $t_2 = T_2 - T_0$ is

$$\Delta t = t_1 - t_2 = T_1 - T_2 = \Delta T\,.$$

Such a difference is not referred to the dynamic temperature T_0 any longer, and thus is not a Celsius temperature according to the definition in the previous equations.

1.3.4.3 Time

The SI unit for time is the second(s). With it, the following derived units are obtained.

1 minute (min) = 60 s
1 hour (h) = 60 min = 3600 s
1 day (d) = 24 h = 86400 s

Prefixes as mentioned in section 1.3.2 are not used with time units.

1.3.4.4 Frequency

The derived unit for frequency is the Hertz (Hz), often denoted as cycle. 1 Hertz equals the frequency of an oscillation with a 1 s period. The prefixes listed in section 1.3.2 are again applicable.

In addition to astronomy, the units of time and frequency are used in surveying in connection with gyro instruments and electronic distance meters.

1.4 Error Theory

1.4.1 The Scope of Error Theory

Depending on their respective purposes, geodetic measurements have to be performed to a certain accuracy and reliability. Because of the shortcomings of measuring instruments and human senses, it is impossible to obtain truly error free measurements. As a rule, the measurements are therefore repeated several times and, if possible, supported by additional measurements. For example, one measures the hypotenuse in addition to the sides of a rectangular triangle, or one observes the third angle of a triangle in addition to the two that are needed.

The evaluation of the measurements leads to the following tasks:

1) to derive the most probable mean value of the desired unknown quantity;

2) to provide a figure for the accuracy or the dispersion of a single measurement;

3) to estimate the accuracy or dispersion of the mean and of its confidence region.

1.4.2 Types of Errors

In reference to their occurrence, errors are divided into blunders, systematic, and random errors.

Blunders (or gross errors) are caused by mistakes in reading measuring instruments, misidentification of targets and similar occurrences. They can be spotted by check measurements and then eliminated.

Systematic Errors influence the results always in the same sense. They are caused by poor calibration or one sided use of measuring instruments as well as by external influences on an instrument or measuring object, such as temperature, pressure, etc. These errors can be eliminated to a large extent for all standard tasks by instrument calibration and by proper selection of the measuring procedure as well as by mathematical compensation for external influences.

Random Errors are all the remaining unknown "elementary errors" after elimination of blunders and systematic errors. They are caused by limitations of the human senses and of the measuring instruments, by uncontrollable changes of the enviroment or of the object to be measured. Random errors will equally often be positive and negative. In mathematical statistics, they are considered as *independent stochastic variables*. In spite of their apparent irregular behaviour, they are accidental and thus have some probability attached to them.

Fig. 1.4.1 represents the distribution of true errors ε_i (section 1.4.3.2) which were obtained from 160 observations of the same angle. According to their size, the ε_i are sorted into groups of 1″ as shown on the abscissa. The height of the rectangles plotted for each group is proportional to the number of errors within the respective group. The resulting histogram shows that the frequency with which an error ε occurs is a function of its magnitude. This phenomenon has been described by C. F. Gauss in his law of errors

$$\varphi(\varepsilon) = \frac{h}{\sqrt{\pi}} e^{-h^2 \varepsilon^2} \tag{1.4.1}$$

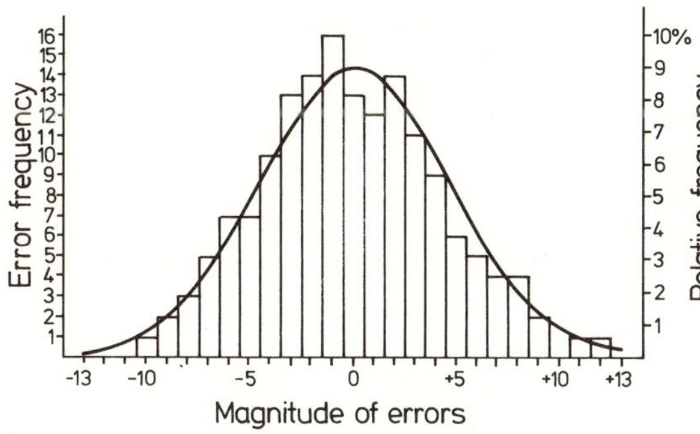

Fig. 1.4.1 Histogram for random errors

where $\varphi(\varepsilon)$ characterizes the relative frequency (i.e. in percent), e denotes the exponential function, which is the base of natural logarithms, and h represents a constant, related to the measuring accuracy.

The theoretical error distribution curve expected from this law is plotted in Fig. 1.4.1 and corresponds quite well with the histogram derived from the measurements. This holds for all larger series of observations which include primarily random errors. In statistical terms, such series correspond to the "*normal distribution*". These measurement discrepancies, caused primarily by random errors, can be rigorously adjusted using the "least-squares method", proposed by C. F. Gauss.

1.4.3 Means and Measures of Dispersion

1.4.3.1 Means

The single results l_i, which would be obtained if a quantity were to be determined by an infinite number of equally precise independent measurements ($n \to \infty$) with random errors, would oscillate around a *mean value* μ. This mean is also called the *mathematical expectation* or the *"true" value of the quantity*. Since usually only a limited number of measurements is available (a random sample of size n), the *sample mean (arithmetic mean)* is used to estimate the true value.

$$\hat{x} = \frac{1}{n}(l_1 + l_2 + \ldots + l_n) = \frac{1}{n} \sum_{i=1}^{n} l_i \,. \tag{1.4.2}$$

The *residuals* or *corrections* which remain after formation of the arithmetic mean

$$v_1 = +\hat{x} - l_1; \quad v_2 = +\hat{x} - l_2; \quad \ldots \quad v_n = +\hat{x} - l_n$$

are such that the sum of their squares (Σv^2) reaches a minimum,

$$\Sigma v^2 = v_1^2 + v_2^2 + \ldots + v_n^2 = \min \,. \tag{1.4.3}$$

This also is the basic requirement for the method of least squares, where the arithmetic mean represents a special case.

1.4.3.2 Measures of Dispersion

In order to obtain a measure of the dispersion of a single measurement, the deviations of the observations l_i from the true value μ are considered. For $n \rightarrow \infty$ this leads to the so-called *"true errors"*

$$\varepsilon_1 = \mu - l_1; \ \varepsilon_2 = \mu - l_2; \ \ldots \ \varepsilon_n = \mu - l_n .$$

The measure of the accuracy of a single observation l_i is then defined by the *"theoretical standard deviation"*

$$\sigma = \pm\sqrt{\frac{\Sigma \varepsilon^2}{n}} \quad \text{for} \quad n \rightarrow \infty . \tag{1.4.4}$$

Usually μ and ε_i are unknown. Then, the sample mean $\hat{x}$ and the residuals v_i (section 1.4.3.1) have to be used instead. Based on statistical estimation, the *"sample standard deviation of a single observation"* is then obtained as an approximation or estimation of σ as:

$$s = \pm\sqrt{\frac{\Sigma v^2}{n-1}} . \tag{1.4.5}$$

The *sample standard deviation of the mean of n observations* (section 1.4.4.1) becomes

$$s_x = \pm\frac{s}{\sqrt{n}} . \tag{1.4.6}$$

1.4.4 The Law of Error Propagation

Often the standard deviation of a function of measured quantities is required in addition to the standard deviation of a single observation. This can be obtained by applying the law of error propagation.

1.4.4.1 Linear Functions

The standard deviation σ_x of the function

$$x = l_1 + l_2 \tag{1.4.7}$$

is to be found, using the given observations l_1 and l_2 with their standard deviations σ_1 and σ_2. In order to compute σ_x, one has to utilize the definition given in equation

(1.4.4) with the assumption that the values l_i in equation (1.4.7) are the means of v original observations with true errors $\varepsilon'_1, \varepsilon''_1 \ldots \varepsilon^{(v)}_1$ or $\varepsilon'_2, \varepsilon''_2 \ldots \varepsilon^{(v)}_2$ where v is infinitely large. This represents v equations of the form $\varepsilon_x = \varepsilon_1 + \varepsilon_2$. By squaring, adding, and dividing by v, one obtains

$$\frac{\Sigma\varepsilon_x^2}{v} = \frac{\Sigma\varepsilon_1^2}{v} + \frac{\Sigma\varepsilon_2^2}{v} + \frac{2\Sigma\varepsilon_1\varepsilon_2}{v}.$$

The first three expressions represent σ_x^2, σ_1^2, σ_2^2 (according to equation (1.4.4)). The last term contains the sum of the mixed products. Since, however, only random errors are considered, i.e. independent stochastic variables, which on the average are as often negative as positive, they practically cancel each other when adding them up. This, together with the division by v, which is very large, indicates that this term approaches zero. Therefore, we can write:

$$\sigma_x^2 = \sigma_1^2 + \sigma_2^2 .$$

If one replaces σ by its estimator s, similar to the transition from equations (1.4.4) and (1.4.5) the following rule is obtained for estimating the standard deviation of the sum in equation (1.4.7)

$$s_x^2 = s_1^2 + s_2^2 . \tag{1.4.8}$$

Applying the same principle to the following function

$$x = \alpha_1 l_1 + \alpha_2 l_2 + \ldots + \alpha_n l_n$$

leads to

$$s_x^2 = \alpha_1^2 s_1^2 + \alpha_2^2 s_2^2 + \ldots + \alpha_n^2 s_n^2. \tag{1.4.9}$$

The following special cases should be noted:

a) For $s_1 = s_2 + \ldots + s_n = s$

$$s_n^2 = s^2 \Sigma\alpha^2 . \tag{1.4.10}$$

b) If in addition all α_i are either $+1$ or -1, then

$$s_x^2 = ns^2 \quad \text{or} \quad s_x = s\sqrt{n} . \tag{1.4.11}$$

This case occurs when levelling or measuring distances with tapes. The rule is expressed as:

If several single observations of equal accuracy are combined to a sum or difference, then the standard deviation of the result increases with the square root of the number of single observations.

c) If equation (1.4.2) is written as

$$\hat{x} = \frac{l_1}{n} + \frac{l_2}{n} + \ldots + \frac{l_n}{n}$$

and all s_i equal s, then the following value of s_x is obtained according to equation (1.4.9)

$$s_{\bar{x}}^2 = \frac{s^2}{n^2} + \frac{s^2}{n^2} + \ldots + \frac{s^2}{n^2} \quad \text{or} \quad s_{\bar{x}} = \frac{s}{\sqrt{n}}$$

which reaffirms the value given in equation (1.4.6). This rule is expressed as:

If the same object is measured n times with the same accuracy, then the standard deviation of the arithmetic mean is reduced by the square root of the number of repetitions.

The reader is advised to clearly separate cases b) and c). See also examples 1 and 2 in section 1.4.4.2.

1.4.4.2 Nonlinear Functions

These functions are usually linearized with the aid of a *Taylor series expansion*. Then equation (1.4.9) can again be applied. The law of error propagation in its most general form is therefore:

for $x = f(l_1, l_2 \ldots l_n)$

$$\sigma_x^2 = \left(\frac{\partial f}{\partial l_1}\right)^2 \sigma_1^2 + \left(\frac{\partial f}{\partial l_2}\right)^2 \sigma_2^2 + \ldots + \left(\frac{\partial f}{\partial l_n}\right)^2 \sigma_n^2 . \qquad (1.4.12)$$

Examples

1) When measuring a distance with a 20 m tape, the standard deviation is 2 mm per tape length. What is the standard deviation of a distance of 400 m?

According to equation (1.4.11)

$$s_{400} = 2\sqrt{20}\ \text{mm} = 9\ \text{mm} .$$

2) A distance of 160 m has been measured four times independently. Each measurement has a standard deviation of 12 mm. What is the standard deviation of the arithmetic mean of the four measurements?

According to equation (1.4.6)

$$s_x = \frac{12}{\sqrt{4}} = 6\ \text{mm} .$$

3) Given is a quantity $a = 87.46$. What is the standard deviation of $x = \log a$ when $s_a = 0.04$?

According to equation (1.4.12)

$$s_x = \frac{d\log a}{da} s_a = \frac{\text{Mod}}{a} s_a = \frac{0.4343}{87.46} 0.04 = 0.00020 .$$

4) The sides of a rectangle and their standard deviations are $a = 39.12$ m, $b = 71.38$ m, $s_a = 0.02$ m, and $s_b = 0.04$ m. What is the area of the rectangle and its standard deviation?

According to equation (1.4.12) with $A = ab$:

$$s_A^2 = (b s_a)^2 + (a s_b)^2 .$$

This leads to $A = 2792\ \text{m}^2$ and $s_A = 2.1\ \text{m}^2$.

5) In a triangle ABC, the sides b and c and the angle α were measured. The side a and its standard deviation are to be computed, using the following values: $b = 221.41$ m, $c = 166.14$ m, $\alpha = 48°47'24''$, and $s_b = 0.06$ m, $s_c = 0.05$ m, and $s_\alpha = 13''$.

With $a = \sqrt{b^2 + c^2 - 2bc\cos\alpha}$, we obtain according to equation (1.4.12)

$$s_a^2 = \left(\frac{b - c\cos\alpha}{a} s_b\right)^2 + \left(\frac{c - b\cos\alpha}{a} s_s^2\right) + \left(\frac{bc\sin\alpha}{a}\frac{s_\alpha}{\varrho}\right)^2$$

or numerically: $a = 167.80$ m and $s_a = 0.04$ m.

1.4.5 Adjustment of Direct Observations with Equal Accuracy

If a quantity has been measured several times, the measurements have to be adjusted according to equations (1.4.2), (1.4.5) and (1.4.6). In order to be able to compute with small figures, an approximate value x_0 is introduced for the unknown x, such that $x = x_0 + \delta x$. If one denotes the measured quantities as L_i instead of l_i, the following basic equation is obtained:

$$L_i + v_i = x_0 + \delta x . \tag{1.4.13}$$

Rearrangement leads to the following *observation equation*:

$$v_i = \delta x - (L_i - x_0) = \delta x - l_i , \tag{1.4.14}$$

in which the l_i have a new meaning.

This leads to the following computational scheme

1) absolute terms $\quad -l_i = -(L_i - x_0)$

2) mean $\quad \delta\hat{x} = \frac{1}{n}\Sigma l_i$

3) residuals $\quad v_i = \delta\hat{x} - l_i$

4) Σv check $\quad \Sigma v = 0$

5) Σv^2 checks $\quad \Sigma v^2 = \Sigma l^2 - \Sigma l\delta\hat{x} = \Sigma l^2 - \frac{(\Sigma l)^2}{n}$

6) standard deviation of a single observation $s = \sqrt{\frac{\Sigma v^2}{n-1}}$

7) standard deviation of mean $s_{\hat{x}} = \frac{s}{\sqrt{n}}$

As a final check, the results are substituted into equation (1.4.13).

Example

In order to determine an angle, a number of observations were made and entered in column 1 in the table below. As an approximate value $x_0 = 52.350°$ was used.

L_i	$l_i = L_i - x_o$	$v_i = \delta\hat{x} - l_i$ +	$v_i = \delta\hat{x} - l_i$ −	v^2	l^2
1	2	3	3	4	5
52.356	$+0.6 \cdot 10^{-2}$		$0.53 \cdot 10^{-2}$	$0.28 \cdot 10^{-4}$	$0.36 \cdot 10^{-4}$
52.348	$-0.2 \cdot 10^{-2}$	$0.27 \cdot 10^{-2}$		$0.07 \cdot 10^{-4}$	$0.04 \cdot 10^{-4}$
52.346	$-0.4 \cdot 10^{-2}$	$0.47 \cdot 10^{-2}$		$0.22 \cdot 10^{-4}$	$0.16 \cdot 10^{-4}$
52.347	$-0.3 \cdot 10^{-2}$	$0.37 \cdot 10^{-2}$		$0.14 \cdot 10^{-4}$	$0.09 \cdot 10^{-4}$
52.352	$+0.2 \cdot 10^{-2}$		$0.13 \cdot 10^{-2}$	$0.01 \cdot 10^{-4}$	$0.04 \cdot 10^{-4}$
52.355	$+0.5 \cdot 10^{-2}$		$0.43 \cdot 10^{-2}$	$0.18 \cdot 10^{-4}$	$0.25 \cdot 10^{-4}$
	$+0.4 \cdot 10^{-2}$	$1.11 \cdot 10^{-2}$	$1.09 \cdot 10^{-2}$	$0.90 \cdot 10^{-4}$	$0.94 \cdot 10^{-4}$

$\hat{x} = +\frac{0.4}{6} = 0.07° \cdot 10^{-2}; \quad \Sigma v = 0.02 \cdot 10^{-2}$

(theoretical value: zero).

$$v^2 = 0.94 \cdot 10^{-4} - 0.4 \cdot 10^{-2}; \hat{x} = 0.94 \cdot 10^{-4} - \frac{0.4^2 \cdot 10^{-4}}{6} = 0.91 \cdot 10^{-4}$$

(directly obtained: $0.90 \cdot 10^{-4}$).

Adjustment results:

$$x = x_0 + \delta\hat{x} = 52.351°$$

$$s = \pm\sqrt{\frac{0.90 \cdot 10^{-4}}{6-1}} = 0.004°; \quad s_{\hat{x}} = \frac{0.004}{\sqrt{6}} = 0.0013°.$$

1.4.6 Adjustment of Direct Observations of Different Accuracy

1.4.6.1 Introduction of Weights

If a measured quantity has been observed several times with different accuracies, these accuracy relations (*"weights"* in terms of adjustment calculus) have to be considered when computing the mean. An observation is given the weight p if it has the same standard deviation as the arithmetic mean of p real or fictitious standard observations with unit weight. For the measurements $L_1, L_2, \ldots, L_n$ with weights $p_1, p_2, \ldots, p_n$ follows from equation (1.4.6):

$$p_1 : p_2 : \ldots : p_n = \frac{\sigma^2}{\sigma_1^2} : \frac{\sigma^2}{\sigma_2^2} : \ldots : \frac{\sigma^2}{\sigma_n^2} . \tag{1.4.15}$$

This means that the weights are inversely proportional to the squares of the standard deviations. The standard deviation of a measurement with weight 1 is called the *standard error of unit weight* and is usually denoted as σ_0.

In order to compute the weight p_i of an observation L_i with a standard deviation σ_i the following equation is used:

$$p_i = \frac{\sigma_0^2}{\sigma_i^2} . \tag{1.4.16}$$

The unit weight is chosen such that the individual weights deviate as little as possible from 1. The following weights are often used for unity:

Weight of a distance of 100 m, of a levelling loop of 1 km, or of an angle observed in direct and reverse.

With the values derived from σ_i^2 in equation (1.4.16) the *law of propagation of weights* follows directly from the law of error propagation (1.4.9):

$$\frac{1}{p_x} = \frac{\alpha_1^2}{p_1} + \frac{\alpha_2^2}{p_2} + \ldots + \frac{\alpha_n^2}{p_n} . \tag{1.4.17}$$

If an observation L_i is multiplied by the square root of its weight, then, according to equation (1.4.17),

$$L_i \sqrt{p_i} \text{ has weight } 1 . \tag{1.4.18}$$

This expression is referred to as the *standardized or normed variable.*

Example:

The angles α and β are observed in a triangle, with

$s_\alpha = 6''$ and $s_\beta = 4''$.

a) What is p_β if $p_\alpha = 1$?
b) What is p_γ for $\gamma = 180° - \alpha - \beta$?

to a): $p_\alpha : p_\beta = \frac{1}{6^2} : \frac{1}{4^2} = \frac{36}{36} : \frac{36}{16}$; $p_\beta = 2.2$

to b): $\frac{1}{p_\gamma} = \frac{1}{p_\alpha} + \frac{1}{p_\beta} = \frac{1}{1} + \frac{1}{2.2} = 1.45$; $p_\gamma = 0.7$.

1.4.6.2 The Weighted Mean

In order to obtain a mean from measurements with different weights, the minimum condition from equation (1.4.3) is replaced by

$$\Sigma v^2 p = \min . \tag{1.4.19}$$

The simple mean (equation 1.4.2) is then replaced by the weighted mean, as

$$\hat{x} = \frac{L_1 p_1 + L_2 p_2 + \ldots + L_n p_n}{p_1 + p_2 + \ldots + p_n} = x_0 + \frac{\Sigma l_p}{\Sigma_p} = x_0 + \delta \hat{x} . \tag{1.4.20}$$

According to equations (1.4.5) and (1.4.18) the standard deviations of unit weight and of weight p_i are then:

$$s_0 = \sqrt{\frac{\Sigma v^2 p}{n-1}} \quad \text{or} \quad s_i = \frac{s_0}{\sqrt{p_i}} . \tag{1.4.21}$$

The standard deviations of the weighted mean and its weight are:

$$s_{\hat{x}} = \frac{s_0}{\sqrt{\Sigma p}} \quad \text{or} \quad p_x = \Sigma p . \tag{1.4.22}$$

1.4.6.3 The Computational Path

The *computational path* corresponds to the one given in section 1.4.5. In addition, each observation equation has its own weight. Instead of items 2, 5, and 7 in section 1.4.5, equations (1.4.20) to (1.4.22) have to be applied, thus replacing the v^2 and l^2 columns by $v^2 p$ or $l^2 p$, and adding lp and vp columns in the numerical example. The Σv check and the Σv^2 check (section 1.4.5, items 4 and 5) now become

$$\Sigma vp = 0 , \tag{1.4.23}$$

$$\Sigma v^2 p = \Sigma l^2 p - \Sigma lp \, \delta \hat{x} = \Sigma l^2 p - \frac{\Sigma lp}{\Sigma p} . \tag{1.4.24}$$

Example

An angle was measured on four days, namely 8 times on day 1, 4 times on day 2, 12 times on day 3, and 8 times on day 4. The daily means were in order: 40.1714°, 40.1718°, 40.1721° and 40.1725°. Find the mean and its standard deviation. The value 40.17° was chosen as the initial approximation. This leads to (all values in 10^{-4} degrees!)

p	$l_i =$ $L_i - \alpha_0$	lp	$v_i =$ $\delta_{\hat{x}} - l_i$	vp +	vp −	$v^2 p$	$l^2 p$
1	2	3	4	5		6	7
2	14	28	+6	12		82	392
1	18	18	+2	2		4	324
3	21	63	−1		3	3	1323
2	25	50	−5		10	50	1250
8		159		14	13	129	3289

$$\delta\hat{x} = \frac{159}{8} = 19.88 \cdot 10^{-4}; \quad \Sigma vp = 1 \cdot 10^{-4}$$

$$\Sigma v^2 p = 3289 - 159 \cdot \delta\hat{x} = 3289 - \frac{159^2}{8} = 129 \quad \text{(check!)}$$

(Note: For the $\Sigma v^2 p$ check, $\delta\hat{x}$ has to be carried to 1 to 2 more place than actually necessary.)

Adjustment results:

$$x = 40.17 + 20 \cdot 10^{-4} = 40.1720^\circ$$

$$s_0 = \sqrt{\frac{129}{4-1}} = 7 \cdot 10^{-4} = 2.3''$$

$$s_{\hat{x}} = \frac{s_0}{\sqrt{\Sigma p}} = \frac{7}{\sqrt{8}} = 2.4 \cdot 10^{-4} = 0.8''.$$

1.4.7 Adjustment of Direct Observations with a Sum Condition

The adjusted values of several measurements L_i often have to fulfil a mathematical condition. For instance, the sum of the angles completing the horizon has to be 360°, or a levelling loop returning to its starting point has to close with zero. In order to adjust such measurements, the weighted mean provides the following approach.

If S is the theoretical value, and ΣL is the sum of the n measurements forming it, then the discrepancy $w = \Sigma L - S'$ is equally divided among the measurements, provided they have equal weight. The standard deviations of an original measurement L_i, or of the adjusted measurement x_i are then

$$s_i = \pm \frac{W}{\sqrt{n}} \quad \text{or} \quad s_{x_i} = \pm s_i \sqrt{1 - \frac{1}{n}}\,. \tag{1.4.25}$$

If the measurements have different weights, then the discrepancy is distributed proportionally to the reciprocal weights $1/p_i$.

The standard deviations of a measurement with unit weight and of an adjusted measurement x_i are then

$$s_i = \frac{W}{\sqrt{n}} \quad \text{or} \quad s_{x_i} = \frac{s_0}{\sqrt{p_i}} \sqrt{\left(1 - \frac{1}{p_i}\right)\left(\Sigma \frac{1}{p}\right)^{-1}} . \tag{1.4.26}$$

1.4.8 Computation of Standard Deviations Using Double Measurements

As a safety measure, n similar quantities (angles, distances, elevation differences, areas, etc.) are often observed twice with equal accuracy. Let L_i and ε_i be the observations and true errors of the first series, and L_i' and ε_i' of the second series. Then according to the definition, observation plus true error yield the true value of a quantity, therefore

$$L_i + \varepsilon_i = L_i' + \varepsilon_i' \quad \text{or} \quad L_i - L_i' = d_i = -\varepsilon_i + \varepsilon_i' .$$

This means, that $d_i = -\varepsilon_i + \varepsilon_i'$ are known as the differences $L_i - L_i'$. If the n possible equations for d_i are first squared then added and finally divided by n, then one obtains

$$\frac{\Sigma d^2}{n} = s^2 + s'^2$$

because the mixed terms approach zero as in section 1.4.4. If the same method of observation has been used for both, s and s' can be considered as being equal, and the standard deviation of a single observation becomes, for *measurements of equal weight*:

$$s = \sqrt{\frac{\Sigma d^2}{2n}} . \tag{1.4.27}$$

The standard deviation of the mean of both observations becomes

$$\bar{s} = \frac{1}{2}\sqrt{\frac{\Sigma d^2}{n}} . \tag{1.4.28}$$

The standard deviation for unit weight for *measurements with different weights* is then

$$s_0 = \sqrt{\frac{\Sigma d^2 p}{2n}} . \tag{1.4.29}$$

For levelling, the weight is inversely proportional to the distance (equation (1.4.17)) because the standard deviation increases with the square root of the levelled distance, according to equation (1.4.11). Therefore, for a distance r_i, the weight becomes $p_i = 1/r_i$.

1.4.9 Adjustment Algorithm for Observation Equations

Adjustment with observation equations is used when several unknowns are to be determined simultaneously and when the number of observations is larger than the number of unknowns. In most cases, the unknowns are not directly observed, but other quantities, which are functions of the unknowns, are. For instance, angles are measured for trigonometric sections, while the coordinates of the new point are the unknowns to be determined. To solve this, observation equations are formed which express the observed quantities as functions of the unknowns. Then the residuals V are adjusted such that the sum of V^2 is minimized (least-squares adjustment).

A simple example for the formation of observation equations is given in section 1.4.5. A solution for the determination of several unknowns is described below. An observation equation is formed for each observation L_i. For several unknowns x, y, z, these equations can be written as

$$\left.\begin{aligned} L_1 + v_1 &= a_{11}x + a_{12}y + a_{13}z \\ L_2 + v_2 &= a_{21}x + a_{22}y + a_{23}z \\ &\vdots \\ L_m + v_m &= a_{m1}x + a_{m2}y + a_{m3}z \end{aligned}\right\} \tag{1.4.30}$$

provided that a linear functional relationship

$$L_i + v_i = f_i(x, y, z)$$

is given.

The elements a_{ij} describe known coefficients (equation 1.4.39). The index i denotes observations, while j represents unknowns.

Equation (1.4.30) can be written in matrix notation as (section 1.5):

$$\mathbf{L} + \mathbf{v} = \mathbf{A}\mathbf{x}$$

$$\begin{bmatrix} L_1 \\ L_2 \\ \vdots \\ L_m \end{bmatrix} = \mathbf{L} \quad \begin{bmatrix} v_1 \\ v_2 \\ \vdots \\ v_m \end{bmatrix} = \mathbf{v} \quad \begin{bmatrix} x \\ y \\ z \end{bmatrix} = \mathbf{x}$$

$$\begin{bmatrix} a_{11} & a_{12} & a_{13} \\ a_{21} & a_{22} & a_{23} \\ \vdots & \vdots & \vdots \\ a_{m1} & a_{m2} & a_{m3} \end{bmatrix} = \mathbf{A}\,. \tag{1.4.31}$$

If the observations have *different weights*, the following equations replace equation (1.4.32), (1.4.33) and (1.4.34):

$$(\mathbf{A}^T\mathbf{P}\mathbf{A})^{-1}\mathbf{x} - \mathbf{A}^T\mathbf{P}\mathbf{L} = \mathbf{0} \tag{1.4.32}$$

The solution of this system of equations provides the unknowns:

$$\hat{\mathbf{x}} = (\mathbf{A}^T\mathbf{P}\mathbf{A})^{-1}\mathbf{A}^T\mathbf{P}\mathbf{L} \tag{1.4.33}$$

(^: estimated)

The cofactor matrix $(\mathbf{A}^T\mathbf{P}\mathbf{A})^{-1}$ is of special interest in (1.4.33). It is used for the computation of the variance-covariance matrix of the unknowns:

$$\begin{bmatrix} \sigma_x^2 & \sigma_{xy} & \sigma_{xz} \\ \sigma_{yx} & \sigma_y^2 & \sigma_{yz} \\ \sigma_{zx} & \sigma_{zy} & \sigma_z^2 \end{bmatrix} = \sigma_0^2(\mathbf{A}^T\mathbf{P}\mathbf{A})^{-1}. \tag{1.4.34}$$

The values along the main diagonal represent the variances of the unknowns, while the off diagonal values are covariances. The factor σ_0^2 is a constant. It is estimated by the empirical variance

$$s_0^2 = \frac{\mathbf{v}^T\mathbf{P}\mathbf{v}}{n-u}, \tag{1.4.35}$$

where n = the number of observations
u = the number of unknowns.

If there is a sufficient number of overdeterminations $(n-u) > 5$, s_0^2 can be used in place of σ_0^2 in (1.4.34).

The residuals are obtained by back substitution into (1.4.31):

$$v = A\hat{x} - L \tag{1.4.36}$$

If the original observation equations

$$L_i + v = f_i(x, y, z)$$

are nonlinear, they are linearized with the aid of a Taylor series expansion. After introducing approximate values

$$x = x_0 + dx; \; y = y_0 + dy; \; z = z_0 + dz \tag{1.4.37}$$

the *linearized observation equations* become

$$L_i + v_i = f_i(x_0, y_0, z_0) + \left(\frac{\partial f_i}{\partial x}\right)_0 dx + \left(\frac{\partial f_i}{\partial y}\right)_0 dy + \left(\frac{\partial f_i}{\partial z}\right)_0 dz + \ldots \tag{1.4.38}$$

with

$$\left(\frac{\partial f_i}{\partial x}\right)_0 = a_{i_1}; \; \left(\frac{\partial f_i}{\partial y}\right)_0 = a_{i_2}; \; \left(\frac{\partial f_i}{\partial x}\right)_0 = a_{i_3} \tag{1.4.39}$$

and

$$L_i - f_i(x_0, y_0, z_0) = l_i \tag{1.4.40}$$

the rearranged observation equations become

$$v_i = a_{i_1} dx + a_{i_2} dy + a_{i_3} dz - l_i \,. \tag{1.4.41}$$

Examples

a) The original observation equations for distances measured from fixed points (x_i, y_i) to a new point (x, y) are

$$s_i + v_i = q\sqrt{(x - x_i)^2 + (y - y_i)^2} \tag{1.4.42}$$

and after introducing approximate coordinates for the new point N:

$$s_i + v_i = q\sqrt{(x_0 - x_i + dx)^2 + (y_0 - y_i + dy)^2}$$

where q is a scale factor, and q_0 its approximate value.

The individual terms of equation (1.4.38) become

$$f_i(x_0, y_0) = q_0\sqrt{(x_0 - x_i)^2 + (y_0 - y_i)^2} = s_i^0$$

$$\left(\frac{\partial f}{\partial x}\right)_0 = a_{i_1} = q_0 \frac{x_0 - x_i}{s_i^0}; \quad \left(\frac{\partial f}{\partial y}\right)_0 = a_{i_2} = q_0 \frac{y_0 - y_i}{s_i^0}$$

$$\left(\frac{\partial f}{\partial q}\right)_0 = a_{i_3} = s_i^0$$

which leads to the rearranged observation equations

$$v_i = q_0 \frac{x_0 - x_i}{s_i^0} dx + q_0 \frac{y_0 - y_i}{s_i^0} dy + s_i^0 \, dq - (s_i - s_i^0) \,. \tag{1.4.43}$$

b) The original observation equations for directions observed from fixed points (x_i, y_i) to a new point (x, y) are

$$(r_N^0)_i + v_i = \arctan \frac{x - x_i}{y - y_i} \tag{1.4.44}$$

where $(r_N^0)_i$ are the oriented directions to the new point. After introducing approximate values, the linearized observation equations become

$$(r_N^0)_i + v_i = \arctan \frac{x_0 - x_i}{y_0 - y_i} + \left(\frac{\partial \arctan g}{\partial x}\right)_0 dx + \left(\frac{\partial \arctan g}{\partial y}\right)_0 dy$$

with

$$g = \frac{x_0 + dx - x_i}{y_0 + dy - y_i}$$

$$\left(\frac{\partial \arctan g}{\partial x}\right)_0 = a_{i_1} = \frac{y_0 - y_i}{(s_i^0)^2} \cdot \frac{180^\circ}{\pi}$$

$$\left(\frac{\partial \operatorname{arctang}}{\partial y}\right)_0 = a_{i_2} = -\frac{x_0 - x_i}{(s_i^0)^2} \cdot \frac{180^\circ}{\pi} \,; \quad (s_i^0)^2 = (x_0 - x_i)^2 + (y_0 - y_i)^2 \,.$$

If $t_1^0 = \arctan \dfrac{x_0 - x_i}{y_0 - y_i}$, then the following rearranged observation equations are obtained:

$$v_i = \frac{y_0 - y_i}{(s_i^0)^2} \cdot \frac{180^\circ}{\pi} dx - \frac{x_0 - x_i}{(s_i^0)^2} \cdot \frac{180^\circ}{\pi} dy - [(r_N^0)_i - t_i^0] \,. \qquad (1.4.45)$$

c) If the directions r_i are measured to several fixed points (x_i, y_i) from the new point (x, y), the observation equations are

$$r_i + v_i = \arctan \frac{x_i - x}{y_i - y} - \gamma \,. \qquad (1.4.46)$$

The arc tan function differs from the one in equation (1.4.44) by having different signs in both numerator and denominator. Thus the coefficients of the observation equations have a different sign than in equation (1.4.45), while their amount remains the same. γ is an orientation unknown. Instead of equation (1.4.45) we get

$$v_i = -\frac{y_0 - y_i}{(s_i^0)^2} \cdot \frac{180^\circ}{\pi} dx + \frac{x_0 - x_i}{(s_i^0)^2} \cdot \frac{180^\circ}{\pi} dy - (r_i - t_i^0) \qquad (1.4.47)$$

with

$$t_i^0 = \arctan \frac{x_0 - x_i}{y_0 - y_i} + 180^\circ \,.$$

In the case of the resection, the coordinates of the new point and the orientation γ of the measured directions are unknown. Thus the system of observation equations can only be solved if an observation equation is present for the unknown γ. Since it is not observed, a fictitious equation has to be introduced:

$$v_{n+1} = \left(\sum_1^n a_{i_1}\right) dx + \left(\sum_1^n a_{i_2}\right) dy - \sum_1^n l_i$$

with weight

$$p_{n+1} = -\frac{1}{\sum_1^n p_i} \qquad (1.4.48)$$

where

$$l_i = (r_i - t_i^0) \,.$$

Individual problems can be solved according to the following schema:

1) Approximate coordinates	x_0, y_0
2) Write an observation equation for each observation as a function of the unknowns	$L_i + v_i = f_i(x, y, z)$
3) Coefficients	$a_{i_1}, a_{i_2} \ldots$
4) Absolute terms	$l_i = L_i - f_i(x_0, y_0, z_0)$
5) Coefficient matrix	$\mathbf{A}$
6) Vector of absolute terms	$\mathbf{l}$
7) Unknowns	$\hat{\mathbf{x}} = (\mathbf{A}^T\mathbf{A})^{-1}(\mathbf{A}^T\mathbf{l}) = \mathbf{N}^{-1}\mathbf{n}$
8) Residuals	$\mathbf{v} = \mathbf{A}\hat{\mathbf{x}} - \mathbf{l}$
9) Check on adjustment	$\mathbf{v}^T\mathbf{v} = -\mathbf{l}^T\mathbf{v}$
10) Empirical variance	$s_0^2 = \mathbf{v}^T\mathbf{v}(n-u)$
11) Variance of unknowns	$((\mathbf{A}^T\mathbf{A})^{-1})s_0^2$
12) Final result	$\hat{\mathbf{x}}$

1.4.10 Error Limits and Confidence Regions

The area under the curve in Fig. 1.4.1 represents the totality of all occurring errors. The probability of obtaining an error for an observation with a magnitude bounded by $\varepsilon = a$ and $\varepsilon = b$, is obtained by integrating equation (1.4.1)

$$\Phi(\varepsilon) = \frac{h}{\sqrt{\pi}} \int_{\varepsilon=a}^{b} e^{-h^2\varepsilon^2} d\varepsilon . \qquad (1.4.49)$$

One can select the boundaries $a = -u_s\sigma$ and $b = +u_s\sigma$, such that u_s represents a number, and σ the theoretical value of the standard deviation for $n \to \infty$ (equation 1.4.4). Table 1.4.1 shows the certainty S in %, that for certain values of u_s the observations fall within the *confidence region*

$$x \pm u_s\sigma . \qquad (1.4.50)$$

From the last row of Table 1.4.1 it is evident that the value of 3σ is only exceeded in 0.3% of all cases.

Table 1.4.1 Standard deviation and probability

u_s	S%
1	68.3
1,96	95
2	95.4
2,58	99
3	99.7

Based on this and experience, most surveying agencies require that observations which exceed the value of three times the theoretical standard deviation be discarded or repeated. This, however, often leads to the misconception that measurements which do not exceed this confidence region, using the computed standard deviation, are always safe. Caution has to be exercised because, from the limited amount of observations, one does not obtain the theoretical standard deviation σ but rather estimations which are the sample standard deviations a and s_x. The smaller the number f of redundant observations, the more uncertain are these estimated values.

In mathematical statistics, the confidence region with its upper and lower boundaries is described by the formula

$$x \pm t_s s_x . \tag{1.4.51}$$

The factor t_s is a function of the number f of redundant measurements used to determine s_x and of the safety percentage S, depending on the specific case. In Table 1.4.2 some values of this so-called "Student's" distribution are given. This table is entered with f and gives the values for t_s for $S = 95\%$ and $S = 99\%$.

Table 1.4.2 Student's distribution value (t_s)

f	Values for t_2 $S = 95\%$	$S = 99\%$
1	12.71	63.66
2	4.30	9.92
3	3.18	5.84
4	2.78	4.60
5	2.57	4.03
8	2.31	3.36
10	2.23	3.17
20	2.09	2.85
50	2.01	2.68
∞	1.96	2.58

The so-called expectancy or true value of the measured quantity lies with the *selected certainty percentage* within the region described in equation (1.4.51).

Examples

1) In the example to section 1.4.5, the following results were obtained using $f = 5$ redundant observations

$$\hat{x} = 52.351° \quad \text{and} \quad s_x = \pm 0.0013°.$$

The confidence region for 95% certainty is then

$$\hat{x} \pm 2.57 \cdot 0.0013 \quad \text{or} \quad 52.351° \pm 0.003°$$

and for 99% certainty:

$$\hat{x} - 4.03 \cdot 0.0013 \quad \text{or} \quad 52.351° \pm 0.005° .$$

The uncertainty of the result is thus 3 or 5 times larger than assumed on the basis of the conventionally computed standard deviation.

2) In the example to section 1.4.6.3, the following results were obtained with $f = 3$, $\hat{x} = 40.1720°$, $s_x = \pm 0.00024°$.

This leads to a 95% confidence region of

$$\hat{x} \pm 3.18 \cdot 0.00024° \quad \text{or} \quad 40.1720° \pm 0.0008°$$

or for the 99% confidence region

$$\hat{x} \pm 5.84 \cdot 0.00024° \quad \text{or} \quad 40.1720° \pm 0.0014° .$$

It is surprising at first to note that the boundaries of the confidence region have expanded with increasing demands on certainty which gives the appearance that the result x is associated with a larger uncertainty than for lower demands. There is, however, a simple explanation: If under identical circumstances a higher percentage of certainty is to be placed within the boundaries of the confidence region, then the boundaries have to expand to encompass a larger region. If it is desired to limit the final result x within narrower boundaries, then either a smaller standard deviation is to be obtained using more precise measurements, or else the number of redundant observations has to be increased in order to obtain a smaller value for t_s in Table 1.4.2.

1.5 Short Introduction to Matrix Algebra

A matrix is a two-dimensional array of elements a_{ij} with i rows and j columns. Value and location of each element a_{ij} is important. A matrix is not a determinant but an array of coefficients, which form an independent mathematical unit for which certain operations are defined.

The elements may be complex numbers, functions, etc. For adjustment calculus, they contain real numbers.

Special types of matrices:

1) Quadradic matrix: number of columns equals number of rows $n = r$
2) Diagonal matrix: $a_{ij} = 0$ for $i \neq j$
3) Identity matrix: diagonal matrix with $a_{ij} = 1$ for $i = j$
4) Null matrix: all $a_{ij} = 0$
5) Column vector: $j = 1$

Rules for matrix operations:

1) $\mathbf{A} = \mathbf{B}$ if both matrices have the same number of rows and columns (matrices are compatible) and all $a_{ij} = b_{ij}$.
2) $\mathbf{A} + \mathbf{B} = \mathbf{S}$ if matrices are compatible, and

$$s_{ij} = a_{ij} + b_{ij} \quad \text{for} \quad \begin{array}{l} i = 1 \ldots n \\ j = 1 \ldots r. \end{array}$$

Under these conditions, the following laws apply:
Commutative law:

$$\mathbf{A} + \mathbf{B} = \mathbf{B} + \mathbf{A}.$$

Distributive law:

$$(\mathbf{A} + \mathbf{B}) + \mathbf{C} = \mathbf{A} + (\mathbf{B} + \mathbf{C}) = \mathbf{A} + \mathbf{B} + \mathbf{C}$$

thus

$$\mathbf{A} + \mathbf{0} = \mathbf{0} + \mathbf{A} = \mathbf{A}.$$

3) $\mathbf{A} - \mathbf{B} = \mathbf{D}$ if matrices are compatible and

$$d_{ij} = a_{ij} - b_{ij} \quad \text{for} \quad \begin{array}{l} i = 1 \ldots n \\ j = 1 \ldots r. \end{array}$$

4) Multiplication by a constant λ

$$\lambda \cdot \mathbf{A} = \mathbf{A}\lambda$$

each element a_{ij} is multiplied by λ.

5) Matrix multiplication

$$\mathbf{A} \cdot \mathbf{B} = \mathbf{P}.$$

This is only possible if the number of columns of **A** equals the number of rows of **B**. Then $p_{ij} = a_{11}b_{11} + a_{12}b_{21} + \ldots + a_{ir}b_{rj}$, i.e. the elements of the row i of **A** are multiplied by the elements of column j of **B** and added up.
When multiplying a matrix with the identity matrix, the matrix remains unchanged

$$\mathbf{A} \cdot \mathbf{I} = \mathbf{I} \cdot \mathbf{A} = \mathbf{A}.$$

The commutative law is not generally valid $\mathbf{AB} \neq \mathbf{BA}$.
However, the associative law

$$(\mathbf{AB})\mathbf{C} = \mathbf{A}(\mathbf{BC}) = \mathbf{ABC}$$

and the distributive law

$$(\mathbf{A} + \mathbf{B})\mathbf{C} = \mathbf{AC} + \mathbf{BC}$$

hold.

6) Multiplication with diagonal matrices **D**
AD means multiplication of all column elements with the diagonal elements (constant factors)

DA means multiplication of all row elements with the diagonal elements (constant factors).

7) The inverse $\mathbf{A}^{-1}$ is obtained from

$$\mathbf{A}\mathbf{A}^{-1} = \mathbf{A}^{-1}\mathbf{A} = \mathbf{I}.$$

This works under the conditions that $\mathbf{A}$ is quadratic and $\operatorname{Det}\mathbf{A} \neq 0$ (no rank deficiency).

8) The transpose $\mathbf{A}^T$ is obtained by interchanging rows and columns of $\mathbf{A}$

$$a_{ij} = a^T_{ji}.$$

Thus $(\mathbf{A}^T)^T = \mathbf{A}$.
If $\mathbf{A}^T = \mathbf{A}$, the matrix is symmetric.

9) Matrices can be divided into submatrices, which can then be treated as matrix elements.

Rules for transpose and inverses:

1) $(\mathbf{A} + \mathbf{B})^T = \mathbf{A}^T + \mathbf{B}^T$
2) $(\mathbf{A}\mathbf{B})^T = \mathbf{B}^T\mathbf{A}^T$
3) $(\mathbf{A}^T)^{-1} = (\mathbf{A}^{-1})^T$
4) If $\mathbf{A}$ is symmetric (i.e. $\mathbf{A} = \mathbf{A}^T$) and if $\mathbf{A}^{-1}$ is defined (i.e. $\operatorname{Det}\mathbf{A} \neq 0$) then $\mathbf{A}^{-1}$ is also symmetric.
5) If $\mathbf{A}^{-1}$ and $\mathbf{B}^{-1}$ are defined, then $(\mathbf{A}\mathbf{B})^{-1} = \mathbf{B}^{-}\mathbf{1}\mathbf{A}^{-1}$.

2. Elements of Surveying Instruments

2.1 Spirit Levels

In order to be able to *refer geodetic position and elevation surveys to the reference surfaces* mentioned in section 1.2, the axes of the measuring instruments have to be brought into the plumb line or normal to it. This is usually accomplished with the aid of level bubbles which are produced in two types, bull's-eye bubbles for rough settings, and tubular levels for precision measurements.

2.1.1 The Bull's-eye Level

The bull's-eye level consists of a round glass container encased in metal. Its lid is milled spherically on the inside, while its bottom is melted together. Except for a small remainder – the bubble –, the container is filled with ether or alcohol. *The centre of the container* is marked by one or more circles (Fig. 2.1.1).

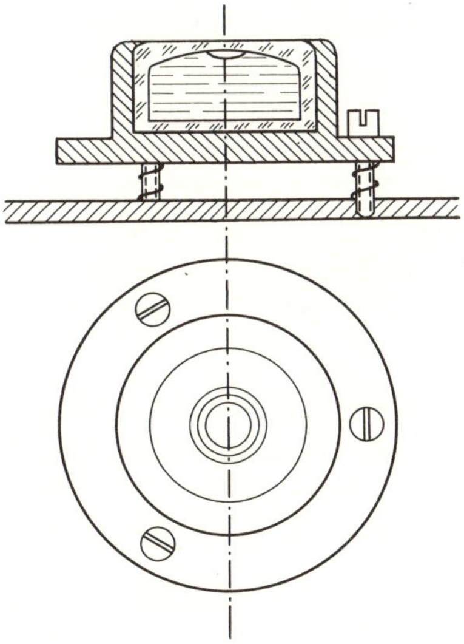

Fig. 2.1.1 Bull's-eye level

A plane, resting on setting screws, such as a plane table, can be levelled with the aid of an adjusted bull's-eye level. Similarly, a vertical axis, held by setting screws can be brought into the plumb line by centring the level bubble using the setting screws.

A bull's-eye level is adjusted if a tangential plane to the spherical cap at its centre is parallel to the support surface of the level or if it is normal to the vertical axis.

In order to *adjust a bull's-eye level*, its bubble is first carefully centred with the aid of the setting screws. Then the level is very slowly moved 180° (by transposing it on the table or rotating the vertical axis). If the bubble wanders, half of the amount each is eliminated with the aid of the setting screws and with the adjustment screws visible on the level casing.

2.1.2 The Tubular Level

The tubular level consists of a cylindrical glass tube encased in metal. Its inside surface is milled in such a way that its longitudinal section represents a circular arc. The tube is melted closed on both ends and filled with ether, except for the elongated bubble, consisting of ether vapour, which always wanders to the highest spot of the level. A graduation is engraved on the exterior wall of the glass body. The graduation lines are separated by 2.26 mm (one Parisian line) for older levels, or 2 mm for newer ones (Fig. 2.1.2).

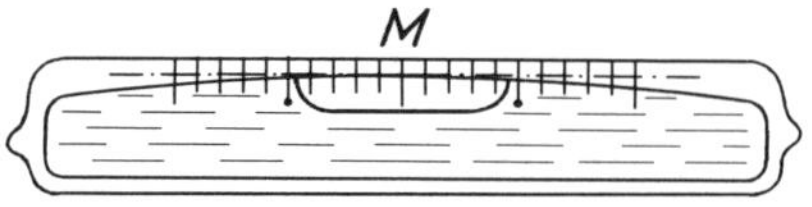

Fig. 2.1.2 Tubular level

The separation of two graduation lines is called one *pars*. The central point of the graduation is known as the *normal point* or *centre mark*. The longitudinal tangent passing through the centre mark at the interior wall is referred to as the *level axis*. Its projection onto the bottom part of the casing – i.e. the plane in which the casing is supported – is called the *setting line*. The angle by which the level has to be tilted to cause a movement of the bubble by one pars is called *level sensitivity*. It is usually given in seconds of arc and amounts to about 5″ for precision levels, 20″ for medium levels, and 45″ for simple ones. The corresponding milling radii are 82.5 m, 20.6 m and 9.2 m, respectively.

A setting level (Fig. 2.1.3) is utilized for levelling straight lines and planes. It is encased such that it can be moved and reversed. If the spirit level is directly or indirectly connected to the vertical axis of an instrument (Fig. 2.1.5), it is used to set the latter in the vertical direction.

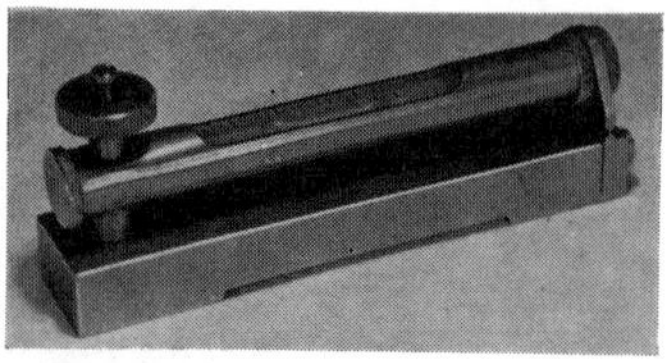

Fig. 2.1.3 Setting level

2.1.3 Adjustment and Use of Tubular Levels

2.1.3.1 The Setting Level

This level is adjusted if the bubble axis is parallel to the setting line. If an adjusted setting level is placed onto a horizontal plane, its bubble will be centred around the centre mark.

To appreciate the behaviour of a setting level, the misadjusted level is placed on a tilted plane. This can be done with the use of a "setting board", i.e. a small plane board which can be tilted around a horizontal axis with the aid of a setting screw. As shown in Fig. 2.1.4, α is the tilt angle of the plane, while β represents the angle of misadjustment of the setting level.

The bubble reading L_1 is the result of both misadjustment and tilt, and amounts to $(\alpha + \beta)$ (left half of Fig. 2.1.4). After the level is reversed (right half of Fig. 2.1.4), the bubble wanders in the opposite direction to L_2 by $(\alpha - \beta)$. Thus the total path of the bubble from L_1 to L_2 represents an angle of 2α. If one denotes the centre of the arc $L_1 L_2$ with S, then $L_1 S = SL_2 = \alpha$. Therefore, α equals zero if the bubble is centred around S, which means that the plane is horizontal. The arc MS corresponds to the angle of misadjustment β.

Therefore, to adjust the level, the arc MS has to be brought to zero, or in other words, the centre of the run has to be brought to the centre mark with the aid of the adjustment screws.

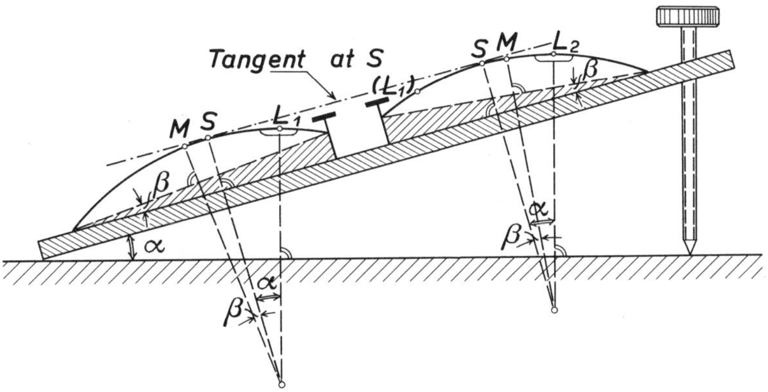

Fig. 2.1.4 Use of setting level

The adjustment becomes rather simple if one uses $\alpha = \beta$. This can be obtained automatically if the bubble is centred to the centre mark with the aid of the setting screw of the setting board. After reversing the level, the bubble wanders by $2\alpha = 2\beta$. Then one half of the reading is eliminated with the setting screw of the setting board, and the setting line is horizontal. The other half of the reading is eliminated by returning the bubble to the centre mark with the aid of its adjustment screws. The level is then adjusted. If the bubble still wanders when the level is again reversed, the procedure has to be repeated.

If the setting line has to be levelled without adjustment of the level, then the centre of the run is determined by reversing the level. Either the centre or one end of the bubble can be used. The bubble is then centred around S with the aid of the setting screw of the setting board.

2.1.3.2 The Vertical Axis Level

This level is not designed as a separate instrument but rather as part of a theodolite or other instrument whose vertical axis is to be set into the plumb direction. Such a level is adjusted if its bubble axis – i.e. the tangent at M (Fig. 2.1.5) – is normal to the vertical axis. The centre of the run S in respect to the vertical axis is obtained by rotating and thus changing the orientation of the bubble axis by 180° and then finding the centre between L_1 and L_2.

To set the vertical axis into the plumb direction and to adjust the vertical axis level, the following steps are necessary, again based on the situation $\alpha = \beta$ as in section 2.1.3.1:

1) Set the vertical axis approximately with the aid of a bull's-eye level.

2) Rotate until the tubular level is parallel to two foot screws, then centre the bubble.

3) Rotate axis with bubble by 180°, and read bubble position.

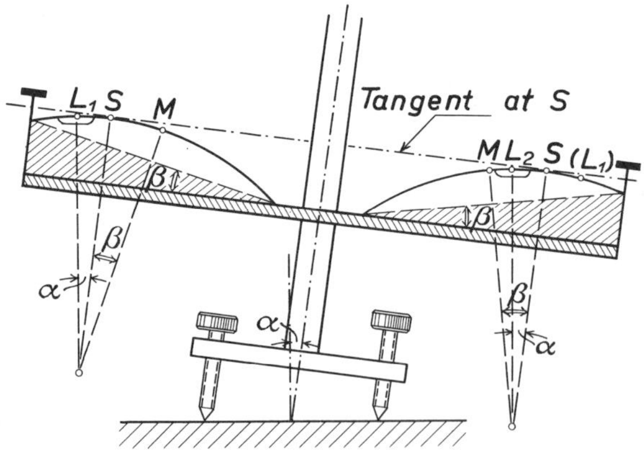

Fig. 2.1.5 Use of vertical axis level

4) Eliminate half of the reading with the foot screws; the bubble is now centred at S, and the axis is vertical in this direction.

5) Rotate axis with bubble by 90° and bring bubble to the previously determined

centre of run S, using the third foot screw; the axis is now also level in the second direction.

6) Check by rotating the axis slowly; if the bubble does not remain stationary in S, repeat steps 2 to 5.
7) If the bubble remains stationary, the axis is correctly placed in the plumb direction. If one centres the bubble around the centre mark with the aid of its adjustment screws, it is properly adjusted.

2.1.4 Determination of the Bubble Sensitivity

This is important for judging a bubble instrument and can be achieved in two ways: *Bubbles which are rigidly connected to a telescope* can be studied with the aid of a graduated ruler (e.g. levelling rod) placed at a distance d. The instrument is set up such that one of the three foot screws points towards the rod. The position of one bubble end is read using the bubble graduation, while the reading l_1 is taken on the rod using the telescope. Afterwards, the bubble is moved by n pars with the aid of the foot screw pointing towards the rod, and a second rod reading l_2 is taken. Now the tilt change in pars can be related to the angle derived from the rod readings, which leads to

$$n\,[\text{pars}] = \frac{l_1 - l_2}{d} \cdot \varrho''$$

or

$$1\,[\text{pars}] = \frac{\varrho''}{n \cdot d}(l_1 - l_2)\,.$$

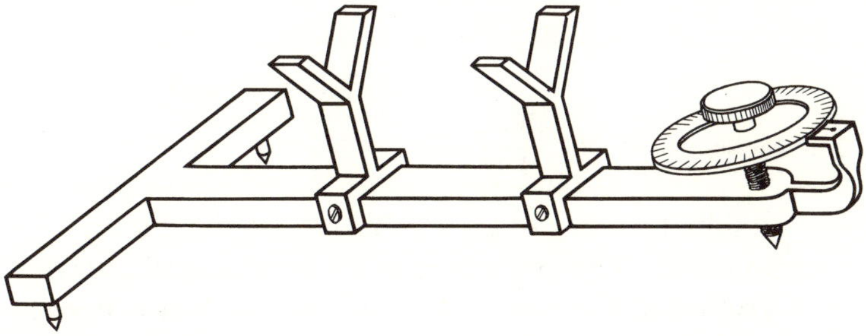

Fig. 2.1.6 Level trier

Example: $n = 20 \quad d = 18.80\text{ m} \quad l_1 = 1.397\text{ m} \quad l_2 = 1.361\text{ m}$

$$1\,[\text{pars}] = \frac{206265 \cdot 0.036}{20 \cdot 18.80} = 20'' .$$

Levels without casing or connection to an instrument are studied with the aid of a level trier (Fig. 2.1.6). It consists of a T-shaped beam with two Y-supports for the level and can be tilted with a precision measuring screw with 0.5 mm pitch whose drum is divided, e.g. into 360°, or 100 parts.

With a pitch h of the screw and an effective length l of the level trier, one complete revolution of the screw represents a tilt change $k'' = \varrho'' \cdot h/l$, which is called the constant of the level trier. This constant is determined by placing a telescope on the level trier and following the procedure outlined above.

drum reading	t	bubble readings left	right	centre	n
0.00	0.25	1.7	15.2	8.45	5.10
0.25		6.8	20.3	13.55	
0.02 0.27	0.25	2.4	16.0	9.20	4.80
0.27		7.2	20.8	14.00	
0.05	0.25	3.0	16.6	9.80	4.75
0.30		7.7	21.4	14.55	
0.13	0.25	4.6	18.3	11.45	4.90
0.38		9.5	23.2	16.35	
0.23	0.25	6.3	20.0	13.15	5.05
0.48		11.4	25.0	18.20	
				Mean	4.92

Therefore the sensitivity is $\frac{0.25}{4.92} \cdot 400'' = 20.3''$

Once k is known, an initial reading of the bubble is taken, and a second reading is obtained after t revolutions. This procedure is repeated several times using slight changes in the initial setting. If the bubble has wandered by a mean value of n pars, the sensitivity of the bubble is calculated as

$$n\,[\text{pars}] = tk''$$

or

$$1\,[\text{pars}] = \frac{t}{n} k'' .$$

Example

Determination of the sensitivity of a tubular level. With the pitch of the screw of 0.5 mm and the effective length of 250.0 mm, k becomes 400″. Readings to 0.01 revolutions of the screw can be taken at the drum.

Therefore the sensitivity is $\frac{0.25}{4.92} \cdot 400'' = 20.3''$.

A setting board can be used as a make-shift level trier. In this case, the length l has to be directly measured in order to determine the constants. The pitch of the screw is determined by pressing the screw against a piece of paper and measuring the separation of several threads with a millimetre ruler.

2.1.5 Peculiarities of Tubular Levels

a) The *bubble graduation* for general instruments usually extends to both sides from the centre mark, while it is continuously numbered for precision levels. When reading levels with centre graduation it is helpful to denote one direction (e.g. towards the objective lens) as positive.

b) *Reversible levels* are milled and graduated on the top and bottom so that they can be used in both positions. They therefore have an upper and a lower bubble axis, which have to be parallel. This requirement is fulfilled to within a few arc seconds for modern levels.

c) *Chamber levels* have at one end a chamber, separated by a wall. Liquid from the level can be transferred into it by raising and lowering the chamber end. This facilitates control of the bubble length, which depends on the temperature of the liquid. It is preferable however to set the level into an insulated casing such that temperature changes have little influence.

d) *Coincidence bubbles* permit precise settings of the level by the observer. This is accomplished by means of a prism system, which splits the bubble ends in longitudinal direction and images them side by side (Fig. 2.1.7).

By moving the foot screws of the tribrach, the bubble ends can be coincided to a symmetrical figure, which represents a centred bubble. Usually, coincidence bubbles do not have a graduation. Their sensitivity can be obtained by measuring the bubble path with a millimetre ruler at the glass body, and finding the tilt angle for a bubble path of 2 mm.

e) The *striding level*, a special form of setting level, is used for levelling cylindrical axes. Instead of a plane base surface, it has two supports which usually are milled in a circular shape. The setting line corresponds to the line connecting the vertex points.

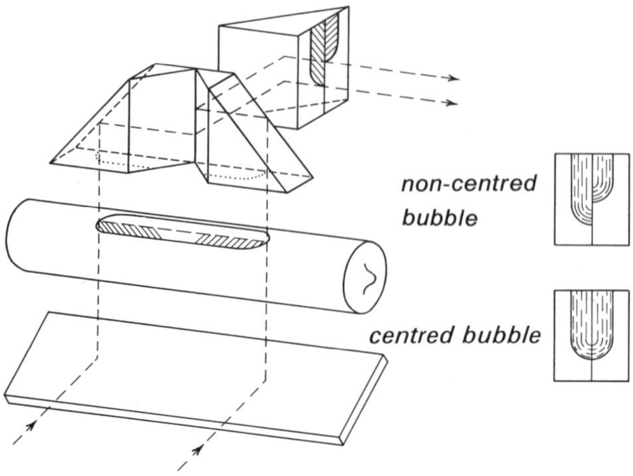

Fig. 2.1.7 Coincidence bubble

f) *Cross divergence of a level* occurs when the horizontal projection of the axis of the telescope bubble is not parallel to the collimation axis, or when the axis of a striding level is not parallel to the cylindrical axis. Cross divergence can be recognized if the bubble wanders in different directions when rotating the telescope back and forth, or tilting the striding level back and forth. It can be eliminated with sideways acting adjustment screws.

2.1.6 Levels and Compensators

Up until about 1950, nearly all surveying instruments were equipped with spirit levels. Spirit levels, especially the precision tubular levels for levelling instruments and height index bubbles, are influenced significantly by one sided solar heating. Furthermore, they react strongly to mechanical disturbances caused by passing vehicles or even by the observer when he walks around the instrument. Tubular levels for precision levelling have to be carefully centred using a setting screw (tilting screw) prior to taking a reading, even with a previously levelled instrument. This is time consuming and unfortunately is occasionally forgotten, especially when measuring vertical angles.

For this reason, the instrument manufacturers tend to produce instruments where the fine levelling is accomplished automatically after rough levelling with a bull's-eye bubble. This is accomplished by utilizing an optical-mechanical element subject to gravity, which is commonly referred to as a *"compensator"* (Chapter 9). While spirit levels are connected with simple devices to levelling telescopes or reading devices for the vertical circle, the type of compensator often influences the total optical-mechanical design of the whole instrument. For some levelling instruments, the compensator represents the key element of the whole instrument. This is the reason why compensators are not dealt with in this section but rather in connection with the levelling instruments (Chapter 9) and the tacheometric instruments (section 4.2).

2.2 Imaging with Lenses

2.2.1 Geometric-optical Fundamentals

Although geometric optics is considered to be a known prerequisite, the most important facts as they relate to telescopes are summarized below:

A lens is a glass body bounded by two spherical surfaces. The line connecting the two centres of the spheres represents the optical axis of the lens, which contains the optical centre as well.

Lenses which become thinner towards the edges are called *convex or collecting lenses*; while lenses which are thinner at the centre than on the edges are referred to as *concave or dispersing lenses*. Fig. 2.2.1 illustrates, in sequence, a bi-convex, a plane-convex, a concave-convex as well as a bi-concave, a plane-concave and a convex-concave lens.

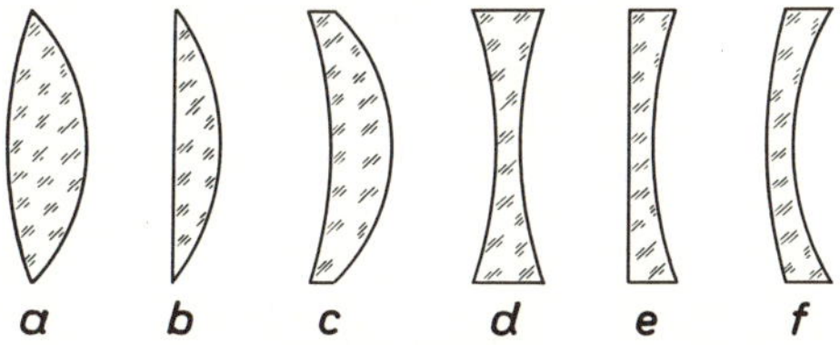

Fig. 2.2.1 Collecting and dispersing lenses

Most optical instruments contain several lenses which have to be centred. A lens system is centred properly if the optical axes of all lenses lie on the same straight line.

The laws governing the path of light rays through lens systems become very simple if one restricts oneself to near axial rays and assumes that the lenses are extremely thin – which is not usually the case in practical applications.

Imaging with convex lenses follows the following laws:

a) Light rays which penetrate the lens at the optical centre continue without being refracted, except for a small parallel shift for inclined rays.

b) Rays entering the lens parallel to its optical axis combine in the focal point after exiting. If inclined parallel rays penetrate the lens, they intersect at a point where the ray passing through the optical centre hits the focal plane.

c) The path of the rays is reversible. The following cases are important for telescope optics, where a denotes the object distance and b the image distance:

Case 1: $a > 2f$: Imaging with a collecting lens (Fig. 2.2.2).

To derive the lens equation, it is evident in the figure that: $y : y' = (a - f) : f$ which leads to

$$\frac{1}{a} + \frac{1}{b} = \frac{1}{f}. \tag{2.2.1}$$

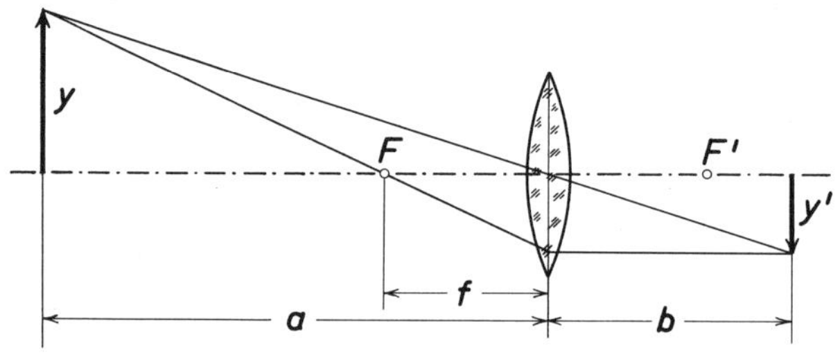

Fig. 2.2.2 Imaging through collecting lenses

The lens creates a real reduced and inverted image of the object, which is generated behind the focal plane for a finite a. The image dimension size y' is obtained as

$$y' = \frac{b}{a} y.$$

Case 2: $a \leqq f$: The collecting lens as magnifying glass (Fig. 2.2.3).

If $a < f$ then the rays themselves do not intersect, but rather their backward extensions. This generates a virtual upright and enlarged image, and the lens becomes a

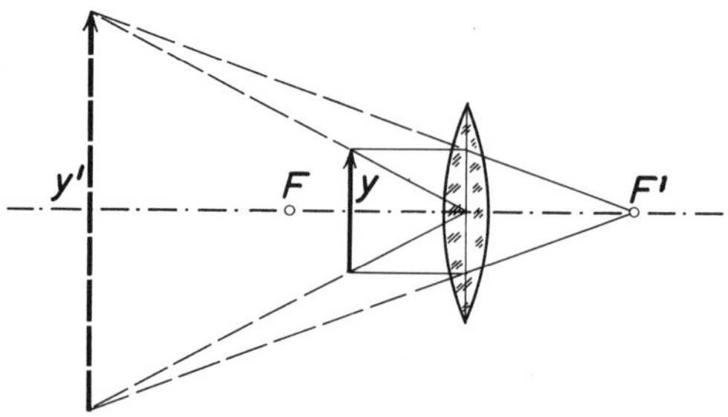

Fig. 2.2.3 Collecting lenses as magnifying glass

magnifying glass. Since the image distance is negative, the equation for the magnifying glass is

$$\frac{1}{a} - \frac{1}{b} = \frac{1}{f}. \tag{2.2.2}$$

If the object falls into the front focal point, then a equals f. The rays from the object leave the lens as a parallel bundle, which permits a relaxation of the eye of the observer. He therefore tends to hold the magnifying glass in such a manner that the object is situated in the front focal plane. For this case, the magnification, which is the ratio of the visual angles as seen through the lens, or with the naked eye, can easily be found: The observer places the object at the most suitable distance w which for viewing with the naked eye amounts to about 25 mm. Thus the visual angle is $y : w$. The visual angle when viewing through a magnifying glass equals $y : f$ (Fig. 2.2.4)

Thus the magnification of the magnifying glass becomes

$$M = \frac{y:f}{y:w} = \frac{w}{f}. \tag{2.2.3}$$

Therefore, magnifying glasses must have short focal lengths in order to be effective.

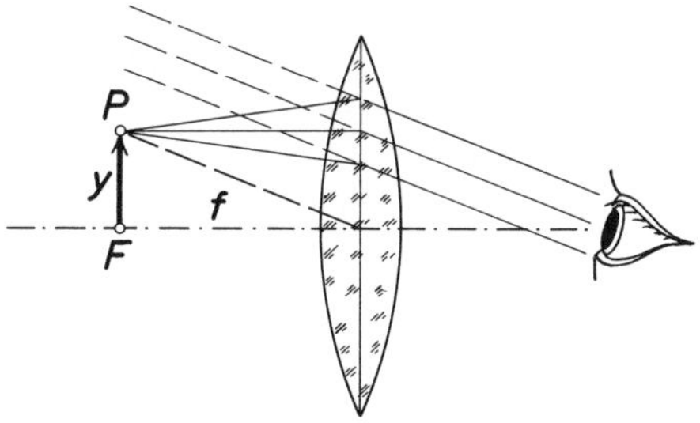

Fig. 2.2.4 Visual angle

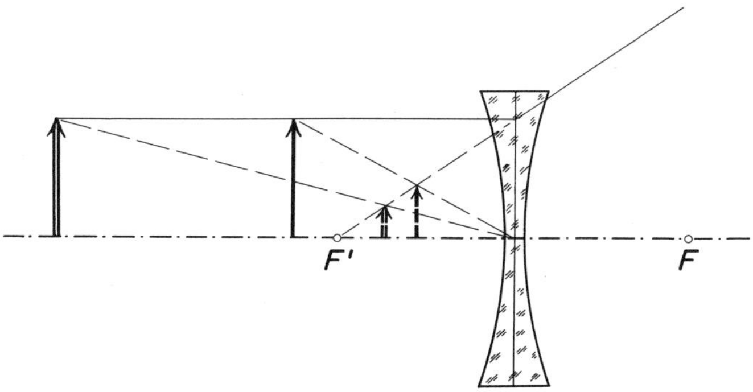

Fig. 2.2.5 Imaging with concave lenses

Imaging with concave lenses results in virtual, reduced and upright images (Fig. 2.2.5). With image distance and focal length becoming negative, the lens equation is then

$$\frac{1}{a} - \frac{1}{b} = -\frac{1}{f}. \tag{2.2.4}$$

All rays parallel to the axis are refracted such that their backward extensions point towards the rear focal point F'.

2.2.2 Imaging Errors

The imaging laws described earlier apply for mono-chromatic light only, and for rays within a narrow space along the optical axis. For all other cases, imaging errors occur which either influence the image quality or else cause distortions of the image positions.

Bundles of rays, parallel to the axis, are affected by chromatic and spherical aberrations. For inclined bundles, astigmatism and curvature of field are also present, and if the bundles are wide, there is coma as well. Finally, imaging outside the axis leads to lens distortions.

a) *Chromatic aberration* (Fig. 2.2.6) is caused by the fact that for white light the component with a shorter wavelength (violet) is refracted more than the component with a longer wave length (red). Therefore, the image positions depend on the color of the light and result in images with coloured edges.

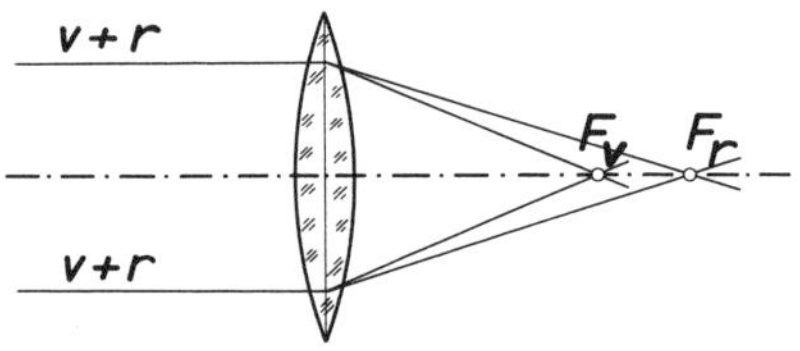

Fig. 2.2.6 Chromatic aberration

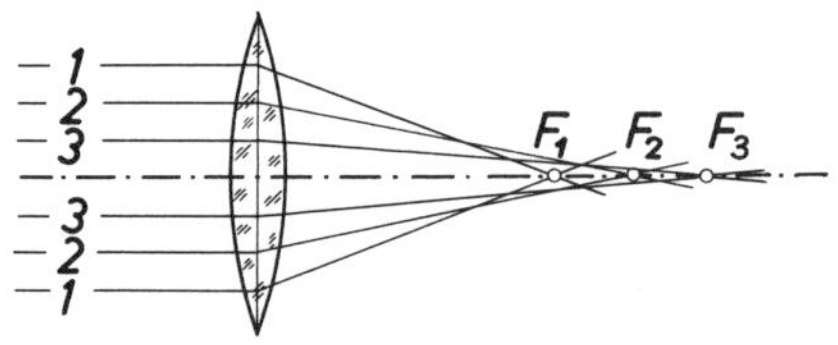

Fig. 2.2.7 Spherical aberration

b) *Spherical aberration* (Fig. 2.2.7) is caused by the fact that the farther the rays are from the axis, the more they are refracted. Thus the image positions on the axis depend on the width of the incident bundle. The observer sees a sharp centre, surrounded by a circle of confusion.

c) *Astigmatism* for inclined rays is obtained, because the focal length for object points outside the optical axis is different in the different planes having the lens axis in common (Fig. 2.2.8).

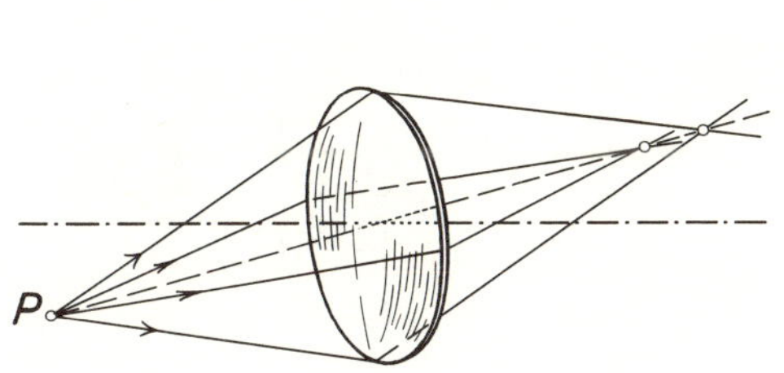

Fig. 2.2.8 Astigmatism

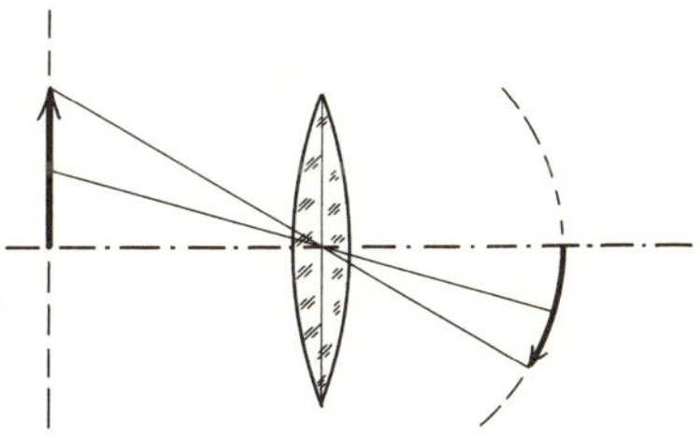

Fig. 2.2.9 Curvature of field

d) *Curvature of field* describes the fact that object planes normal to the optical axis are not imaged as planes but as flat dishes (Fig. 2.2.9).

e) The *coma* represents spherical aberration of inclined bundles; the object point is imaged as a fuzzy centre with a comet-like tail.

f) *Lens distortion* occurs when the principal rays of inclined bundles with increasing angle of incidence are distorted such that in the image plane they appear shifted towards or away from the image centre. A square in the object space, for instance, is distorted to a barrel or pillow shape.

For surveying instruments, distortion is usually insignificant. It is, however, very significant in photogrammetry, where it can be practically eliminated with the use of rather complicated lens systems.

2.3 Measuring Telescopes

2.3.1 Structure of a Measuring Telescope

In its simplest form, the *telescope* consists of two centred collecting lenses as stated in 1611 by *J. Kepler*. The objective lens has a large focal length, while the eyepiece has a short one. The objective lens produces a reversed and reduced image which is viewed through the eyepiece. In the upper part of Fig. 2.3.1, the rays for a finite object are traced while the lower part shows the imaging of an ∞ object, located off the axis.

To convert the telescope to a *measuring telescope* and also to keep imaging errors as small as possible, several additions to and changes from the original type are necessary.

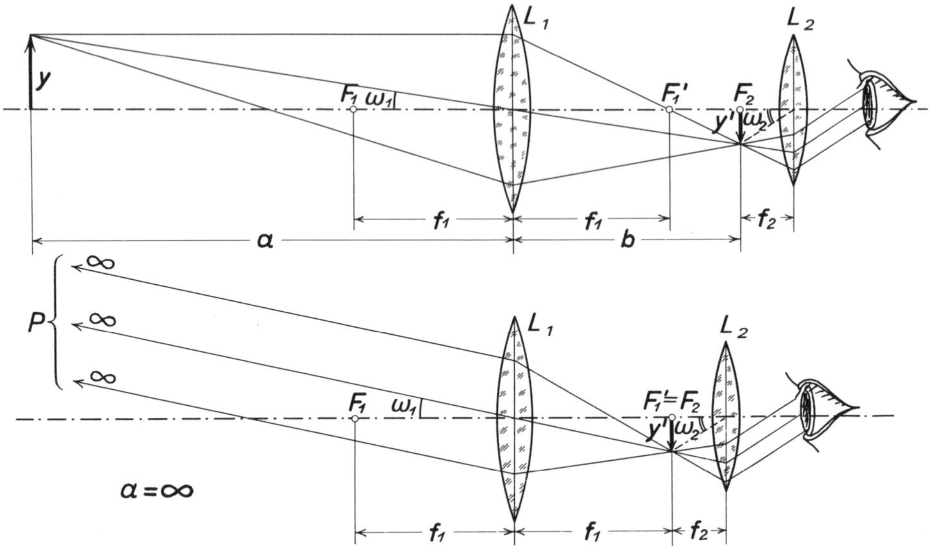

Fig. 2.3.1 Kepler's telescope

2.3.1.1 Cross Hairs

Cross hairs consisting of spider threads in older instruments are inserted at the rear end to enable proper sighting with the telescope. In newer instruments, they are replaced by a glass plate – the reticule – with fine cross hairs, which is held by 3 or 4 adjustment screws (Fig. 2.3.2). As a rule, the cross hairs are located just in front of the eyepiece, which can be slightly moved for sharper vision.

The optical centre of the objective lens and the intersection of the cross hairs define the line of sight (collimation axis). In order to set it parallel to the axis of a telescope bubble or normal to the horizontal axis of a theodolite, the cross hair plate is held in a ring which can be slightly shifted using adjustment screws (adjustment of the collimation axis).

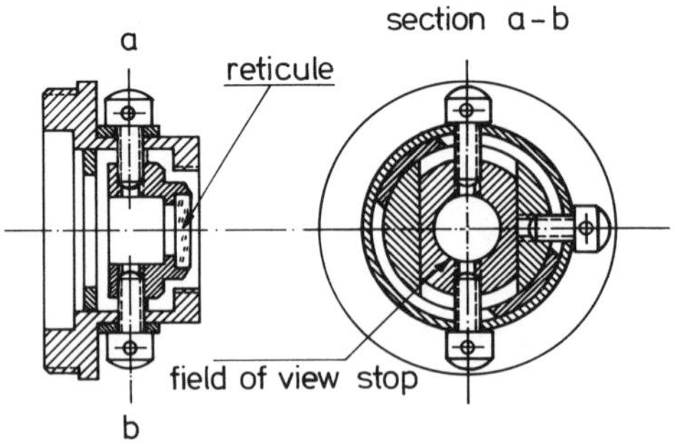

Fig. 2.3.2 Reticule with cross hairs

2.3.1.2 Focussing Lens

If the target is located at a finite distance, the image is formed behind the focal plane of the objective lens, moving farther the closer the object. The cross hairs have to be brought to exactly this location. After having set the eyepiece such that a sharp image of the cross hairs is obtained, this can be achieved with the aid of a focussing lens.

In the *telescope with focussing lens*, sometimes referred to as telescope with constant length, the position of the cross hairs cannot change (Fig. 2.3.3). Therefore the image, generated by the objective lens, has to be shifted into the cross hair plane. For this, a – usually slightly negative – focussing lens is placed between the objective lens and its rear focal point. This lens disperses somewhat the rays gathered by the objective lens. Therefore, the rays will intersect a bit later, i.e. the image is generated farther back, and moves farther away, when the focussing lens gets closer. This solution avoids the disadvantages of the eyepiece slide. It also permits a shortening of the telescope by selection of suitable focal lengths, even though the image is shifted along.

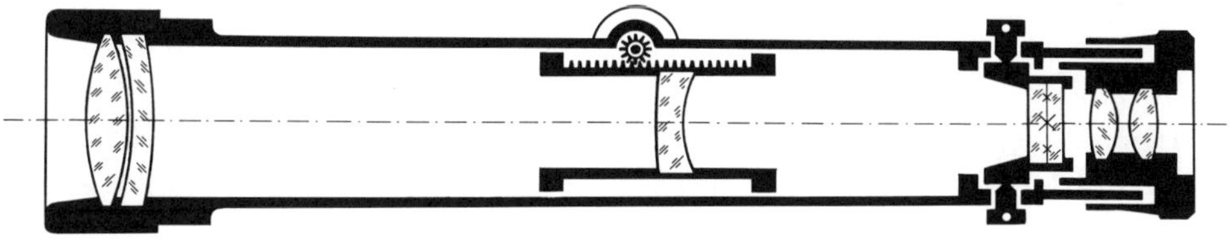

Fig. 2.3.3 Telescope with internal focussing lens.

2.3.1.3 Objective Lens and Eyepiece

These are composed of several lenses with different refractive indices and radii in order to reduce chromatic and spherical aberrations. To reduce deviations in colour, the individual lenses of the eyepieces are separated.

The *objective lens* of geodetic instruments used to be predominantly of the type designed by Frauenhofer. This objective lens system consists of a biconvex lens of

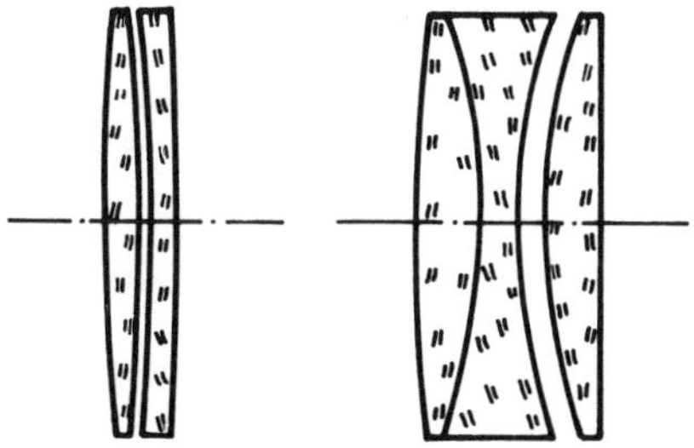

Fig. 2.3.4 Frauenhofer's and apochromatic objective lenses

relatively weak refracting crown glass ($n = 1.52$) and of a concave-convex lens of strongly refracting flint glass ($n = 1.62$), shown on the left side of Fig. 2.3.4. Newer instruments are often equipped with apochromats.

The *eyepiece*, as designed by Ramsden, consists in its simplest form of two plane-convex lenses with equal focal lengths. They are separated by $^2/_3 f$ with the curved sides facing each other (Fig. 2.3.5, left). In order to accommodate an observer with abnormal eyesight so that he can see both image and cross hairs sharply, the eyepiece can be slightly moved in respect to the cross hair plate. In modern eyepieces, e.g. orthoscopic ones, the collecting lens and eye lens consist of two lenses each.

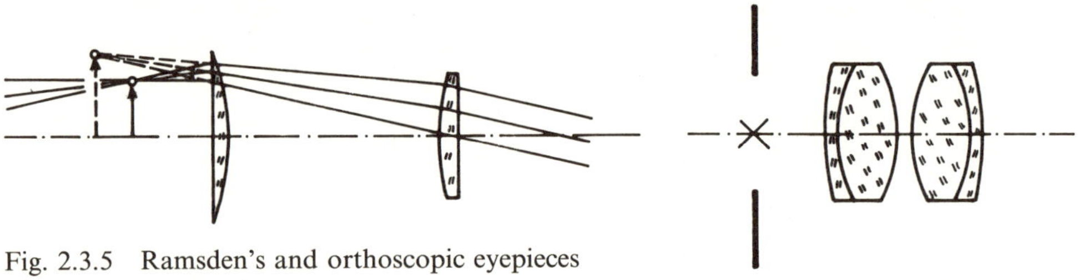

Fig. 2.3.5 Ramsden's and orthoscopic eyepieces

A terrestrial eyepiece is obtained, if an additional reversing lens is inserted. Telescopes with terrestrial eyepieces generate upright images. The insertion of compound prisms has the same effect (section 9.1.3).

2.3.1.4 Stops

In order to keep astigmatism, curvature of the field, and coma small, the incoming bundles of rays are restricted by opening stops; furthermore, bundles whose principal ray is sharply inclined are cut out by field of view stops. Usually the housing of the objective lens acts as an opening stop. The ring which carries the cross hairs serves as a field of view stop. It has the effect that an inclined bundle of rays will only enter the eyepiece if not more than half of the bundle is cut off by the eyepiece housing (Fig. 2.3.6).

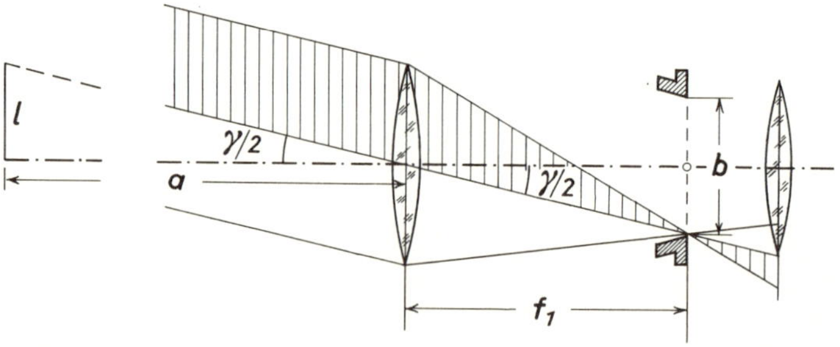

Fig. 2.3.6 Opening and field of view stop

If one imagines the field of view stop as being imaged by the objective lens into the object space, it would create a sharp boundary of the object resembling a mask. This virtual mask is called the *entrance opening*.

The images of the opening stop are referred to as *pupils*. If the housing of the objective lens acts as an opening stop, it also is the *entrance pupil*. Its image, generated by the eye – perhaps by means of a focussing lens – is the *exit pupil*, as shown in Fig. 2.3.7. It

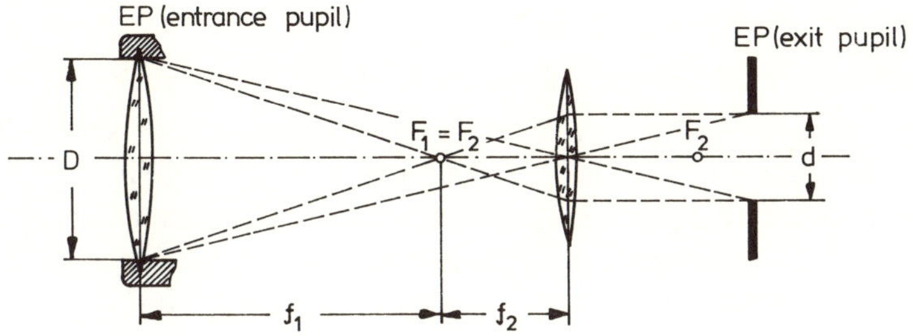

Fig. 2.3.7 Entrance and exit pupil

appears as a bright circle near the rear focal point of the eyepiece and can be caught on a ruler and thus measured.

2.3.2 Magnification, Field of View, Brightness and Resolution

These characteristics primarily identify the value of a telescope. They are defined for a telescope focussed to infinity. In order to keep the illustration reasonably simple, the case shown at the lower part of Fig. 2.3.1 is utilized; f_2 is the focal length of the eyepiece, while f_1 represents the focal length of the system consisting of objective plus focussing lenses at ∞ setting.

The *telescope magnification* is the ratio of the visual angles ω_2 and ω_1 with which a distant object is seen through the telescope or with the naked eye. As evident in Fig. 2.3.1, the angles ω_1 and ω_2 also appear in the space between the two lenses L_1 and L_2. Therefore

$$\omega_1 = \frac{y'}{f_1}; \quad \omega_2 = \frac{y'}{f_2}$$

and

$$M = \frac{\omega_2}{\omega_1} = \frac{f_1}{f_2}.$$

In addition, with D being the diameter of the entrance pupil (housing of the objective lens), and d the diameter of the exit pupil, $M = D : d$, because the ratio $f_1 : f_2 = D : d$

is evident in Fig. 2.3.7. To determine M according to the latter formula, the telescope is focussed at ∞. Then D is measured directly, while d is caught and measured with a ruler behind the eyepiece. The latter can be improved by using a dynameter, instead of the ruler, which is a magnifying glass with a fine graduation in its field of view.

The magnification can be approximately determined with the aid of a levelling rod set up at some distance. It is viewed with one eye through the telescope and with the other one looking straight ahead to determine how many directly seen units cover one unit as seen through the telescope.

Telescope magnification for surveying instruments lies between 15 and 50 times.

The *field of view* is the conical space with the opening angle γ, which is bounded by the principal rays passing through the field of view stop at ∞ setting (Fig. 2.3.6). In order to approximately determine γ, a levelling rod is set up at a rather long distance, a, from the instrument. The farthest possible rod readings at the edges of the field of view are taken, the difference of which represents the rod section l.

This leads to

$$\gamma = \frac{\mathrm{b}}{f_1}\varrho \approx \frac{1}{\mathrm{a}}\varrho \ .$$

For surveying telescopes, γ amounts to about 1° to 2°.

The *telescope brightness* is the ratio of area brightnesses H and H_0 with which an object is seen through the telescope and with the naked eye. With d and p being the diameters of the exit and eye pupils, respectively, and c a coefficient characterizing the light transmission of the telescope in terms of the original light intensity, then

$$h = \frac{H}{H_0} = c\left(\frac{d}{p}\right)^2 .$$

This holds if $p \geqq d$. Thus the images generated through the telescope at the retina are only as bright as without the telescope in the ideal case $c = 1$ and $d = p$. However, c is always smaller than 1. It can be estimated using the rule that each refracting surface reflects 5% of the light, and that each centimetre of glass path absorbs 0.5% of the light. The losses due to reflection are reduced to approximately 1%, if the lenses are coated with the reflex reducing "T-film", which can be noticed by its violet glimmer. A telescope, consisting of objective lens, focussing lens, cross hair plate and two eyepiece lenses, suffers a loss of about 40% of the light unless the reflex reducing T-film is used which cuts the loss to about 10%. Thus c equals 0.6 for the first case and 0.9 for the second one. Especially when measuring in poor light conditions, the T-film is very advantageous.

Optical companies often state that the telescope brightness is simply the value of D^2.

The *resolution of a telescope* is its ability to separately image two distinct object points. Due to the wave-nature of light, the light is diffracted at the edges of the lens housing and the stops. The optical image of an object point is thus not a geometric point but rather a small diffraction disk. Its diameter depends on the opening angle ω_2 and the wave length of the light λ and becomes

$$b_2 = \frac{1.22\,\lambda}{n_2 \sin \omega_2} .$$

This diffraction disk can be transformed back into the object space. However, only half the size of this fictitious object can be utilized, because the eye can only resolve half of the diffraction disk for psychological reasons.

The limit of resolution of a telescope can only be represented in angular units with f_1 as focal length of the objective lens, $2 f_1 \sin \omega_1$ as effective objective opening, and $\lambda = 0.000555$ mm. The resolution of the telescope in object space becomes

$$R'' = \frac{1.2\lambda \cdot 206265''}{\text{objective opening}} = \frac{140\ [\text{mm}]}{\text{objective opening [mm]}} \cdot 1'' .$$

The suitable magnification $\bar{M}$ is designed such that the visual angle behind the eyepiece becomes equal to the psychological limit for viewing which amounts to approximately $2' = 120''$. This is the case if 140/objective opening = 120, which means that the suitable magnification of the telescope is

$$\bar{M} \approx \text{objective opening.}$$

The manufacturers of surveying instruments usually restrict themselves to about 70% of this value.

2.3.3 Use of the Telescope

Prior to the actual observation, a possible cross hair parallax has to be removed, i.e. if image and cross hair planes do not coincide. This is achieved with the following steps:

1. Aim the telescope against the sky and rotate the eyepiece until sharp cross hairs are seen.
2. Aim the telescope towards the object and move the eyepiece slide or focussing lens until a sharp image of the object is obtained as well.
3. Check this by moving your head sideways several times. The position of the object with respect to the cross hairs has to remain unchanged.

After elimination of the parallax, the cross hairs are centred onto the target with the aid of the fine motion screws of the instrument. The standard deviation of setting onto the target has been determined empirically for telesopes with 10 to 50 times magnification as

$$s_t = \pm \frac{1.0}{\sqrt{M}} \quad \text{to} \quad \pm \frac{1.5}{\sqrt{M}} \text{[mgon]}$$

or

$$s_t = \pm \frac{3''}{\sqrt{M}} \quad \text{to} \quad \pm \frac{5''}{\sqrt{M}} \,.$$

For usual surveying instruments, this error can be taken as 2″ or 0.6 m gon.

2.4 Tripods and Instrument Base

2.4.1 Instrument Base

For most surveying instruments, the upper part is distinguished from the instrument base. The latter usually consists of a tribrack with a centrally milled opening (axis socket) directed towards the top. It is fastened to a tripod or a pillar and remains stationary during measurements. The upper part, which carries the measuring telescope and reading devices, fits with its axis plug into the socket of the base and is rotated by the observer towards each target. Plug and socket have conical shapes for older instruments with the plugs resting on a spring for support. Since this arrangement requires frequent monitoring, cleaning and lubricating of the axis, most manufacturers – facilitated by progress in mechanical precision work – have changed to cylindrical axes.

2.4.2 Tripods and Tripod Connections

A tripod usually consists of three legs of wood or light metal, which are fastened together by a tripod top. Each leg consists of two props, which become narrower towards the bottom. They are reinforced by cross braces and end with an iron tip. For several decades, tripods with telescopic legs have been manufactured (Fig. 2.4.1).

At the top, the legs are fastened either with bolts and wing nuts or with ball joints or with roller joints such that they can be moved in planes normal to the top plate (Fig. 2.4.2) without wobbling.

For older instrument types, the tripod is connected with the instrument base by means of a screw bolt, which, after penetrating the circular opening of the tripod plate, is screwed into the instrument base. A strong spiral spring pulls the instrument onto the tripod plate. The three (or four, for some older instruments) supports of the instrument base rest on support plates or else in notches which are cut into the tripod plate.

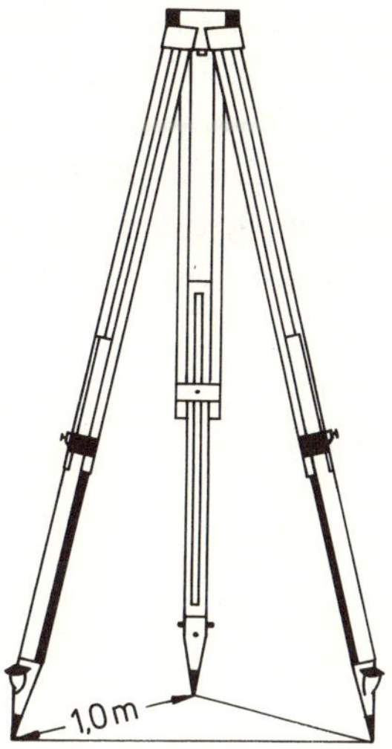

Fig. 2.4.1 Tripod with telescopic legs

In modern instruments, the supports of the base rest on a base plate, which is pressed against the tripod plate with a tightening screw (Fig. 2.4.3). A second – spring loaded – plate, which is lightly connected to the base plate with screws, surrounds the supports and holds the instrument base in place.

The spiral spring, or the spring plate, have to be elastic so that the vertical axis of the instrument can be brought to the plumb direction with the aid of the foot screws, i.e. they have to permit proper levelling of the instrument.

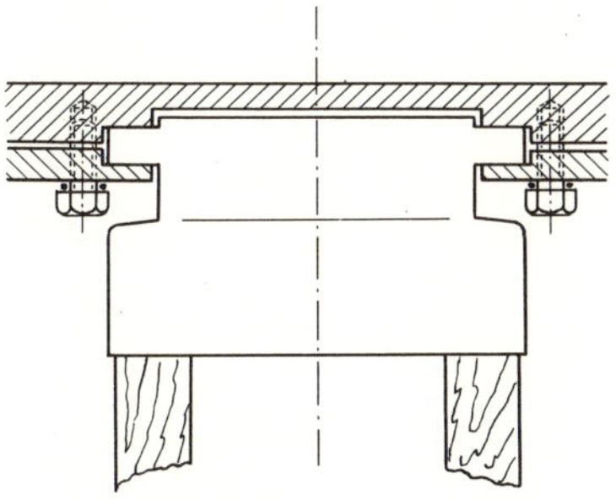

Fig. 2.4.2 Tripod tops

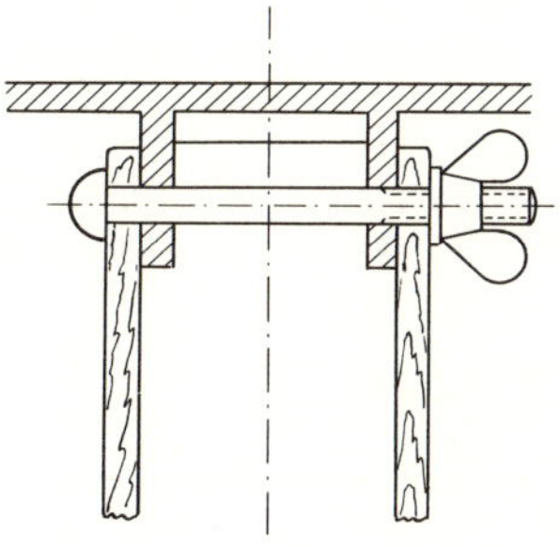

Fig. 2.4.3 Tripod connection

2.4.3 Levelling Tripods

Levelling tripods have been developed for faster set-ups of theodolites and levels. Kern (Switzerland), for instance, manufactures levels with a cylindrical foot which is milled conically on the bottom surface. This support surface can be shifted along the spherical tripod head until the bull's-eye bubble is centred (Fig. 2.4.4). The tightening screw under the spherical tripod head is tightened, and fine levelling with the aid of a tilting screw can proceed. Kern developed a similar tripod for levelling theodolites (section 3.5.2). Levelling tripods have also been developed by other manufactures, such as Breithaupt (W. Germany), and Zeiss (W. Germany).

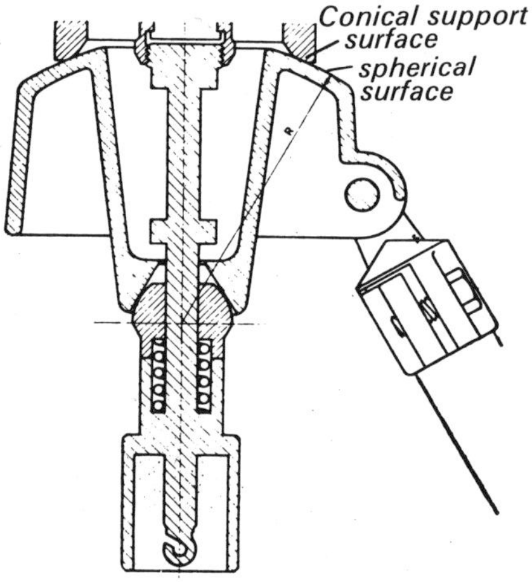

Fig. 2.4.4 Levelling tripod by Kern

3. Theodolites and Angular Measurements

3.1 Horizontal, Vertical and Space Angles

In many surveying tasks, points P_1, P_2 ... are to be determined in a rectangular coordinate system using measured directions and distances (Fig. 3.1.1).

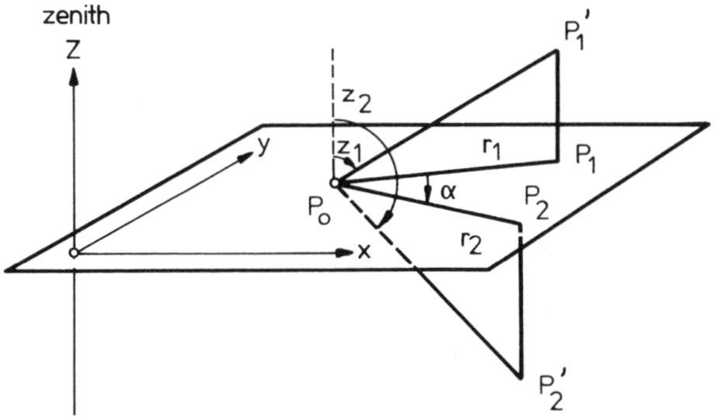

Fig. 3.1.1 Horizontal and vertical angles

The measured directions should be simply defined with respect to the coordinate axes. As it is rather simple to realize the plumb line in surveying instruments, the z-axis is selected in this direction which generates a horizontal x, y plane as the reference surface.

An angle is obtained as the difference between two directions. Depending on whether a horizontal or vertical plane is utilized, one distinguishes between horizontal and vertical angles. The projections of the lines $P_0 P_1'$ and $P_0 P_0'$ onto the horizontal plane define the horizontal angle α (Fig. 3.1.1). The rays from P_0 to P_1', P_2' ... and their projection onto the horizontal plane through P_0 form the vertical angles $(90° - z)$. Frequently the zenith direction is used as a reference, thus providing the complement to 90° which is called the zenith angle z.

The space angle $P_1' P_0 P_2'$ can be measured with a sextant but is of no practical use in surveying.

3.2 The Theodolite (Transit)

3.2.1 External Structure

The theodolite, sometimes referred to as a transit, is used to measure both horizontal and vertical angles (*Cooper* 1982). Its structure is illustrated in Fig. 3.2.1.

The theodolite consists of a fixed base and a top part which can be rotated around the vertical axis. This top part consists of standards which connect the vertical and horizontal axes. The latter is supported by the axis bearings of the standards and

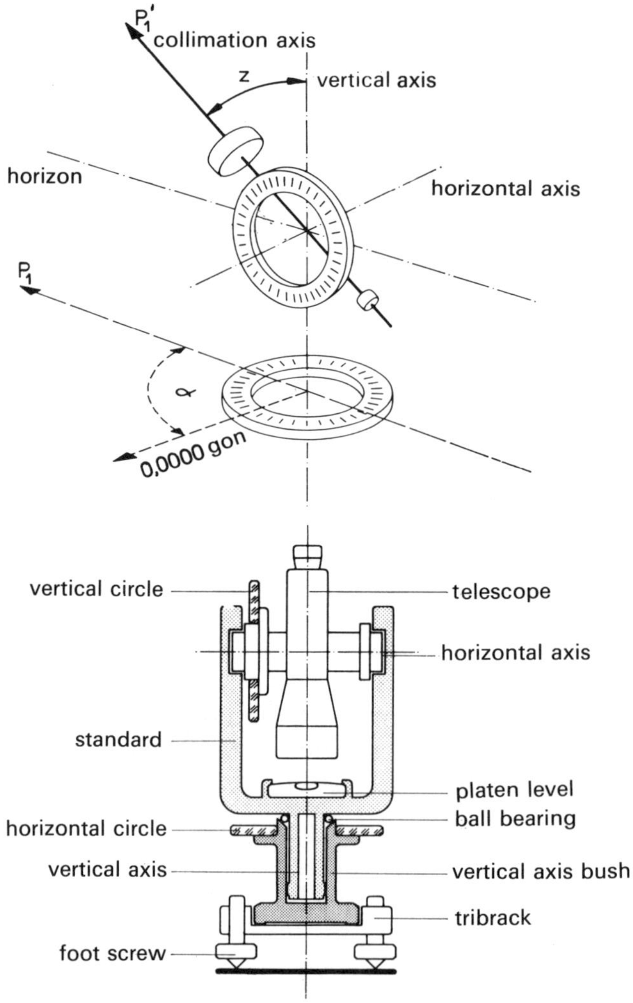

Fig. 3.2.1 Structure of a simple theodolite

carries telescope and vertical circle, which is graduated in degrees or gon. The vertical axis is part of the standards and fits onto the vertical axis bush of the base, which carries the horizontal circle. Again it is graduated in either degrees or gon. The standards are equipped with reading devices for the horizontal and vertical circles. The base can be levelled with the aid of foot screws (three or four) and the platen level. The connection between the vertical axis socket and tribrack can be either fixed or exchangeable for forced centring (section 3.5.4).

3.2.2 The Axes

3.2.2.1 The Vertical (or Azimuth) Axis

This axis has two purposes, namely, to support the weight of the standards and to ensure that the rotational axis of the top part coincides with the centre of the graduation of the horizontal circle. Various designs have been used for this axis (see Figs. 3.2.2 to 3.2.5), which is generally produced of hardened steel. Older theodolites have conical axes, which have to be fitted individually and require frequent cleaning and lubrication. Modern instruments are nearly exclusively equipped with cylindrical axes, which are more suited to mass production and practically need no maintenance.

The reduction of contact area between axis and bush leads to favourable friction conditions, which are further improved by supporting the axis with ball bearings. For some designs a lower and upper guidance ring defines the centre of the axis as well as its angular position. The weight of the standards is then supported by a ball bearing at the bottom end (Fig. 3.2.2). Other designs define the angular position and centre of axis by a lower guidance ring and a ball bearing at the upper end (Fig. 3.2.3). The latter also carries the weight.

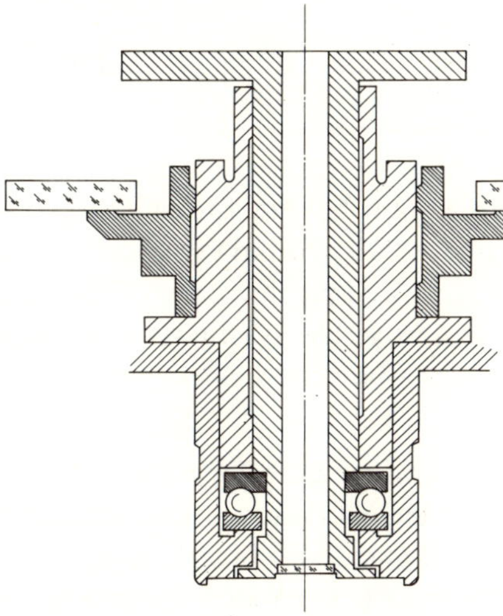

Fig. 3.2.2 Axis system with two guidance rings (*Zeiss*, Oberkochen Th 42)

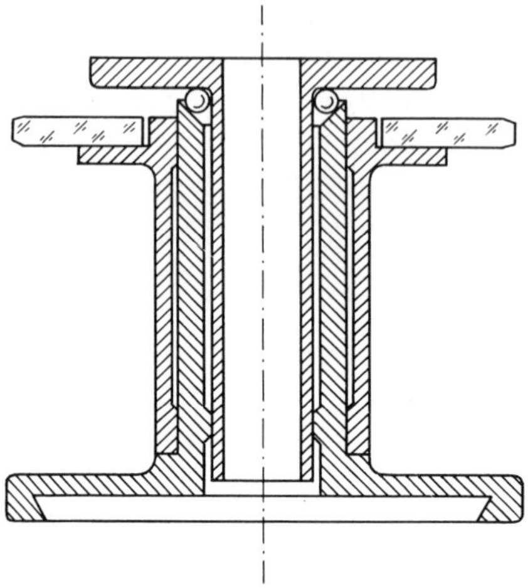

Fig. 3.2.3 Axis system with one guidance ring and ball bearing (*Wild* T 2)

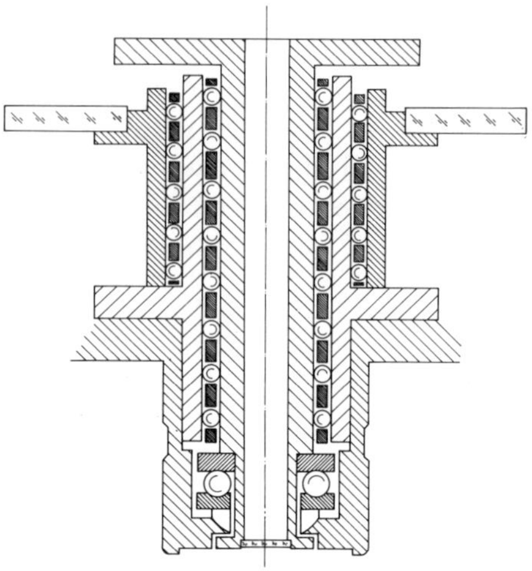

Fig. 3.2.5 Axis system with ball guidance (*Zeiss*, Oberkochen)

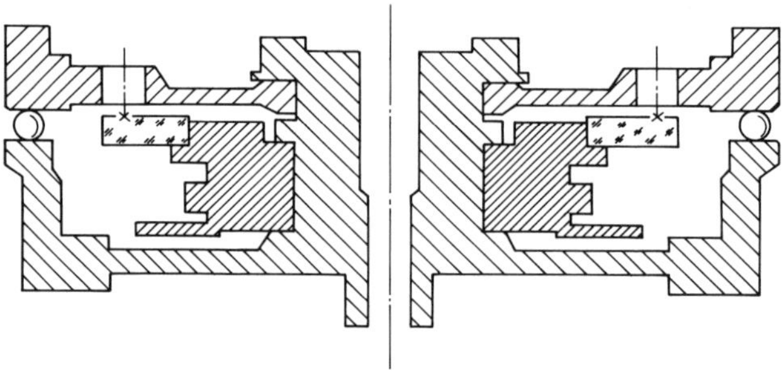

Fig. 3.2.4 Axis system with plane bearing and axis plug (*Kern* DKM 2)

The axis system illustrated in Fig. 3.2.4 has a plane bearing for definition of the angle position and an axis plug for the centre with a ball bearing to support the weight of the standards. The ball guidance axis (Fig. 3.2.5) represents a rolling axis system rather than a sliding one as for the cylindrical axes. The system is free of play because the diameter of the balls is only a few μm less than the space between the axis and bushing.

A distinction between single and biaxial systems is made depending on the arrangement of the vertical axis. Single axis systems have a rigidly connected axis bush and graduated circle (Fig. 3.2.1), while this is not the case for biaxial systems (Figs. 3.2.2 to 3.2.5). The circle has a circle bush which can be rotated around the axis bush without coming in contact with the vertical axis. This circle bush has to ensure that the vertical axis and circle centre coincide. In conjunction with specific measuring procedures, a movable horizontal circle offers the following advantages:

- short periodic graduation errors are reduced and blunders detected when observing in sets (section 3.7.3);
- unavoidable estimation errors when reading are reduced by repetition (section 3.7.4);
- specific values can be set for staking.

Optical theodolites usually are equipped with a movable circle. In electronic theodolites it is possible to substitute electronic switching for the mechanical circle rotation, thus electronic precision theodolites do not always have a movable horizontal circle.

Standard circle and axis system form a uniformly connected element in instruments for forced centring. This can be fastened to the base with the aid of a clamping device, and lifted after loosening the clamp (see section 3.5.5).

3.2.2.2 Horizontal (or Elevation) Axis

This axis facilitates up and down tilting of the telescope and is arranged normal to the optical axis of the telescope. The cylindrical axis plugs of the horizontal axis rest normally in V-supports (Fig. 3.2.6) which are two contact surfaces at 45° from the vertical. The axis is without play, because the relatively high pressures concentrated on a small support surface equally distribute the lubrication film.

One of the supports can be raised or lowered by a small amount such that the horizontal axis can be adjusted normal to the vertical axis. This adjustment is possible for

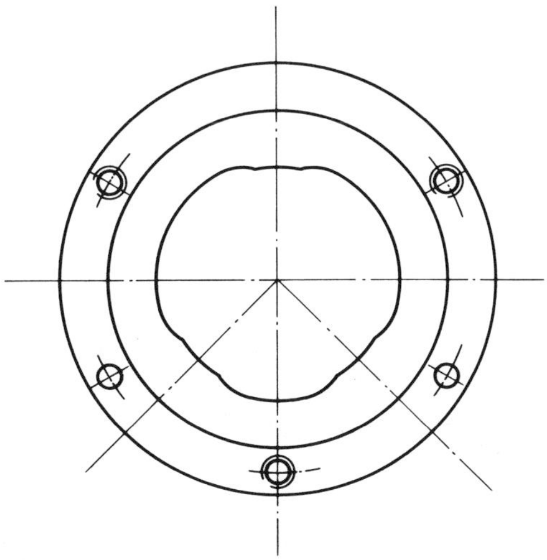

Fig. 3.2.6 Horizontal axis support (*Kern* DKM 2 A)

many theodolites, while others facilitate a change of the angle between these axes by using a wedge shaped flange.

The measuring telescope is positioned in the centre of the horizontal axis. The standards should be high enough to permit plunging of the telescope.

Instruments designed primarily for vertical angle measurements occasionally have the telescope at one end of the horizontal axis outside the supports.

The horizontal axes are normally hollow and thus provide room for the optical rays used for circle readings.

3.2.2.3 The Collimation Axis

The collimation axis of the telescope is (with sufficient accuracy) the straight line passing through the cross hair centre and the centre of the objective lens at ∞ setting (section 2.3.1.3). Its location can be changed somewhat by a mechanical adjustment.

3.2.2.4 Vertical Axis Setting

One or several spirit levels are attached to the top part primarily for setting the vertical axis into the plumb direction.

Electronic levels may fulfil the same function for electronic theodolites. For an angular measuring accuracy of 0.1 to 0.2 mgon ($\approx 0.5''$), the vertical axis has to be vertical within a tolerance of 1 mgon ($3''$). Normally rough levelling is accomplished with the aid of a bull's-eye level. A precision level, arranged normal or parallel to the horizontal axis, is used for fine levelling of the instrument. Theodolites with electronic levels require fine levelling with the foot screws only up to a certain limit provided an internal computer applies corrections for the influences of residual inclination of the vertical axis.

3.2.3 Graduated Circles

3.2.3.1 The Horizontal Circle

The horizontal circle can consist of either *glass* or *metal* (brass), or *plastic* for simple instruments. Its graduation increases clockwise.

Circles *graduated in degrees* are used in most countries and for astronomic observations. Circles with *gon graduations* are preferred for surveying instruments in central Europe.

The graduation is produced as positive or negative with the aid of special graduation machines. Production circles are manufactured by contact copying from this original, which is sometimes referred to as the "mother circle". The following steps are given for a negative: A light sensitive lacquer layer on the top surface of the circle is illuminated through the negative, after which graduation and numbering is dissolved from the lacquer layer. Subsequently, the free glass spots are vaporized with chrome and the remaining lacquer is wiped off. As a result, a positive image of graduation and numbering is obtained. The graduation lines on glass circles have a thickness of several μm.

The influence of possible circle subdivision errors becomes smaller for larger circle diameters. Therefore, the diameter of the circle is often viewed as a sign of the quality of the theodolite. Field instruments have circle diameters of 6 to 10 cm, with circle *subdivisions* of usually 20′ or 30′, or 0.5 gon.

More accurate readings are obtained by *interpolation* between neighbouring graduation lines using reading microscopes or electronic scanners. Errors in the graduation lines can be kept small by selected scanning procedures and special measuring arrangements.

Sections of graduated *circles of optical theodolites* are illustrated in Fig. 3.2.7 (a) to (c). There are simple graduations (a), double circle graduations (b), and double circle graduations with single and double graduation lines (c).

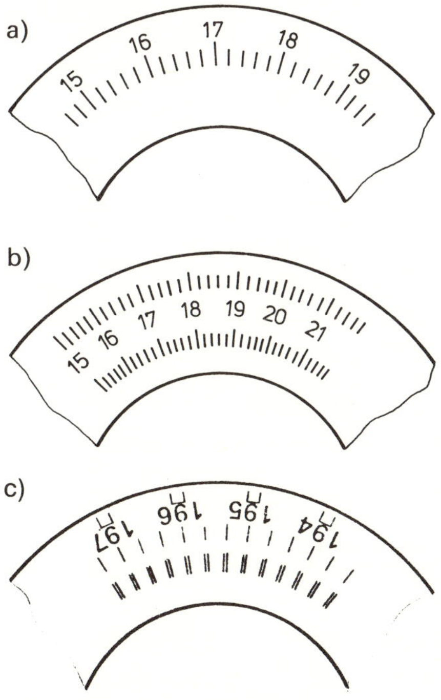

Fig. 3.2.7 Circle sections of theodolites with optical reading (*Kern*)

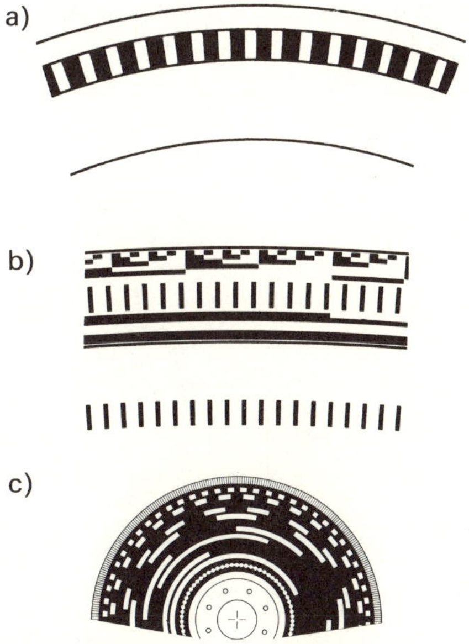

Fig. 3.2.8 Circle sections of theodolites with electronic circle scanning
a) *Wild* T 2000, Kern E 1
b) *Zeiss-Oberkochen*, Elta 2
c) *Hewlett-Packard* HP 3820

Sections of graduated *circles of electronic theodolites* are illustrated in Fig. 3.2.8 (a) to (c). They carry binary coded numbers or none at all instead of the customary Arabic numerals.

3.2.3.2 The Vertical Circle

This is made of the same material as the horizontal circle and is subdivided similarly, although it is occasionally a bit smaller. While the circle remains stationary for horizontal angle measurements when the alidade is rotated, the vertical circle is attached to the horizontal axis and moves with the telescope.

The reading device is rigidly attached to the standards. Since vertical angles are referred to the zenith direction, the reading index and centre of the circle are on a line pointing along the plumb direction. A manually adjustable reading device is illustrated schematically in Fig. 3.2.9. In this particular case the reading device consists of a frame that carries the index as well as a height index bubble. A fine motion screw rotates this frame slightly around an axis that coincides with the horizontal axis. After the height index bubble is centred by the observer using the fine motion screw, both index and centre of the circle point into the zenith direction, provided that the index bubble is properly adjusted. Most newer instruments are equipped with an automatic index, where a compensator acting on gravity automatically sets the index into its zero position (section 10.2.3). As illustrated in Fig. 3.2.9, the vertical circle is graduated such that zenith angles, rather than vertical angles, are read.

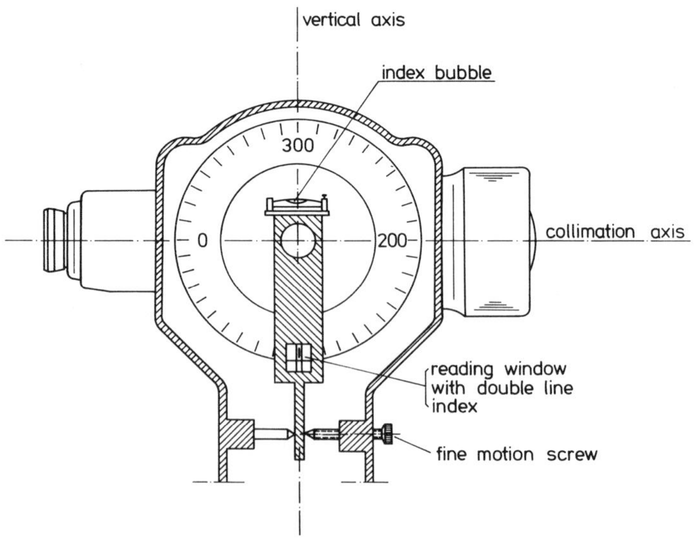

Fig. 3.2.9 Vertical circle (schematic)

It is evident from Figs. 3.1.1 and 3.2.9 that the vertical angle is referred to the point of intersection between vertical, horizontal and collimation axes. Thus the vertical distance between the ground point and horizontal axis has to be considered for elevation calculations.

3.2.4 Clamps, Fine Motion and Circle Motion

In order to keep and accurately set the circle and/or telescope in a certain position, the manually rotatable parts have to be solidly linked. An additional rotation by a small angle must be possible utilizing a fine motion device. Among others, these tasks can be achieved using

- clamps and fine motion screws
- circle clamp
- circle drive with cogwheel arrangement.

The top part and vertical axis socket are temporarily locked together during the measurement of horizontal angles. A simple design is illustrated in Fig. 3.2.10.

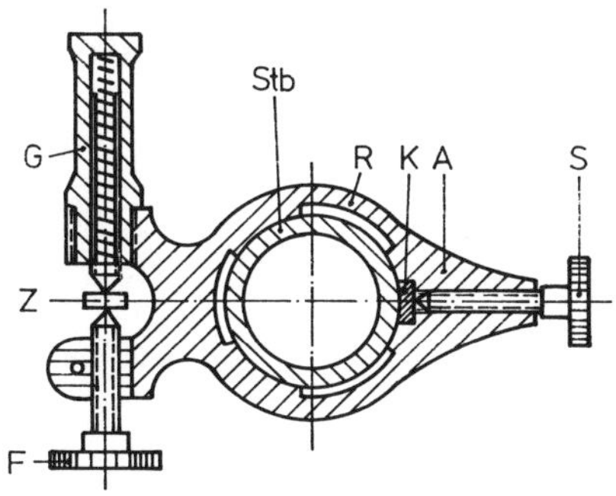

Fig. 3.2.10 Clamp with lateral fine motion screw

The ring *R* is pressed against the axis socket using the screw *S* and a clamping pad *K*. The fine motion screw *F* and its counteraction spring *G* hold a peg *Z* which is connected to the standards. This fine motion facilitates small rotations of the standard. The setting accuracy can be improved for precision instruments by an additional lever system.

In newer instruments the clamp and fine motion are arranged coaxially, as shown in the example in Fig. 3.2.11. A rotation of the clamping screw *S* presses the clamping pad *K* against the vertical axis socket *Stb*. When turning the fine motion screw *F*, a bolt *B* presses against the lever *H* whose motion is transferred onto the standards with a second bolt *Bo*.

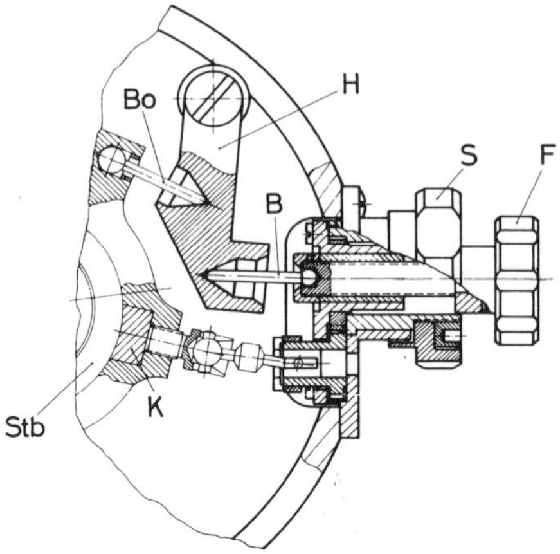

Fig. 3.2.11 Clamp with coaxial fine motion screw (F. W. Breithaupt & Son)

The height clamp fixes the horizontal axis in respect to the standards. The fine motion screw facilitates the exact setting of the vertical directions. Their design principles correspond to the ones for horizontal angle measurements. Figs. 3.2.12 and 3.3.7 illustrate different constructions.

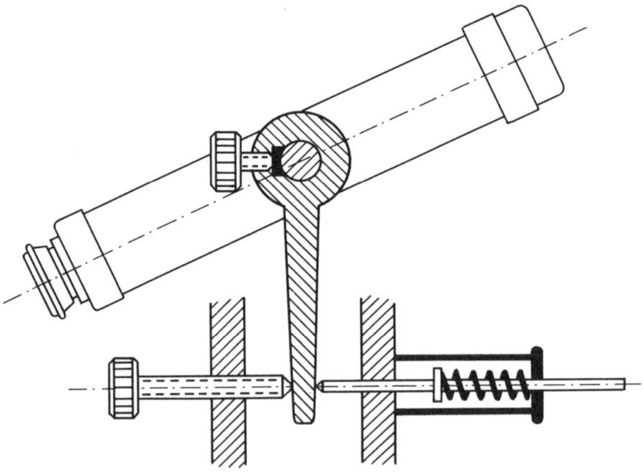

Fig. 3.2.12 Height clamp and fine motion

Biaxial instruments offer additional possibilities for circle motion. In simple theodolites the circle bush is resting under friction on the vertical axis bush. In order to move the circle, it is temporarily linked to the standards with the aid of a circle clamp. This clamp is located on the standard as shown in Fig. 3.2.13. A disk membrane is connected to the circle which can be temporarily fastened to a lever situated on the top part. The circle will then follow the motion of the top part.

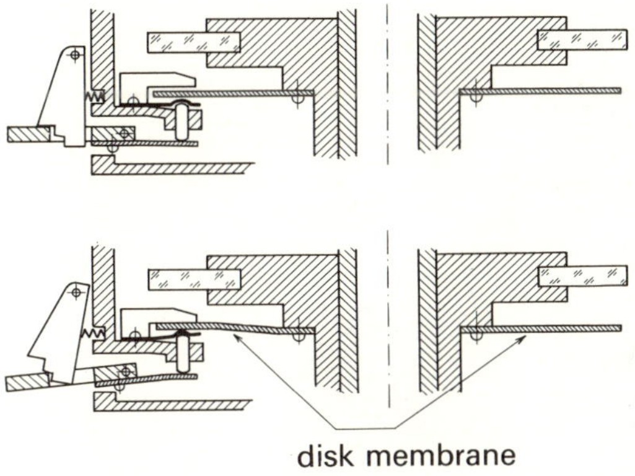

Fig. 3.2.13 Circle clamp (*Zeiss*, Jena)

More sophisticated theodolites have two clamps, each equipped with a fine motion screw. Fig. 3.2.14 illustrates possible arrangements. At instruments of type (a), the observer can couple the top part with the circle support, as well as with the axis bush, and shift each with a fine motion. At the design (b), the top part can be coupled with the circle support and the latter with the axis bush. Again, the individual elements can be moved with respect to each other. These designs with circle clamps facilitate

- orientation of the horizontal circle (zero setting)
- repetition measurements.

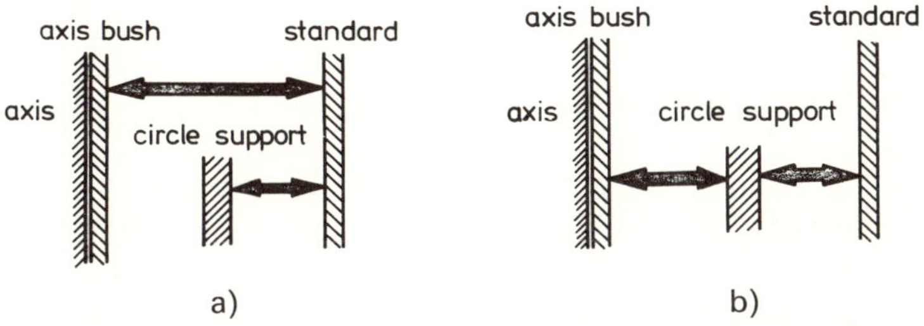

Fig. 3.2.14 Different arrangements for the fine motion between the top part and the circle (e.g. a) *Zeiss*, Oberbochen Th 42, *Kern* KiS, KiM; b) *Wild* T16, T1

Theodolites of medium and higher accuracy have a circle drive with a cogwheel arrangement which permits rotation of the horizontal circle even though the top part and the vertical axis bush are rigidly connected. The design as illustrated in Fig. 3.2.15 has a circle bush with cogs which is activated by the cogwheel of the setting knob. A spring pushes the cogwheel off the cogs when it is not activated.

The advantages of mechanical circle motion can be replaced by electronic scanning

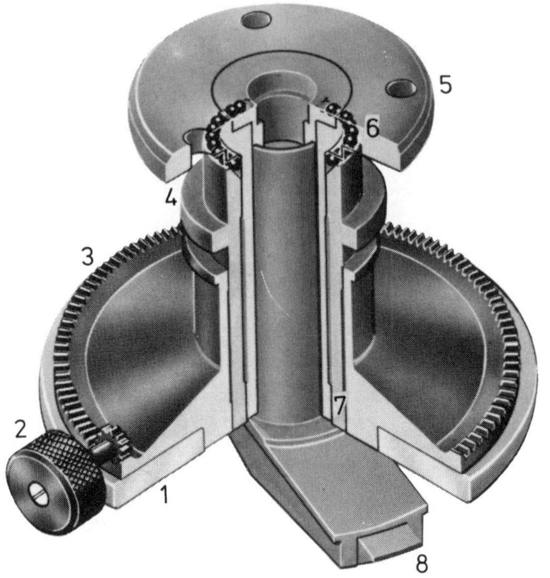

Fig. 3.2.15 Circle drive with cog wheel arrangement (*Wild* T2)
1 horizontal circle, 2 setting knob, 3 circle bush, 4 axis bush, 5 vertical axis, 6 ball bearing, 7 slide-bearing, 8 reading prism

procedures in electronic theodolites. Therefore most electronic theodolites have fixed circles.

3.2.5 Circle Reading and Scanning Devices

The direction of the collimation axis can be either read at the circles or else electronically scanned. The type of data gathering is of prime importance for subsequent data handling. Therefore, one distinguishes between *optical theodolites* (utilizing optical circle readings) and *electronic theodolites* (utilizing electronic circle scanning), even though their mechanical design is rather similar. With electronic theodolites an automatic data flow can be generated from the gathering of the data to data handling and documentation. They are especially suited in conjunction with electronic distance meters for extended areas and large amounts of data. Optical theodolites are far less expensive, but they are also less suited for automation. They are thus more economical for smaller tasks.

3.3 Optical Theodolites

Circle reading devices are primarily built as vernier or reading microscopes. Only very primitive instruments have a simple line index with a magnifying glass. The type of reading device influences the structure and accuracy of the theodolite.

3.3.1 Vernier and Vernier Theodolites

The vernier is a short auxiliary graduation attached to the alidade such that it is in one plane with the limbus. The basic idea of its design is as follows: $(n-1)$ units of the circle graduation C face n units V of the vernier graduation such that

$$(n-1)\,C = nV\,.$$

The difference

$$C - V = \frac{C}{n} = a$$

is called the vernier constant. If the xth vernier line coincides with a line of the circle graduation, the separation of the zero line of the vernier from the previous graduation line of the circle is $x \cdot a$.

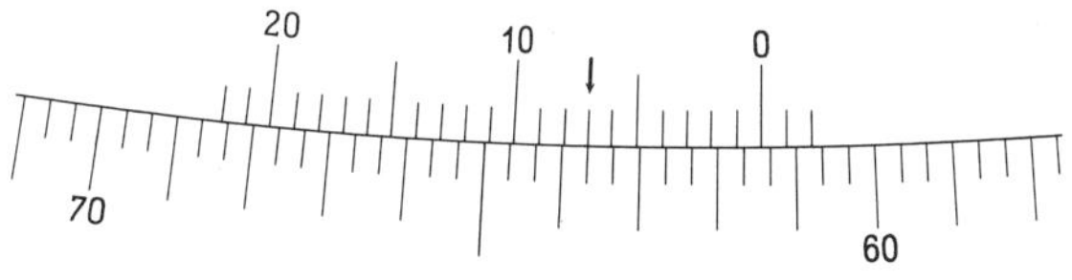

Fig. 3.3.1 Vernier

The circle shown in Fig. 3.3.1 is graduated into ${}^{1}/{}_{3}{}^{\circ} = 20'$, i.e. $C = 20'$. The vernier has 20 intervals, which means that $a = 20'/20 = 1'$. When reading, one first has to find and record the graduation line preceding the zero mark of the vernier, e.g. $61^{\circ}20'$ in the example at hand. Then one moves toward the left to the vernier until a vernier line "coincides" with a graduation line, i.e. they match in such a manner that they appear as one straight line. This is the case for the 7th vernier line, which thus provides the total reading as $61^{\circ}27'$.

Above the reading locations there are magnifying glasses which can be shifted somewhat within their casings to generate a sharp image. It is advisable to observe the position of the neighboring line pairs in addition to the coinciding one. To make this possible, even when the first or last line of the vernier coincides, one or two lines are usually provided beyond the end lines of the vernier. If no line pair provides exact coincidence, the mean of the two best fitting line pairs is taken.

Vernier theodolites are rarely built any more. The available instruments are either one axial or else repetition instruments, usually having two verniers. Some instruments do not even have vertical circles. Depending on the accuracy, there are construction, engineer's, and precision theodolites.

The construction theodolites are simple instruments with circle subdivisions of whole or half degrees and vernier accuracies of 1 to 2 minutes of arc.

The engineer's theodolites normally are equipped with a repetition arrangement and with a vertical circle. Both circles are divided into $^1/_2{}^\circ$ or $^1/_3{}^\circ$ (1 gon or 0.5 gon). The verniers provide 30″ or 20″ (1 cgon or 0.5 cgon). The common transit belongs to this category.

Verniers are found on older precision theodolites but only in combination with microscopes.

3.3.2 Reading Microscopes and Microscope Theodolites

Reading microscopes appear in several types of design, namely

- line microscopes,
- scale microscopes,
- line microscopes with optical micrometer, and
- coincidence microscopes with optical micrometer.

While the first three cover only one circle position for reading, the coincidence microscope permits simultaneous reading on two diametrically opposite circle positions.

The optical systems are usually placed inside a tube next to the telescope (Figs. 3.3.2 and 3.3.3).

Both aiming and reading are simple and fast, if the eyepiece of the microscope is next to that of the telescope, and if the images of both circles are simultaneously displayed in every observation position. The circle images are colour coded for some theodolites. The readings of horizontal and vertical circles are then clearly distinguished which practically eliminates reading the wrong circle. Usually a rotatable mirror on one side of the standards provides illumination. For night observations or underground an artificial light source is required. Many theodolites are equipped with an electric lamp attachment for these purposes.

Most modern theodolites have only one reading microscope.

In an optical sense, the microscope can be considered as a telescope with an objective lens with a short focal length, the object being located between the single and double focal lengths. This object, of course, is a section of the circle graduation. The image plane of the microscope carries a reading index in place of the reticule, whose separation to the previous circle graduation line is to be determined. Depending on

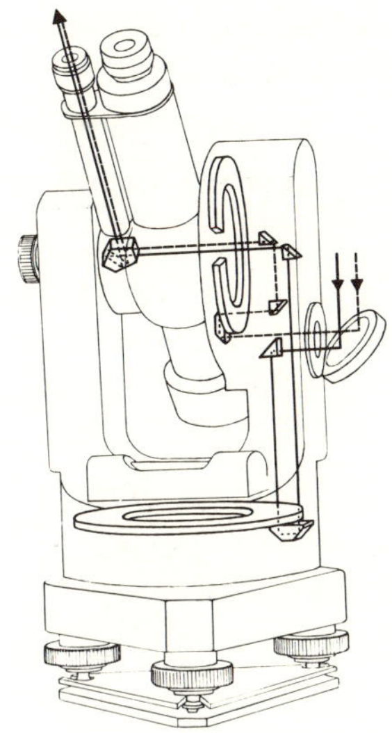

Fig. 3.3.2 Optical train of a reading microscope in a theodolite with one circle reading position

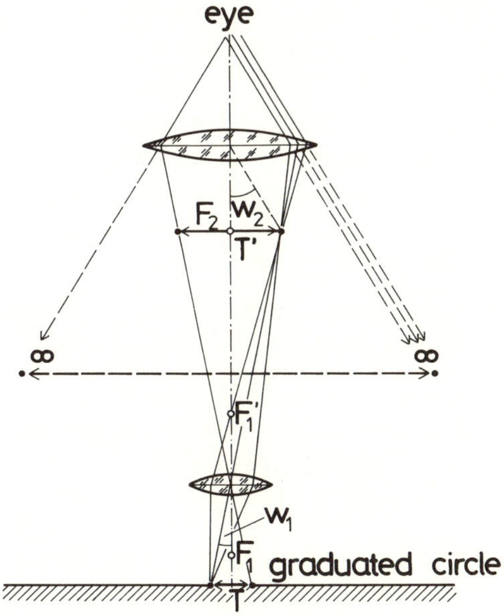

Fig. 3.3.3 Optical train within the reading microscope

how this is accomplished, one distinguishes between line, scale and coincidence microscopes.

In order to provide correct readings, the microscopes have to fulfil several optical conditions. Therefore, appropriate adjustment possibilities are necessary.

a) The separation between image plane and objective lens – i.e., the focal length – must be slightly variable in order to obtain a sharp image of the circle sections in the index plane.

b) If there is a scale or similar feature in addition to the zero line in the image plane, then the separation between microscope and circle – i.e., the object distance – must be variable, so that the scale fits exactly to the image of the corresponding graduation interval.

A possible difference between the scale dimensions and the image of the corresponding graduation interval is denoted as the *run*. For modern compact instruments, deficiencies in a) or b) can be corrected only in special workshops.

In addition to these objective settings, necessary for adjustment, there is a subjective demand as well. The observer has to be able to shift the eyepiece in its housing to bring the index into the most suitable position for his eye.

3.3.2.1 The Line Microscope

The simple line microscope (Fig. 3.3.4) uses a line as the index which is realized by a spider thread or a line engraved into a reticule. Only the first of the conditions mentioned in section 3.3.2 has to be fulfilled for proper adjustment. The circle of a line microscope theodolite is normally subdivided into 10′ or 10 cgon. The reading is obtained by estimating the separation of the index from the nearest graduation line to within a tenth of the graduation interval. The reading accuracy is not very high, but reading is fast and safe. Fig. 3.3.5 shows the field of view of a more modern theodolite with both horizontal and vertical circle readings.

The vertical circle in example b) of Fig. 3.3.5 carries an additional percent graduation between -15% and $+15\%$ which permits the direct determination and control of slopes.

If a higher reading accuracy is desired, the separation of index and graduation line is increased with the aid of a mechanical or optical micrometer.

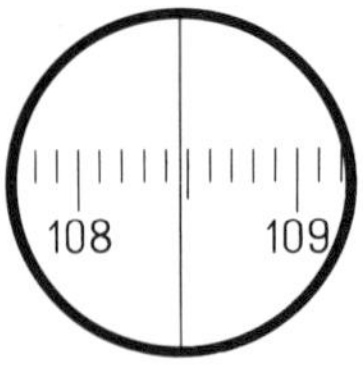

Fig. 3.3.4 Reading of a simple line microscope

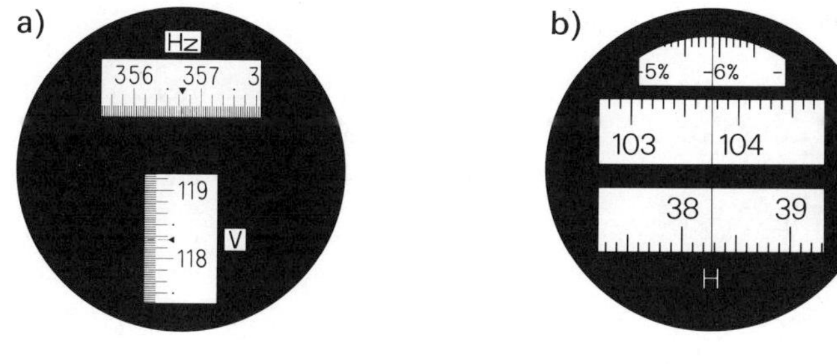

Fig. 3.3.5 Fields of view of line microscopes with horizontal and vertical circle readings
a) *Wild* T05 b) *Kern* K0-S

3.3.2.2 The Scale Microscope

This microscope is equipped with a scale in the image plane of the microscope whose length corresponds to the line separation of the graduated circle, provided that it is properly adjusted. Numbering originates at the zero index, and usually increases in the opposite direction to the circle graduation. Thus the observer reads directly the separation between the zero line and the previous graduation line.

For newer theodolites, the circles are subdivided into 1 gon or 1° with a scale of 10 mgon or 0.5′ or 1′ graduation. Depending on the instrument, 1 to 2 mgon or 0.1′ are estimated. Different examples are illustrated in Fig. 3.3.6.

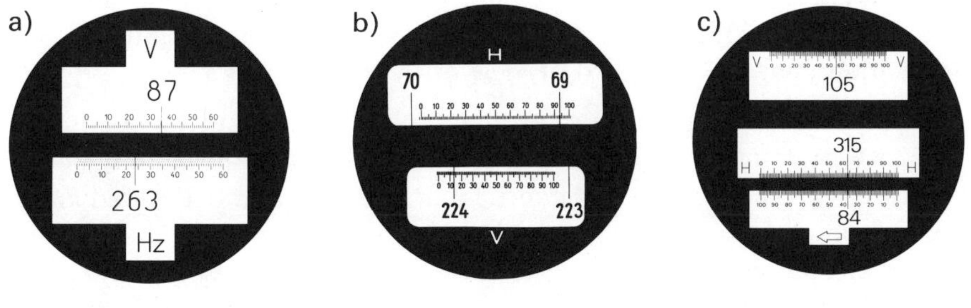

Fig. 3.3.6 Fields of view of scale microscopes
a) Zeiss, Oberkochen, TH42; b) *Zeiss*, Jena Theo 020B; c) *Kern* Kl-S

The images of horizontal and vertical circles in a) and b) are colour coded to avoid additional reading errors. The horizontal circle of instrument c) is numbered both clockwise and counter-clockwise. Thus the time consuming and error prone calculation of complementary angles as needed for staking is not necessary.

3.3.2.3 Line Microscope with Optical Micrometer

In addition to the reading system, a movable deflection element is utilized. This permits a shift of the image of the circle section relative to the reading index. This deflection element usually consists of a plan-parallel plate which can be rotated around a fixed axis (Fig. 3.3.7).

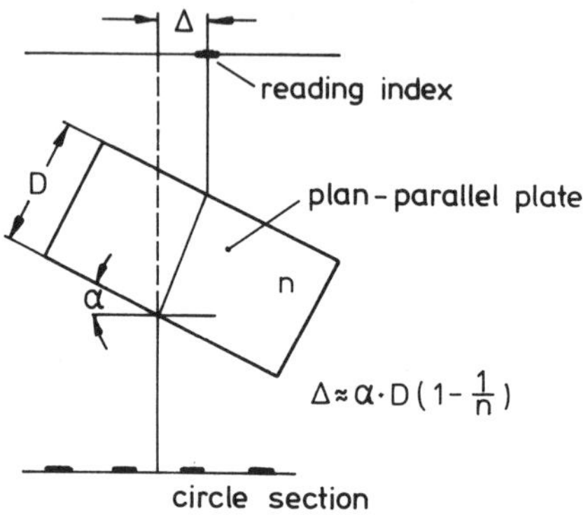

Fig. 3.3.7 Plan-parallel plate used as deflection element in optical micrometer, (n = refractive index; D = plate thickness)

The shift Δ of the image of the circle segment is approximately proportional to the rotation angle α of the plan-parallel plate. The observer moves the image until the previous graduation line and the reading index coincide. A double line is particularly suited as a reading index. The shift amount Δ is read in angular units on a micrometer graduation imaged in the field of view of the micrometer.

The functional principle of an optical plan-parallel plate micrometer is illustrated in Fig. 3.3.8. A rotating mirror is used to direct light onto the reading position of the vertical circle. The segment of the vertical circle is imaged into the graduation plane of the horizontal circle through prisms, a liquid compensator and a system of lenses. The images of both circles are projected into the index plane by additional prisms and lenses. After a further deflection, this image is observed in a microscope located parallel to the telescope. There are two windows, each with a double line index to facilitate the reading of both circles. A scale, rigidly connected to the plan-parallel plate, is observed through a third window with a wedge shaped index. The parallel plate is connected to the micrometer screw with a lever. If the parallel plate is rotated with the micrometer screw, the third window with its wedge shaped index provides the shift amount of the circle images in angular units. The horizontal and vertical circles can be consecutively read using the same micrometer.

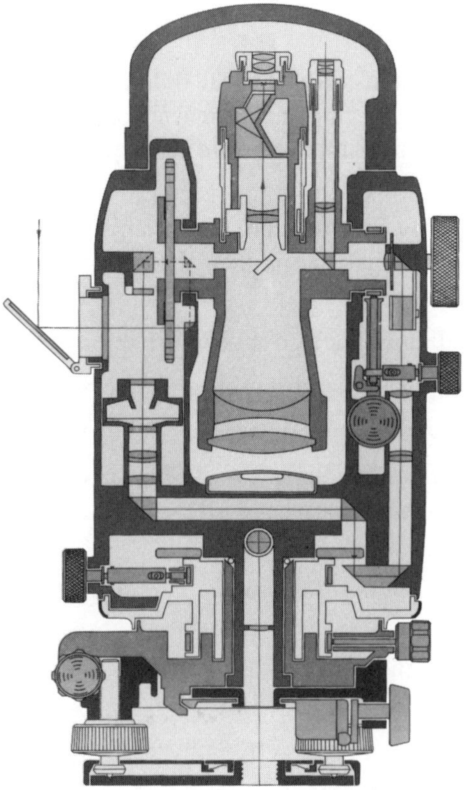

Fig. 3.3.8 Optical train in a theodolite with optical micrometer (*Wild* T1)

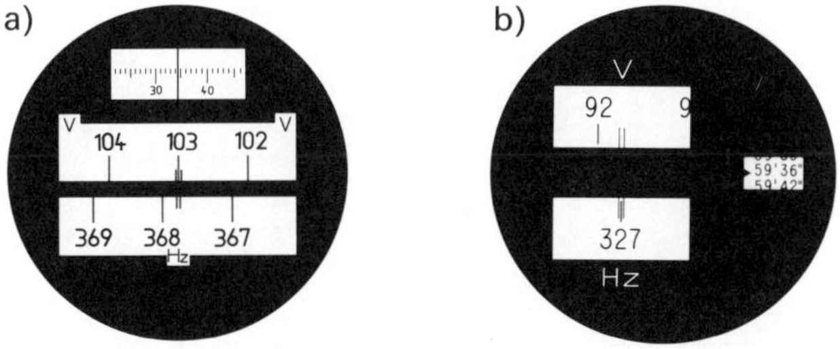

Fig. 3.3.9 Fields of view of line microscope with optical micrometers
a) *Breithaupt* TEAUT; b) *Wild* T1;

The graduation interval of the circles is usually 1° or 1 gon. Intervals of 5 mgon to 1 cgon or 10″ to 20″ can be read directly, while smaller intervals of about 1 mgon or 3″ are obtained by estimation. In Fig. 3.3.9 different readings are illustrated.

3.3.3 The Coincidence Microscope

This was first built by Carl Zeiss in Jena according to H. Wild's design. It brought about a complete change from previous design principles. The basic principle of this design is to optically combine two diametrically opposite circle positions such that they can be read at once.

Since then different imaging systems have been utilized. Depending on the theodolite, the graduation lines are combined in such a manner that their images in the microscope

- meet along a separating edge while moving in opposite directions (Fig. 3.3.11);
- are superimposed and move in opposite directions (Fig. 3.3.16b);
- are superimposed and move in the same direction (Fig. 3.3.16a).

The opposite sense of motion is achieved when using a roof prism in the optical train, while simple deflecting prisms lead to the same sense of motion. Basically, two imaging processes can be distinguished

- both images are symmetrically matched and then imaged in the microscope (Figs. 3.3.10 and 3.3.12);
- one reading position is imaged onto the diametrically opposite one, and both are then imaged through the microscope (Figs. 3.3.14 and 3.3.15).

The circle graduations consist either of single or double lines (Fig. 3.2.7).

The schematic illustration in Fig. 3.3.10 provides the basic idea. The symmetrical imaging system (1) images two sections of diametrically opposite circle segments (2) in the image plane of the measuring microscope.

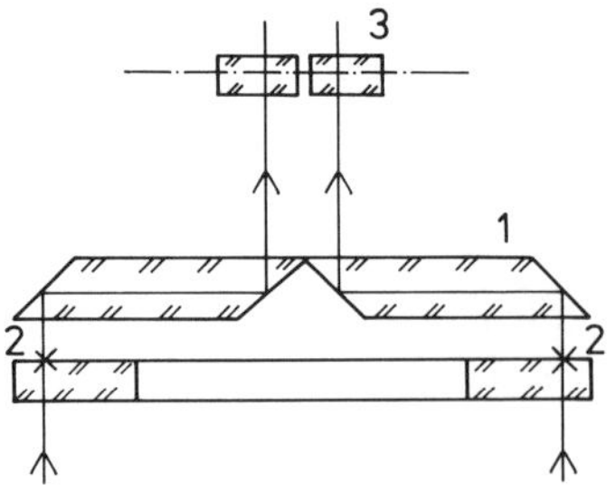

Fig. 3.3.10 Imaging of two circle segments with a symmetrical imaging system

As shown in Fig. 3.3.11, one circle section appears upright, the other upside down along a separation edge. If a separate index is available (dotted line in Fig. 3.3.11 (a)), then it has to have equal separation from the corresponding numbers which differ by 200 gon (180°).

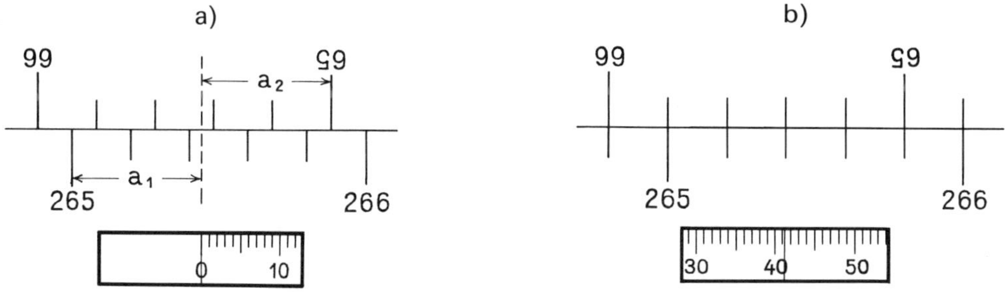

Fig. 3.3.11 Coincidence microscope before and after operating

The reading is obtained as follows. Starting with the upright gon value in the left half of the reading window, and assuming for a moment that the dotted reading index is there, the following reading would be obtained:

on the upright circle section:	265	$+a_1$
on the inverted circle section:	65	$+a_2$
which gives a mean of:	265	$+\dfrac{a_1+a_2}{2}$.

Therefore, decimals of a gon are obtained by dividing the apparent distance between the graduation lines 265 and 65 in half. In order to get this halved amount automatically, one should read as follows: Take the graduation line which differs by 200 gon as the reading index, and value every 20 cgon interval as 10 cgon. Then the reading becomes 265.44 gon, with the last digit being estimated. The dotted zero mark is therefore not necessary for the reading and thus is often omitted. With this, a coincidense microscope without micrometer is obtained.

Most coincidence theodolites, however, are equipped with optical micrometers. In this case, the last two digits are not estimated. The interval between direct and reflected graduation lines is measured with a micrometer. In its simplest form, the micrometer consists of two plan-parallel plates, (3) in Fig. 3.3.10, which rotate in opposite directions. They are rotated with the aid of a screw drum until the graduation lines of both circle sections coincide. Then the full and decimal gons are obtained from the main graduation as mentioned before, while the amount exceeding full 10 cgon units is halved and then displayed on an auxiliary scale. The exact reading is then 265.4412 gon.

The reading system of the theodolite Th2 by Zeiss Oberkochen is based on the described principle, and the Wild T2 has a similar optical train (Fig. 3.3.12). The fields of view of both theodolites are now digitized to such an extent that only 0.1 mgon for 400 gon graduation, or 1″ intervals for 360° graduation, are read on a scale (Fig. 3.3.13). Horizontal and vertical circles carry two concentric graduations in

Fig. 3.3.12 Optical train for horizontal circle reading (*Wild* T2)

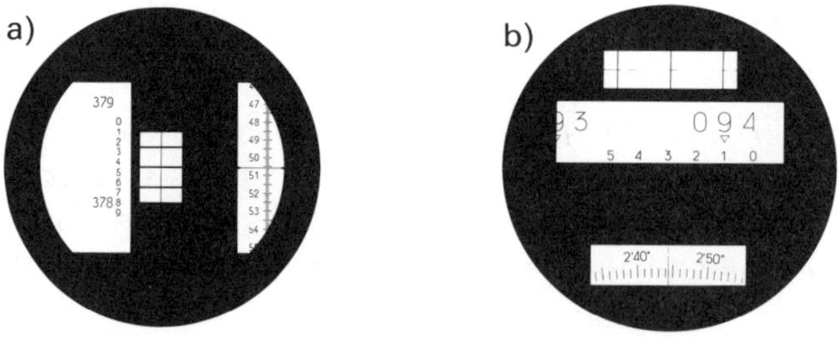

Fig. 3.3.13 Fields of view of coincidence microscopes
a) *Zeiss*, Oberkochen Th2 b) *Wild* T2

some theodolites (Figs. 3.3.14 and 3.3.15). In such double circle theodolites, one segment of one graduation is imaged onto the diametrically opposite one such that both are superimposed. This image appears in the field of view of the microscope after some additional imaging processes.

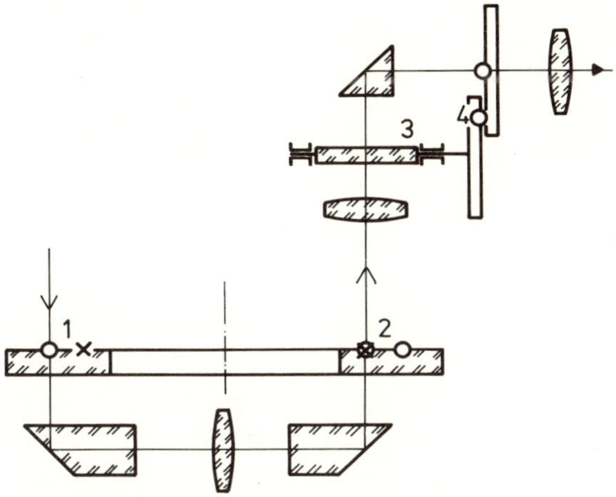

Fig. 3.3.14 Double circle theodolite with simple graduation (*Kern* DKM 1)
1 outer circle graduation, 2 inner circle graduation, 3 plan-parallel plate of micrometer, 4 micrometer scale

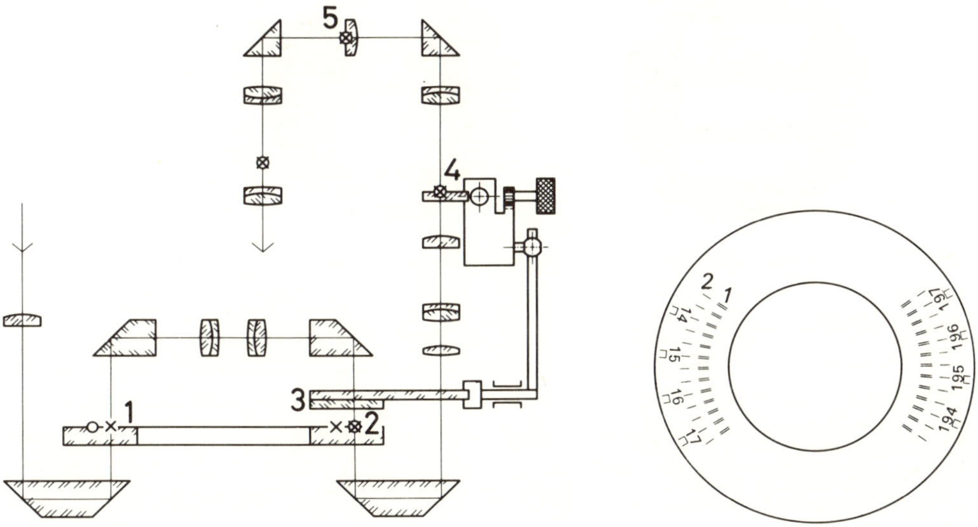

Fig. 3.3.15 Double circle theodolite with circle graduation of single and double lines (*Kern* DKM 2A)
1 inner graduation, 2 outer graduation, 3 plan-parallel plates of micrometers, 4 micrometer sled with scale, 5 graduation of scale microscope, ⊗ intermediate image

There are different solutions for achieving the imaging and coincidence, as indicated in the following examples. The imaging system images the lines of the double circle with simple graduation so that they are shifted but move in the same sense (Fig. 3.3.14). The lines appear then as rigidly corresponding double lines. A fine reading is obtained when they are symmetrical to a fixed index line after moving them with a plan-parallel micrometer (Fig. 3.3.16a). This rather simple system permits smaller dimensions and lower weight of the theodolite while providing a reading accuracy of 0.5 mgon (1″).

The small travelling and expedition theodolite Kern DKM 1 is equipped with such a device.

Another instrument (Fig. 3.3.15) has one graduation with single and the other with double lines. One segment of one graduation is imaged onto the diametrically opposite segment through an optical system generating an opposite sense of motion. The micrometer consists of a measuring parallel plate in the optical train between the diametrically opposite circle positions. A second plan-parallel plate between the numbered circle position and the field of view stop is used for the rough reading. The observer moves the measuring parallel plate until the graduation lines of one side appear symmetrically between the ones of the other side (Fig. 3.3.16b).

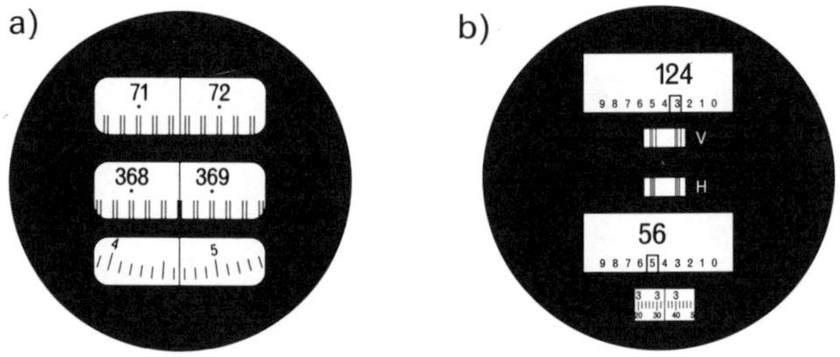

Fig. 3.3.16 Fields of view of coincidence microscopes in which graduation lines are superimposed
a) *Kern* DKM 1 b) *Kern* DKM 2A

The reading optics also serves as a scale microscope which captures 10 cgon or 10′ figures. The graduation of the scale microscope consists of one row of figures only. The second parallel plate shifts the common image of the diametrically opposite circle position across a row of numbers such that the desired tenth of a degree of the rough reading stands directly below the numbered graduation line (Fig. 3.3.16b). The frame around the numbers to be read is an extension of the numbered graduation line. The image shift by means of the auxiliary parallel plate need only be such that the frame is centred approximately around the number to be read.

With a bit of practice, the graduation lines can be coincided for most instruments to within 1″ or 0.2 to 0.3 mgon. Since two diametrically opposite circle sections are used for the reading, the result is free of influences of possible circle eccentricities (section 3.6.2.1).

3.3.4 Classification of Optical Theodolites

Based on the accuracy, one distinguishes between theodolites of low, medium, high and highest accuracy (see Table 3.1).

3.3.4.1 Low Accuracy Theodolites

Low accuracy theodolites, or *construction or minute theodolites* are designed for simple construction measurements and rough topographical work. If these instruments have microscope readings at all, then only simple line or scale microscopes without micrometers. They have circle diameters of 50 to 80 mm and telescope magnifications of 15 to 20 times. Some instruments do not have a vertical circle.

Usually the observer can directly read intervals of 10′ or 10 cgon and estimate to 30″ or 0.5 to 1 cgon (Figs. 3.3.17, 3.3.4 and 3.3.5).

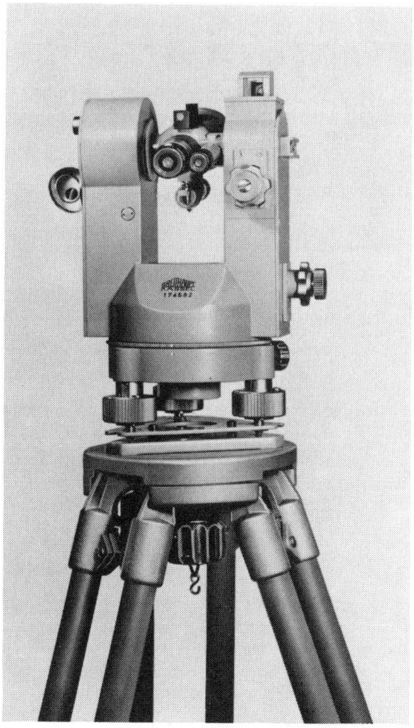

Fig. 3.3.17 Theodolite with line microscope (*Breithaupt* TEKAT)

Table 3.1 Optical Theodolites

Manufacturer	Type	Telescope Magnification	Telescope Obj.-aperture mm	Circle Diam horiz. mm	Circle Diam vert. mm
1	2	3	4	5	6
(1) Low Accuracy Theodolites					
Breithaupt, W. Germany	Tekat	18	30	65	47
Ertel, W. Germany	BTU	24	30	80	60
Fennel, W. Germany	FTS	19	25	80	60
Kern, Switzerland	KO-S	19	24	79	69
Wild, Switzerland	T 05	19	25	67	50
Zeiss, E. Germany	Theo 080 A	18	32	61	61
(2) Theodolites of Medium Accuracy					
Breithaupt, W. Germany	TEAUT	28	40	78	50
Fennel, W. Germany	FT 1 A	30	40	90	70
Kern, Switzerland	K 1 S	30	45	95	75
Kern, Switzerland	K 1 M	30	45	89	70
Kern, Switzerland	DKM 1	20	30	56	56
Nikon, Japan	NT-2 A	30	45	84	84
Nikon, Japan	NT-2 C	30	45	84	84
Sokkisha, Japan	TS 6	30	45	95	85
Sokkisha, Japan	TM 20 E	30	40	80	70
Wild, Switzerland	T 1	30	40	73	65
Wild, Switzerland	T 16	30	40	79	79
Zeiss, W. Germany	Th 4	25	35	98	85
Zeiss, W. Germany	Th 42	30	40	98	85
Zeiss, E. Germany	015 B	30	40	86	86
(3) Theodolites of High Accuracy					
Kern, Switzerland	DKM 2 A	32	45	80	74
Nikon, Japan	NT-5	30	45	94	74
Sokkisha, Japan	TM 1 A	30	40	94	80
Wild, Switzerland	T 2	30	40	90	70
Zeiss, W. Germany	Th 2	30	40	100	85
Zeiss, E. Germany	010 B	30	40	86	86
(4) Theodolites of Highest Accuracy					
Kern, Switzerland	DKM 3	45	68	104	104
MOM, Hungary	Te-B 23	30	45	93	60
Wild, Switzerland	T 3	24–40	60	135	90
Wild, Switzerland	T 4	60–80	70	240	135
Zeiss, E. Germany	Theo 002	38–75	65	250	200

System	Reading [gon] horiz. Circle		horiz. Circle		[° ′ ″] horiz. Circle		horiz. Circle	
	directly	esti-mated	directly	esti-mated	directly	esti-mated	directly	esti-mated
7	8	9	10	11	12	13	14	15
Opt. scale	5	0.5	5	1	5′	0.5′	5′	0.5′
Direct	10	1	10	1	5′	0.5′	5′	0.5′
Opt. scale	2	0.5	5	1	5′	0.5′	5′	0.5′
Direct	10	1	10	1	5′	0.5′	5′	0.5′
Direct	10	1	10	1	5′	0.5′	5′	0.5′
Direct	10	1	10	1	5′	0.5′	5′	0.5′
Opt. micro.	1	0.1	1	0.1	20″	2″	20″	2″
Opt. scale	1	0.2	1	0.2	1′	10″	1′	10″
Opt. scale	1	0.2	1	0.2	0.5′	0.1′	0.5′	0.1′
Opt. micro.	0.2	0.1	0.2	0.1	6″	3″	6″	3″
Coinc. micro.	0.1	0.05	0.1	0.05	10″	1″	10″	1″
Opt. scale	0.5	0.05	0.5	0.05	10″		10″	
Opt. scale	0.5	0.05	0.5	0.05				
Opt. scale	1	0.1	1	0.1	1′	0.1′	1′	0.1′
Opt. micro.	0.5	0.1	0.5	0.1	20″		20″	
Opt. scale	0.2	0.1	0.2	0.1	6″	3″	6″	3″
Opt. scale	1	0.1	1	0.1	1′	0.1′	1′	0.1′
Opt. scale	1	0.2	1	0.2	20″		20″	
Opt. scale	1	0.2	1	0.2	20″		20″	
Opt. micro.	0.2	0.1	0.2	0.1	6″		6″	
Coinc. micro.	1	0.5	1′	0.5	1″	0.1″	1″	0.1″
Coinc. micro.	1″		1″					
Coinc. micro.	1		1		1″	0.1″	1″	0.1″
Coinc. micro.	1		1		1″		1″	
Coinc. micro.	2		2		1″		1″	
Coinc. micro.	1		1		1″	0.1″	1″	0.1″
Coinc. micro.	0.5	0.1	0.5	0.1	0.5″	0.1″	0.5″	0.1″
Coinc. micro.					1″		1″	
Coinc. micro.	1	0.5	1	0.5	0.2″	0.1″	0.2″	0.1″
Coinc. micro.					0.2″	0.05″	0.2″	0.1″
Coinc. micro.					0.2″		0.2″	

3.3.4.2 Theodolites of Medium Accuracy

Medium accuracy theodolites, often referred to as *engineer's or stadia theodolites*, are instruments with medium reading accuracy, designed primarily for traversing, mini-triangulations and layout surveys (Table 3.1 (2)). They are repetition instruments equipped with stadia hairs for optical distance measurements. The diameter of the horizontal circle is 70 to 100 mm, while the vertical circle is occasionally smaller. Newer types usually have an automatic height index (section 10.2.3). According to the reading microscopes, the following distinctions are made:

a) *Line microscope theodolite with optical micrometer.* This design provides readings to 1 mgon or 3″ (Figs. 3.3.9 a) and b)). As a rule, however, a fully digitized reading is provided.

b) *Scale microscope theodolites.* Full minutes of arc can be directly read on the scales (1′ or 1 cgon), and 0.1′ or 1 mgon can be estimated (Fig. 3.3.6 and 3.3.18).

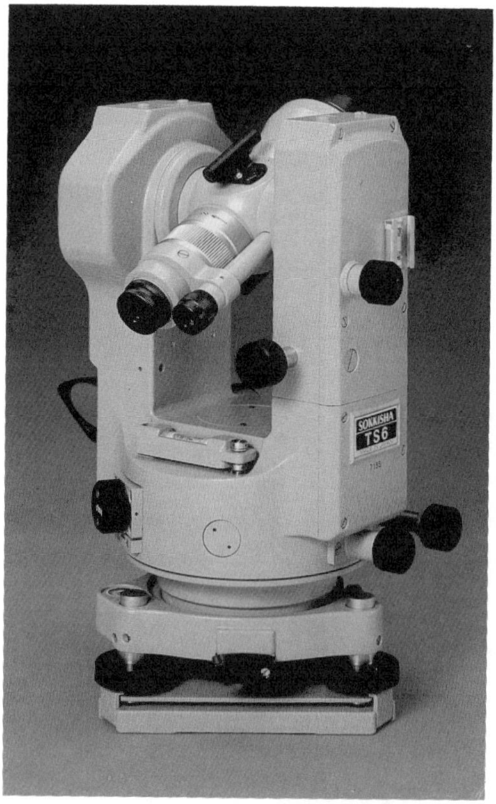

Fig. 3.3.18 Theodolite with scale microscope (*Sokkisha* TS 6)

c) *Coincidence theodolites with simple micrometer.* Kern, Aarau, used this principle to build the very compact and light double circle theodolite DKMl, which allows for direct readings of 0.1′ or 1 cgon and estimation of the fractions (Fig. 3.3.16a).

3.3.4.3 Theodolites of High Accuracy

High accuracy theodolites, often referred to as *precision or second theodolites*, are designed primarily for second- and third-order triangulation and precision staking. They do not have a repetition capability. The diameters of the horizontal and vertical circles are 80 to 100 mm, and 70 to 90 mm, respectively.

Practically all modern precision theodolites have coincidence microscopes with optical micrometers. The latter utilize either graduations rising in opposite directions which have to coincide without an index, or else graduations where an index is to be centred between the double lines of the graduation. A few instruments of new design have a largely digitalized reading.

The reading system of the vertical circle usually has a compensator so that the directions are automatically referred to the zenith.

The standard deviation for a single circle reading, dominated by the standard deviation of the coincidence with the micrometer, is approximately 1″ or 0.2 to 0.3 cgon for modern instruments, and slightly higher for older models.

3.3.4.4 Theodolites of Highest Accuracy

Theodolites of highest accuracy are specialized instruments designed for first-order triangulation, tunnel staking, etc. (e.g. Wild T 3, Kern DKM 3, Zeiss, Jena Theo 002). The standard deviations can be reduced to ±0.05 to 0.01 mgon or 0.1″ for a single reading.

Thus the reading accuracy reaches the magnitude of refraction influences. It is therefore not wise to improve it further.

3.4 Electronic Theodolites

3.4.1 Electronic Circle Scanning Devices

Based on their mechanical structure, the angular measuring systems of conventional and electronic theodolites are quite similar. The fundamental element for both groups of instruments is a graduated circle with one or two graduations centred to its axis. The graduation consists of graduation lines or other patterns, i.e., traces bordered by sine lines (Fig. 3.2.8).

The reading microscopes are replaced by scanning systems consisting of optical electronic elements. *Code and incremental methods* are utilized for the scanning process.

The measured quantities are obtained in binary form, are handled and stored by a computer, and can be displayed in digital form if desired.

For binary operation two symbols are needed, i.e., the figures 0 and 1 or a light and dark field (Figs. 3.4.1 and 3.4.2). In measuring instruments they are represented by two states, i.e., voltage and no voltage, current and no current.

	2^3	2^2	2^1	2^0
0	0	0	0	0
1	0	0	0	1
2	0	0	1	0
3	0	0	1	1
4	0	1	0	0
5	0	1	0	1
6	0	1	1	0
7	0	1	1	1
8	1	0	0	0
9	1	0	0	1
10	1	0	1	0
11	1	0	1	1
12	1	1	0	0
13	1	0	0	1
14	1	1	1	0
15	1	1	1	1

Fig. 3.4.1 Binary code

	2^3	2^2	2^1	2^0
0	0	0	0	0
1	0	0	0	1
2	0	0	1	0
3	0	0	1	1
4	0	1	0	0
5	0	1	0	1
6	0	1	1	0
7	0	1	1	1
8	1	1	1	0
9	1	1	1	1

Fig. 3.4.2 1-2-4-2 code

For ease of use, it is necessary to convert such results into decimal numbers, which can be achieved by means of a code. Depending on the measuring devices, several codes are utilized. The simplest one is the binary code shown in Fig. 3.4.1. As illustrated in Fig. 3.4.1, four binary places, representing powers of two, cover 16 decimal numbers. The 1-2-4-2 code (Fig. 3.4.2) is advantageous for electronic number operations. The instrumental operation of such number operations is described in *Kahmen* (1978).

3.4.2 Control and Operation of Electronic Measuring Instruments

A microprocessor or process computer represents the central unit for controlling and operating modern surveying instruments and directly influences their operational mode. The *functions of a microprocessor* and process computer are thus briefly discussed in order to explain the functional operation of the surveying instruments.

A microprocessor consists of buffer memory, data and command registers, the arithmetic unit (computer), and the command decoder (Fig. 3.4.3). Nowadays, data and commands can be accepted and handled by the microprocessor at a rate of one word per microsecond (1 000 000 words per second). Data and commands are first stored in buffers. The command decoder decodes the commands sitting in the command register and then controls the individual units of the microprocessor depending on the meaning of the commands, i.e. the data from the data register are correlated in the computer when acting on arithmetic commands. This procedure is illustrated in Fig. 3.4.3 for addition. The results are subsequently stored in buffers and transferred to data storage via the data bus. The command decoder causes a data transfer between buffer and input/output registers for input/output commands. From there the data are further transferred to a peripheral module (angular measuring system, distance meter, direction sensor for checking of vertical axis), or else information is obtained from there. Input/output commands are not used for transfer of measuring data, but for logic information used for monitoring the measuring process.

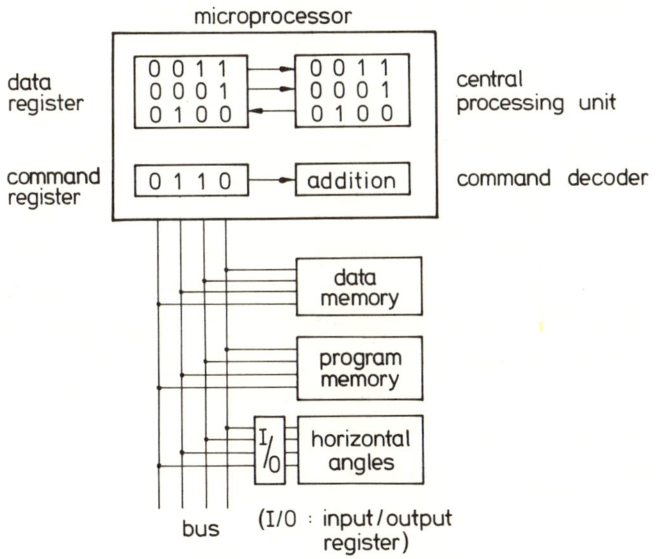

Fig. 3.4.3 Simplified functional diagram of a microprocessor

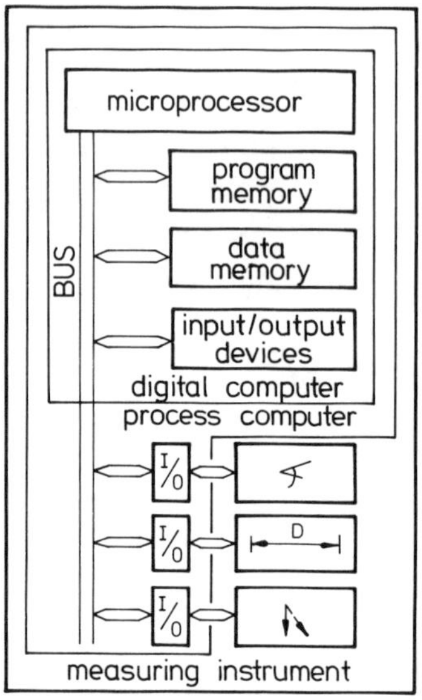

Fig. 3.4.4 Simplified functional diagram of a measuring instrument, controlled by a microprocessor

A *digital computer* (or microcomputer) is basically a combination of microprocessor, memory, and input/output devices. The example in Fig. 3.4.4 illustrates how the peripheral modules – memory, input/output units – can be connected to the microprocessor via a bus. A bus is a multichannel connection for information transfer. The functional diagram also shows how additional peripheral units – horizontal and vertical angle measuring system, distance meter and direction sensor for the vertical axis – can be coupled with the microprocessor via the bus. The digital computer is thus directly linked to the process and fulfils the important additional functions of a process computer.

The term *process computer* is presently applied to digital computers the design of which permits direct coupling to a process (i.e. measuring process) and which can handle the process data in real time. The possibility of process interrupt, which can redirect the programmed operation, is another typical feature of a process computer.

Process computers are used in modern surveying instruments to control the whole measuring process, including input/output, as well as to determine corrections and compute desired values (i.e. horizontal distance, elevation difference, etc.) from field measurements.

3.4.3 Analogue-Digital Conversion of Angles

The measured angles are analogue values at first. An analogue-digital converter transforms them into binary numbers so that the process computer can handle them.

The *analogue-digital conversion* of angles is based on code or incremental methods. *Code methods* utilize coded graduated circles where each graduation line is uniquely limited to a coded signal (Figs. 3.2.8 and 3.4.5). Newer instruments utilize electro-optical scanning systems.

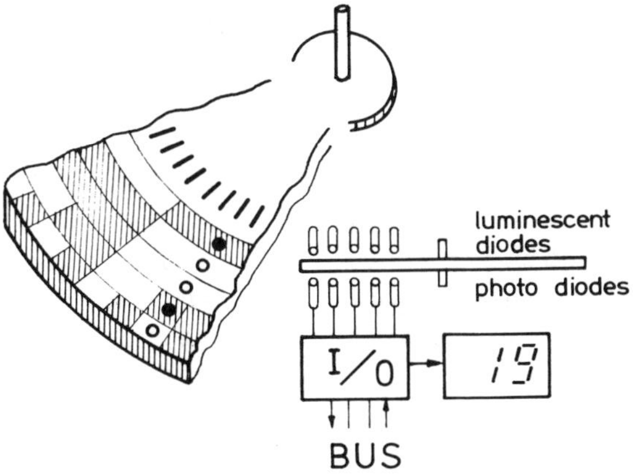

Fig. 3.4.5 Angular scanning using coded circles

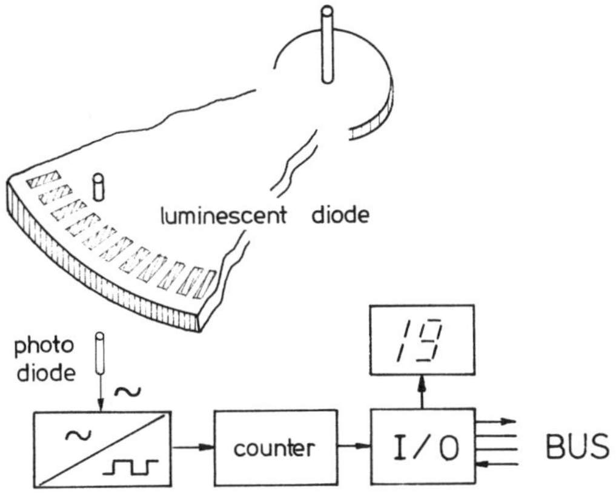

Fig. 3.4.6 Angular scanning using incremental methods

A technically possible solution is illustrated in Fig. 3.4.5. The code tracks consist of transparent and nontransparent fields. A luminescent diode is arranged above each track, emitting infrared light. Photo diodes below the tracks convert the light signals into electrical ones. Depending on whether a transparent field is above the photo diode or not, voltage or a zero signal is generated. A bus transfers to computer and storage the binary figure, which corresponds to the particular graduation line. The computer has to manipulate this data prior to the decimal display. The code disk shown in Fig. 3.2.8 has a resolution of 0.25 gon ($\sim 1\%$).

Incremental methods utilize graduated circles with radial line rasters which represent an increasing series of light/dark fields (increments). These fields (Fig. 3.4.6) are also denoted as "yes/no elements". This method does not permit the determination of absolute directions, but only direction changes by correcting the number of light/dark fields covered by the scanning system. Newer instruments utilize electro-optical scanners. It is also possible to produce the yes/no elements of translucent and non-translucent fields. In the example in Fig. 3.4.6 a luminescent diode is placed above the raster which produces a different light intensity below it depending on whether it is directly, partially or not above a transparent field. If the graduated circle moves in relation to the diode, then its radiation is subjected to an approximate sine-type intensity modulation. Each yes/no element generates one period of such a sine signal. The photo diode (or a photo transistor) below the circle and a luminescent diode convert the light signal to an electrical signal. An additional electronic component transforms the sine signal into a rectangular signal. The counter then counts the number of signal periods generated by the rotation. The result reaches the computer and storage via a bus and is subsequently displayed digitally after additional manipulations.

It is difficult to sense each individual line with the scanning system, especially if the raster is quite fine. It is thus more advantageous to generate the intensity modulation with a group of graduation lines. There are different solutions for this averaging approach in electronic theodolites.

A raster of 25000 lines can be produced for circles with diameters of 70 to 100 mm, providing a least count of 0.16 gon ($\sim 8'$).

The accuracy of the described angular measuring systems is generally not sufficient for geodetic tasks. A sufficient accuracy can be achieved with circle diameters of 70 to 100 mm, if

- electronic interpretors are used, and
- diametrically opposite readings are taken because of possible circle eccentricities (*Kahmen* 1978).

3.4.3.1 Electronic Interpolators of Medium Accuracy

Electronic theodolites of medium accuracy generally use the incremental method for scanning the graduated circles. The scanning head generates one period of an electric sine-signal (Fig. 3.4.7) if the top part rotates across one light/dark field. An electronic counter counts the number of signal periods for rough measurements.

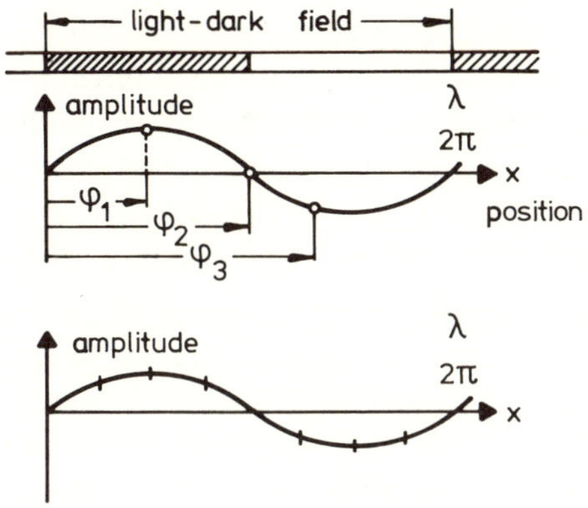

Fig. 3.4.7 Rough and fine measurement for angle scanning

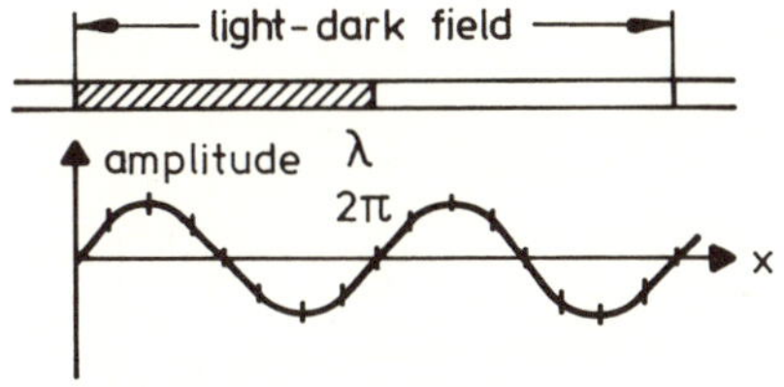

Fig. 3.4.8 Interpolator 1:16

The length of a signal period is 2π or λ. Phase angles ϕ_1, ϕ_2, ϕ_3 ..., which are fractions of 2π, correspond to all states of oscillation. These phase angles are used for fine measurements. An additional electronic interpolator in a simple scanning system can distinguish 8 signal sections by varying phase angles within one signal period (Fig. 3.4.7). With this interpolation in a 1 : 8 ratio a resolution of 0.002 gon ($\sim 6''$) can be achieved for a line raster of 25000 increments.

Two signal periods can be generated for one light/dark field using special imaging methods (Fig. 3.4.8). If the electronic interpolator again distinguishes 8 different states within one period, then the light/dark field can be subdivided in a 1 : 16 ratio. Thus the number of graduation intervals is magnified by a factor of 32, or the zero paths of both signals are shifted by $^1/_{32}$ of a signal period. This arrangement provides a resolution of 0.001 gon ($\sim 3''$) for a line raster of 12500 increments. A microprocessor controls the measuring process and the generation of the final result.

3.4.3.2 Electronic Interpolation of High Accuracy

Electronic interpolation of high accuracy is obtained either with a coincidence method or by phase comparison or by mathematical interpolation.

The *coincidence methods* are based on the principles illustrated in Figs. 3.3.14 and 3.3.15. The positions of the graduation lines, however, are determined by the scanning of code traces. An electro-optical measuring system performs symmetric coincidence of the graduation lines. Thus a subdivision of each interval into more than 1000 steps is possible. The partial results of rough and fine measurement reach the microprocessor through a connection system and are combined into the final result.

The *phase comparison method* is an incremental approach, where partial steps within a signal period are generated by phase comparison. The graduated circles of the HP 3820 tacheometer by Hewlett Packard carry an analogue trace at the outer and inner edges with code traces in between (Fig. 3.2.8c). The rough measurement is accomplished by scanning the code trace, while the fine reading is generated by scanning the analogue traces.

When turning the theodolite standards, each analogue trace is scanned by four diodes, distributed evenly over a signal period. Four electric sine-signals i_1, i_2, i_3, i_4 are generated having a phase shift of 90° against each other (Fig. 3.4.9).

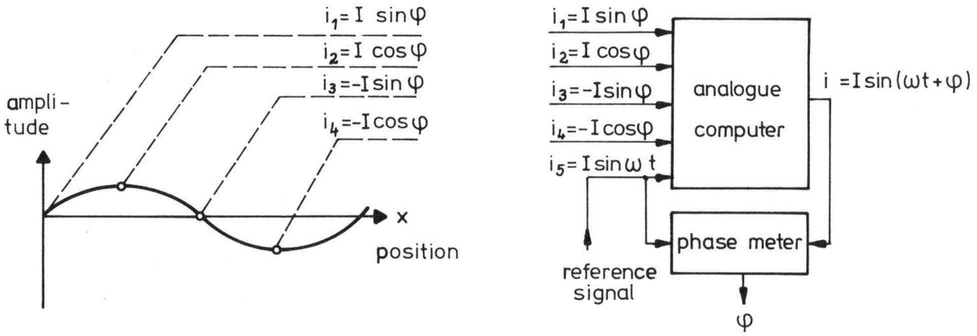

Fig. 3.4.9 Fine scanning of circles using phase comparison (*Hewlett Packard* HP 3820)

These four signals are sent to an analogue computer together with a reference signal. There the trigonometric functions are combined in such a manner that an output signal with the frequency of the reference signal and a zero phase angle ϕ is obtained. This output signal reaches the phase difference meter together with the reference signal. With the two analogue traces the phase difference $\Delta\phi$ is finally determined twice. ϕ permits a subdivision of a graduation interval into about 1000 separate intervals. The rough reading is obtained by scanning the code traces. A microprocessor within the instrument combines the partial results (two fine and one rough reading) into the final result. The latter is subsequently displayed and stored.

The electronic Theomat T2000 theodolite by Wild also utilizes a phase comparison method. The angular scanning systems consist each of a turning glass circle, two diametrically opposite fixed light barriers L_S, and two diametrically opposite light barriers L_R which rotate with the theodolite standards. The circle contains 1024 evenly distributed graduation lines. For simplicity, a system with just one each movable and fixed light barrier is being considered (Fig. 3.4.10). The light barrier L_S represents the zero direction of the "circle orientation". The angle ϕ between this zero direction and the second light barrier L_R corresponds to the direction shown on the display.

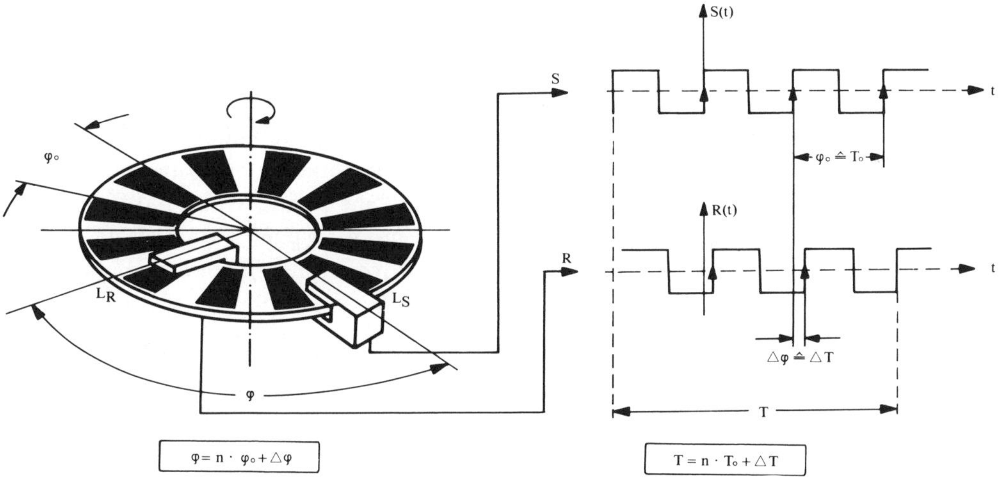

Fig. 3.4.10 Fine scanning of circles using phase comparison (*Wild* Theomat T 2000) (φ: phase angle; *T*: Time)

The main components of a light barrier are a transmitter diode and a receiver diode. A rotation of the graduated circle causes a sinusoidal modulation of the light of the transmitter diode. The receiver diode converts the light signal to an electrical signal. The analogue signal is then converted into a rectangular one for further processing. The light barriers L_R and L_S generate one such rectangular signal each. The angle ϕ is then given by the phase difference of the two signals

$$\phi = n \cdot \phi_0 + \Delta\phi \; .$$

The direction measurement is performed in two steps: in the rough measurement, the integer number n of signal periods $\varphi_0 = 2\pi$ is determined, while the fine reading is obtained from the phase difference between two inclined branches of the signals.

A special microprocessor (angle processor) monitors rough and fine readings. It initiates the rotation of the graduated circle and triggers the measurements once the proper number of revolutions per time unit is reached. Special reference marks on the circle facilitate the rough reading. As soon as a light barrier recognizes a reference mark, a counter starts to count the signal periods. As a check, two rough measurements are carried out. 512 phase comparisons comprise one fine reading, which means that every second graduation line is involved. After the elementary reading is computed, the action switches to the diametrically opposite light barrier. The fine reading is the arithmetic mean of all elementary readings and thus is free of eccentricity errors. Once the angle processor has combined rough and fine readings into a complete measurement, the result is transferred to a central processor.

The fine reading can also be derived from *mathematical interpolation* as, for instance, in the E2 precision theodolite by Kern. The method is based on incremental angle measurements.

The glass circles carry 20000 radial lines. An imaging system superimposes a section of 200 lines onto its diametrically opposite section. Since the imaging process includes a magnification factor of 1.005, a Moire pattern with sinusoidal light/dark variation (Fig. 3.4.11) is obtained. Thus a direction is not determined by two individual lines, but rather by groups of lines, which is statistically better. The Moire pattern is optically magnified by a factor of two and imaged onto a system of four photo diodes, distributed evenly over one Moire period. When turning the theodolite, these diodes generate four electrical signals shifted by 90° each.

For the rough reading, the number of Moire periods generated by rotation of the standards is counted. The phase angle φ is determined for the fine reading, which represents the position of the diode system within the imaged Moire period. The

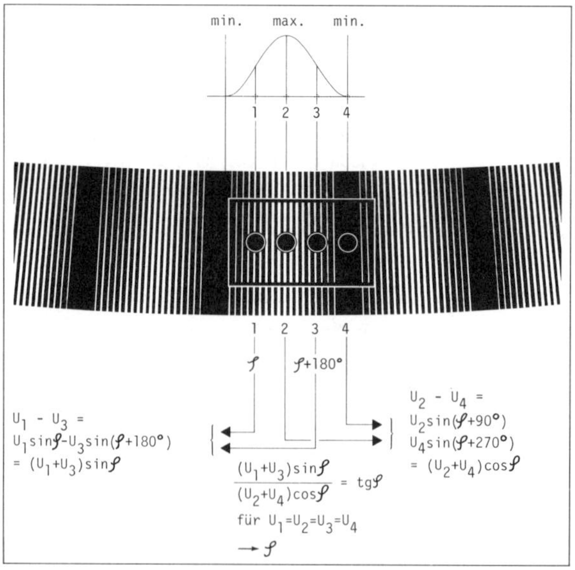

Fig. 3.4.11 Fine scanning of circles using mathematical interpolation (*Kern* E2)

phase angle is obtained by evaluating the voltage signals u_1, u_2, u_3, u_4 as generated by the photo diodes. As illustrated in Fig. 3.4.11, the different signals $u_1 - u_3$ and $u_2 - u_4$ are first formed and then divided in a microprocessor after an analogue/digital conversion.

This leads to the function $\tan\varphi$ from which the phase angle φ is extracted within the microprocessor.

The quadrant of φ still needs to be determined. This can be done uniquely by using the signal $u_2 - u_4 = (u_2 + u_4)\cos\varphi$.

One full revolution of the standards generates 40000 signal periods because of the diametrically opposite imaging. This means that a rough reading has a directional resolution of 0.01 gm ($\approx 30''$). The interpolation of the E 2 theodolite has a resolving power of 1 : 100.

With electronic interpolators of high accuracy, directions can be resolved to an accuracy of 0.0001 ... 0.0002 gon ($0.3''$ ... $0.6''$).

3.4.4 Classification of Electronic Theodolites

Similar to optical theodolites, it is possible to distinguish between instruments of medium and high accuracy. The latter have a resolution of 0.1 to 0.2 mgon ($\sim 0{,}5''$) per direction while the resolution of medium accuracy instruments is 1 to 2 mgon ($\sim 5''$).

Generally, electronic theodolites are not built as single instruments, but rather are combined with an EDM to an electronic stadia or total station type instrument. Further details are covered in section 12.2.

3.5 Levelling and Centring of Measuring Instruments

The extension of the rotational axis of a theodolite or distance meter is in the plumb direction and passes through the centre of a ground point.

Directions, horizontal and vertical angles, as defined in section 3.1, or distances are only obtained correctly if these conditions are fulfilled. Levelling and centring constitutes proper set-up of the measuring instruments. Spirit bubbles are used for levelling, while lateral shifts are necessary for centring (Fig. 3.5.1).

Depending on the required accuracy, the connection between the rotational axis and the ground point can be as plumb bob, centring rod or optical plummet.

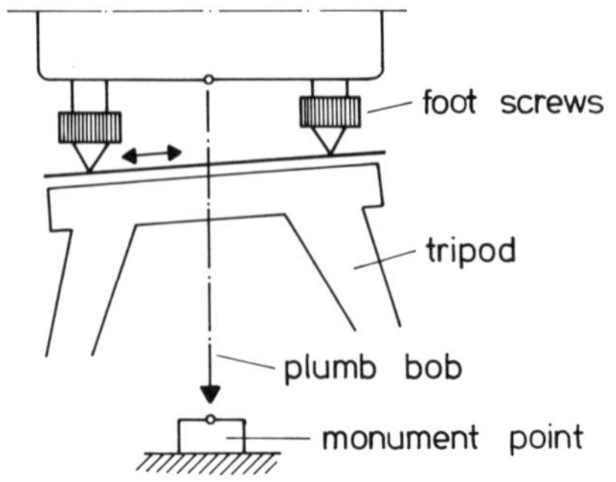

Fig. 3.5.1 Centring of a tribrack by shifting laterally

3.5.1 Levelling and Centring with a Plumb Bob

For simple field surveys a plumb bob is used. The plumb string is attached to a point positioned at the extension of the vertical axis of the instrument. The tribrack is first centred by shifting laterally, and then levelled with the platen level of the theodolite.

As illustrated in Fig. 3.5.2, the bubble is set parallel to two foot screws A and B and then centred. After rotation by 90° it is again centred, using foot screw C. This procedure is repeated for a check.

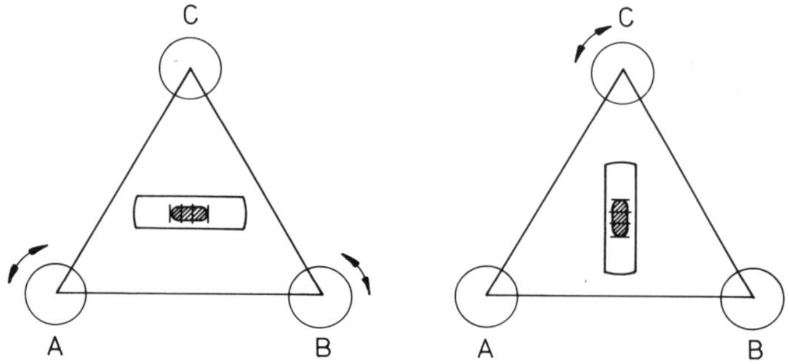

Fig. 3.5.2 Levelling of a tribrack

Changes of the foot screws result in centring errors, whose magnitude can be estimated as shown in Fig. 3.5.3.

If the foot screw C is turned, then the point 0, located in the extension of the vertical axis, follows a circular arc rather than moving vertically. This creates a centring error e.

$$e = \frac{d}{2} - \frac{d}{2}\cos\alpha = \frac{d}{2}\left(1 - 1 + \frac{\alpha^2}{2!} - \ldots\right) = \frac{d}{2} \cdot \frac{\alpha^2}{2!}$$

with

$$\alpha = \frac{2x}{3d}; \qquad e = \frac{x^2}{qd}.$$

A centring accuracy of ± 3 to 5 mm can be achieved with a plumb bob provided that there is no wind and the ground point is well defined. Thus this centring error can usually be neglected.

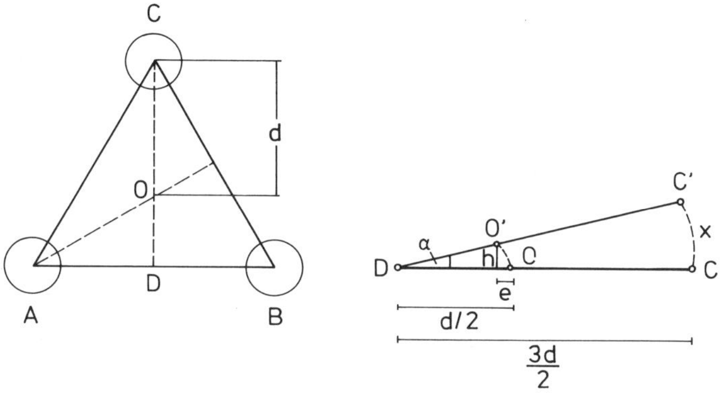

Fig. 3.5.3 Estimation of centring errors caused by levelling of the instrument

3.5.2 Levelling and Centring with a Centring Rod

For precise measurements it is desirable – especially on windy days – to utilize a centring rod, i.e. a telescopic steel pipe with pointed end, screwed onto the tripod and carrying a bull's-eye bubble. Fig. 3.5.4 illustrates such a design. The observer has to roughly level the tripod head (3) and then places the point of the centring rod (6) onto the centre of the ground point. At the upper end of the centring rod there is a carrier plate (1) for the instrument sliding along a spherical head, which in turn can move relative to the tripod head.

The carrier plate is shifted along the spherical head which in turn is shifted along the tripod head until the centring rod is vertical as indicated by the bull's-eye bubble (5). Carrier plate, spherical head and tripod head are then fastened together with the aid of a clamping handle (4). The observer subsequently centres and fastens the measuring instrument and tribrack onto the carrier plate with the aid of a locking lever. Normally only a slight levelling is required for the instrument using the foot screws. If the bull's-eye level is not adjusted, its centre has to be determined by rotating the rod by 180°.

The centring accuracy for a centring rod is approximately ± 1 mm.

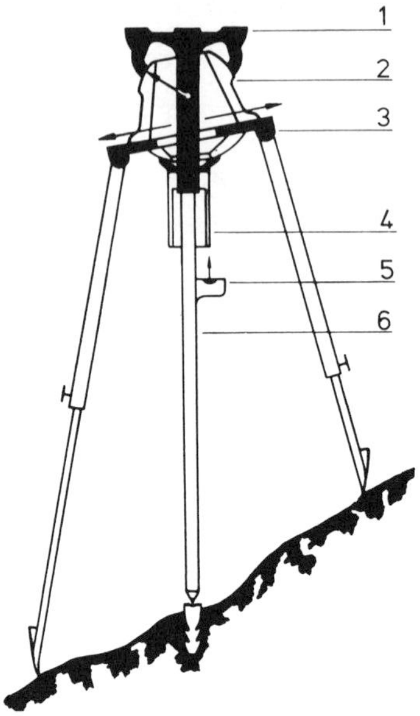

Fig. 3.5.4 Design principle of the *Kern* centring rod

3.5.3 Ball Base with Parallel Motion Device

The spherical foot, which is used as support for several of the instruments built by Zeiss, Oberkochen, permits freehand levelling to such an extent that two easily accessible setting knobs suffice for fine levelling. This type of support also offers better protection against undesirable turning of the instrument. Furthermore, a "rectangle slide" is fitted between instrument and tripod to accelerate centring of the instrument. This slide permits movements on the tripod in two perpendicular directions only, although this is hardly noticed by the observer.

3.5.4 Levelling and Centring Using an Optical Plummet

Modern theodolites usually are equipped with an optical plummet. It consists of a small telescope placed in the vertical axis viewed from the side by means of a reflecting prism (Fig. 3.5.5).

The tripod is approximately levelled and centred above the ground point. Then the observer views into the plummet and sets its cross hair onto the ground point by turning the foot screws. Then the bull's-eye bubble is centred by extending or reducing the length of the tripod legs. After levelling with foot screws and platen level, the centring is corrected by shifting the instrument along the tripod head.

Tribracks and target supports are often equipped with optical plummets, and some

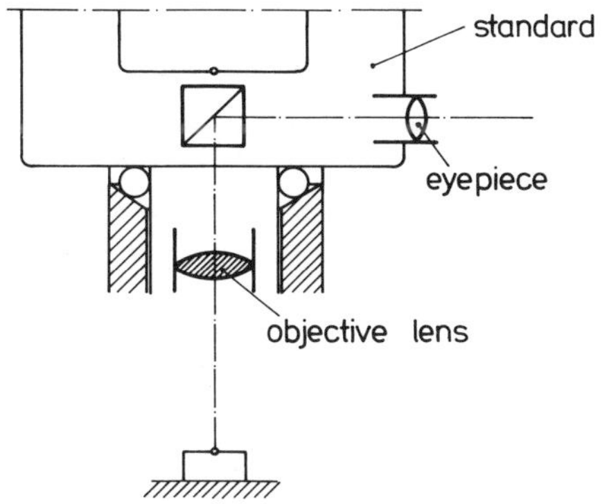

Fig. 3.5.5 Optical plummet located in the upper part of the theodolite

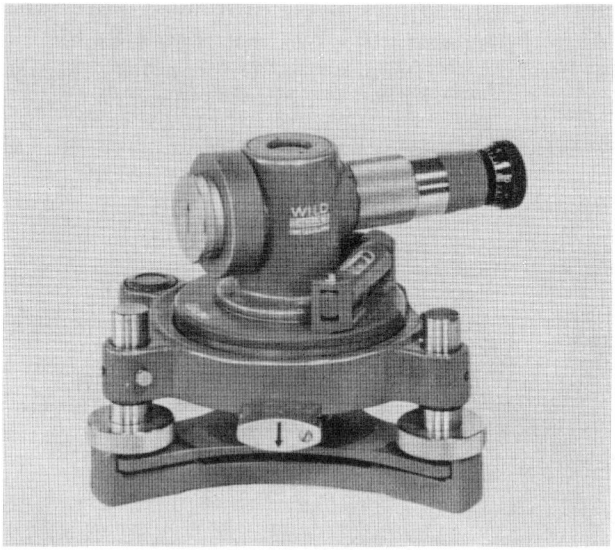

Fig. 3.5.6 Universal Zenith and Nadir Plummet with tribrack (*Wild* ZBL)

manufacturers offer optical plummets as separate units, such as zenith plummets, nadir plummets and universal plummets (Fig. 3.5.6). They are primarily used for tasks like centring a tribrack above a ground point or below a ceiling point, shaft plumbing, and plumbing of high structures, such as smokestacks, bridges, and power dams.

Simple optical plummets provide a centring accuracy of ± 0.5 mm, while individual precision units reach accuracies of ± 0.5 mm/100 m.

It is necessary to regularly check optical plummets, especially the ones in tribracks because the observation takes place in one position only. For this the outline of the clamped tribrack plate is marked with pencil onto the tripod head. After levelling the bull's-eye bubble, the plumbed point is marked on graph paper on the ground. Then the tribrack is loosened and rotated by 120° and carefully placed inside the outline. After clamping and levelling, the plumbed point is again marked. This is repeated once more. The plummet is properly adjusted if all points coincide. If not, the cross hair is shifted to the centre of gravity of the three plumbed points.

3.5.5 Forced Centring

This method guarantees the same position even in cases when instrument, target, distance meter and reflector have to be exchanged at one point, e.g. when measuring a precision traverse.

The centred tribrack remains at the station, solidly connected to the tripod until all measurements are completed. The bottom of the instruments carry one or three centring plugs, which fit into precision fittings of the tribrack and can be solidly fastened during the measurements. Depending on the design, the fastening is achieved either with a clamp on the side or else with a centrally located rotational lock. Fig. 3.5.7 illustrates forced centring with one centring plug and a clamp.

Fig. 3.5.7 Forced centring with the centring plug and clamp

Instruments manufactured, by for instance, Breithaupt, Zeiss-Jena and Zeiss-Oberkochen are equipped with this type of forced centring. The plug is pressed into a *V* bearing with a high accuracy. Instruments manufactured by Wild and Kern operate with a central locking system. When turning the locking lever for forced centring of Kern instruments (Fig. 3.5.8), three claws of the central lock grasp the base plate of the instrument and fasten it onto the carrier plate. For Wild instruments (Fig. 3.5.9), a centrally located rotational lock locks the three plugs of the centring connection into the centring bush of the tribrack. The accuracy depends on how well the plug fits into the centring bush.

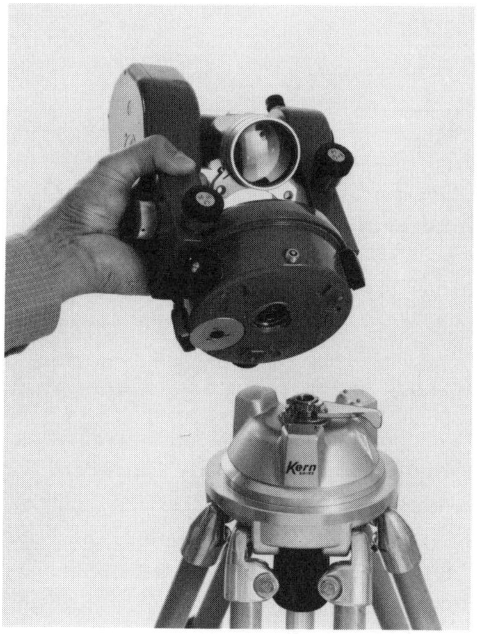

Fig. 3.5.8 Forced centring with central lock (*Kern*)

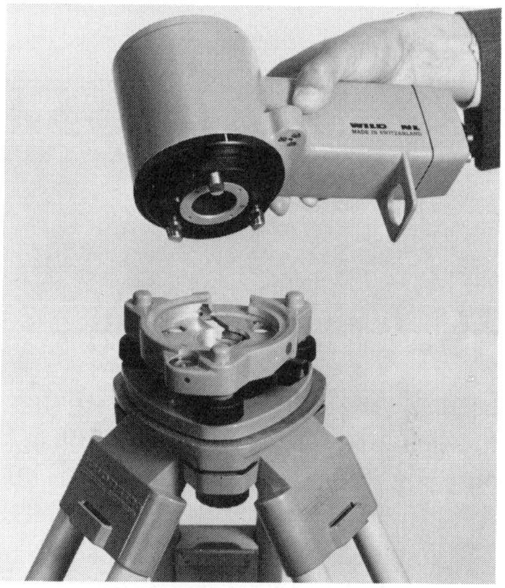

Fig. 3.5.9 Forced centring with rotational lock (*Wild*)

Centring accuracies of ± 0.03 to 0.1 mm are common for forced centring. Forced centring with constant instrument height is necessary when measuring three-dimensional precision networks. This requirement is fulfilled if one of the three foot screws remains unchanged whenever an instrument is being levelled.

Since it is cumbersome to level an instrument without touching one of the foot screws, special devices have been developed. Fig. 3.5.10 illustrates a Kern design. In order to assure a constant instrument height when levelling, one of the three support points is placed into the vertical axis and held fixed. This mechanical design leads to a joint within the vertical axis, consisting of a ball in the bottom part of the instrument and a cone in the tribrack. The observer can tilt the vertical axis around the centre of the sphere by operating two foot screws, arranged normal to each other. A special pressure spring provides lateral guidance (*Aeschlimann* 1975).

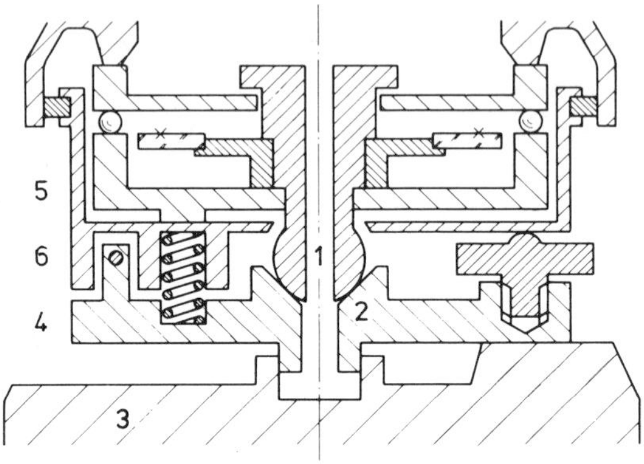

Fig. 3.5.10 Forced centring with constant instrument height
1 sphere, 2 cone, 3 support plate with forced centring device, 4 base plate of tribrack, 5 theodolite base, 6 pressure spring

3.6 Testing and Adjusting a Theodolite

Even the most efficient and experienced mechanic cannot perfectly transform the mathematical design idea into the actual instrument. Furthermore, certain elements may slightly change their position due to use, temperature influence, etc. Both facts result in errors which have to be eliminated either by instrument adjustment or by selected measuring procedures. The most important errors are *axis errors, eccentricities,* and *circle graduation errors.*

3.6.1 Axis Errors

A theodolite has three principal axes, mainly the vertical, or azimuth, axis (V_1 V in Fig. 3.6.1), the horizontal, or elevation, axis H_1 H, and the collimation axis, or line of sight, ZZ, as well as the axes of the levels. If horizontal angles are to be measured, then the collimation axis has to move in a vertical plane when tilting the telescope. This is the case if:

a) the collimation axis ZZ is normal to the horizontal axis HH;

b) the horizontal axis HH is normal to the vertical axis VV;

c) the vertical axis is exactly vertical.

If one of these conditions is not fulfilled, then one refers to a collimation, horizontal or vertical axis error.

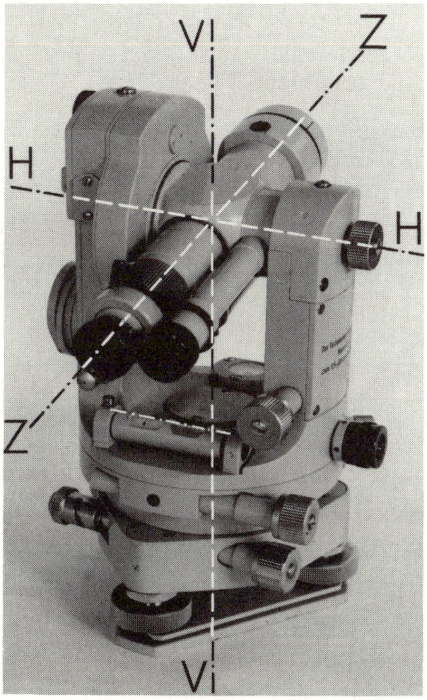

Fig. 3.6.1 Theodolite axes (schematic)

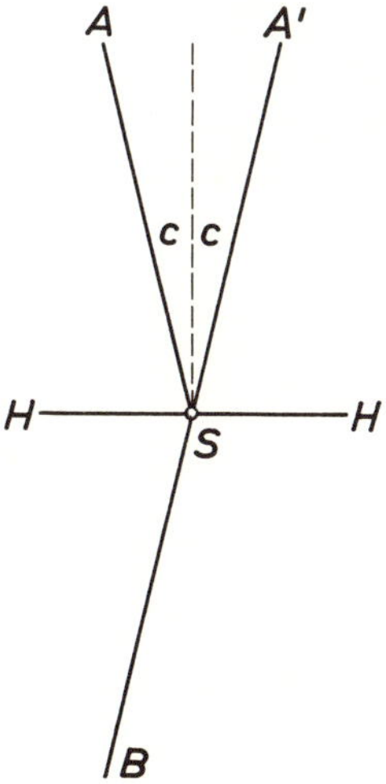

Fig. 3.6.2 Determination of a collimation error

3.6.1.1 The Collimation Error

A point A, located at about the same height as the station S, is sighted and the circle is read (Fig. 3.6.2). Then the telescope is plunged pointing towards B. It is then rotated by exactly 180° (200 gon) pointing towards A′. The angle ASA′ $= 2c$ represents twice the collimation error. It is adjusted by removing half of the angle USA′ with the aid of adjustment screws on the side of the reticule.

When sighting to an elevated target appearing at a vertical angle h, the directional error caused by a collimation error c is $(c) = \frac{c}{\cos h}$. As evident from the earlier deliberations, a direction is free of the influence of the collimation error, if it consists of the mean of the readings in direct and reverse telescope positions. For $h = 0$, (c) becomes c.

3.6.1.2 The Horizontal Axis Error i

This occurs when the horizontal axis deviates by an angle i from the horizontal plane when the vertical axis is plumb (Fig. 3.6.3). It can usually be recognized if the cross hair, set onto the vertical edge of a house, moves away from it when the telescope is being tilted. There are two ways to determine this error.

a) After the collimation error has been eliminated, the telescope is pointed to a high point A. It is then tilted into the horizontal plane. Then a reading a_1 (Fig. 3.6.4) is taken at a horizontal scale placed normal to the line of sight. The telescope is then plunged and again sighted to A. After tilting it down, the reading a_2 is taken at the scale. The separation $(a_2 - a_1)$ is caused by twice the horizontal axis error.

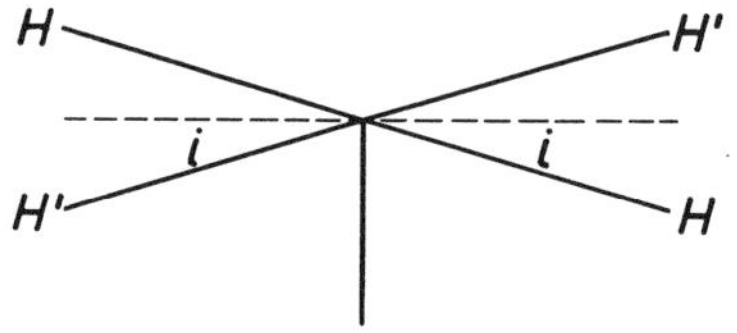

Fig. 3.6.3 Horizontal axis error

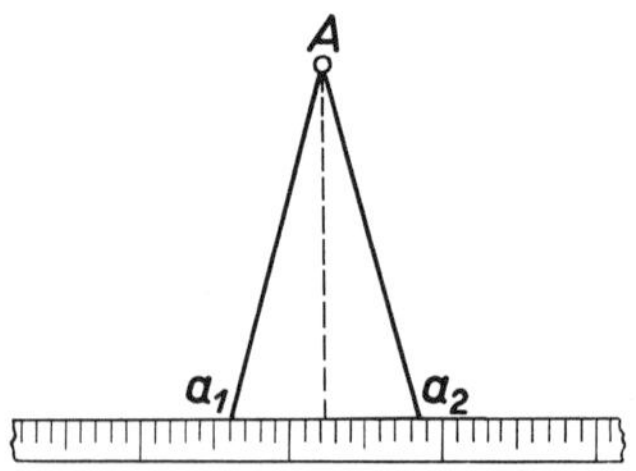

Fig. 3.6.4 Determination of the horizontal axis error

To eliminate the horizontal axis error, the telescope is diverted to the mid point $^1/_2(a_1 + a_2)$ at the scale. The telescope is tilted until A appears in the field of view. Then one axle support is raised or lowered with the aid of adjustment devices, described in section 3.2.2.2, until point A appears exactly in the centre of the cross hairs. For modern instruments, where the horizontal axis is inside the housing, this has to be done in a mechanical workshop.

b) If point A is sighted in both telescope positions, then the difference between the two readings (reduced by 180°) represents twice the influence of the horizontal axis errors, provided that there is no collimation error. By setting the telescope into the mean position of both circle readings, the horizontal axis error can be adjusted as mentioned in a).

The effect (i) on an observed direction caused by the horizontal axis error when sighting at a vertical angle h equals $(i) = i \tan h$. It thus has no effect on directions where the target is in the horizon ($h = 0$). Using this formula it is possible to calculate the amount of the horizontal axis error from its effect on an inclined sight as in a) or b). From this it is also obvious, that the effect of the horizontal axis error is eliminated when taking the mean of direct and reverse readings.

3.6.1.3 The Vertical Axis Error v

This is not an instrument error but rather a setting up error because the vertical axis is not plumb. It can be practically avoided if the vertical axis is brought to the plumb position with determination of the bubble centre, according to the procedure given in section 2.1.3.2.

The effect (v) of a vertical axis error v on a measured direction is a function of the vertical angle h to the target as well as the angle representing the separation of the measured direction from the vertical plane containing the inclined vertical axis. This effect is

$$(v) = v \sin \alpha \tan h \, .$$

Unlike the collimation and horizontal axis errors, this effect (v) cannot be eliminated when measuring in both telescope positions. Since (v) increases with the tangent of the vertical angle, the vertical axis has to be plumbed extremely carefully when measuring inclined directions.

Some theodolites and electronic tacheometers are additionally equipped with a horizontal axis micrometer (*Aeschlimann* 1979) or an electronic level (*Bullock and Warren* 1976) to measure the remaining tilt of the vertical axis. The measuring results obtained with these additional devices permit correction of the horizontal directions for vertical axis errors.

3.6.2 Eccentricities

On the graduated circle, in order to obtain correctly the horizontal angle according to the rotation of the collimation axis, two conditions have to be fulfilled:

- the centre of the graduation has to coincide with the vertical axis, and,
- a line normal to the vertical axis has to intersect with the collimation axis of the telescope.

If this is not the case, then a circle, or telescope, eccentricity is present.

3.6.2.1 Circle Eccentricity

If the centre of the circle graduation (L) is not identical with the centre A of the rotation of the platen in its bush axis, then the "circle readings" for the two sides of an angle are not identical to the actual rotation of the platen. Fig. 3.6.5 shows the angle φ left side of which happens to coincide with the directions AL of the eccentricity with respect to its extension CC'. Its right side points in the direction AZ_1. The platen is rotated by the angle φ when pointing to Z_1.

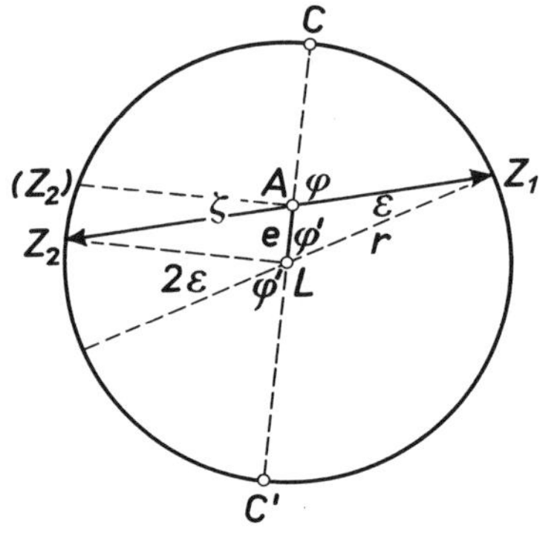

Fig. 3.6.5 Circle eccentricity

The circle readings for φ are then at Z_1:

$$CLZ_1 = \varphi' = \varphi - \varepsilon, \text{ instead of } CAZ_1,$$

and at Z_2:

$$C'LZ_2 = \varphi'' = \varphi' + 2\varepsilon = \varphi' + \varepsilon \text{ instead of } C'AZ_2\,.$$

When taking the mean, we obtain therefore

$$\frac{\varphi' + \varphi''}{2} = \varphi\,.$$

Thus when reading at two diametrically opposite circle positions and taking the mean value, the rotational angle ϱ is correctly obtained. If there is only one circle reading as in some theodolites of medium accuracy (section 3.3.4.2), then the circle eccentricity is only eliminated when taking the mean of direct and reverse readings.

Instruments with two reading positions often have index bending ξ, which is illustrated in Fig. 3.6.6 where the second index points to (Z_2) instead of to Z_2.

ξ remains constant, while the error ε caused by the eccentricity depends on the angle φ and thus can assume all values between zero and $\varepsilon_{max} = e \cdot \varrho / r$ according to

$$\sin \varepsilon = \frac{e}{r} \sin(180° - \varphi) \text{ or } = \frac{e \cdot \varrho}{r} \sin \varphi \,.$$

ξ influences the readings for both sides of an angle by the same amount and thus is eliminated when finding the difference of right minus left side of an angle. Furthermore, it is also eliminated when plunging the telescope. The same holds for modern theodolites with coincidence reading.

Even a small eccentricity has a large influence on the reading. For instance, an eccentricity $e = 0.01$ mm for a theodolite with 10 cm circle diameter causes an error according to $b : r = \alpha : \varrho$ of

$$\varepsilon_{max} = \frac{0.01 \cdot 63\,662}{50} = 12.7 \text{ mgon} \qquad (\text{or } +2'') \,.$$

Therefore the manufactures try to reduce the eccentricity to below 1 μm.

3.6.2.2 Eccentricity of the Collimation Axis

In Fig. 3.6.6, the direction to a target P is to be determined. L represents the centre of the circle. If the collimation axis has an eccentricity e, then it is always tangent to a

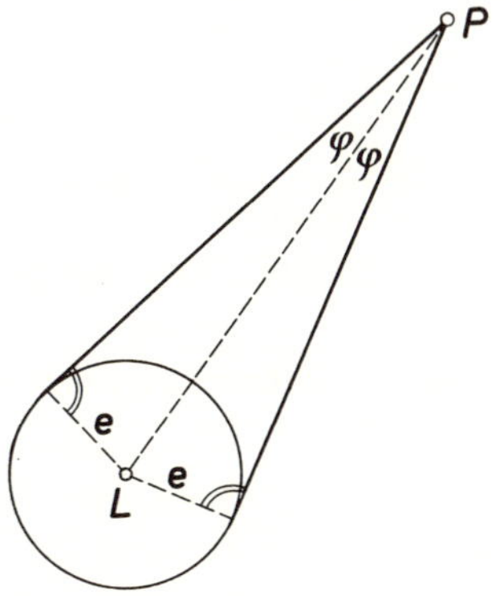

Fig. 3.6.6 Eccentricity of collimation axis

circle around *L* with radius *e*. Thus in the direct and reverse telescope position it causes the same directional deviation but with opposite sign, therefore the mean of direct and reverse readings is not influenced by eccentricity of the collimation axis. This holds strictly even if the whole telescope is located eccentric and not centred on the horizontal axis (section 3.2.2.2).

3.6.3 Circle Graduation Errors

If a theodolite with a circle diameter of 10 cm is graduated to 20 cgon, it has 2000 graduation lines, separated by 0.16 mm. In order to guarantee an accuracy of 0.5 mgon, each of these has to be placed correctly to within 0.4 μm. In spite of the finest graduation machines and extreme care of the mechanics, this accuracy is not always achieved. Both regular and irregular graduation errors occur, whose exact determination falls outside the scope of this book. They are commonly treated by measuring the same angle on several symmetrically distributed locations of the circle. This leads to elimination of the regular graduation errors and reduces the irregular ones to such an extent that, for common practical tasks, they can be neglected.

3.6.4 Mechanical Errors in Practice

Eccentricities and graduation errors are usually negligibly small in modern instruments. It is necessary, however, to check the instruments for collimation and vertical axis errors, especially if they are not eliminated through the measuring procedure, i.e., when observing in one telescope position only.

3.7 Horizontal Angle Measurements

3.7.1 General Rules

When considering horizontal angle measurements, one distinguishes between simple angle measurements and direction measurements. For certain tasks, repetition angle measurements, angle measurements in all combinations, and the sector method are used. Each of these methods utilizes direct and reverse readings because of collimation, horizontal axis, and eccentricity errors.

3.7.2 Simple Angle Measurement

This is the simplest way of observing the angle between two directions. First the instrument is centred and levelled. Then the left target is sighted and recorded. The same is repeated for the right target. The difference of mean right minus mean left represents the angle in the direct telescope position. Subsequently, the telescope is plunged and the observations are carried out in the same manner for the reverse telescope position, except that the right target is observed first. The mean value for both telescope positions is formed, which concludes the first measurement of the angle. If the angle is to be observed 2, 3, or n times to improve the accuracy, then the circle is rotated by $180°/n$ prior to each new measurement. The mean of all observations is used as the final value.

Example

target	readings		angle		mean
	direct	reverse	direct	reverse	
1	2	3	4	5	6
71	0°31′09″	180°31′16″			
22	50°03′22″	230°03′39″	49°32′13″	49°32′23″	49°32′16″

3.7.3 Direction Measurement in Sets

This method is used when more than two rays originate from the observation station. Without changing the circle, all targets are observed clockwise in order with readings being taken at both indices. Then the telescope is plunged and all targets are observed in the reverse telescope position and in reversed order. Finally the mean directions are reduced to the starting direction (zero-direction) by subtracting the value obtained for this direction from all others. Such a series of measurements is referred to as a set, occasionally as a full set, while the observations in only one telescope position are considered as a half set. As in section 3.7.2, the circle is shifted by $180°/n$ when measuring n sets. The final directions are obtained as mean values from the reduced means of the individuals sets. The calculations of means and reduced means are to be checked with sum checks, as indicated in the numerical example.

Example: Direction Measurements

target	readings direct	reverse	mean	reduced mean	final directions	d	v	v^2
1	2	3	4	5	6	7	8	9
	[°]	[°]	[°]	[°]	[°]	[°] · 10^{-4}		
			1. set					
71	0.577	200.579	0.5780	0.0000	0.0000	0	+ 2	4
22	55.618	255.623	55.6205	55.0425	55.0420	− 5	− 3	9
2	95.341	295.345	95.3430	94.7650	94.7640	−10	− 8	64
53	266.489	66.491	266.4900	265.9120	265.9127	+ 7	+ 9	81
					187	− 8	0	
					$[d]_1 : s =$	− 2		
			2. set					
	67.606	267.513	67.5095	0.0000		0	+ 2	4
	122.548	322.553	122.5505	55.0410		+10	+12	144
	162.270	362.279	162.2745	94.7650		−10	− 8	64
	333.422	133.424	333.4230	265.9135		− 8	− 6	36
						− 8	0	
					$[d]_2 : s =$	− 2		
			3. set					
	134.150	334.158	134.1540	0.0000		0	− 4	16
	189.194	389.199	189.1965	55.0425		− 5	− 9	81
	228.913	28.919	228.9160	94.7620		+20	+16	256
	0.064	200.069	0.0665	265.9125		+ 2	− 2	4
	592	652	6220	560	3 = *187*	+17	+ 1	763
				4 × 780 = 120	$[d]_3 : s =$	+ 4		
				4 × 095 = 380				
		622		4 × 540 = 160				
				220				

$$s(r) = \sqrt{\frac{763}{(3-1)(4-1)}} = 11^\circ \cdot 10^{-4} = 3.6'' \qquad s(\hat{r}) = \frac{11.3}{\sqrt{3}} = 6^\circ \cdot 10^{-4} = 2''$$

The reversal of the order for the reverse telescope position serves to eliminate possible torsion effect of tripod or pillars. For precision measurements, the top part should be rotated clockwise only, and the last twist on the fine setting screws is to go to the right, i.e. against the pressure of the spring. After three sets at the most, one has to check whether the instrument has remained level.

To compute standard deviations, a bundle of directions with s rays and n sets is considered. First the difference d between the observed direction and the final mean direction is found. With the sum of these differences Σd, the correction $v_i = d_i - \Sigma d/s$ is computed for each direction. The fact that the sum of the corrections v has to be zero, within rounding accuracy, is used as a check for the computations. With this, the standard deviation of a direction observed in one set becomes

$$s_r = \sqrt{\frac{[vv]}{(n-1)(s-1)}}$$

and for the mean direction R, determined in n sets:

$$s_{\hat{R}} = \frac{s_r}{\sqrt{n}} = \sqrt{\frac{[vv]}{n(n-1)(s-1)}}.$$

3.7.4 Repetition Angle Measurements

Vernier and simple microscope theodolites usually have a sighting accuracy of 0.3 to 0.5 mgon, while the reading at the indices has an accuracy of 1 to 2 mgon. Therefore, the sighting accuracy gets lost because of the uncertainty of the reading. To counter this disadvantage, n measurements of an angle are mechanically added on the graduated circle (repetitions). The difference between the initial and final readings equals n times the angle. The single angle is thus obtained by dividing by n, which means that the reading errors are also reduced to one n-th of their original amount thus reaching the magnitude of the sighting error. Therefore, for the above mentioned accuracies, 4 to 6 repetitions in each telescope position are appropriate. A 5-times repetition requires the following steps in each telescope position:

a) set telescope to left target and record initial reading A_i,

b) loosen platen clamp and sight to right target (record a rough reading),

c) clamp circle to top post and set telescope to left target always moving clockwise,

d) repeat b) and c) until the fifth setting of the right target, then read A_e. $^1/_5(A_e - A_i)$ equals the single angle in the direct telescope position,

e) plunge the telescope, and repeat steps a) to d) starting with the right target,

f) add the n-times angles of both telescope positions and divide by $2n$ (in this case $2n = 10$).

Once precision theodolites (section 3.3.4.3) became available, the repetition method lost its importance. It remains, however, a valuable approach if one has to obtain high accuracies with simpler instruments, i.e. for measuring a parallactic angle (section 4.3.2.3). The value of the concept remains without question.

3.7.5 Special Methods for Angular Measurements

If the stations are far apart, all targets are rarely visible at the same time. Thus single angles have to be measured instead of directions in sets. For this purpose several methods have been developed.

3.7.5.1 Angles with Closure of Horizon

Individual angles are measured between two consecutive targets. This is done systematically until the total horizon is covered. Then the discrepancy of the sum of these angles to 400 gon is divided evenly into all observed angles. This approach is also recommended for shorter sights if the observation pillars are affected by torsion due to the influence of the sun or are unstable for other reasons.

3.7.5.2 Angles in All Combinations

To achieve a better error distribution than in section 3.7.5.1, all the $^1/_2 \cdot s(s-1)$ possible angles within a bundle of s rays are measured. In the situation constructed in Fig. 3.7.1, this means six angles, namely 12, 13, 14, 23, 24, and 34. To eliminate discrepancies, the most probable value for the angles 12, 13, and 14, which share the initial direction as one side, is computed from the direct observations with weight 2, and from the sums and differences, each with weight 1.

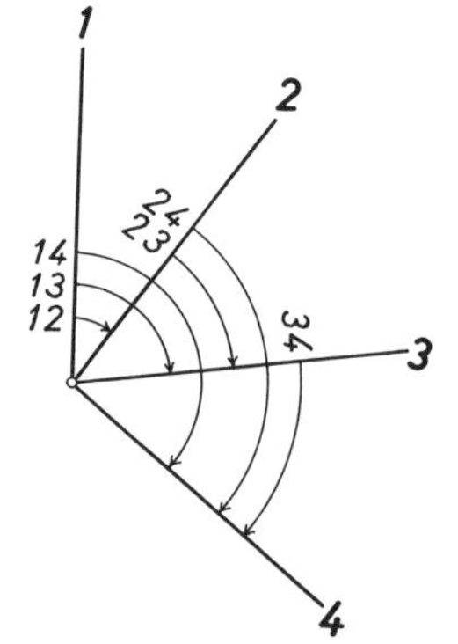

Fig. 3.7.1 Angles in all combinations

If, for instance, 12 is the observed angle, and (12) its adjusted value, then the following is obtained:

$$(12) = {}^1/_4[2 \cdot 12 + 1 \cdot (13 - 23) + 1 \cdot (14 - 24)]$$
$$(13) = {}^1/_4[2 \cdot 13 + 1 \cdot (12 + 23) + 1 \cdot (14 - 34)]$$
$$(14) = {}^1/_4[2 \cdot 14 + 1 \cdot (12 + 24) + 1 \cdot (13 + 34)]$$

As a rule, this method is used only in primary triangulation networks.

3.7.5.3 The Sector Method

This method is used if rays of different order originate at the same station. In the well-known Swiss sector method three or five primary directions, distributed evenly over the horizon, are selected. For primary triangulation, these angular sectors are determined with a weight of about 8, i.e. 8 fold observations of each angle. The mean values are adjusted to the horizon, i.e. 400 gon, according to the method in section 3.7.5.1. Subsequently, the angles of the 1st, 2nd, and 3rd densification are observed with weights 6, 4, and 2, respectively, and adjusted according to the same principle within the sector formed by two primary directions.

3.8 Orientation with Gyro Instruments

3.8.1 Fundamentals

Based on investigations by *Anschütz* (1915) and *Schuler* (1922), the gyro has been used in mining surveying since the 1920s as a gyro-compass or for meridian reference. Within the last two decades, gyros have also been applied for other surveying tasks, especially for the orientation of long or open traverses and in engineering surveys. Several companies build *"gyro theodolites"*, i.e. instruments where theodolite and gyro form one unit. More recently, for financial reasons, mainly *gyro attachments* are marketed which can be mounted onto common theodolites, either directly or with an adapter.

Any solid body rotating around an axis can be considered as being a gyro, usually driven by a fast running motor. Physically important characteristics of a gyro are its inertia and precession, a movement caused by an external force acting upon the rotational axis.

For technical purposes, the body of the gyro is usually symmetric. For surveying purposes, the following gyro type is most suitable.

3.8.2 The Pendulum Gyro

This gyro hangs on a thin but strong and elastic metal ribbon and swings freely in the air.

A gyro, which is supported symmetrical to its centre of gravity and with as little friction as possible, is called a "free gyro" because gravity cannot generate a moment of rotation. Even if the support is rotated around the three spatial axes, the gyro axis retains its orientation in space. If, however, a gyro is suspended above its centre of gravity, then gravity influences the orientation of the gyro axis. This type of gyro is called a "gravity bound gyro".

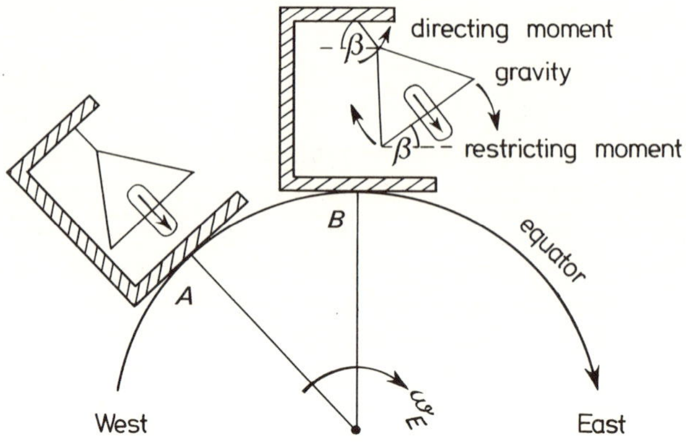

Fig. 3.8.1 Gyro at the equator

The behaviour of such a gyro is especially simple when it is placed at the equator (Fig. 3.8.1). In its initial position A, the gyro axis is horizontal and oriented in an east-west direction. Due to the rotation of the earth, the gyro reaches position B. The gyro axis tries to retain its spatial orientation during this motion, which pushes the centre of gravity outside the vertical. The gyro reacts to the resulting moment M by shifting its axis around the vertical, i.e. by getting into a precession motion. This motion which is forced onto the gyro lasts until the rotational axis of the gyro becomes parallel to the earth axis, and thus points towards geographic north (Fig. 3.8.2).

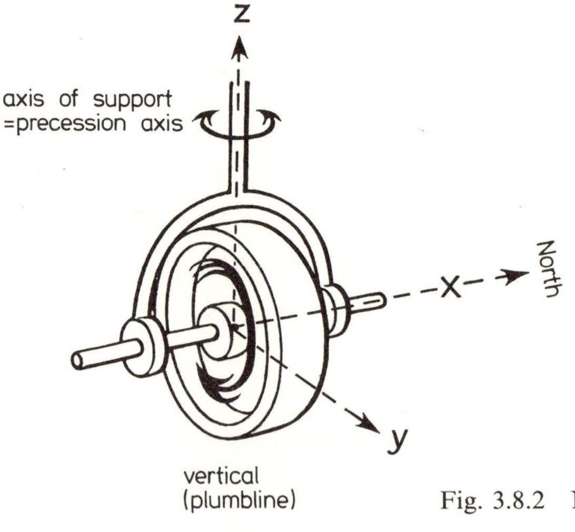

Fig. 3.8.2 North-seeking gyro

The directional moment R of the gyro is

$$R = \Theta_k \omega_k \omega_E \sin\alpha = J_\omega \omega_E \sin\alpha \qquad (3.8.1)$$

where

Θ_k denotes the moment of inertia of the gyro

ω_k the angular velocity of the gyro
ω_E the angular velocity of the earth
α the angle formed by the horizontal projection of the gyro axis onto the meridian, i.e. its azimuth

and

$$J_\omega = \Theta_k \omega_k$$

is the rotational impulse of the gyro. If the gyro is located at latitude φ (Fig. 3.8.3) then the angular velocity of the earth's rotation is split into

$$\begin{aligned} \omega_E \sin\varphi &= \text{rotation around the vertical} \\ \omega_E \cos\varphi &= \text{rotation around the meridian}\,. \end{aligned} \tag{3.8.2}$$

The rotation around the meridian determines the directional moment which, for latitude φ, becomes

$$R_\varphi = \theta_k \omega_k \omega_E \cos\varphi \sin\alpha = J_\omega \omega_E \cos\varphi \sin\alpha\,. \tag{3.8.3}$$

Thus R takes on its maximum value for latitude $\varphi = 0$, i.e. at the equator, and its minimum for $\varphi = 90°$, i.e. at the pole. In the azimuth the directional moment increases with increasing deviation α from the meridian. If the gyro is within the meridian, it does not have a directional moment because of $\alpha = 0$, i.e. it is in the rest position.

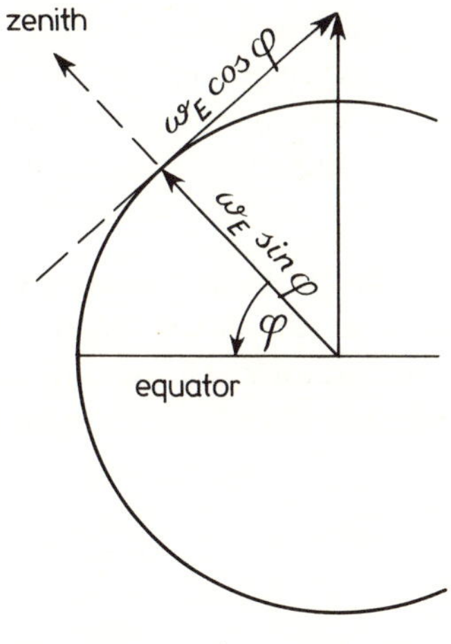

Fig. 3.8.3 Gyro at latitude φ

Theoretically, the rotational axis of the gyro should quickly move to the north-south direction. However, because of the small forces acting on it, it does not come to rest immediately, but oscillates with the following swinging time.

$$T = 2\pi \sqrt{\frac{I_\omega}{M(\omega_E \cos\varphi + D_B/I_\omega)}} \approx \frac{T_0}{\sqrt{\cos\varphi}}\,, \tag{3.8.4}$$

where

M represents the binding moment caused by the gravity acceleration of the path,

D_B the moment of torsion of the suspending ribbon, and

T_0 the swinging time at the equator.

Thus the swinging time increases with increasing latitude. Some practical values for T_0 and T would be:

$$T_0 = 7\text{ min}; \quad T_{60^\circ} = 9.5\text{ min}; \quad T_{75^\circ} = 13.5\text{ min}.$$

The rotational component of the earth around the vertical has the effect that the gyro axis tends to lift up, i.e. become parallel to the earth's axis (Fig. 3.8.1, position B). This elevating moment

$$E = J_\omega \omega_E \sin\varphi \tag{3.8.5}$$

is counteracted by the binding moment M, which means that even though this elevating angle $\beta = E/M$ increases with latitude, it does not exceed a few minutes of arc.

3.8.3 Mechanical Structure

The structural elements of a gyro are usually divided into the *carrying, direction producing,* and *direction receiving systems.* For the gyro theodolite, schematically illustrated in the left part of Fig. 3.8.4, the carrying system consists of the external housing (4) plus a structure which carries the theodolite, i.e. the direction receiving system (3).

The gyro with horizontal axis of rotation, i.e. the direction producing system (1), is suspended like a torsion pendulum with the aid of a thin elastic metal ribbon and oscillates around the meridian because of the earth's rotation. The direction receiving and producing systems are optically connected with the autocollimation mirror (2) at

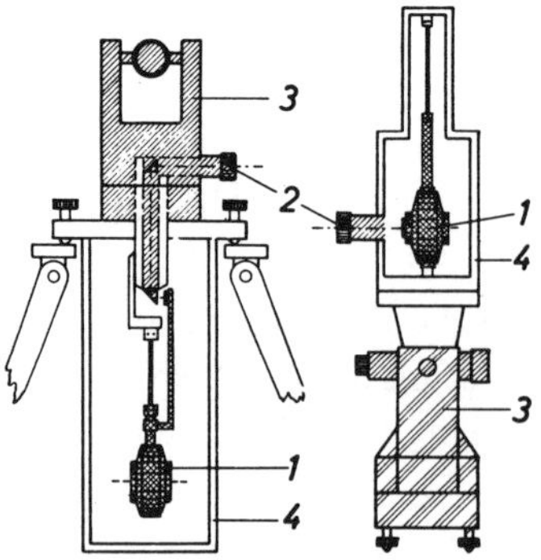

Fig. 3.8.4 Gyro theodolite and gyro attachment

the gyro mast, utilizing several prisms and the hollow axis of the theodolite. The oscillation movements are tracked with the autocollimation device.

The desired central position is derived from the observation results.

The gyro attachment (1), shown in the right part of Fig. 3.8.4 is rigidly connected to the platen of the theodolite. Therefore, there is no clear separation of the three systems in this case. In fact, the theodolite assumes some of the functions of the carrying system by carrying the gyro housing (4). The direction receiving system consists of two separate elements, mainly the device to observe the oscillations (2) and the theodolite (3) itself. Only the gryo as a direction producing system (1) has the same function as in the gyro theodolite.

The ribbon supported compact instruments where gyro and theodolite form one unit have a mass of 10 to 30 kg, not counting tripod and power supply unit. The gyro motors are battery operated by a transistor A/C generator producing a current with approximately 400 Hz and 110 V which provides a rotational velocity for the gyro of approximately 24,000 revolutions per minute. Gyro attachments with a mass of 2 kg are attached to the theodolite with a bridge and powered by cable from a separate battery and converter unit.

Table 3.8.1 gives a selection of gyro instruments of some well-known instrument manufacturers.

Table 3.8.1 Gyro instruments

manufacturer	type	gyro type	half pendulum time $T/_2$	warm-up	measuring time	mass: instrument	mass: total equip.	Std. deviation of one measurem:	Special features
			[sec]	[min]	[min]	[kg]	[kg]	[″]	
a) gyro theodolites									
Fennel. Kassel	KT 3	tape in air	240	2	20	8	42	20	
MOM. Budapest	GiB 1	tape in air	270	2	35	11	55	15	
b) gyro attachments									
Fennel. Kassel	TK 4	tape in air	240	2	25	4	18	30	coincidence position of object. See Fig. 3.8.5
MOM. Budapest	GiC 2	tape in air	220	2	25	9.3	18	20	
Wild AG Heerbrugg	GAK 1	tape in air	240	1.5	20	1.8	13.3	30	Fig. 3.8.6
Wild AG Heerbrugg	ARK 1	tape in air	210	1.5	10	5.6	27.4	20	north seeking gyro

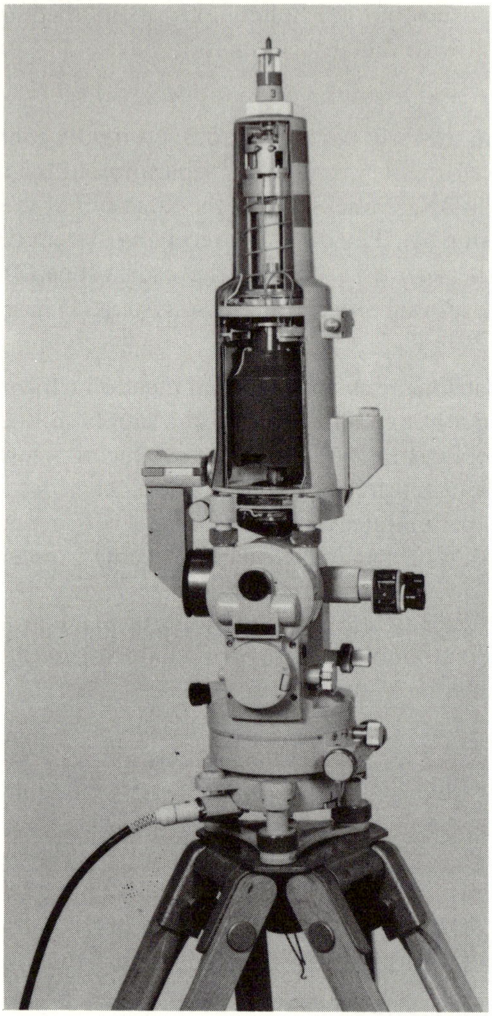

Fig. 3.8.5 Gyro attachment GiC 2 by *MOM*

Fig. 3.8.6 Gyro attachment GAK 1 by *Wild*

3.8.4 Observation Methods for Gyro Attachments

After setting up, the gyro oscillates, as mentioned in equation (4), with little dampening symmetrically to the meridian because of the inertia of its mass. The direction of the gyro axis is projected onto a fixed auxiliary scale with the aid of a light index (autocollimation device). This permits reading of the gyro movements, especially its amplitude.

Prior to the actual measurement, the gyro has to be preoriented to within 10 to 30 cgon. To achieve this, the oscillation of the gyro is followed with the theodolite platen by keeping the light index in the central zero position of the auxiliary scale.

At the instant of reversion, the light index remains stationary for a moment. This moment is used for reading the horizontal circle and recording the results. The same is repeated at the following reversion points U_2 and U_3. These readings provide then an approximation for the north orientation as

$$(N)_1 = \frac{1}{2}(U_1 + U_2) \quad \text{or} \quad (N)_2 = \frac{\frac{1}{2}(U_1 + U_3) + U_2}{2} .$$

The final value for geographic work is obtained either

- by observing a sequence of reversing points plus reading the corresponding horizontal circle positions with subsequent meaning for each point of reversion, or
- by measuring the passing times and amplitudes of the oscillations with respect to a reference direction preoriented towards north.

In both cases, a calibration correction (section 3.8.5) has to be applied. The reader is referred to the operation manual provided by the manufacturer for further details.

3.8.5 Instrument Constant and Meridian Convergence

The *instrument constant* E is a correction to the index reading on the horizontal circle of the theodolite. It is determined by the manufacturer prior to delivery at a calibration line with known azimuths. Basically, E represents a type of collimation axis error, being the angle between the collimation axis of the theodolite and the axis of the autocollimation system of the gyro. The arithmetic sign of E is selected such that E has to be subtracted from the index reading Z for gyro north on the horizontal circle of the theodolite.

Thus the azimuth α from station S to point P as determined by gyro and theodolite is

$$\alpha = \mathrm{r} - (\mathrm{N} - \mathrm{E}) = \mathrm{r} - \mathrm{N} + \mathrm{E}$$

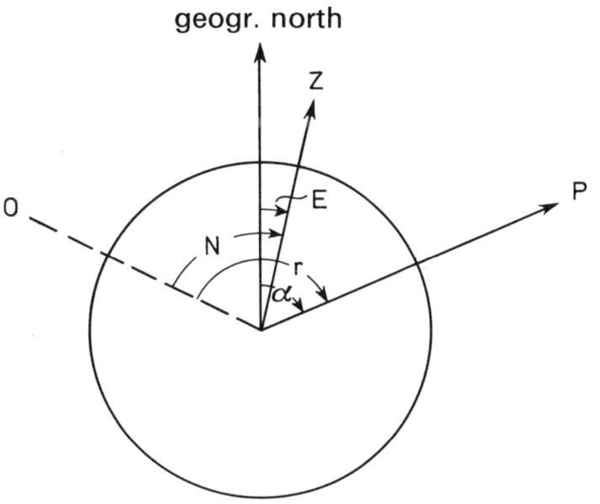

Fig. 3.8.7 Calibration constant E

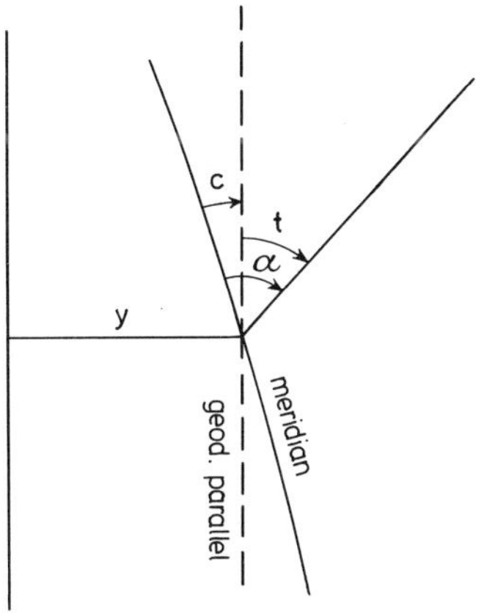

Fig. 3.8.8 Meridian convergence

where r is the observed direction (Fig. 3.8.7). Because of the relatively unstable connection between theodolite and gyro, the instrument constant E has to be checked periodically.

An additional correction, namely for *meridian convergence*, is necessary if the north direction as determined with the gyro is to be used in a rectangular geodetic coordinate system, e.g. in a transverse Mercator system. The abscissa axes of such systems coincide with multiples of 3° longitude, counted from the Greenwick meridian, which are referred to as the principal meridians of the systems.

The reference direction for a geodetic azimuth t to a point P is, unlike for the astronomic azimuth α, not the direction N to geographic north but to grid north, which is parallel to the principal meridian. The angle between the directions to geographic and grid north at the station S is called the meridian convergence. It is denoted as c, increasing clockwise from the local meridian. Thus, according to Fig. 3.8.8,

$$\alpha = t + c \quad \text{or} \quad t = \alpha - c$$

when the station lies east of the principal meridian; c can be computed as

$$c = l \sin\varphi = \frac{(180°/\pi)y}{R} \tan\varphi$$

where l denotes the longitude difference between principal and local meridians, y the geodetic ordinate of the station in reference to the principal meridian (in North America, this is usually referred to as X), and R the radius of curvature of the earth at the meridian. Calibration correction and meridian convergence can be combined as

$$t = r - N + (E - c)\ .$$

As evident from Fig. 3.8.8, the meridian convergence becomes negative if the station is located west of the principal meridian.

4. Distance Measurements

4.1 Simple Distance Measurements

4.1.1 Distance Measurements with Rigid Rulers

Very precise distance measurements in small places can be performed with rigid steel rulers. Measurements of this kind occur quite frequently in connection with industrial and engineering surveys. Steel rulers are also used for the determination of the length of comparators which in turn serve as a calibration standard for tapes and measuring wires.

In the past, 5 m long wooden survey rods have been used for cadastral and detail surveys, especially in mountainous terrain. With the advent of electronic distance meters, however, their use has been declining, such that they are now obsolete.

4.1.1.1 Measurements with Steel Rulers

Steel rulers are recommended for precise measurements in small spaces, as needed in engineering surveys where, for instance, sensitive machinery has to be precisely placed. In such cases, especially when the accuracy requirements are high, two 1 metre long smooth steel rulers with sharp edges at both ends are recommended. Their standard length (at 20°C) and their temperature dependency are determined by calibration. These *"normal metres"* are usually not graduated. Remainders are thus measured with either short rigid steel tapes or, for small distances, with a steel measuring wedge as shown in Fig. 4.1.1.

These measurements are subject to different systematic influences, which can be estimated and corrected for, using the following formulas:

(1) *Calibration correction*

Conventional distance measuring equipment is calibrated with the aid of a length comparator (e.g. a laser interferometer).

This leads to

$$k_E = l \cdot \frac{\delta}{L_0} \tag{4.1.1}$$

where

k_E = calibration correction
L_0 = nominal length in metres

$$l = \text{length to be corrected}$$
$$\delta = L_0 - I_E = \text{difference from } L_0 \text{ at calibration temperature (e.g. 20 °C) in mm}$$
$$\delta / L_0 = \text{calibration correction referred to unit length.}$$

(2) *Temperature correction*

$$k_t = l \cdot \alpha_{st}(t - 20\,°\mathrm{C}) \tag{4.1.2}$$

where

k_t = temperature correction
l = length to be corrected
α_{st} = linear temperature expansion coefficient of steel
t = field temperature during measurement
20 °C = calibration temperature.

Both corrections are combined in one equation, which is representative for the specific measuring tool

$$L = L_0 + k_E + k_t\,. \tag{4.1.3}$$

This equation is also used to describe calibration rulers, which are marketed in pairs. Packed in a wooden crate, they are delivered with a calibration certificate. Usually they are denoted with A and B and are 1000 ± 0.02 mm long. They normally consist of unhardened steel with a temperature expansion coefficient of $11.5 \cdot 10^{-6}$ m per °C.

These steel rulers are not only used to calibrate measuring rods or steel tapes but also directly for calibrating machine parts.

The calibration of 5 m measuring rods is given as an example. For this, two steel edges are placed on a smooth flat beam such that their separation exceeds the desired length by about 1 cm (Fig. 4.1.1). This separation is measured with the steel rulers and measuring wedge as mentioned above. Then the machine part is placed onto the beam between the steel edges. The difference is measured with a short millimetre ruler or tape, or with a measuring wedge.

Example: Calibration of 5 m rods.

Equation of the normal metres:

$$\text{ruler } A = 1\text{ m} - 0.15\text{ mm} - 0.0115(t - 20°)\text{ mm}$$
$$\text{ruler } B = 1\text{ m} + 0.07\text{ mm} + 0.0115(t - 20°)\text{ mm}\,.$$

Ruler length at calibration temperature $t = 16$ °C:

$$A = 1\text{ m} - 0.10\text{ mm} \qquad \mathrm{B} = 1\text{ m} - 0.02\text{ mm}\,.$$

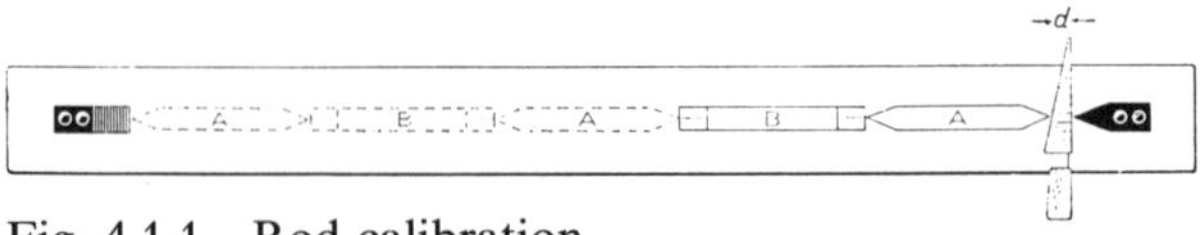

Fig. 4.1.1 Rod calibration

Double measurement of comparator length:

$$1.\ 3A + 2B + d_1 = 5\,\text{m} + 0.34\,\text{mm} + 9.10\,\text{mm}$$
$$2.\ 2A + 3B + d_2 = 5\,\text{m} + 0.26\,\text{mm} + 9.20\,\text{mm}\,.$$

Mean:

Calibration of two 5 m rods:

$$L_r = K_5 - d_r = 5\,\text{m} + (9.45 - 8.30)\,\text{mm} = \underline{5\,\text{m} + 1.15\,\text{mm}}\,,$$
$$L_b = K_5 - d_s = 5\,\text{m} + (9.45 - 8.83)\,\text{mm} = \underline{5\,\text{m} + 0.62\,\text{mm}}\,.$$

As shown in the second example, steel rulers can be used to determine the length of a comparator for the calibration of steel tapes (see also section 4.1.2.4).

Example: Determination of the length of a tape comparator.

A tape comparator is required for an accurate calibration. Usually, a stable measuring bench, which is the length of the tape, is placed on a wall (preferably inside a building) at a height of about 1.2 m. In its simplest form, the comparator has a metal plate with millimetre graduation at each end.

Furthermore, both ends need devices for holding or pulling the tape. The separation between the starting index of the tape comparator (Fig. 4.1.2) and the zero line of the millimetre graduation is measured in the same manner as for the rod comparator (Fig. 4.1.1). The tape is calibrated with the same tension that is used for field measurements.

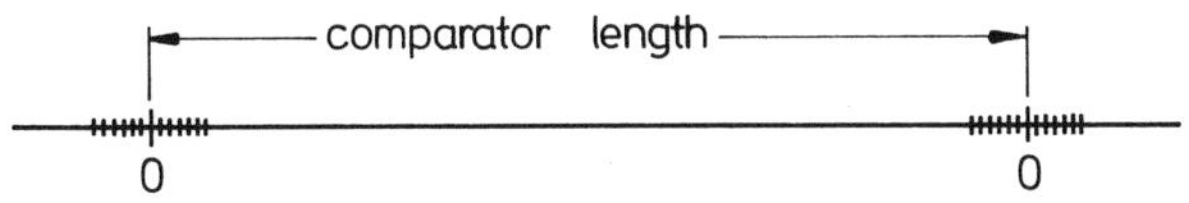

Fig. 4.1.2 Comparator for tape calibration

As in the previous example, the calibration temperature is $t = 16°\text{C}$, i.e.:

$$A = 1\,\text{m} + 0.10\,\text{mm}; \quad B = 1\,\text{m} + 0.02\,\text{mm}\,.$$

Double measurement of comparator length K_{20}:

$$1.\ 10A + 10B + d_1 = 20\,\text{m} + 1.20\,\text{mm} - 0.13\,\text{mm}$$
$$2.\ 10A + 10B + d_2 = 20\,\text{m} + 1.20\,\text{mm} - 0.11\,\text{mm}\,.$$

Mean:

$$K_{20} = 20\,\text{m} + 1.08\,\text{mm}\,.$$

This is valid for the temperature at the time of calibration. As shown in section 4.1.2.4, the survey tapes are to be calibrated for a certain reference temperature.

4.1.1.2 Measurements with Wooden Rods

Wooden distance rods were used in the past for cadastral surveys, especially in mountainous terrain. They were manufactured of well dried and knot-free pine, and usually come in 3 or 5 metre lengths. They had an oval cross section, whose major axis was reduced from about 5 cm in the centre of a 5 m rod to about 3 cm at its ends. The ends were equipped with steel caps which are either flat or else have sharp edges rotated by 90°.

As protection against humidity, the rods were painted with an oil based paint such that black and white, or red and white alternated with each full metre.

Decimetres were either marked with colored rings or with brass knobs, while centimetres were estimated. Usually, one worked simultaneously with two rods (black/white and white/red), thus effectively having a 10 metre ruler. The measuring methods differed according to the terrain.

4.1.2 Distance Measurements with Steel Tapes

Steel tapes are primarily used for distance measurements in small areas, usually in conjunction with the orthogonal or tie-in methods. For longer distances, larger areas or the coordination of planimetric points, nowadays electronic distance meters are used nearly exclusively (section 4.4).

In *flat terrain* the tape is consecutively placed along a straight line between the end points (property corner, traverse point). The resulting distance is then composed of the number of full tape lengths plus the reading within the last length.

In *inclined terrain*, only the horizontal component is utilized in order to obtain a horizontal distance. Either the *step method* or the *reduction method* is utilized in this case.

a) The *step method* (Fig. 4.1.3) is slightly easier when measuring downhill. Then the front end of the tape is lifted until the tape is horizontal. This is often done by estimating the right angle between tape and plumb string, however, using a small level

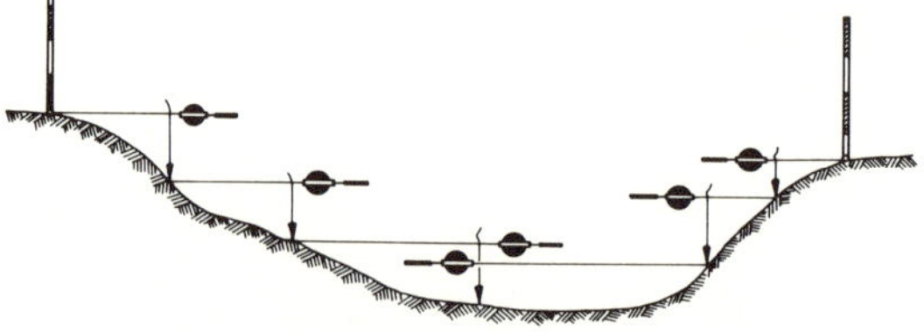

Fig. 4.1.3 Step method with tapes

is better. A plumb bob is used to mark the starting point for the next tape length. This procedure is repeated until the end, where the remainder is read on the horizontal tape, again using the plumb bob. Uphill measurement is a bit more difficult because the tape has to be placed against the free hanging plumb string. For higher accuracy requirements, a distance is measured twice, once up, once down, with the mean used to compensate for systematic errors.

b) The *reduction method* utilizes tape lengths laid along the terrain. Their inclination β (see Fig. 4.1.6) is determined with a pendulum or bubble device. It is then used to reduce the distance to the horizontal. In this case there are no major problems, provided the reductions to horizontal plus for temperature expansion are applied, and the tape is handled carefully. For free hanging use, however, which for practical reasons utilizes either narrow steel tapes or invar wires, additional laws of mechanics have to be considered.

A rope, spanned between two supports P_1 and P_2 with equal height, sags because of gravity if it is homogeneous and flexible but not expanding. Mathematically, the sag curve or chain line is described as a hyperbolic cosine line (cosh). With the origin of the coordinate system lying at a distance m on the normal to the curve, below its lowest point, its equation is (also Fig. 4.1.4):

$$y = m \cosh \frac{x}{m} \tag{4.1.4}$$

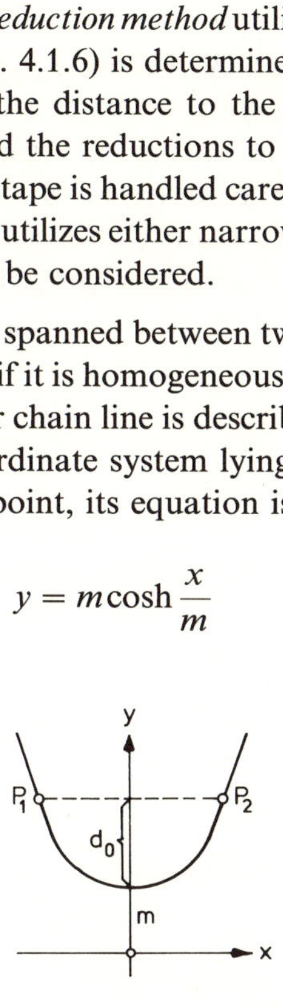

Fig. 4.1.4 Chain line

The constant m is computed using two conditions.

The gravity acting on a small arc segment $\Delta s = PP'$ and the pulling force at the ends of this arc PP' have to be in equilibrium both horizontally and vertically. Rather than presenting a detailed derivation, only the results are presented. With the introduction of $q = p/2H$, both conditions lead to

$$m = y_0 = \frac{H}{p}, \quad \text{and} \quad \frac{x}{m} = \frac{p}{2H} 2x = 2qx \,, \tag{4.1.5}$$

where H is the horizontal component of the pulling force P, and p is the weight of the tape per metre length. Inserting equation (4.1.5) into equation (4.1.4) results in the chain line equation (Fig. 4.1.5)

$$y = \frac{H}{p} \cosh 2qx \,. \tag{4.1.6}$$

For the determination of a correction to the measurement, this equation is developed into a series:

$$y = \frac{1}{2q} + qx^2 + \frac{q^2}{2} x^2 + \frac{2}{45} q^5 x^4 \ldots \,. \tag{4.1.7}$$

In practical surveying, various tapes or invar wires are used instead of a rope. A tape with its weight p in kg/m is pulled with a force P, usually measured in kp (1 kp = 9.806 65 N), with a tension handle or else with the aid of tension weights. In deviations from the theoretical chain line, usually an elevation difference $h_1 - h_2 = h = L^R \cos\beta$ is encountered between the points of support P_1 and P_2, where the angle β represents the vertical angle of the chord. Furthermore, the horizontal projection L of the chord L^R, rather than the arc length, is required (Fig. 4.1.5).

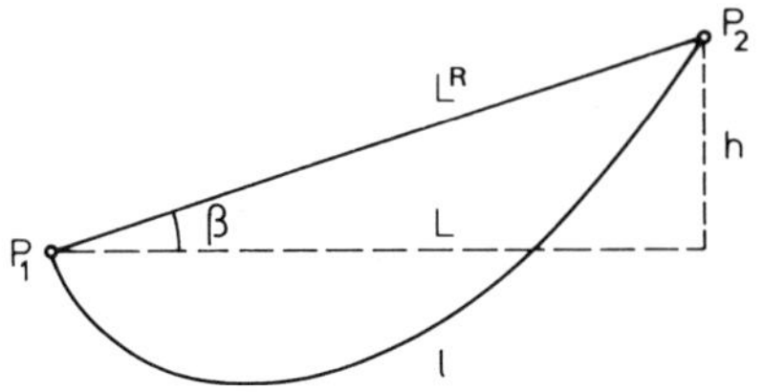

Fig. 4.1.5 Deviations from the theoretical chain line

4.1.2.1 Measurements with Free Hanging Tapes

The following corrections, in metres, are most important for practical work:

(1) *Standardization and temperature correction*

Temperature and standardization corrections can be combined to

$$k_t = l\{\delta/L_0 + \alpha(t - 20\,^\circ\mathrm{C})\} \tag{4.1.8}$$

where

L equals the nominal length of the tape in m,

δ equals the additional length beyond L in metres obtained by tape calibration at 20°C,

δ/L represents the calibration correction referred to 1 metre,

α is the temperature expansion coefficient per metre, i.e. $0.0111 \cdot 10^{-3}$ for steel,

t represents the tape temperature during the measurements in °C. Thus, for a 20 m steel tape, the temperature correction alone for $t = 25$°C amounts to

$$k_t = 0.23\; 10^{-3}(25 - 20)\ \mathrm{m} = 1.15\ \mathrm{mm}\,,$$

which means that k_t for 100 m and $\Delta t = +5\,^\circ\mathrm{C}$ becomes approximately 6 mm. This amount has the magnitude of the desired measuring accuracy and thus cannot be neglected.

(2) *Sag correction*

Because of the difference between chain line and chord, this correction becomes for different heights of the supports (Fig. 4.1.5)

$$r_d = -\frac{l^3 p^2}{24\,P^2}\cos^2\beta = -\frac{lp^2}{24\,P^2}(l^2 - 2h^2) \qquad (4.1.9)$$

where

p = tape weight per unit length (kp/m)
P = pull (N)
β = inclination of chord
h = elevation difference of supports.

If both supports are of equal height, than $\cos^2\beta = 1$ and $2h^2 = 0$. For the mentioned 20 m steel tape with $p = 0.02$ kp/m and $P = 5$ kp, because $\frac{p}{P} = 4.10^{-3}$ this correction becomes

$$r_{d_0} = \frac{-l^3}{24}\,\frac{p^2}{P^2} = \frac{-20^3}{24}\,\frac{16}{10^6} = -5\ \mathrm{m}\ .$$

If p and P are unknown, the sag offset h_0 is determined by levelling. With

$$d_0 \approx \frac{p}{P}\,\frac{l^2}{8} \qquad (4.1.10)$$

the correction becomes

$$r_{d_0} \approx \frac{-8}{3}\,\frac{d_0^2}{l}\,. \qquad (4.1.11)$$

With the values for p, P and l as given above, a sag of $d = 20$ cm is obtained for the 20 m tape (equation (4.1.9)). Inserting this value into equation (4.1.10), again $k_d = -5$ mm is obtained.

For 100 m, this amounts to 2.5 cm and therefore cannot be neglected. If the tape is subjected to a tension of $P = 10$ kg, then both values are reduced to one quarter.

(3) *Correction for pull*

Different pull forces cause different elastic expansion of the tape. If the field pull force differs from the one applied during calibration, then the following correction is needed:

$$k_p = \frac{l(P - P_0)}{Q \cdot E} \qquad (4.1.12)$$

where

P_0 = pull at calibration (e.g. 150 N)
Q = cross section of tape
E = module of elasticity.

If, for example, a 20 m tape with a (20 × 4) mm^2 cross section is used with a pull of 100 N, instead of 150 N, then the following correction is to be applied (E = 20 GPa)

$$k_p = \frac{(100 - 150) \cdot 20 \text{ Nm}}{20 \cdot 0.4 \cdot 20 \text{ mm}^2 \text{ GPa}} = -0.6 \text{ mm} .$$

Thus, for many practical applications it is sufficient to keep the pull force within 50 N of the calibration pull.

(4) *Correction due to inclined tape and mis-alignment*

As shown in Fig. 4.1.6, the corrections are:

$$k_a \cdot (2l - k_n) = h^2$$

$$k_a \approx \frac{h^2}{2L}$$

$$k_a = l(1 - \cos\beta) = 2l\sin^2\frac{\beta}{2} = \frac{h^2}{2l} . \qquad (4.1.13)$$

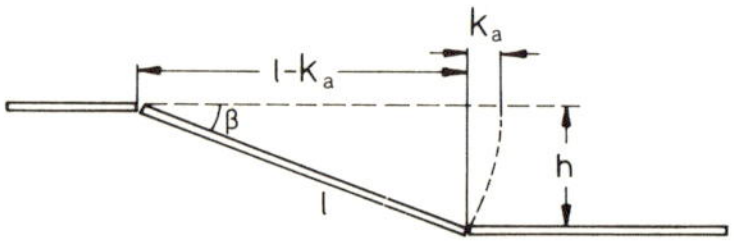

Fig. 4.1.6 Inclined tape and mis-alignment

An elevation difference of 0.2 m causes a correction of 1 mm for a 20 m tape. Thus for many practical purposes, it is sufficient to estimate the horizontal position of the tape.

When computing the corrections, p and P are predetermined, δ is obtained during calibration while l, t, β and h, and d have to be observed during the measurement. By using these corrections, the resulting horizontal distance between two points P_1 and P_2 is then

$$L = L_0 + k_{k,t} + k_d + k_p + k_a . \qquad (4.1.14)$$

4.1.2.2 Measurements with Fully Supported Tapes

Provided that there are no obstacles (e.g. vegetation), the tape can be laid on the ground in flat areas. Thus the k_d correction is not needed. The slope angle β can be determined with a free hand level.

4.1.2.3 Tapes on Reels

These tapes are manufactured of regular or rust-free steel or else of invar and come in 10, 20, 25, 30 and 50 m lengths with a cross section of 13×0.2 mm^2. Recently they have been coated with polyamide (nylon). Usually the tapes are subdivided into metres, decimetres, and centimetres with millimetres for the first metre. Since each decimetre is numbered, reading errors are practically eliminated. The graduation is either etched into or out of the steel or else placed beneath the nylon coating for better preservation. The weight of nylon coated tapes is 18.8 g per metre for tapes by BMI (Bavarian Measuring Industries) and Meywald K.G. Their coefficient of expansion is $11.5 \cdot 10^{-6}$, while their tension strength is 140 kg/mm^2. The tapes are equipped with a ring on one end and fastened to a reel with a crank handle on the other so that they can be easily rolled up after use (Fig. 4.1.7). The zero line of the graduation often lies where the ring and tape meet. However, a graduation which begins 1 to 2 dm inside, is preferable (Fig. 4.1.8).

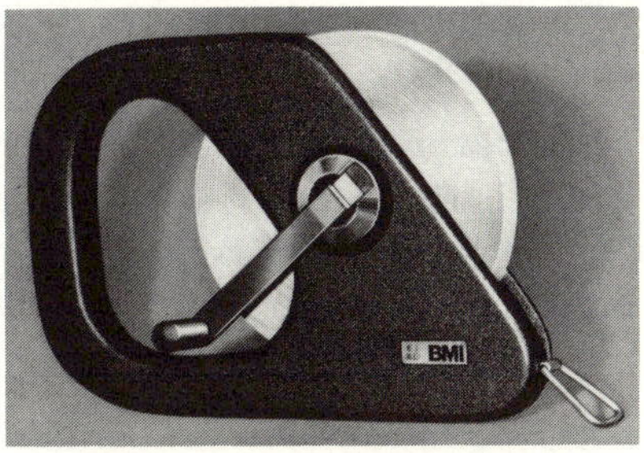

Fig. 4.1.7 Reel for the *BMI* tapes

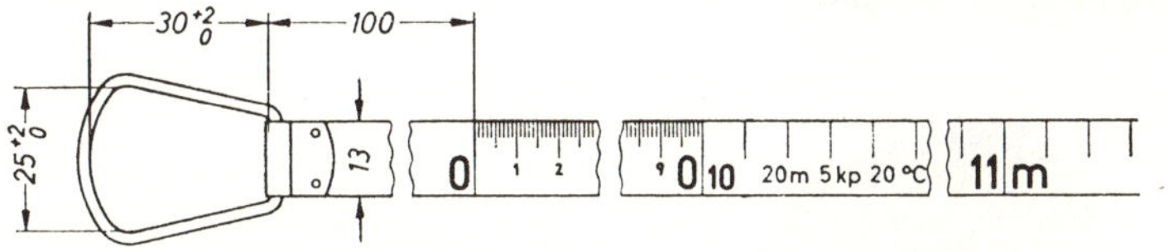

Fig. 4.1.8 Tape with graduation

Rings with 28 mm internal diameters can fit over a range pole. For such tapes, the zero mark is in the centre of the ring which is occasionally quite suitable. The tension (at least 5 kg) is measured with a tension handle utilizing any elastic spring which can be hooked into the ring of the tape. For the BMI tapes (Fig. 4.1.7), the spring is located in a separate handle, while it is built into the lower left corner of the reel for the Maywald tapes (Fig. 4.1.9).

When measuring across rail installations which are conducting low intensity electricity (e.g. for automatic switches), steel tapes have to be insulated. Because of their dimensional instability, cloth and plastic tapes are not suitable for surveying work.

It is not necessary to explain the *measuring procedure for flat terrain*, as one tape length after the other is placed along a straight line, marked if needed, with range

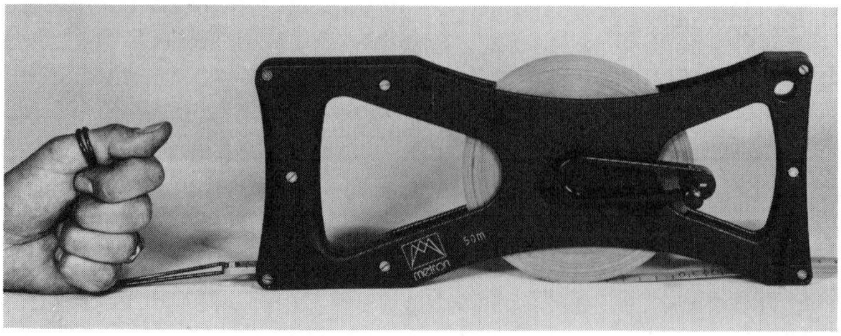

Fig. 4.1.9 Reel with tension handle for the *Meywald* tapes

poles. If, however, there are some larger obstacles on the ground, or the measurements have to cross, for instance, a grain field, then it might be advisable to measure at about 1 m to 1.5 m above the ground. Plumbing is then carried out with range poles and small levels. Usually, 20 m tapes are utilized, however, in urban areas often 30 or 50 m tapes are more suitable. Field measurements are frequently performed without tension handles.

For slopes up to 10% one can measure fully supported and use a slope indicator, i.e. a free hand level. Such an instrument consists of a telescope and a freely rotatable vertical circle which is oriented into the plumbline. As a rule, not only vertical angles but also slopes in percentages or direct reduction values can be read.

The Necli free hand level by Breithaupt, for instance (Fig. 4.1.10), has a broken telescope and a transparent graduated circle which is supported such that it can freely swing and is oriented by a pendulum. Liquid dampening facilitates fast operation.

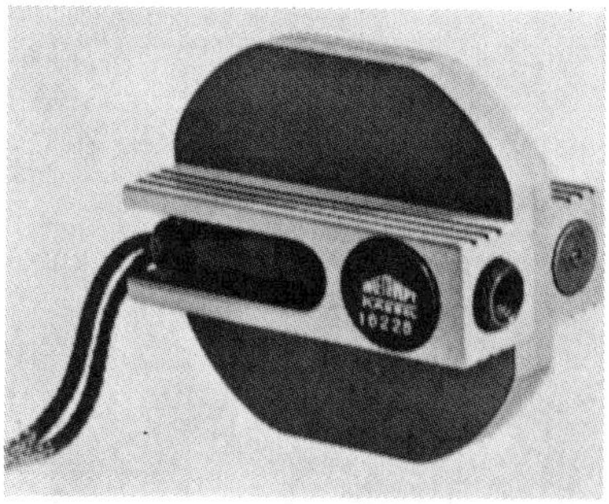

Fig. 4.1.10 Necli free hand level (*Breithaupt*)

Looking through the eyepiece, one sees the target point as well as the vertical angle h in degrees, gon or percent and also the amount $100\,(1-\cos h)$ which is to be subtracted from a 100 m long slope distance for reduction to the horizontal (Fig. 4.1.11). For a 20 m tape, this reduction amounts to 10 cm or 39 cm respectively for $h = 10\%$ or 20 %. An error in the inclination by 1 % causes a 5 cm distance error for a 30 % slope.

For slopes of more than 10 %, surprisingly good results can be obtained using the step method. *For steeper slopes*, the tape is rolled up and shortened such that plumbing can be done without problems using a plumb bob, or range poles. In steep terrain, the head tapeman gives the rear tapeman the respective point on the tape and calls out its reading until the whole 20 m is measured or staked. Only then is the next tape length started.

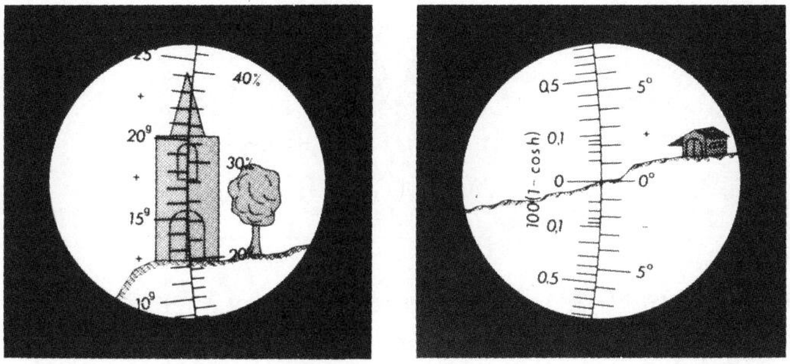

Fig. 4.1.11 Field of view of the Necli free hand level

For highest accuracies, the tapes are put under 5 kg of tension and the corrections mentioned in section 4.1.2.1 are applied. A standard deviation of less than ± 2 cm per 100 m can then be expected.

The low weight, quick and sure reading, and the possibility to reduce the tape length to the approximate length of the desired distance have made the tape on reel one of the most commonly used devices for simple distance measurements.

4.1.2.4 Calibration of Steel Tapes

In most practical cases, steel tapes can be calibrated with sufficient accuracy by direct and return measurement of a 100 m test distance, which had been carefully determined with an electronic meter. The difference provides the standardization correction for the total measuring procedure, i.e. tape and method.

For higher accuracy requirements, a special comparator is used for tape calibration (Fig. 4.1.2). The result thus obtained is valid for the tape temperature during calibration and has to be reduced to the normal temperature of 20 °C. For this it is sufficient to use $11.5 \cdot 10^{-6}$ m per metre and degree Celsius as the temperature expansion coefficient for steel.

Example

Calibration of a 20 m steel tape L_{20}.

Length of comparator:

$$K = 20\text{ m} + 1.08\text{ mm} .$$

Reading at the graduation resulted in $d = 1.5$ mm for the tape with 5 kg pull at $t = 16\,^\circ\text{C}$.

Thus

$$L_{20} = K - 1.5\text{ mm} = 20\text{ m} - 0.42\text{ mm}$$

at $t = 16\,^\circ\text{C}$.

Therefore, for $t = 20\,^\circ\text{C}$.

$$L_{20} = 20\text{ m} - 0.42\text{ mm} - 20 \cdot 0.0115(16 - 20)\text{ m} ,$$
$$L_{20} = 20\text{ m} + 0.50\text{ mm} + 20 \cdot 0.0115(t - 20)\text{ mm}$$

or for $t = t^\circ$.

For a temperature increase of 10 °C, a 20 m tape expands by 2.3 mm, i.e. a 100 m long distance would be short by 11.5 mm. Even for measurements of medium accuracy, these amounts cannot be neglected. Therefore, for practical work it is useful to devise a little table, which gives the tape length or the reductions for a 100 m distance at 0°, 5°, 10° ... 35 °C. It is rather difficult to obtain the tape temperature in the field. Therefore on overcast days, the air temperature, measured with a whirling thermometer, is used for the reductions. On sunny days, better results are obtained if a small thermometer is attached to the tape. Even then temperature errors of ±5 °C have to be expected, which cause distance uncertainties of 6 mm per 100 m.

4.1.3 Accuracy of Simple Distance Measurements

4.1.3.1 Errors in Distance Measurements

Both systematic and random errors are encountered in simple distance measurements. Among the systematic errors are: incorrect length of the measuring tool, deviation from a straight line, deviation from a horizontal plane, sag of the rods, sideways sliding of the rods on hills, etc.

Random errors are due to consecutive placing of rods or tapes, plumbing and reading. According to equation (1.4.11) random errors increase with the square root of the distance, systematic errors proportionally to the distance, while reading errors are independent of the distance. In the worst case, these three error components will add up. Thus the total error of a distance s can be expressed as

$$D = a\sqrt{s} + bs + c .$$

4.1.3.2 Tolerance Limits

Practical experience coupled with the accuracy requirements for certain types of surveys have led to numerical values for the coefficients a, b and c, resulting in tolerance limits which cannot be exceeded by the difference of two independent measurements of the same distance.

In Germany, for instance, legal surveys have to conform to

$$D = 0.008\,s - 0.0003\,s - 0.05$$

while traversing requires tighter limits, namely

$$D_t = 0.006\sqrt{s} - 0.02\,.$$

As mentioned in section 1.4.9, these values represent absolute standards which, for practical purposes, means three times the standard deviation (3σ).

Historically, the tolerance limits were given in three distinct groups, based on the physical circumstances, such as easy, medium and difficult. This idea has been discarded because of the shifting emphasis from mechanical distance measurements to optical, electronic or even photogrammetric distance measurements which are totally independent of the terrain.

In urban areas, the above mentioned limits have been reduced, following the tendency to group tolerance limits according to the different value of the land.

4.2 Tacheometry (Stadia Distances)

4.2.1 Optical Distance Measurements

A tacheometer (stadia) is an instrument which permits the measurement of the distance in addition to horizontal directions and vertical angles by optical or electronic means. Electronic tacheometers are described in section 12.2.

The optical distance measurement is based on the parallactic triangle (Fig. 4.2.1), where the distance s is derived from the base b and the parallactic angle γ.

$$s = \frac{b}{2}\cot\frac{\gamma}{2}\,. \tag{4.2.1}$$

In order to measure only one parameter, either the base or the parallactic angle is held

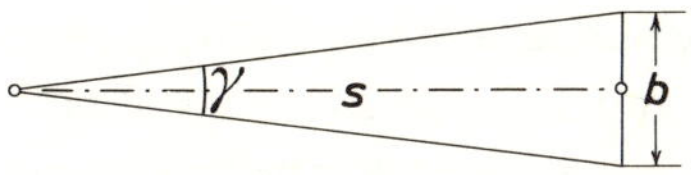

Fig. 4.2.1 Parallactic triangle

constant. Depending on the purpose of the measurement, the base can be located either at the instrument or the target station. The distance meters are divided into simple stadia instruments, which yield slope distances for inclined sights, and self reducing tacheometers which provide horizontal distances and elevation differences. Considering the various optical means for generating a parallactic angle, such as fixed and movable cross hair lines, stops, curves, and wedges, the following design possibilities exist:

a) constant base at observation station, no special targets

b) variable base at observation station, target marked with vertical staff

c) constant base at target station, carrying 2 fixed targets

d) varying base at target station, consisting of a rod with decimal graduation; observation instrument could be:

   i) theodolite with stadia hairs

   ii) self reducing tacheometer with slide device, tangent scale, diagram, or movable cross hair lines

   iii) tacheometer with simple wedge attachment

   iv) self reducing wedge tacheometer.

Designs mentioned in a) and b) are suitable for low accuracy only, and are covered briefly in section 4.2.3. Designs c) or d–iii) and d–iv) are covered in sections 4.3.2 and 4.3.3.

In the following section, the emphasis is placed on design d–i) while d–ii) is covered together with the applications of stadia measurements in Chapter 12.

The oldest type of optical distance meter utilizes the basic stadia principle with two or more horizontal stadia hairs on the reticule of the measuring telescope and a graduated vertical rod at the target station. Early instruments were built by Montanari (around 1670 in Italy) and Green (around 1780 in England). Both required rods with movable indices whose separation had to be read at the rod by an assistant. In 1810, the mechanic G. von Reichenbach from Munich equipped his distance meters with strong Fraunhofer objective lenses and was able to read a graduated rod placed at the target through the telescope. Therefore, simple stadia instruments are often referred to as Reichenbach's stadia theodolites (Fig. 4.2.2).

4.2.2 Reichenbach's Stadia Theodolite

4.2.2.1 Determination of Horizontal Distance

In order to change a measuring telescope of the Kepler type (section 2.3) to a distance meter, two horizontal stadia lines are added symmetrically to the cross hair, consisting of a single horizontal and vertical line. These stadia lines are at a distance of $^1/_2 p$ each from the centre cross hair line, thus are separated by p.

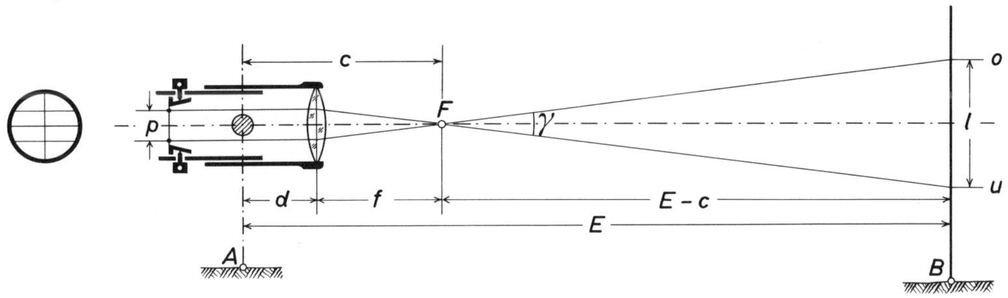

Fig. 4.2.2 Reichenbach's stadia theodolite

To measure the distance $AB = E$ in Fig. 4.2.2, such a telescope is set up and levelled at point A, while a graduated rod, usually a simple levelling rod with centimetre graduation, is held vertically at B.

The horizontal distance $AB = E$ is then obtained as follows: Rays parallel to the optical axis passing through the stadia hairs are refracted at the objective lens and form the angle γ at the front focal point. When extended in straight lines, they hit the rod and define the rod section $l = o - u$. This leads to the following relation:

$$E - (d + f) = \frac{f}{p} l \qquad (4.2.2)$$

where d denotes the horizontal distance between the vertical axis and objective lens, and f is the focal length of the latter. Combining $d + f = c$ and using $\frac{f}{p} = k$, this results in the basic stadia formula

$$E = c + kl\,. \qquad (4.2.3)$$

The small or *addition constant* c amounts to a few centimetres in modern instruments, while the large or *multiplication constant* k is set to 100 by choosing $p = 0.01f$. In a few cases $k = 200$ with $p = 0.005f$.

The point at which the parallactic angle γ is generated is called the anallactic point of the measuring telescope. For Kepler telescopes it is located at the front focal point of the objective lens. Because of $E - c = {}^1/_2 l \cot\frac{\gamma}{2}$, $\gamma = 34'23''$ or 63.66 cgon for $k = 100$.

A *telescope with movable focussing lens* has an effective objective lens consisting of both the objective and focussing lenses. When moving the focussing lens, the equivalent focal length of the system and thus the multiplication constant k change. Furthermore, the focal point and thus the anallactic point are slightly shifted, so that both c and k are functions of the distance.

If one traces the images of the points of intersection of stadia hairs and vertical cross hair line as generated in object space for different positions of the focussing lens, two slightly curved hyperbola branches are obtained. At a 5 to 10 m distance from the objective lens, these branches become straight asymptotes. Thus the horizontal distance E can be computed according to equation (4.2.3) without any problem if $E > 10$ m. It is, however, better to use $k = 100 + dk$, which leads to

$$E = c + (100 + dk)l = 100\,l + (c + dk \cdot l) \tag{4.2.4}$$

or with

$$\begin{aligned} \Delta E &= c + dk \cdot l\,, \\ E &= 100l + \Delta E \end{aligned} \tag{4.2.5}$$

where ΔE is determined empirically. Since today it is possible for the manufacturers to keep c and dk as well as their changes small, ΔE can often be neglected for routine jobs, and need only be known for more accurate work, e.g. stadia traverses. ΔE can be determined empirically as follows.

A horizontal test line is marked with pegs at uneven distances near 20, 40 … 120 m. With calibrated rods, tapes or an EDM instrument the accurate distances E_i of these points from the vertical axis of the instrument are measured. The same distances are measured by stadia, where l is determined 5 to 10 times at the carefully plumbed rod. Using the mean values l_i for each of the known distances E_i, the preliminary differences $(\Delta E_i) = E_i - 100\,l_i$ are found. These values are plotted in a right-handed coordinate system against $100\,l_i$. Subsequently a best fitting straight line is drawn (Fig. 4.2.3) to obtain adjusted values for ΔE which are tabulated. The ordinate at the ΔE axis provides an approximation for c, while dk is given by the slope of the line, which leads to $k = 100 + dk$.

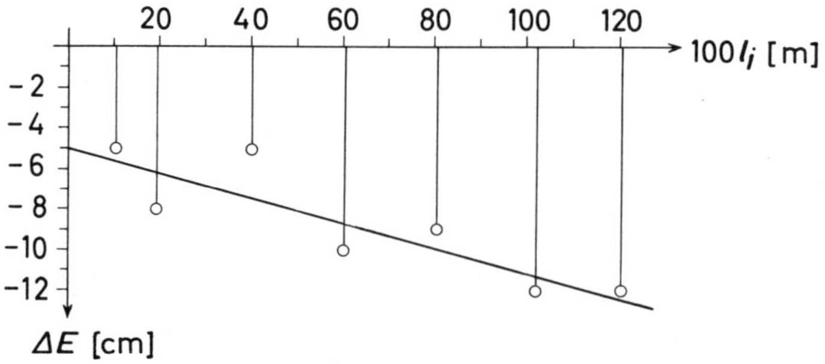

Fig. 4.2.3 Determination of stadia constant

The precision of this calibration depends largely on how well the rod sections were determined. A millimetre ruler for the shorter sights is useful for reading, while nests of 3 or 4 points which are later meaned may also improve the accuracy.

Example

Determination of ΔE, c and k for the Theo 030 No. 121241 theodolite.

After setting up the test line, the distances E_i from the origin to the pegged points were measured optically with the theodolite. For the point $E = 80.75$, for instance, the rod sections $l = o - u$ were obtained as shown in Table 4.2.1 (a), which provided a mean value $l = 0.8064$. Similarly, the l_i values for the other points were determined and used for the calculation of the preliminary values (ΔE_i) shown in Table 4.2.1 (b). After plotting $100\,l_i$ and ΔE_i on a graph, a best fitting straight line was drawn, which was used to obtain the final ΔE_i values, given in Table 4.2.1 (c).

Table 4.2.1 (a)

Determination of l for $E = 80.75$			
$o - u$	$+v$	$-v$	vv
m	mm/10		
0.806	4		16
0.805	14		196
0.806	4		16
0.807		6	36
0.807		6	36
0.806	4		16
0.807		6	36
0.806	4		16
0.807		6	36
0.807		6	36
8.064	30	30	440

$l = 0.8064$

$$\sigma_l = \pm\sqrt{\frac{440}{9 \cdot 10}} = \pm 0.2 \text{ mm}$$

(b)

Preliminary (ΔE)		
E	$100\,l$	(EE)
m	m	cm
10.35	20.40	− 5
19.80	19.88	− 8
40.10	40.15	− 5
60.18	60.28	−10
80.55	80.64	− 9
102.05	102.17	−12
120.09	120.21	−12

$c = -0.05$

$$dk = \frac{(-11) - (-5)}{100} = -0.06$$

$k = 99.94$

(c)

Adjusted ΔE	
$100\,l$	ΔE
m	cm
7	
	− 6
24	
	− 7
40	
	− 8
57	
	− 9
73	
	−10
90	
	−11
107	
	−12
123	

The constant k can be determined faster and more accurately with the aid of a collimator (Fig. 9.1.9) equipped with a 0.1 mm scale in the eyepiece or else with an eyepiece micrometer. In addition to the collimator, the measuring telescope is also focussed to infinity.

Then the separation ϱ_k of the images of the stadia hairs can be measured with the scale or the eyepiece micrometer in the image of the reticule, generated at the focal plane of the collimator. With f_k denoting the focal length of the collimator, $f_k/p_k = f/p = k \cdot {}^1/_2 \cot\gamma/2$.

The value for k, obtained in this manner, is only valid for the infinite focus setting, which practically is reached for $E = 100$ m in stadia theodolites.

In the focal plane of an Askania collimator with $f_k = 2250$ mm, the separation of the stadia hairs for the Theo 030 No. 121 241 theodolite was measured as $\varrho_k = 22.52$ mm, which resulted in $k = 2250 : 22.52 = 99.91$. This value differs from the calibration value using the test line by 0.03%, or 3 cm for 100 m. Generally, an uncertainty of $\pm 0.05\%$ has to be expected.

4.2.2.2 Reduction Formulae for Inclined Sights

The formulae given in section 4.2.2.1 are valid for horizontal sights only, i.e. when using a stadia level. However, stadia hairs are used in theodolites when measuring in hilly terrain as well. This provides two new tasks, namely the reduction of slope distances to horizontal ones and the computation of elevation differences between observation station and target. If a vertical rod is sighted with a stadia theodolite whose collimation axis forms a vertical angle α with the horizontal plane, then the rod section l is not normal to the collimation axis but at the angle 100 gon + α or 100 gon $- \alpha$ (Fig. 4.2.4).

In order to determine the slope distance E', a rod section l' formed by the parallactic angle at a fictitious rod normal to the collimation axis is required. With a small approximation which amounts to less than 0.02 mm for $\alpha = 30$ gon and $l = 100$ cm, this rod section is

$$l' = l \cos\alpha \, , \qquad (4.2.6)$$

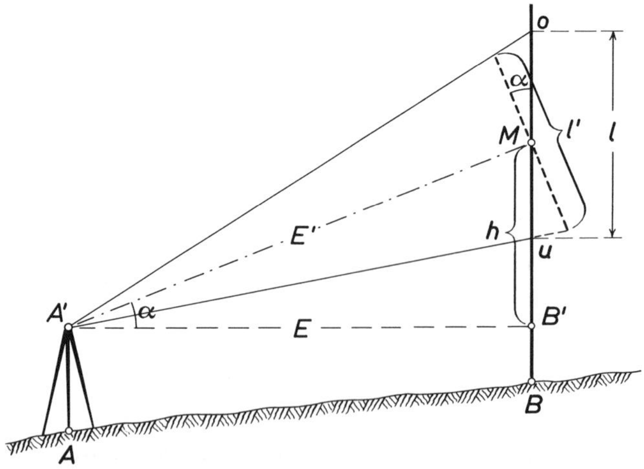

Fig. 4.2.4 Inclined stadia sight

and according to equation (4.2.3), the slope distance $A'M$ becomes

$$E' = o + kl\cos\alpha \,. \tag{4.2.7}$$

The horizontal projection $A'B'$ of this slope distance is

$$E = c\cos\alpha + kl\cos^2\alpha \tag{4.2.8}$$

while the elevation difference $B'M = h$ between the horizontal plane passing through the horizontal axis and the point M at which the centre cross hair is imaged on the rod is

$$h = E'\sin\alpha = c\sin\alpha + kl\sin\alpha\cos\alpha \,. \tag{4.2.9}$$

In order to obtain simpler computation formulae, one uses in equation (4.2.8) $c\cos\alpha \approx c\cos^2\alpha$, and in equation (4.2.9) $c\sin\alpha \approx c\sin\alpha\cos\alpha$, and according to equation (4.2.5)

$$(c + kl) = (\Delta E + 100\,l) = s \,. \tag{4.2.10}$$

Then the following practical formulae are obtained

$$\left.\begin{aligned} E &= s\cos^2\alpha \,, \\ \Delta H &= s\sin\alpha\cos\alpha = {}^1/_2\, s\sin 2\alpha \,. \end{aligned}\right\} \tag{4.2.11}$$

The approximations for $c\cos\alpha$ and $c\sin\alpha$ cause errors of 29 mm in E and 15 mm in h for $c = 0.30$ m and $\alpha = 30$ gon at 100 metres, which can easily be neglected for basic stadia work. These errors are totally eliminated if the addition constant is near zero, as is the case for most modern instruments.

There are numerous auxiliary devices to compute E and h according to equation (4.2.11), such as diagrams, tables, slide rules, etc. Today, pocket calculators are used most frequently.

4.2.2.3 Accuracy of Stadia Measurements

a) *Distance Errors*

The most important error sources for stadia distances are inaccuracies in ΔE or c and k, readings of the rod and vertical angle, refraction influences and non-verticality of the rod.

To cover the last deficiency right away, a rod which is inclined by the angle δ would cause a reduction of l by $\cos(\alpha + \delta)$ instead of by $\cos\alpha$ as in equation (4.2.6).

With

$$l\cos(\alpha + \delta) = l(\cos\alpha\cos\delta - \sin\alpha\sin\delta)$$

and δ being a small angle, we get

$$l\cos(\alpha+\delta) \approx l\cos\alpha(1-\delta\tan\alpha)\ . \tag{4.2.12}$$

According to equation (4.2.8) this leads to

$$E = c\cos\alpha + kl\cos^2\alpha(1-\delta\tan\alpha)\ . \tag{4.2.13}$$

Now the law of error propagation is applied to this equation:

$$\sigma_E^2 = \left(\frac{\partial E}{\partial c}\right)^2 \sigma_c^2 + \left(\frac{\partial E}{\partial k}\right)^2 \sigma_k^2 + \left(\frac{\partial E}{\partial l}\right)^2 \sigma_l^2 + \left(\frac{\partial E}{\partial \alpha}\right) \sigma_\alpha^2 + \left(\frac{\partial E}{\partial \delta}\right) \sigma_\delta^2\ . \tag{4.2.14}$$

Neglecting the δ term in the first four components, the partial derivatives become:

$$\begin{aligned} &\frac{\partial E}{\partial c} = \cos\alpha; \quad \frac{\partial E}{\partial k} = l\cos^2\alpha; \quad \frac{\partial E}{\partial l} = k\cos^2\alpha\ , \\ &\frac{\partial E}{\partial \alpha} = c\sin\alpha + kl2\sin\alpha\cos\alpha \approx kl\sin 2\alpha\ , \\ &\frac{\partial E}{\partial \delta} = -kl\cos^2\alpha\tan\alpha = \frac{1}{2}kl\sin 2\alpha\ . \end{aligned} \tag{4.2.15}$$

To obtain some numerical values, let us consider a 100 m distance, measured with a stadia instrument with $k = 100$, i.e. $k = 100$ and $l = 100$ cm in equation (4.2.15). From experience it is known that $\sigma_c \approx \pm 2$ cm and $\sigma_k \approx \pm 0.05$. A single rod reading can be obtained within ± 1 mm, which means ± 1 mm $\sqrt{2}$ for the rod section. Due to refraction influences (section 4.3.3.1), this amount has to be increased by about 20 %, which leads to $\sigma_l = \pm 0.18$ cm. The vertical angle is obtained to about $\sigma_\alpha \approx \pm 2$ cgon ($\approx 1'$) with a simple stadia theodolite.

Using a bull's-eye level, the rod is plumbed to within about 50 cgon ($\approx 25'$), which leads to $\sigma_\delta = \pm 50$ cgon. Thus, when inserting these values into equation (4.2.14), the standard deviation (in cm) for a 100 m stadia distance becomes:

$$\sigma_{100}^2 = (2\cos\alpha)^2 + (5\cos^2\alpha)^2 + (18\cos^2\alpha)^2 + (3\sin 2\alpha)^2 + (40\sin 2\alpha)^2\ . \tag{4.2.16}$$

The following table is computed with this equation:

α		error in [cm] caused by					
[gon]	[°]	σ_c	σ_k	σ_l	σ_α	σ_δ	σ_{100}
0	0	±2	±5	±18	±0	± 0	±19 cm
10	0	2	5	18	0	12	±22 cm
20	18	2	4	16	2	24	±27 cm
30	27	2	4	15	3	32	±36 cm

This shows that the precision of the determination of the rod section l is not significant, while for inclined sights it is very important that the rod be properly plumbed.

Thus it is advisable to brace the rod with a range pole. On average, a moderately inclined stadia distance of 100 m can be expected to have an uncertainty of about 0.2%. This holds for distances between 20 and 130 metres. The accuracy decreases rapidly for longer distances, which means that for distances of 300 m, used for small scale topographic mapping, errors of $\pm 0.4\%$ or more are to be expected.

b) *Elevation Errors*

When determining the elevation difference h, equation (4.2.9) becomes, with the rod inclination α:

$$h = c\sin\alpha + \frac{1}{2}kl\sin 2\alpha(1 - \delta\tan\alpha)\,. \tag{4.2.17}$$

Similar to the approach in a), the standard deviation of an elevation difference for a 100 m distance is obtained as

$$\begin{aligned}\sigma_h^2 &= (2\sin\alpha)^2 + (2.5\sin 2\alpha)^2 + (9\sin 2\alpha)^2 \\ &\quad + (3\cos 2\alpha)^2 + (80\sin^2\alpha)^2\,.\end{aligned} \tag{4.2.18}$$

This leads to the following table

α [gon]	[°]	h for 100 m	σ_c	σ_k	σ_l	σ_α	σ_δ	σ_h
0	0	0.0 m	0.0	0.0	0.0	3.0	0.0	± 3 cm
10	9	15.4 m	0.3	0.8	2.8	2.8	1.6	± 4 cm
20	18	29.4 m	0.6	1.5	5.3	2.4	7.8	±10 cm
30	27	40.4 m	0.9	2.0	7.3	1.8	15.6	±17 cm

The error influences for the elevation difference correspond – although with different magnitude – to the ones for the distance measurement. Again, careful determination of the rod section and proper verticality of the rod are critical. It is also evident, that for stadia elevations it is quite sufficient to measure the vertical angle to ± 2 cgon ($\approx \pm 1'$). If, in addition, the error of the instrument height is taken to be 1 to 2 cm, then a standard deviation of a stadia elevation over 100 m of ± 5 cm can be expected for sights inclined to about 10 gon.

4.2.3 Tacheometers with Base at Instrument (Range Finders)

4.2.3.1 Instruments with Constant Base

Instruments of this kind have two objective lenses, each of which generates an image of a target object. The two images are diverted into a common eyepiece by means of mirror or prism system and coincided with an optical element (i.e. parallel plate, wedge, lens etc.), which is rotated with a measuring screw. The distance is then read on a scale attached to the drum of the measuring screw. The target area does not have to be entered. For geodetic purposes, instruments with a short base are too inaccurate, and the ones with a long base too cumbersome. Furthermore, neither horizontal nor vertical circles are available.

4.2.3.2 Instruments with Variable Base

These instruments also serve for very limited accuracy requirements only. The first modern design was the Teletop (Fig. 4.2.5) developed by Zeiss, Jena, in the 1930s, which combines a range finder with a telescope compass and a vertical circle.

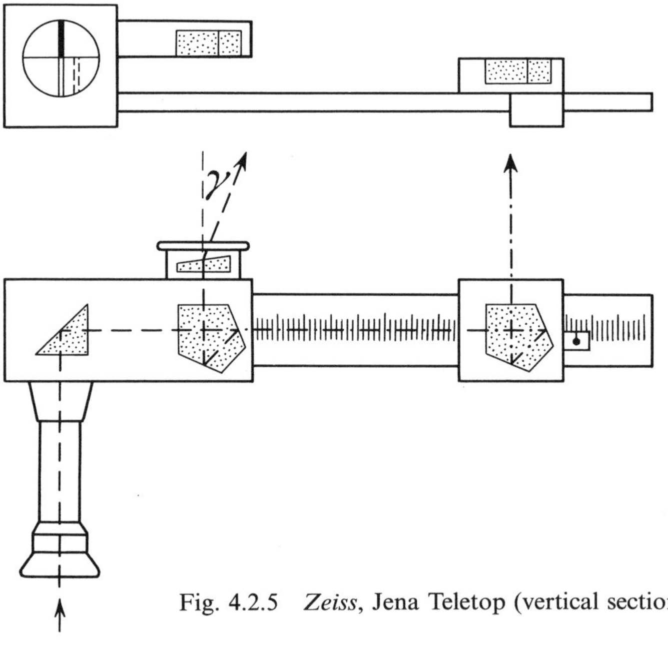

Fig. 4.2.5 *Zeiss*, Jena Teletop (vertical section and plan view)

This instrument was further developed into the BRT006 self reducing tacheometer, and is used primarily in legal surveys, where it is quite useful when encountering inaccessible points. For topographic applications, however, it is too cumbersome.

In the Todis distance meter by Breithaupt, Kassel, a single vernier theodolite is combined with a range finder (Fig. 4.2.6). The theodolite support carries a base extending about 35 cm on two rings which can be rotated.

Each of the arms of the base carries a slide with a rectangle prism and distance wedge which permit deflection of the collimation rays by (100 gon + δ or 100 gon + ε) towards the centre such that the rays intersect with the angle $\gamma = \delta + \varepsilon$. One obtains

Fig. 4.2.6 *Breithaupt* Todis

therefore two symmetric images, of a range pole placed at the target, which have to be coincided. Direction and elevation are set with the telescope of the theodolite and read at the compass or vertical circle. The instrument is equipped with three pairs of wedges which provide the following accuracies under favourable external conditions.

Wedge pair	Range	Accuracy in % of distance
1 : 100	10– 75 m	±0.1
1 : 200	15–150 m	±0.2
1 : 500	< 400 m	±0.8

Under less favourable conditions, distance errors of 1 % to 2 % can be expected for longer ranges. The instrument is quite suitable for expedition work.

4.3 Precision Distance Measurements with Mechanical and Optical Means

The significantly improved methods for distance measurements described in this section were largely developed in the first half of this century. Historically, these methods represent the application to surveying of methods typically used in physics and mechanics. By considering meteorological conditions, weight, expansion, etc., the surveying approaches became more sophisticated.

The rapid development of electronic distance measurements has reduced the use of these methods significantly. However, they are still utilized when EDM instruments break down or are not available. Furthermore, these methods retain their didactic importance. Without the preceding use of the mentioned physical methods, the introduction of electronic distance meters would have faced virtually unbearable difficulties. The methods developed in the first half of this century were and are primarily used in control surveys. Standard deviations which are roughly one-third smaller than the ones mentioned in the tolerance limits in section 4.1.3.2 are required. Thus a standard deviation of the following magnitude is desired:

$$\sigma_s = \pm 0.002\sqrt{s}.$$

This, therefore, is the minimum requirement for the distance measuring methods discussed in the following sections.

4.3.1 Precision Distance Measurements with Wires

For precision measurements it is very important to have a constant measuring system. Invar wires are commonly used. Invar is a nickle-steel alloy whose temperature expansion coefficient is very close to zero. The wires are normally used free hanging, with a constant pull being applied by weights or springs (Fig. 4.3.1). In order to keep the sag small, wires with small cross-section are used. Thus the measuring system is not graduated continually. Two types of arrangements are common, namely the use of wires with known length and with unknown, but constant, length.

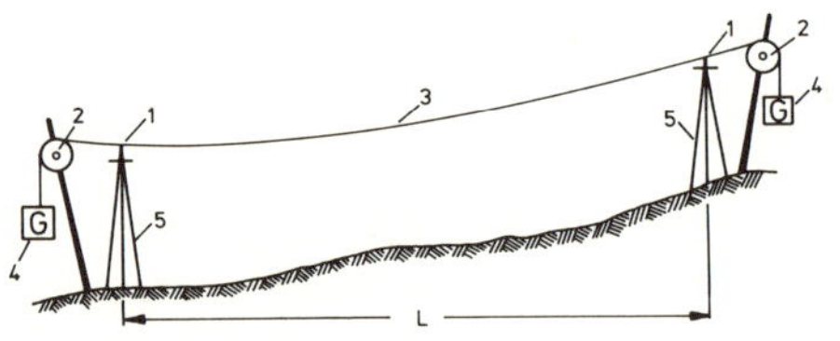

1 = reading index
2 = roll
3 = measuring wire
4 = weight
5 = tripod

Fig. 4.3.1 Distance measurement with free hanging wire

4.3.1.1 Wires with Known Length

The common length of invar wires is 24 metres and their thickness is 1.65 mm. Eight centimetre long scales with millimetre graduation are attached to both ends of the wires.

The proper tension is achieved by using weights, usually 10 kg. The total distance, whose inclination should not exceed 3 gon (or 3°), has to be subdivided into partial distances of about 24 metres, each of which is measured with at least 3 different invar wires.

All corrections mentioned in section 4.1.2.1 have to be carefully applied and supplemented by others. The most important areas of application are:

- Base line measurements. Accuracies of 0.3 ÷ 0.5 mm/km can be obtained.
- Precision engineering surveys. Accuracies of 0.03 mm per wire length can be obtained.

4.3.1.2 Wires with Constant but Unknown Length

These wires are used to determine changes in lengths. The sole purpose of the wire is to connect the terminal points of the distance.

Length changes can then be directly determined with lower relative accuracy requirements (10^{-2} to 10^{-3}), rather than being derived from the differences of long lines. Measuring *systems with and without recording* device are available.

The ISETH distometer (by Kern, Aarau) is a *system without recording*. Fig. 4.3.2 illustrates its operation. The system consists of measuring plugs, the wire, and the distometer as the main measuring part. The distometer can be utilized together with wires of different length (1 to 50 m) consecutively at different projects. As shown in Fig. 4.3.3, the distometer contains two dials: one for distance measurement, the other for tension control. The distances may have any inclination. Length changes can be obtained with an accuracy of 0.02 mm/20 m and $1 \cdot 10^{-6}$ of the distance.

Measuring *systems with a recording* device use mechanical transmission in connection

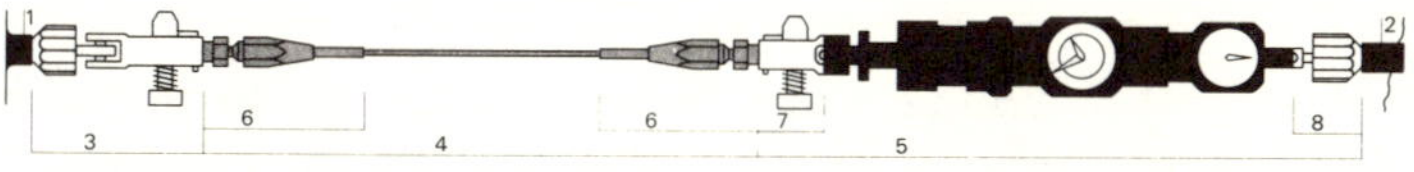

Distometer equipment in operation
1. Gauge pin welded
2. Gauge pin set in concrete
3. Hinge fitting with bracket for wire coupling
4. Invar wire
5. ISETH distometer
6. Wire coupling
7. Bracket for wire coupling on distometer
8. Hinge fitting on distometer

Fig. 4.3.2 Measuring set-up for the ISETH distometer (*Kern*)

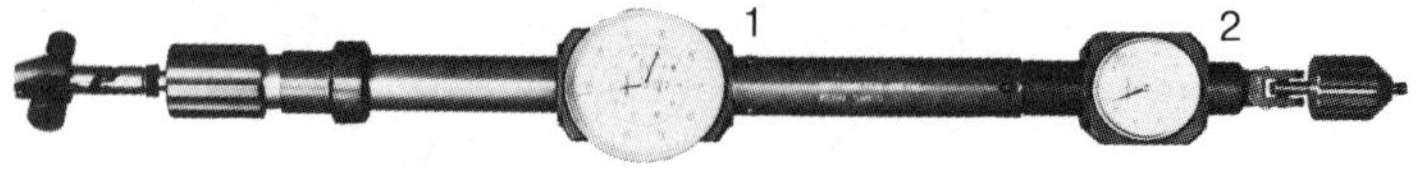

Fig. 4.3.3 Structure of the distometer

with a recording pen or else electronic analog-digital converters in conjunction with electronic data storage.

4.3.2 Indirect Distance Measurements with Subtense Bar

4.3.2.1 Fundamentals

As discussed in section 4.2.1, indirect optical distance measurements are based on the parallactic triangle, where the distance is derived from the base b and the parallactic angle γ (also Fig. 4.2.1). According to equation (4.2.1), the distance is:

$$s = \frac{b}{2} \cot \frac{\gamma}{2} \,. \tag{4.3.1}$$

When applying the theory of errors, one differentiates equation (4.2.1) and divides both sides by s. To consider the worst case, the error components are added which, with neglecting a few insignificant terms, leads to

$$\frac{ds}{s} = \frac{db}{b} + \frac{d\gamma}{\gamma} \,, \tag{4.3.2}$$

or, the relative error of the distance s is equal to the sum of the relative errors of b and γ.

To utilize some figures, let us assume that a distance of 100 m is to be derived from a base $b = 2$ m with a relative error of not more than 1 : 5000, i.e. to within 2 cm. This can be reached if both error components achieve a relative accuracy of 1 : 10000. For $s = 100$ m, γ becomes 1.2732 gon. Therefore the required accuracies are $db \geqq 0.2$ mm and $d\gamma \leqq 0.13$ mgon (0.4″).

To keep the required measuring effort as small as possible, instruments have been developed which have either a constant base or a constant parallactic angle. Thus only the angle or the base have to be observed.

4.3.2.2 The Subtense Bar

Preferably a 2 metre long subtense bar is used as constant base (Fig. 4.3.4). It consists of a light metal pipe which is fastened onto a theodolite tripod by means of a tribrack such that it is horizontal. The bar carries carefully designed targets at both ends whose separation is held as accurately as possible at 2 metres with the aid of an invar tape, invar rod, or similar device for temperature compensation.

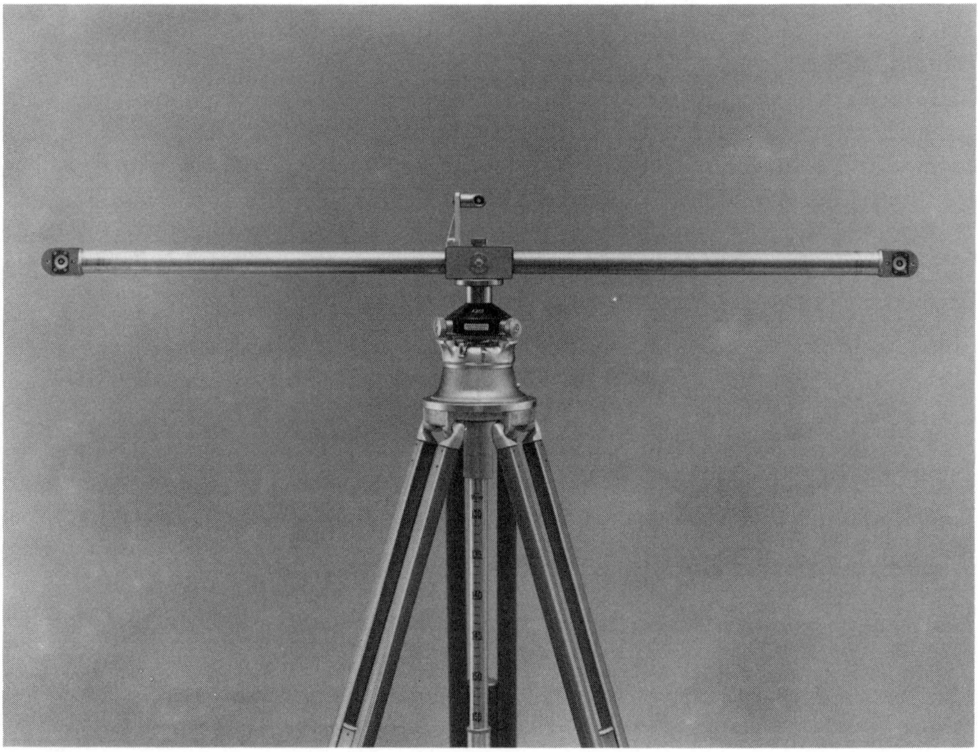

Fig. 4.3.4 Subtense bar (by *Kern*)

The rod is levelled using a spirit bubble, and set normal to the line of sight with the aid of a diopter. For transporting the bar, it can be folded or separated into two, 1 m long sections.

In order to reach an accuracy of 1 : 10000, i.e. 0.2 mm for the separation of the targets, devices for folding or separating and rejoining have to be such that neither the separation changes nor a bend occurs. When setting up, the rod has to be levelled and set normal to the line of sight within 0.3 gon ($\approx 15'$).

4.3.2.3 Measuring the Parallactic Angle

The parallactic angle between the two targets of the subtense bar can be measured with either a precision or an engineer's theodolite. The accuracy requirements for 1:10000 vary with the distance:

s[m]	10	25	50	75	100
$d\gamma$[mgon]	1.3	0.52	0.26	0.20	0.13
$d\gamma$[″]	4	1.7	0.8	0.6	0.4

For an angle measured direct and reverse with a modern precision theodolite, a standard deviation of ± 0.3 mgon ($\pm 1''$) can be expected. To reach $\sigma_\gamma \leqq 0.13$ mgon for a 100 m distance, about 10 sets have to be measured, as the standard deviation is reduced by the square root of the number of additional observations. This is not economical. Therefore, as a rule, distances are limited to 75 m which, with a ± 1 cm tolerance, leads to $\sigma_\gamma \leqq 0.23$ mgon. Two full sets with a precision theodolite, and 8 to 10 repetitions with an engineer's theodolite are required to reach this accuracy. When combining all error sources, a standard deviation of 1 to 2 cm for 75 m can thus be reached, i.e. a relative accuracy of 1 : 5000.

Since measurements of horizontal angles are always independent of the slope of the terrain, distances obtained with a horizontal subtense bar are always horizontal distances.

4.3.2.4 Subtense Bar Measuring Arrangements

a) *Base at one end*

For distances up to 75 m, the arrangement shown in Fig. 4.2.1 is used. Denoting the measured parallactic angle with γ_1, the distance becomes

$$s = \frac{b}{2}\cot\frac{\gamma_1}{2} . \tag{4.3.3}$$

The error due to the angular measurements is, because of equation (4.3.3):

$$\sigma_s = \frac{s^2}{b}\frac{\sigma_\gamma}{\varrho} 50.08\,\sigma_\gamma s^2 \tag{4.3.4}$$

where σ_s in [cm], s in [hm], σ_γ in [mgon], $b = 2$ m and $\varrho = 63\,662$ mgon.

b) *Base in centre*

For distances between 75 m and 150 m the subtense bar is placed normal to the distance in the centre and observed from both end points (Fig. 4.3.5).

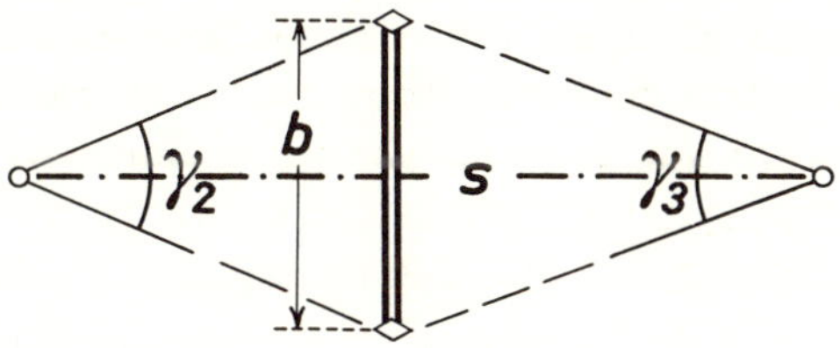

Fig. 4.3.5 Base in centre

The distance is then:

$$s = \frac{b}{2}\left(\cot\frac{\gamma_2}{2} + \cot\frac{\gamma_3}{2}\right). \tag{4.3.5}$$

Assuming that γ_2 and γ_3 were measured with the same accuracy, the standard deviation of the distance becomes:

$$\sigma_s = \frac{s^2}{2.8b}\,\frac{\sigma_\gamma}{\varrho}\;5\,0.028\,\sigma_\gamma s^2 \tag{4.3.6}$$

When comparing this to equation (4.3.4) it is evident that the standard deviation of the distance is reduced to nearly one third when the subtense bar is placed in the centre.

For this arrangement, it is best to have target marks which are visible from both sides. This eliminates the zero error of the bar, which is present when the straight line connecting the end marks does not pass through the vertical axis of the subtense bar.

4.3.3 Accuracy Considerations for Optical Distance Measurements

As evident from the experience gained in the last 50 years, the accuracy of ± 2 cm per 100 m can only be reached with well adjusted instruments and proper operation. In order to compensate for the major error sources, the observer therefore has to be aware of them.

4.3.3.1 Refraction Influences

Most optical distance measurements take place in an airspace about 2 m high directly above the terrain. This space can be considered to consist of layers of air with varying density stacked parallel to the surface. This varying density causes variations in the refractive index. Therefore, a light ray passing from instrument to rod is not straight, especially when it passes through several layers of air. The curvature caused by this is referred to as the *influence of ground refraction*.

The ideal layers of air, however, are frequently disturbed. Sun radiation mostly heats the air directly on top of the terrain surface, so that the temperature decreases and the density increases when moving upwards (negative temperature gradient). After the sun's radiation has stopped (at night) the trend slowly reverses until the layer closest to the terrain surface becomes the coldest one (positive temperature gradient).

This temperature gradient plus the wind cause exchanges of air masses with varying velocity. If the exchange is very rapid, then the air movement is visible to the observer and often prohibits proper measurements, especially at noon (heat waves). If layered bodies of air move slowly, a floating condition occurs. This is a rather abnormal refraction behaviour, which usually cannot be noticed by the observer and thus creates a dangerous error source.

Ground refraction, *heat waves* and *floating* represent the most important error sources for optical distance measurements. Since these phenomena are primarily a function of the distance to the terrain surface, horizontal rods are used for precision distance measurements. However, similar temperature and air layers can occur at walls and hedges during sunshine, which causes a lateral deflection of the ray (lateral refraction). Therefore, it is advisable to always stay several metres away from such objects.

4.3.3.2 Instrument Errors

The addition and multiplication constants, as well as index bubble and reduction devices, are subject to change with extended use. It may therefore be necessary to adjust these according to the instructions given by the manufacturer, which vary for the different instrument types.

Experience has shown that the addition and multiplication constants change so little that an adjustment is not usually necessary. Periodic checks every one to two weeks, and calibrations which also cover regular rod and human errors (section 4.3.4.5) are more advisable. For this, a 100 m long test line with marks at every 20 m is most suitable. The test results yield correction values which are usually tabulated for every 10 m up to 130 m and added to the field measurements.

Errors of the reduction device become significant in steep terrain only. Their influence is cancelled if the mean of measured forward and backward distances is taken.

Calibration results using an inclined test line can again be used to generate a correction table.

Errors of the height index bubble have to be adjusted according to the manufacturer's instructions.

4.3.3.3 Rod Errors

Rods can have regular and irregular graduation errors. Numerous investigations have shown that the irregular ones remain below 0.05 mm. Regular graduation errors and long periodic changes in the length of the rod are covered in the calibration with test lines, as recommended above. Furthermore, it is advisable to compare the rods from time to time with calibrated normal metres.

Bends of subtense bars, and warping of horizontal distance rods are more dangerous. They have the effect that the measuring targets get closer together, thus giving distances which are too long. Furthermore, all distances are incorrect by a constant amount, namely the separation of the centre of the rod from the straight line connecting the target marks.

If, for instance, a 1 m long rod section is warped by 10 mm, the targets are closer by 0.25 mm. For $k = 100$, a distance error of 25 mm occurs. In addition a constant error of ± 10 mm depending on the direction of the deviation is present. It is therefore important to return warped or bent rods to the manufacturer for repair.

4.3.3.4 Errors in Setting Up

This concerns mainly the rod and covers the following errors:

a) rod is not horizontal
b) rod is not normal to the direction of the distance
c) the rod section is not symmetric to the target position.

For errors mentioned in a) and b) the resulting distance is too long. The accuracy of 1 : 10000 is obtained as long as each of these errors above does not exceed 0.64 gon ($\approx 35'$). Since both errors act in the same sense, a rule of thumb limits the deviation to 0.3 gon ($16'$). The error c) remains below the accuracy limit if the rod section at least touches the centre of the target at one end.

4.3.3.5 Human Errors

Human errors are primarily caused by the setting and reading habits of the observer, and occasionally by eye deficiencies. The most important of these errors are:

a) Decimal preference, i.e. when estimating, the observer prefers certain decimals, e.g. 0.2 and 0.8. By repeating the measurement at a different rod position and by exercising self control, this error can be counteracted. If an optical micrometer is used, as with most wedge tacheometers, this error is non-existent.

b) Coincidence errors, i.e. the observer coincides differently when approaching the line from left or from right. This error is eliminated if one coincides once from left and once from right and then uses the mean of the two values.

c) Eye deficiencies can be caused by asymmetry of the cornea or the glass body of the instrument. This error is counteracted with double image tacheometers superimposing the exit pupils of the direct and deflected images. Large deficiencies, such as scars on the cornea, can render an observer unsuitable for optical precision distance measurements.

d) Irregular human errors can be reduced by properly designing target marks and by repeated sightings.

4.4 Electronic Distance Measurements

4.4.1 Fundamental Concepts of Distance Measuring Devices Using Electromagnetic Waves

When utilizing electro-magnetic waves for distance measurements, either *impulse or phase comparison methods* are applied. Instruments used for tasks in cadastral, engineering, or geodetic surveying operate mainly with the phase comparison approach, while impulse and phase comparison methods can be used for satellite positioning.

4.4.1.1 Principle of Impulse Methods

The instruments consist basically of a transmitter, receiver, propagation timer and microprocessor (Fig. 4.4.1). If all these components are located at the observation station, then a reflector is needed at the target station.

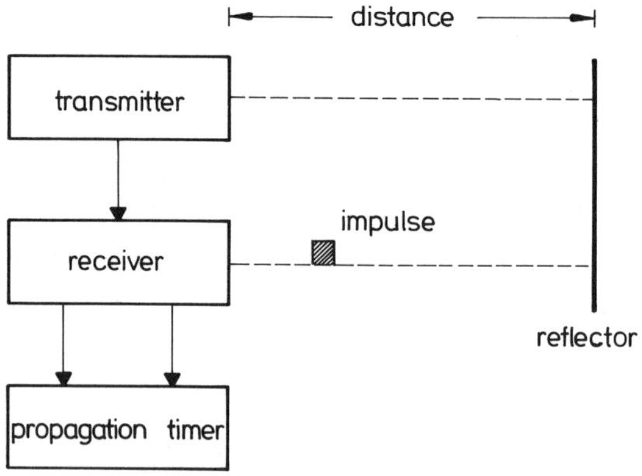

Fig. 4.4.1 Simplified principle of an impulse method

The transmitter sends an impulse (i.e., light or microwave impulse) which propagates to the other endpoint and returns. The receiver receives a small fraction of the emitted impulse immediately and also the impulse which has travelled along the line and back. The propagation timer, which is controlled by an extremely accurate clock, determines the difference in propagation time for the two impulses. This difference corresponds to the time t needed by the external impulse to twice travel the distance D. This distance is then

$$D = k_A + c\frac{t}{2} \tag{4.4.1}$$

where c denotes the light velocity for the average refractive index along the line. If the electric origin of the signal does not correspond to the vertical axis of the measuring instrument, then a zero correction K_A is to be added. Several methods can be used to determine the zero correction (section 4.4.2).

Since the light velocity c is very large, the time interval t is extremely short. Thus, in order to determine D to within 0.5 cm, the time has to be measured to within 0.03 ngon seconds (ns).

Distance measuring systems with the transmitter at one end of the line and the receiver at the other are also used in positioning. Highly accurate clocks are required at both stations in this case. The distance is then obtained by

$$D = K_A + c \cdot t\,. \tag{4.4.2}$$

Impulse methods have long ranges, as individual impulses can be emitted with a high intensity.

4.4.1.2 Principle of Phase Comparison Methods

The instruments consist of a transmitter, receiver, phase meter and microprocessor (Fig. 4.4.2). If these components are located at one station, then a reflector is needed at the target station.

The transmitter generates a sinusoidal measuring signal

$$y_s = \hat{A} \cdot \sin \omega t \tag{4.4.3}$$

with a wavelength

$$\lambda = \frac{c_0}{n \cdot f}. \tag{4.4.4}$$

This wavelength depends on the velocity of electro-magnetic waves in a vacuum ($c_0 = 299\,792.5$ km/s), the mean index of refraction n along the line, and the frequency f. The receiver immediately receives a small fraction of the emitted signal and also the signal which has travelled along the line. The latter signal suffers a phase shift which is

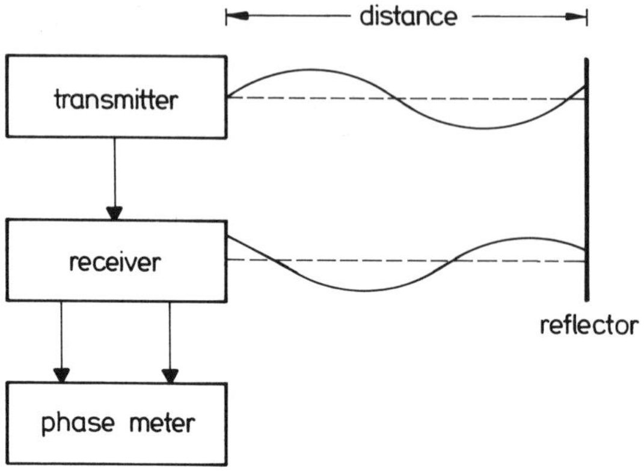

Fig. 4.4.2 Simplified principle of a phase comparison method receiver

dependent on the length of the travelled distance. At the receiver it has the following form:

$$y_E = \hat{A}\sin(\omega t + \Phi) \tag{4.4.5}$$

where

$$\Phi = a \cdot 2\pi + \Delta\varphi\,. \tag{4.4.6}$$

2π corresponds to the length of a complete wave, a denotes the number of complete waves, while $\Delta\varphi$ represents the phase difference between the end of the last complete wave and the receiver. Thus, the residual distance becomes

$$r = \frac{\Delta\varphi}{2\pi} \cdot \lambda\,. \tag{4.4.7}$$

The distance D can be uniquely determined if it is less than half the wavelength. Phase meters have a limited accuracy of about $\pm\gamma/4000$ to $\pm\lambda/8000$. If, for example, an instrument operates with a wavelength of 20 m and has a phase meter with $\pm\lambda/4000$ accuracy, then the residual distance is resolved with an accuracy of ± 5 mm. Unique results and high accuracies can only be obtained if the measurements are carried out in steps, each step with a larger wavelength. The measurement with the smallest wavelength is the *fine reading*, the others are *rough readings*. These steps are illustrated in the following example:

	Measuring Frequency	Wavelength	Residual Distances	
3rd reading	150 KHz	2000	789	rough readings
2nd reading	1.5 MHz	200	882	
1st reading	15 MHz	20	8437	fine reading
			788.437 m	

Thus, the final result is obtained by adding the first digit of each of the rough readings to the fine reading. It is apparent that it is not detrimental when the accuracy of the rough readings decreases by a factor of ten for increasing wavelengths.

Since the signal travels twice along the line, twice the distance is obtained. The distance is obtained after dividing by 2 or by replacing n in equation (4.4.4) by $n\,2 = U$. U is usually referred to as 'scale'. Depending on the measuring frequency, the smallest scale is referred to as 'fine scale' while the others are called 'rough scales'.

The phase position of the signals drifts because the parameters of the components are temperature dependent. Thus no fixed starting point for the measured distance can be defined for phase measurements. Therefore, the phase comparison is repeated along an internal calibration distance which is obtained by inserting a reflector into the optical train. If the time interval between external and internal measurements is very short, and if the distance D' is computed from the difference

$$D' = D'_2 - D'_1\,, \tag{4.4.8}$$

then the drift influence is practically eliminated. D'_2 and D'_1 denote external and internal distances, respectively.

If the effective reflecting surfaces of the internal and external prisms deviate by c_1 or c_2 from the vertical axis of the measuring instrument, then the zero correction (Fig. 4.4.3) becomes:

$$k_0 = c_1 + c_2\,. \tag{4.4.9}$$

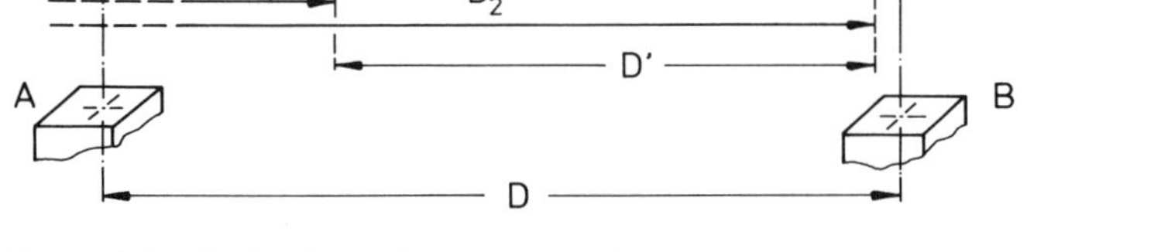

Fig. 4.4.3 Definition of zero correction

The determination of the zero correction is described in section 4.4.2.

A microprocessor usually controls the measuring steps, computes the result from the readings, and applies zero and scale corrections. The final result is then displayed in numerical form.

4.4.1.3 Carrier Waves and Their Modulation

In order to measure distances with an accuracy of a few millimetres or centimetres, it is not feasible to directly transmit the measuring signals because their propagation characteristics can only be determined inaccurately. However, the propagation characteristics of high frequency electro-magnetic waves can be determined with sufficient accuracy.

Therefore, the measuring signals are superimposed on high frequency waves, called carrier waves. This can be done by *amplitude, frequency, or impulse modulation.* The amplitude or frequency of the carrier changes in the rhythm of the measuring signals for amplitude or frequency modulation. Impulse modulation is achieved when the transmitter sends the carrier at constant time intervals for a very short period (Fig. 4.4.4).

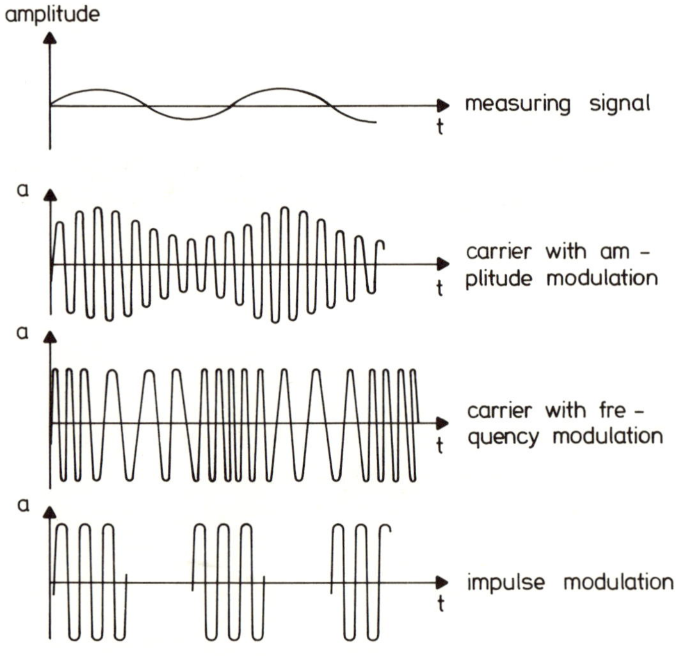

Fig. 4.4.4 Modulation of carrier wave

Electro-magnetic waves of the infrared, visual light, and microwave ranges are suitable as carrier. Their geometric behaviour is only influenced by refraction. The signal path can be described sufficiently accurately as a straight line for distances of up to 10 km, and as a circular arc for longer distances.

The velocity of the carrier waves within the optical range can be determined with higher accuracy than for microwave carriers. However, the range of instruments using these optical carriers is influenced by haze, rain, and air turbulence. The range of microwave instruments is largely independent of the meteorological conditions. Since the carriers of the measuring signals have a significant influence on the characteristics of the measuring instruments, a distinction is made between electro-optical and microwave distance meters.

4.4.1.4 Simplified Models of Electro-optical Distance Meters

As mentioned in section 4.4.1.3, infrared or visual light waves are suitable as carriers of the measuring signals. By choosing infrared radiation as the carrier and limiting the range to 15 km, a technically simple instrument can be obtained. If a laser is used as the carrier, then the range can be extended to 60 km. Such instruments, however, are heavier, larger and more expensive.

The block diagram (Fig. 4.4.5) illustrates the major elements of a distance meter operating with infrared carrier waves. The infrared radiation is generated with a Ga-As diode.

A high frequency (HF) oscillator generates the measuring signal in the form of AC voltage. This high frequency (e.g. 15 MHz) voltage controls the Ga-As diode. This changes the intensity of the infrared radiation such that it increases and decreases with the beat of the AC voltage (light-dark effect).

With the aid of the transmission optics, the light is subsequently directed towards a reflector set up at the end point of the distance. The reflector sends the light back to the transmitter. The receiving optics directs the incoming light toward a photodiode which transforms the changes in intensity into voltage changes of the same frequency. These are fed into a phase meter consisting of an impulse giver, a measuring gate and

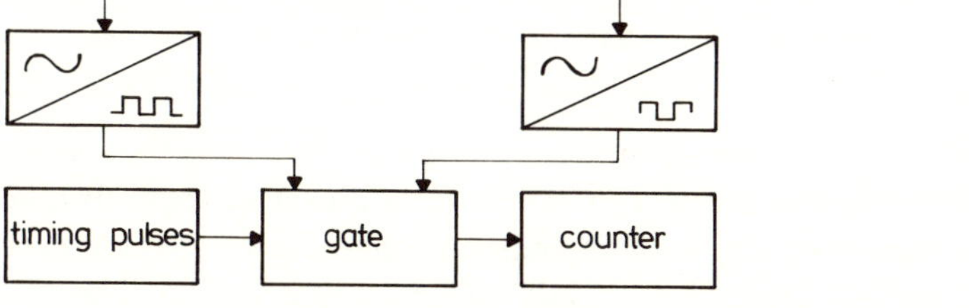

Fig. 4.4.5 Simplified block diagram of a distance meter

a counter. Furthermore, a small portion of the original oscillator signal is directly fed into the phase meter. However, the phase meter cannot directly use the high frequency signals. They have to undergo several mixing steps to become low frequency signals and are then changed into rectangular signals in an auxiliary circuitry.

Usually, there is a phase lag between the reference and measuring signals because the latter has travelled the line back and forth. Therefore, the phase difference $\Delta \varphi$ represents the actual measuring result (section 4.4.1.2) and is determined by the phase meter. This measuring process is repeated internally and with different frequencies until the distance can be uniquely derived from phase differences.

The schematic structure illustrated in Fig. 4.4.5 varies for different instrument types. It is possible, for instance, to omit impulse giver, modulation oscillator and the left mixing step if the counting frequency can be obtained directly from the HF oscillator by means of additional circuitry.

Fig. 4.4.6 is a simplified block diagram for a distance meter with long range. A laser generates the carrier wave. The measuring signal is indirectly superimposed onto the laser through a KDP modulator (see section 4.4.1.6). The laser signal travels along the line. Afterwards, a secondary electron multiplier (SEM) modifies it into an electric signal.

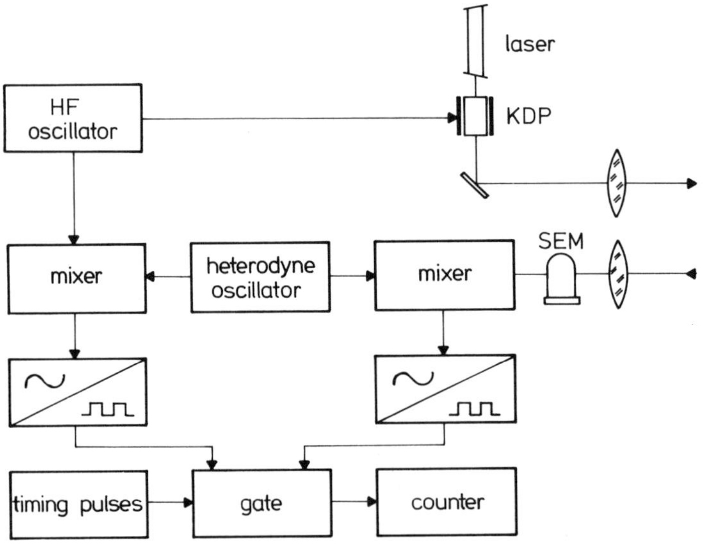

Fig. 4.4.6 Simplified block diagram of a long range distance meter

The further handling of the signal and the phase comparison is the same as for the distance meters conceptualized in Fig. 4.4.5. In addition to these distance meters there are several ones operating with an analogue phase meter. It is not necessary for these to change the low frequency signals into rectangle signals. The phase meter consists of a resolver circuit plus a phase detector.

4.4.1.5 Simplified Model of a Microwave Distance Meter

Distance meters utilizing microwaves as carriers require a master station and an active remote station, each of which is equipped with transmission and receiving capability. Fig. 4.4.7 provides a simplified illustration of the functional principle of a microwave distance meter.

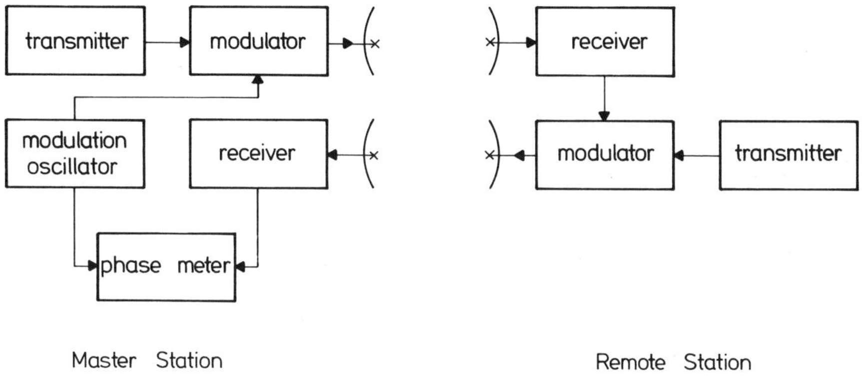

Fig. 4.4.7 Simplified block diagram of a microwave distance meter

The carrier is generated at the transmitter by a reflex clystron or a gun-diode. A measuring signal is superimposed by frequency modulation. A guartz oscillator monitors the modulation.

The modulated carrier signal is emitted by means of an antenna, received at the remote station and demodulated. This demodulated measuring signal is amplified and superimposed on the carrier of the remote station, again by frequency modulation. It is sent back to the master by means of the antenna. Upon reception it is again demodulated. The demodulated signal is sent to a phase meter where its phase is compared to a reference signal obtained directly from the oscillator. Depending on the instrument, this phase comparison is obtained either by an analogue phase meter (resolver circuit) or a digital phase meter. The process is repeated with different modulation frequencies to facilitate the unique determination of long distances from the phase differences, as in section 4.4.1.2

4.4.1.6 Elements of Electronic Distance Meters

As already mentioned in section 4.4.1.3, modulation superimposes periodic signals onto shortwave light such that their propagation time is proportional to the distance to be measured. To achieve this, either the light waves emitted by the light source are modulated (indirect modulation), or else the light is influenced at its origin, namely the light source itself, and thus modulated (direct modulation). The optical measuring signals of distance meters with ranges up to 15 km are generated by *direct modulation with a luminescent diode* (Ga-As diode, Fig. 4.4.8). The elements can be housed in a very small space, and the voltage supply is simple. Thus, measuring instruments of this type can be built with low weight and small dimensions.

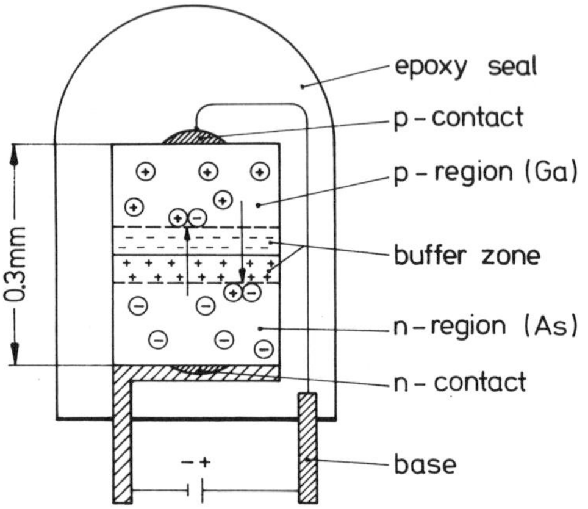

Fig. 4.4.8 Operation principle of a luminescent diode

Semiconductors (gallium arsonide) are placed between two electrodes inside the Ga-As diode. By implanting foreign atoms, an electron surplus is generated in one half of the diode, and an electron deficiency in the other. In the approximate 10 mm thick buffer zone, electrons and holes are combined to form a barrier. If an external voltage is applied, electrons and holes obtain higher energy and are driven into the barrier zone, a phenomenon commonly denoted as energy injection.

These carriers of an electric charge unit in the barrier zone and release the energy of the voltage source as heat and – more important here – as photons. Thus a luminescent radiation is generated which is in the near infrared band with about a 90 nm wavelength.

If a different voltage is applied to the diode, then different injection currents flow through it. Depending on the intensity of the injection current, different radiation intensity is obtained. This correlation is evident from the characteristic line, illustrated in Fig. 4.4.9. This characteristic line is straight within a certain range. If the applied voltage generates injection currents falling within the straight section of the characteristic line, then the voltage is transformed linearly into intensity changes of the infrared carrier. Thus, sinusoidal voltage oscillations generate sinusoidal current oscillations, which in turn generate a sinuoidal amplitude modulation of the carrier.

Unfortunately, this radiation is invisible which makes the alignment of instrument and reflector more difficult. Long range electro-optical distance meters utilize a laser beam. The laser radiation is *indirectly modulated with a KDP modulator*. A KDP

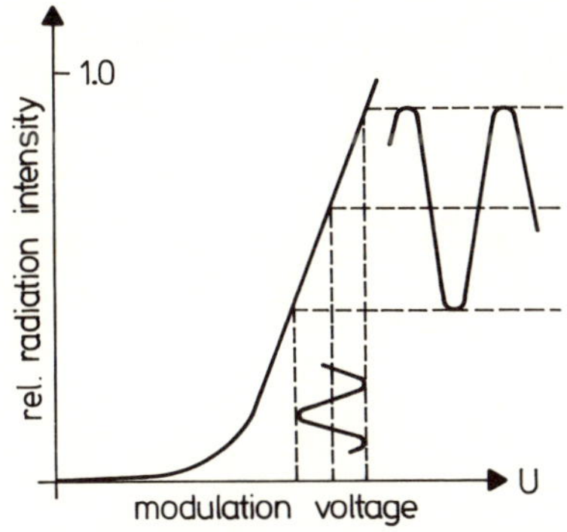

Fig. 4.4.9 Characteristic line of a luminescent diode

modulator consists of a KDP crystal, polarizer and analyser (Fig. 4.4.10). The KDP crystal is located between two electrodes. If a voltage is applied to these, the crystal becomes double refracting; i.e., incoming light is separated into two components, oriented normal to each other. This electro-optical effect is used for light modulation in the following manner.

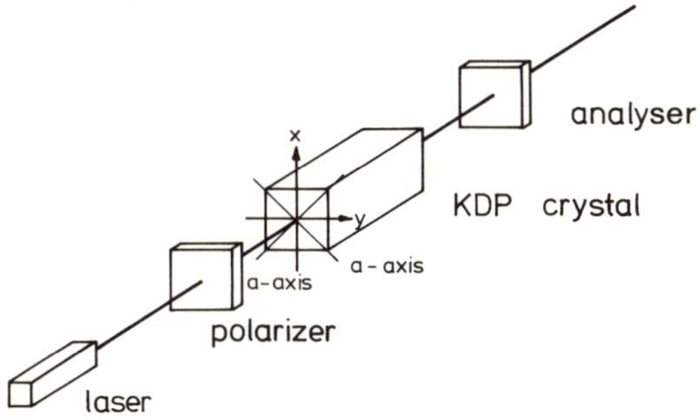

Fig. 4.4.10 KDP modulator

The light emitted from the laser is linearly polarized parallel to the *a*-axis by the polarizer. It then hits the KDP crystal, where the light is split by the electro-optical effect into two components which coincide with the main axes x and y of the crystal. The light oscillating parallel to the x-axis propagates through the crystal with a velocity that differs from the velocity of the light that is polarized parallel to the y-axis. The axes of the polarization ellipse coincide with the *a*-axes. If a weak AC voltage is superimposed onto the DC voltage at the electrons, then the shape of the ellipse changes in the rhythm of the voltage changes. Only light of one polarization direction can penetrate the analyser. If the analyser with its polarization direction lies parallel to one axis of the polarization ellipse, then the light intensity reaches a maximum behind the analyser. The characteristic line (Fig. 4.4.11) describes the properties of the modulator. If the AC voltage is a sine-voltage then a light signal is generated at the exit from the modulator with a sinusoidal amplitude modulation.

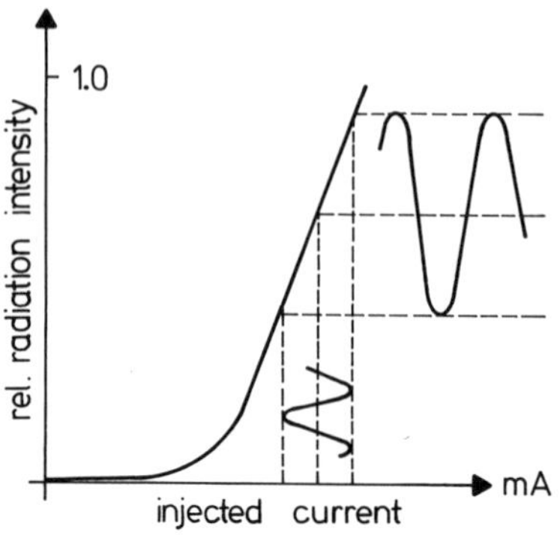

Fig. 4.4.11 Characteristic line of a KDP modulator

The measuring signal that has travelled along the line is separated from the carrier by *demodulation* and then converted into an electrical signal. *Photodiodes* and secondary electron multiplexes are primarily utilized as demodulators for short and medium range distance meters. Photodiodes are elements of semiconductors (Fig. 4.4.12). The voltage is applied to the *pn* elements in the direction of the barrier. The diode has a large resistance if there is no incoming light and thus virtually no current will flow.

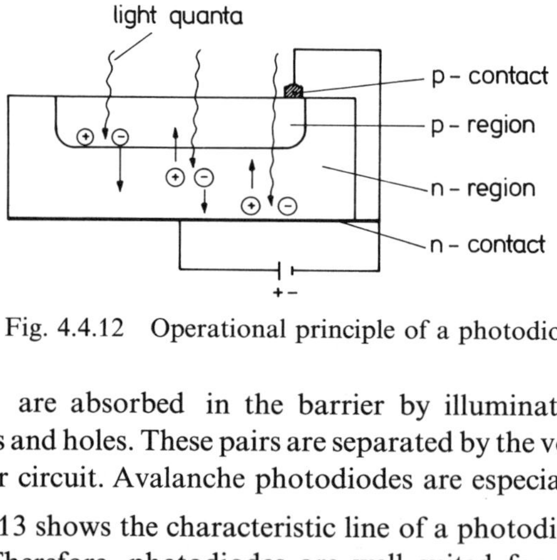

Fig. 4.4.12 Operational principle of a photodiode

Photons are absorbed in the barrier by illumination, which generates pairs of electrons and holes. These pairs are separated by the voltage and generate a current in the outer circuit. Avalanche photodiodes are especially effective.

Fig. 4.4.13 shows the characteristic line of a photodiode, which is linear for a large range. Therefore, photodiodes are well suited for converting linearly sinusoidal amplitude modulated light signals into electrical signals.

Secondary electron multipliers (SEM) are used as demodulators for long range distance meters. They consist of a vacuum tube with cathode, anode and diodes (Fig. 4.4.14).

Electrons are freed from the cathode by incoming light. These electrons move from diode to diode, freeing additional electrons in the process. Thus an amplification is

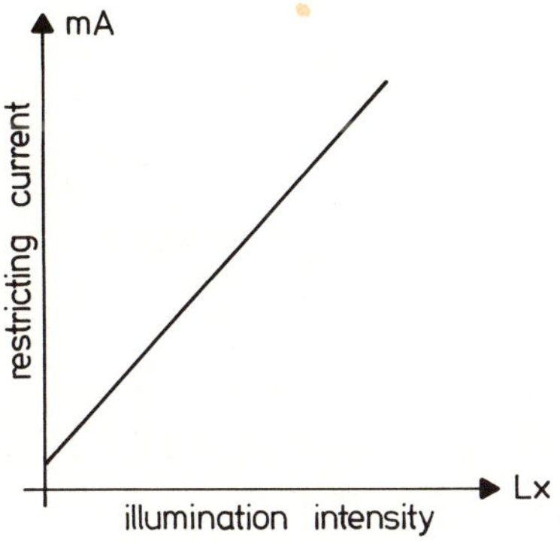

Fig. 4.4.13 Characteristic line of a photodiode

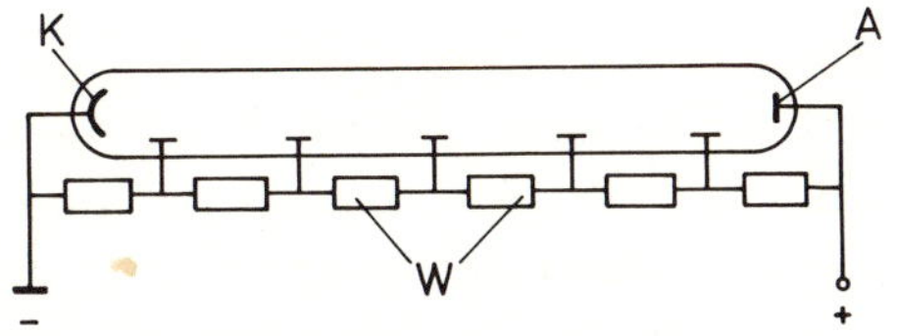

Fig. 4.4.14 Secondary electron multiplier
A = anode
K = cathode
W = resistors

achieved. The electrons finally reach the anode and generate a current in the outer circuit. The SEMs have a characteristic line similar to the one for photodiodes, which explains the similarity in the effect.

Once the measuring signal has travelled along the line, its *phase difference $\Delta\varphi$ is measured* with respect to a reference signal. According to section 4.4.1.2, the final distance is computed by applying equation (4.4.7) several times.

In the *analogue method*, it is technically extremely difficult to determine the phase differences for the high frequencies of about 15 MHz used for modulation. Therefore the high frequency is transformed into a low frequency of equal phase with between 1.5 and 150 kHz by superimposing a slightly different auxiliary frequency. Low frequency resolvers are used as phase shifters. Such a resolver is similar to an AC motor. It consists of two fixed starter coils normal to each other and a rotating coil (rotor). The phase shift generated by turning the rotor is proportional or "analogue" to the angle of rotation. This rotational angle is displayed by means of a counter.

For setting the resolver, an "electrical reference point" is needed. It is supplied by a phase detector with zero-instrument. Photodetectors indicate a zero voltage if two incoming signals have a phase difference of 90° or 270°. Therefore, the resolver has to be coupled with a phase detector (Fig. 4.4.15) for phase measurements. For phase comparison the rotor is turned until a zero voltage is shown at the indicator. The

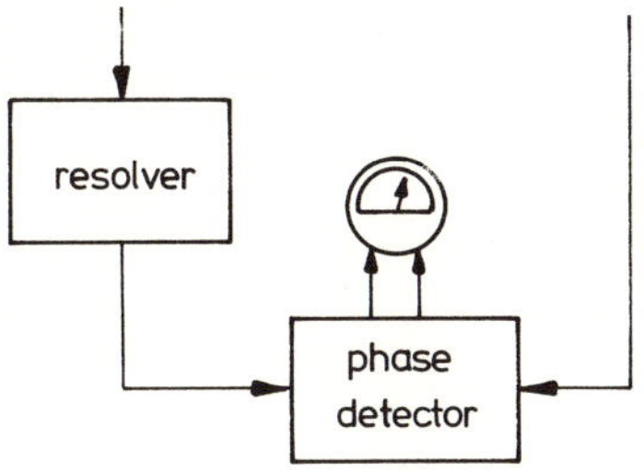

Fig. 4.4.15 Analogue phase measurements

phase difference is then

$$\Delta\varphi' = 90° + \Delta\varphi\,, \quad \text{or} \tag{4.4.10}$$

$$\Delta\varphi'' = 270° + \Delta\varphi\,. \tag{4.4.11}$$

It is important to work with only one equation, i.e., with one reference point. The residual distances r for rough and fine readings are normally obtained by measuring differences, which eliminates the constant segment of 90° or 270°. The accuracy of phase measurements with resolvers is about 10^{-3} and 10^{-4}. Therefore the phase angle is usually given with three digits. As shown in the example in section 4.4.1.2 the phase measurements are arranged such that ambiguities disappear.

If, however, the measuring results are to be obtained in *digital form* or registered automatically, then it is advisable to perform *digital phase measurements* (Fig. 4.4.16).

Both low frequency measuring signal and reference signal, available in sinusoidal form, are triggered, i.e., transformed to rectangular signals. These triggered signals monitor the gate between an impulse generator, which generates impulses in rapid sequence, and an impulse counter. The front of the reference signal opens the gate, while the front of the measuring signal closes it. The number of impulses passing through the open gate and reaching the impulse counter provides a means of measuring the phase angle between the reference and measuring signals. This measurement can be repeated at will, starting with the next front of the reference signal. The mean

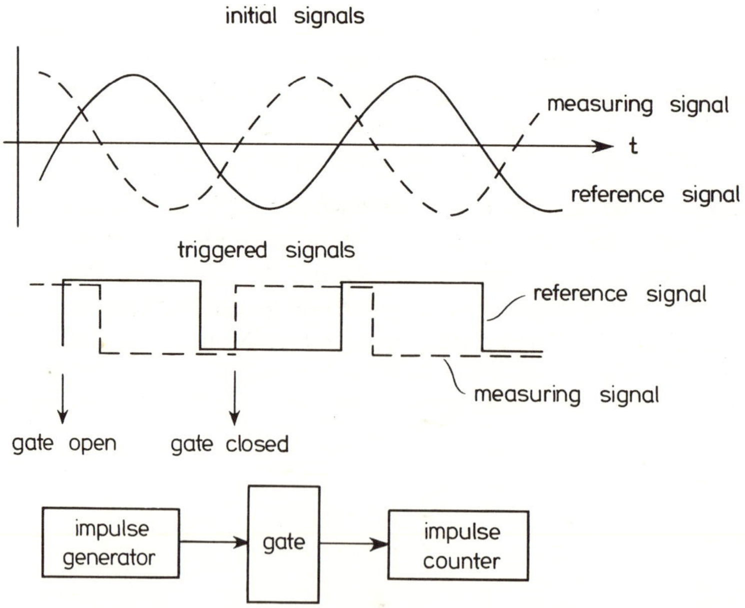

Fig. 4.4.16 Digital phase measurements

of many measurements provides a rather high accuracy of about 10^{-4}. Therefore, most modern short range EDM use digital phase measurement with a display of the result at the instrument.

4.4.2 Instrumental Errors: Calibration

Basically, the electronic distance determination is based on an equation of the form $D = k_0 + kD^A$ like other distance determinations. This means that the rough reading D^A has to be multiplied with a *scale constant* k and an *addition correction* k_0 in order to yield the desired distance D. The modulation frequency f_0 of an electro-optical distance meter is usually selected such that the wavelength becomes a round figure for a certain normal average value of the refractive index n_0. Thus the scale constant would be equal to 1 for normal conditions. If, however, the actual values for air temperature and pressure differ from the normal values, k changes and the reading D^A has to be corrected accordingly. However, the influence of such differences is rather small for electro-optical distance meters. As will be seen in equation (4.4.19), a temperature difference of 10 °C or a pressure difference of 25 Torr causes a change in the refractive index, and thus in k, of 1×10^{-5} which corresponds to a distance error of 1 cm/km. Section 4.4.4 describes the correction for the measured distances when the instrument constant n_0 differs from the actual refractive index.

The scale constant k can also deviate from 1 if the modulation frequency f does not correspond to its design value f_0. Generally, the stability of the frequencies is quite high, however, from time to time it is advisable to check them with a frequency tester. The *frequency correction*, due to deviation from the design frequency, amounts to:

$$k_f = D^A \frac{f_0 - f}{f_0}. \tag{4.4.12}$$

The *zero correction* (equation (4.4.9) and Fig. 4.4.17) is to be checked as well as the scale constant. It takes into account the difference between electrical and mechanical zero positions at the instrument as well as the separation of the reflection point and the physical centre of the reflector and the propagation time reduction inside the glass body. While the latter reflector components remain constant, unless the reflector is changed, the instrumental components may change. This is caused mainly by phase inhomogenities of the transmitters and of the photodetectors. The mutual influence of the signals of transmitting and receiving parts has to be mentioned in this connection as well. The latter leads to cyclic changes in k_0 just like possible errors of the phase meter. In most cases, however, these changes can be kept sufficiently small.

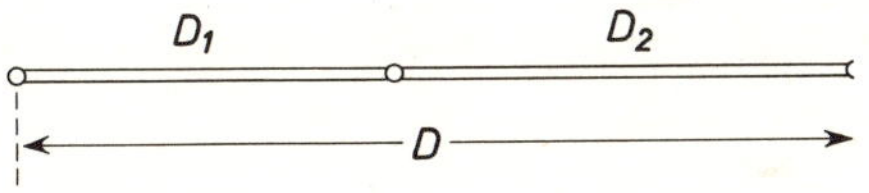

Fig. 4.4.17 Determination of the zero correction

It is therefore advisable to regularly check the zero correction k_0. As a rule, a very simple arrangement is sufficient: If the scale constant k equals 1, the following equations are valid:

$$D = k_0 + D^A; \quad D = D_1 + D_2 = k_0 + D_1^A + k_0 + D_2^A$$

which leads to

$$k_0 = D^A - (D_1^A + D_2^A) \; . \tag{4.4.13}$$

An independent determination of the zero correction together with its accuracy requires a large number of calibration measurements. These are carried out most efficiently on a subdivided testline (Fig. 4.4.18). Such a testline should fulfill the following requirements:

- The partial distances should uniformly cover the total measuring range used in practical work, so that influences correlated with the distance can be detected.
- The lengths of the partial lines should be such that they are evenly spread over the fine scale of the instrument.
- The reference distances are to be measured with a distance meter of highest accuracy.

Normally a total distance of less than 1 km is selected because of local terrain difficulties and in order to keep the influence of the refractive index small. This arrangement is usually sufficient to obtain the characteristic lines of the distance meters with the needed accuracy. If the measurements are carried out in all combinations (Fig. 4.4.18), a larger number of comparative observations is available. The evaluation is best performed in two steps:

1) graphical representation of the differences ΔD (true minus measured D);

2) determination of the parameters of the characteristic line and their accuracy by least-squares adjustment.

Some typical results are presented in Fig. 4.4.19. Example (a) represents a very carefully adjusted distance meter. Its characteristic line is horizontal and coincides with the zero line. In example (b) the zero correction has a discontinuity in the area of 100 m.

Fig. 4.4.18 Testline and measuring arrangement for the determination of the zero correction of a distance meter

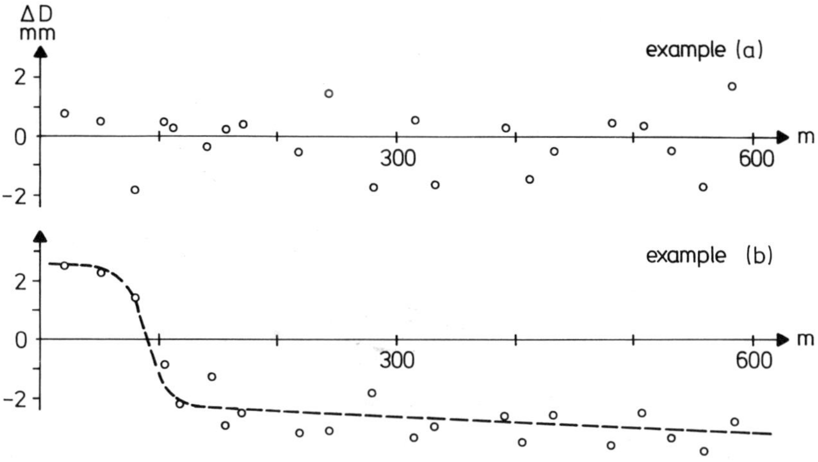

Fig. 4.4.19 Characteristic lines of different distance meters

4.4.3 Refractive Index, Refraction Coefficient

Both velocity of propagation and path for light or microwaves depend on the atmospheric conditions. The refractive indices n_L and n_M affect the velocity, while the refraction coefficients k_L and k_M cause a curvature of the path.

The *refractive index* for microwaves can be computed with the *Essen* and *Froome* (1951) formula

$$N_M = (n_M - 1) \cdot 10^6 = \frac{77.64}{T}(p - e) + \frac{64.68}{T}\left(1 + \frac{5748}{T}\right)e\,. \tag{4.4.14}$$

The refractive index n_{Gr} for the group of visual light radiation is computed according to *Barrel* and *Sears* (1939) at 0 °C and 760 Torr:

$$(n_{Gr} - 1)10^7 = 2876.04 + 3 \cdot \frac{16.288}{\lambda^2} + 5 \cdot \frac{0.136}{\lambda^4}\,, \tag{4.4.15}$$

where the wavelength λ has the dimension µm.

The actual refractive index n_L for t °C, p Torr total pressure, and e Torr vapour pressure is obtained from

$$(n_L - 1) = 98.7 \cdot 10^{-5}\,\frac{n_{Gr} - 1}{1 + \alpha t} \cdot p - \frac{4.1 \cdot 10^{-8}}{1 + \alpha t}\,e;\ \alpha = 0.003\,661\,, \tag{4.4.16}$$

with

$$e = E' - D \cdot p(t - t') \tag{4.4.17}$$

where

D is a constant (0.000 662 for measurements across water, and 0.005 83 for measurements across ice),

E' is the saturation vapour pressure for the temperature of the wet thermometer, and
t' is the wet temperature.

The saturation vapour pressure is usually computed according to Magnus-Tetens as:

$$\log E' = \frac{\alpha t'}{t' + \beta} + \gamma \tag{4.4.18}$$

where

$\alpha = 7.5$ and $\beta = 237.3$ for across water, and
$\alpha = 9.5$ and $\beta = 265.5$ for across ice
γ remains the same at 0.7857.

Deviations of the computed refractive indices n_M and n_L from the instrument constant n_0 require corrections, as discussed in section 4.4.4.

By substituting equation (4.4.17) into equations (4.4.14) and (4.4.16) and differentiating these equations one obtains the following error formulas for the scale error in parts per million:

$$dn_L \cdot 10^6 = -1.00\, dt + 0.28\, dp \tag{4.4.19}$$

$$dn_M \cdot 10^6 = -4.35\, dt + 6.67\, dt' + 0.26\, dp \tag{4.4.20}$$

for $t = 15\,°\mathrm{C}$, $t - t' = 4\,°\mathrm{C}$, $p = 1000$ mbar.

The wet temperature is of secondary importance for the computation of n_L. Its influence is less than $1 \cdot 10^{-6}$ for temperatures $t < 30\,°\mathrm{C}$ and a relative humidity of less than 60%. However, dry and wet temperatures have to be measured extremely carefully for microwave measurements, i.e., with an Assmann aspiration psychrometer. Errors of 0.23 °C of the dry temperature or 0.15 °C of the wet temperature cause a scale error of 1 mm per km because of the error in e.

The *refraction coefficient* k is defined as

$$k = \frac{R}{\varrho} = -\frac{\text{curvature of path}}{\text{earth curvature}}$$

where R = radius of the earth and

$$\frac{1}{\varrho} = -\frac{\operatorname{grad} n}{n} \cos\alpha \approx \frac{dn}{dh}$$

where α denotes the angle of slope of the path.

Thus $\frac{1}{\varrho}$ is approximately proportional to the value of the refraction coefficient k which decreases with increasing height. k therefore actually denotes the relative curvature of the path.

Average values for unperturbed atmosphere are $k_L = 0.13$ and $k_M = 0.25$, which means that the radius of curvature of the light path is approximately $8\,R$, while it is about $4R$ for microwaves. Above grassland the following mean values can be expected at 40 to 100 m above the terrain:

$k_M = 0.25$ at day during the whole year and on overcast winter nights
$k_M = 0.50$ at summer nights and clear winter nights
$k_L = 0.20$ at day and on overcast nights during the whole year
$k_L = 0.20$ at clear nights
$k_L = 0.13$ at days without clouds.

Differences in t, p, and t' from the assumed standard require the corrections discussed in section 4.4.4.

4.4.4 Correction Because of Velocity of Propagation

The distance D' as read at the instrument is obtained based on a fixed value n_0 for the refractive index. Considering n as a mean refractive index valid for the whole path of the rays, it can be approximated by introducing into equations (4.4.14) or (4.4.16) the mean values for the meteorological data as observed at both end points of the distance. The difference $(n - n_0)$ is then used to compute a *first velocity correction* k_n.

A *second velocity correction* $k_{\Delta n}$ is primarily influenced by the refraction coefficient. These corrections amount to:

$$k_n = D^A \frac{n_0 - n}{n} \approx D^A (n_0 - n) \tag{4.4.21}$$

$$k_{\Delta n} = -(k - k^2) \frac{(D^A)^3}{12 R^2} \, . \tag{4.4.22}$$

Thus the *observed and corrected distance* becomes, using equations (4.4.12), (4.4.13), (4.4.21) and (4.4.22):

$$D = D^A + k_0 + k_f + k_n + k_{\Delta n} \, . \tag{4.4.23}$$

Note that the correction (4.4.22) is applicable for distances of more than 10 km in length only. When measuring long distances in alpine terrain or across water, special refraction conditions are encountered and have to be considered (section 4.4.6).

4.4.5 Geometric Reductions

The value D obtained with equation (4.4.23) represents a circular arc with radius ϱ between the points with sea level elevations H_1 and H_2. This arc has to be reduced in steps (see Fig. 4.4.23), first for the curvature of the path to the (inclined) chords S^R with r_k; then with r_H to the chord S^0 of the arc of the earth's surface passing through the reference horizon, and finally with r_E to S because of curvature of the earth.

The final distance becomes with the *known elevation difference* $\Delta H = H_2 - H_1$ for the end points:

$$S = D + r_k + r_H + r_E \tag{4.4.24}$$

with

$$r_k = -k^2 \frac{D^3}{24\,R^2} \tag{4.4.25}$$

and

$$r_H = \left(\sqrt{\frac{1 - \left(\frac{\Delta H}{S^R}\right)^2}{\left(1 + \frac{H_1}{R}\right)\left(1 + \frac{H_2}{R}\right)}} - 1 \right) S^R$$

see equation (4.4.32) where

$$S^R = D + r_k \approx D$$

and also

$$r_E = \frac{D^3}{24\,R^2}\,. \tag{4.4.26}$$

The corrections $k_{\Delta n}$, r_K and r_E can be neglected for distances of less than 10 km. The magnitude of the sum of these corrections is evident from the following table, where

$$k_{\Delta n} + r_k + r_E = (1 - k^2)\frac{D^3}{24\,R^2} \tag{4.4.27}$$

k	$D = 20$ km	40 km	60 km	80 km	100 km
0	0.008				
0.125	0.006	0.050	0.168	0.398	0.880
0.20	0.005	0.042	0.141	0.333	0.740
0.30	0.004	0.032	0.108	0.255	0.564
0.50	0.002	0.016	0.055	0.130	0.288
1.00	0.000				

ΔH has to be determined with a high accuracy for short distances with large elevation differences as shown in the diagram in Fig. 4.4.20.

If zenith angles were measured for the geometric reduction, then

$$S^0 \approx S = S^R \cdot \sin z' \left[1 + \frac{1}{R}\left(-H_1 - S^R \cos z' \left(1 - \frac{k}{2} \right) \right) \right]$$

(see equation (4.4.37)) for highest accuracy requirements, and

$$S^0 \approx S = S^R \cdot \sin z' \left(1 - \frac{H_1}{R} \right)$$

(see equation (4.4.39)) for lower accuracy requirements and shorter distances.

In order to keep the refraction influence small, the reduction with zenith angles is used for distances only up to 3 km. Then the reductions according to equations (4.4.22), (4.4.25) and (4.4.26) can be omitted. Then

$$S^R = D^A + k_0 + k_f + k_n \,. \tag{4.4.28}$$

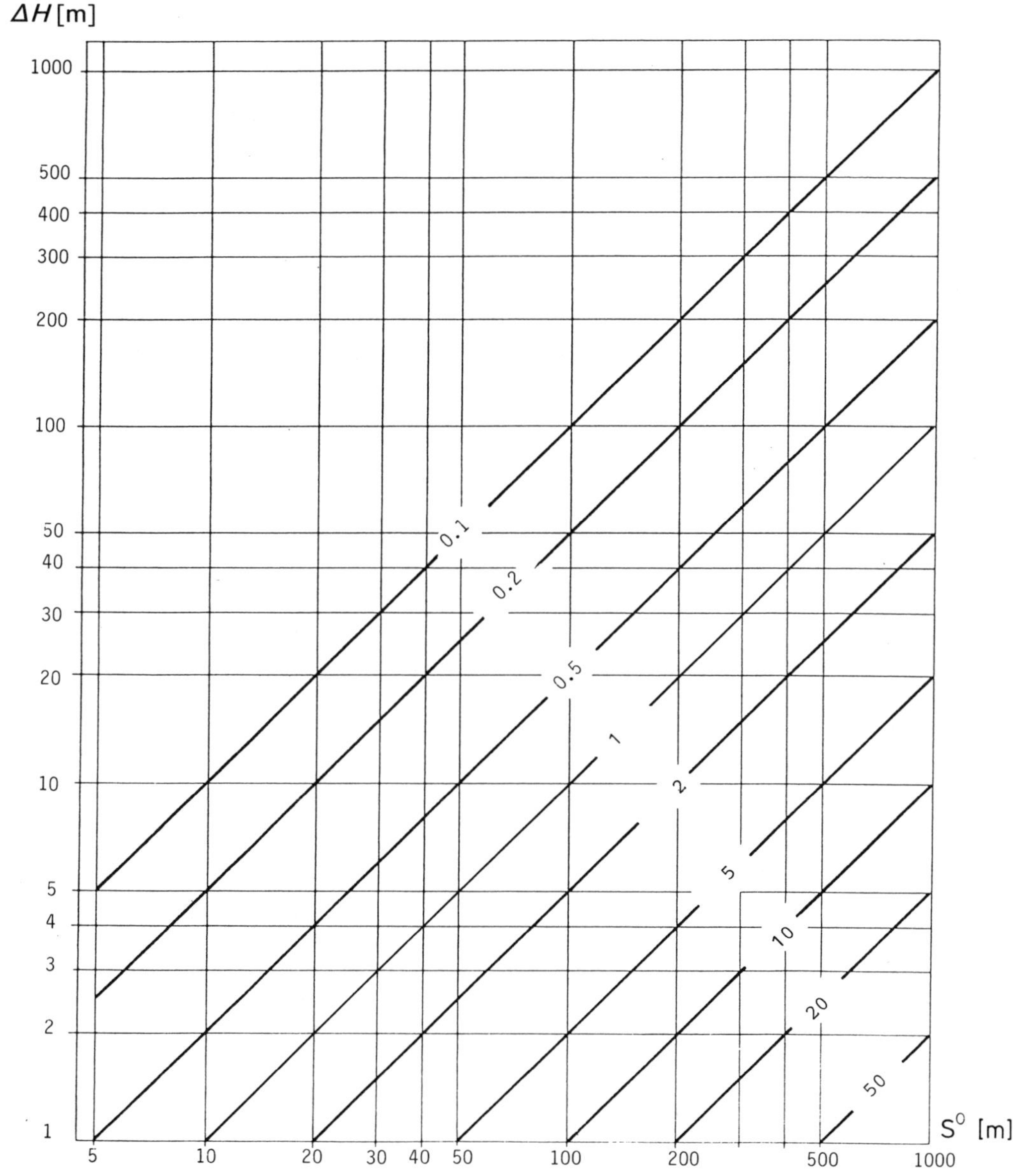

Fig. 4.4.20 Precision of elevation differences in mm for reduction of slope distance

Again, the accuracy of the zenith angle has to increase for larger elevation differences (Fig. 4.4.21). Fig. 4.4.22 provides some guidance as to when equation (4.4.37) can be replaced by the simplified equation (4.4.39).

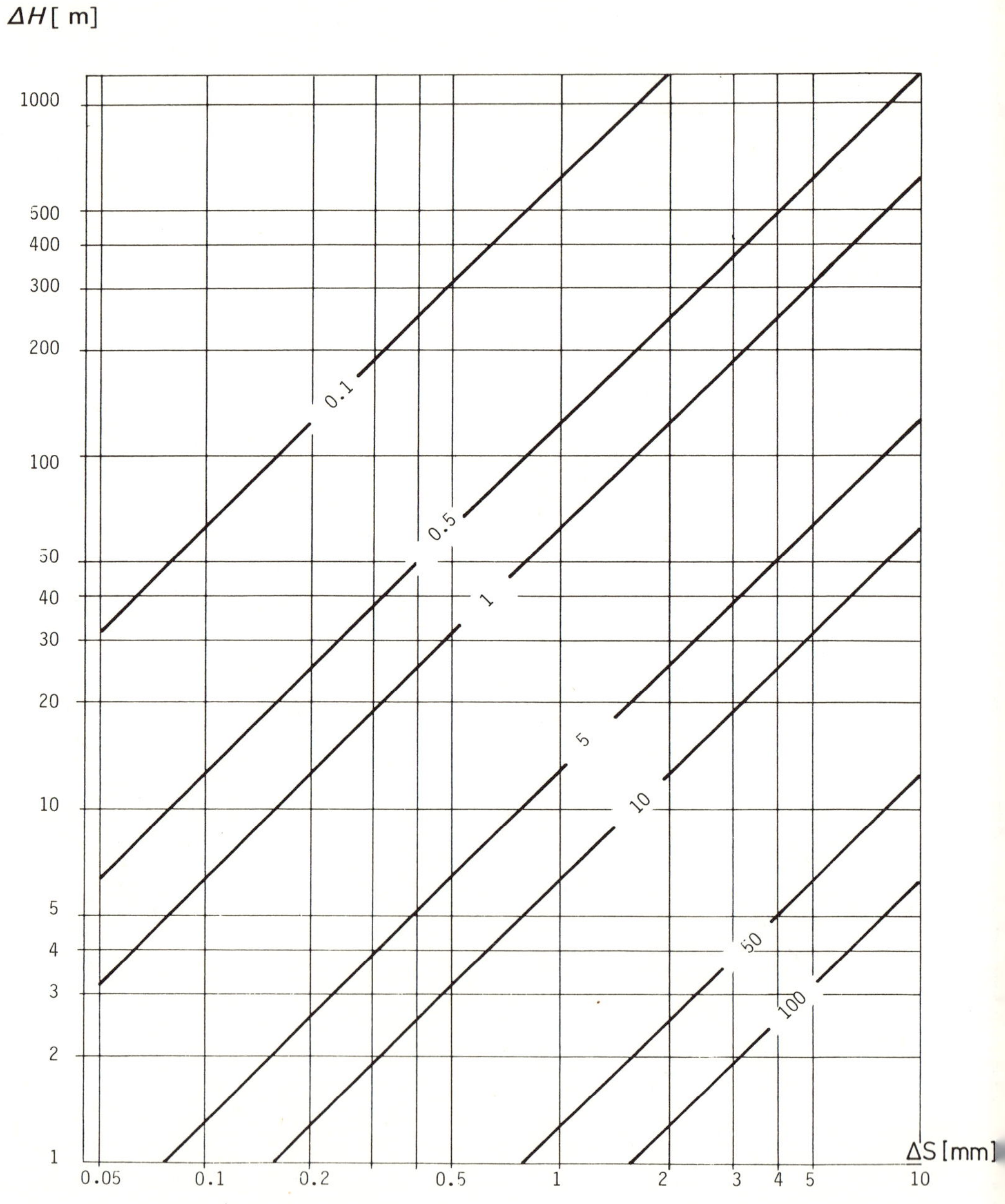

Fig. 4.4.21 Precision of vertical angles in mgon for reduction of slope distance

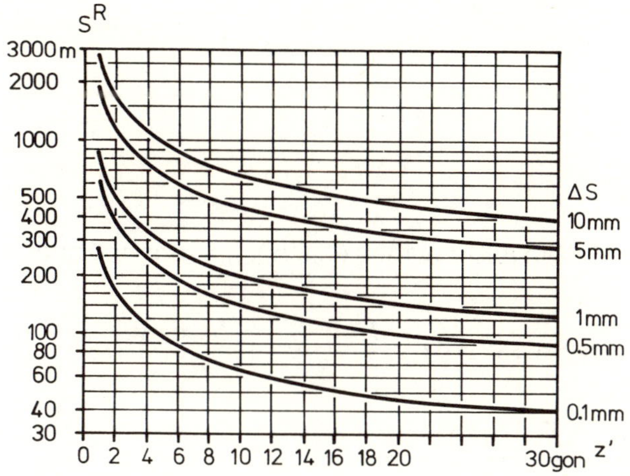

Fig. 4.4.22 Diagram of reduction errors as functions of distance and slope

4.4.5.1 Reduction Formula for Known Elevation Difference

Distances measured on the earth's surface represent arcs on the surface of the reference ellipsoid within a vertical section. Usually these distances are less than 100 km long, thus the arcs can be approximated by circular arcs.

In Fig. 4.4.23, P_1 and P_2 represent terrain points with elevations H_1 and H_2 above the reference sphere with radius R. Applying the cosine rule to triangle $P_1 P_2 M$ we get

$$(S^R)^2 = (R + H_1)^2 + (R + H_2)^2 - 2(R + H_1)(R + H_2)\cos\gamma \,. \qquad (4.4.29)$$

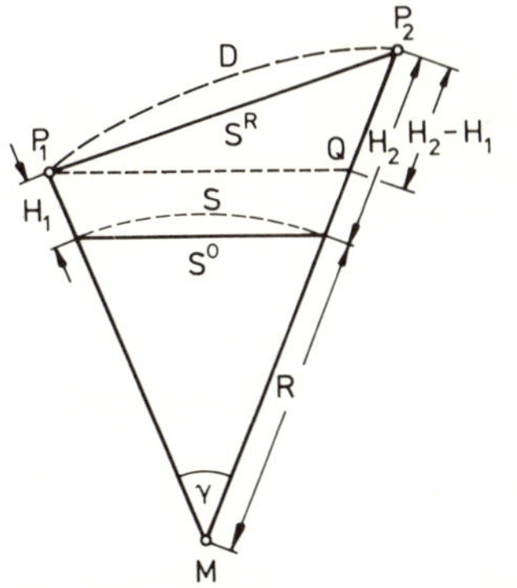

Fig. 4.4.23 Reduction of slope distance using elevation difference

Since

$$\cos\gamma = 1 - 2\sin^2\frac{\gamma}{2}$$

and

$$\sin\frac{\gamma}{2} = \frac{S^0}{2R},$$

we obtain

$$\cos\gamma = 1 - \frac{(S^0)^2}{2R^2}. \tag{4.4.30}$$

After inserting equation (4.4.30) into equation (4.4.29), squaring and multiplying, we get

$$(S^R)^2 = (H_2 - H_1)^2 + (S^0)^2\left(1 + \frac{H_1 + H_2}{R} + \frac{H_1 H_2}{R^2}\right)$$

or

$$(S^0)^2 = \frac{(S^R)^2 - (H_2 - H_1)^2}{1 + \frac{H_1 + H_2}{R} + \frac{H_1 H_2}{R^2}} = \frac{(S^R)^2 - \Delta H^2}{\left(1 + \frac{H_1}{R}\right)\left(1 + \frac{H_2}{R}\right)}$$

with $\Delta H = H_2 - H_1$, we obtain

$$S^0 = \sqrt{\frac{(S^R)^2 - \Delta H^2}{\left(1 + \frac{H_1}{R}\right)\left(1 + \frac{H_2}{R}\right)}}. \tag{4.4.31}$$

This leads to the correction term for slope and height correction:

$$r_H = \left(\sqrt{\frac{1 - \left(\frac{\Delta H}{S^R}\right)^2}{\left(1 + \frac{H_1}{R}\right)\left(1 + \frac{H_2}{R}\right)}} - 1\right) S^R \tag{4.4.32}$$

For longer distances (10 km) it is necessary to consider the ellipsoid as a reference surface. The radius of curvature of the normal section of the ellipsoid at the azimuth A can be computed with Euler's formula (*Torge* 1980):

$$\frac{1}{R} = \frac{\cos^2 A}{M} = \frac{\sin^2 A}{N} \tag{4.4.33}$$

where

M = radius of curvature in the meridian
N = radius of curvature normal to the meridian
A = azimuth.

The elevations H_i above the reference ellipsoid are computed as MSL elevation plus geoidal height. A mean value for R can be utilized for shorter distances.

For specific tasks, e.g., staking of structures, the distances have to be reduced to a horizon that is located at the elevation H_r above the reference sphere.

$$S^{H_r} = S^0 \frac{R+H_r}{R} = S^0\left(1+\frac{H_r}{R}\right)$$
$$= \left(1+\frac{H_r}{R}\right)\sqrt{\frac{(S^R)^2 - \Delta H^2}{\left(1+\frac{H_1}{R}\right)\left(1+\frac{H_2}{R}\right)}}\,. \qquad (4.4.34)$$

By analysing equations (4.4.31) or (4.4.34) with the aid of the law of error propagation and neglecting the earth's curvature and reduction to the reference sphere, one obtains:

$$(S^0)^2 = (R^R)^2 - \Delta H^2\,;$$
$$2S^0 dS^0 = -2\Delta H d\Delta H$$
$$\lg S^0 = \lg \Delta H + \lg d\Delta H - \lg dS^0\,. \qquad (4.4.35)$$

This equation (4.4.35) is used for the construction of a diagram which relates the required accuracy for an elevation difference for a certain reduction accuracy, i.e., $dS^0 = 0.1$ mm (Fig. 4.4.20). For a reduction accuracy of $\Delta S^0 = 1$ mm, all numerical values on the lines have to be multiplied by 10. Steep sights and short distances require a high accuracy for ΔH.

4.4.5.2 Reductions of the Slope Distance Using Zenith Angles

Given are the points P_1 and P_2 at elevations H_1 and H_2 above the reference sphere, the distance D and the zenith angle z' which contains the angular refraction error $\delta/2$ (Fig. 4.4.24). Wanted are the distances S^{Hr} at a selected horizon above the reference sphere and the horizontal distance S^H at the reference sphere:

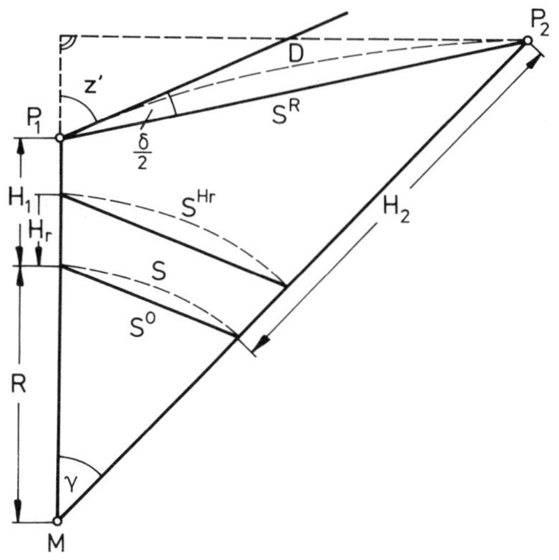

Fig. 4.4.24 Reduction of slope distance using zenith angle

$$\tan\gamma = \tan\frac{S^{Hr}}{R+H_r} = \frac{S^R \sin(z'+\delta/2)}{R+H_1+S^R\cos(z'+\delta/2)}.$$

Since the use of zenith angles for the reduction of slope distances is limited to distances of 3 km or less, the following applies:

$$\frac{S^{Hr}}{R+H_r} = \frac{S^R \sin(z'+\delta/2)}{R+H_1+S^R\cos(z'+\delta/2)}.$$

Applying the addition theorem leads to

$$\frac{S^{Hr}}{R\left(1+\frac{H_r}{R}\right)} = \frac{S^R(\sin z'\cos\delta/2+\cos z'\sin\delta/2)}{R\left(1+\frac{H_1}{R}+\frac{S^R}{R}(\cos z'\cos\,\delta/2-\sin z'\sin\delta/2)\right)}.$$

Since $\cos\delta/2 \approx 1$, $\sin\delta/2 \approx \delta/2$, and $\sin z'\cdot\sin\delta/2$ is negligibly small, we get:

$$S^{Hr} = \left(1+\frac{H_r}{R}\right) S^R(\sin z'+\delta/2\cos z')\left(1-\frac{H_1}{R}-\frac{S^R}{R}\cos z'\right).$$

The refraction angle is (see section 10.6):

$$\frac{\delta}{2} = \frac{1}{2}\frac{S^R}{R}\sin z' k\ .$$

Therefore

$$S^{Hr} = \left(1+\frac{H_r}{R}\right) S^R\sin z'\left(1+\frac{1}{2}\frac{S^R}{R}\cos z' k\right)\left(1-\frac{H_1}{R}-\frac{S^R}{R}\cos z'\right),$$

$$S^{Hr} = S^R\sin z'\left[1+\frac{1}{R}\left(H_r-H_1-S^R\cos z'\left(1-\frac{k}{2}\right)\right)\right]. \tag{4.4.36}$$

H_r becomes zero for horizontal distances in the reference surface, which leads to:

$$S = S^R\sin z'\left[1+\frac{1}{R}\left(-H_1-S^R\cos z'\left(1-\frac{k}{2}\right)\right)\right]. \tag{4.4.37}$$

By analysing equation (4.4.37) with the aid of the law of error propagation and neglecting earth curvature, refraction and reduction to the reference sphere, one obtains:

$$\begin{aligned} S &= S^R\sin z' \\ dS &= S^R\cos z'\,dz' \\ dS &= \Delta H dz' \\ \lg dS &= \lg\Delta\mathrm{H}+\lg dz'\ . \end{aligned} \tag{4.4.38}$$

This equation (4.4.38) is used for the construction of a diagram which relates the required accuracy for a zenith angle to a certain reduction accuracy ΔS (Fig. 4.4.21) Increasing elevation differences require higher accuracy for the zenith angles.

Simplification of equation (4.4.37) leads to

$$S = S^R \sin z' \left(1 - \frac{H_1}{R}\right) \tag{4.4.39}$$

where the influence of earth curvature and refraction is not considered. The factor $(1 - H_1/R)$ reduced the horizontal distance from the horizon of the horizontal axis of the theodolite to the reference surface. If one makes the refraction coefficient k zero and expands the expression in parentheses in equation (4.4.37), then it becomes obvious that

$$|r_{\text{ecur}}| = \frac{(S^R)^2}{2R} \sin 2z' \tag{4.4.40}$$

represents the influence of the earth curvature on the distance reduction.

The error of the reduction increases with the square of the slope distance, and linearly with the zenith angle. The lines in Fig. 4.4.22 indicate the range of validity for the simplified relation (equation (4.4.38)). The curves were constructed using a rearrangement of equation (4.4.40):

$$S^R = \sqrt{\frac{|r_{\text{ecur}}|2R}{\sin 2z'}}\,.$$

4.4.6 Special Refraction Models for Microwaves

a) The mean value N for the refractive index is used as $N = \frac{1}{2}(N_1 + N_2)$, where N_1 and N_2 are determined at the end points. The following assumptions are needed in this case:

$$\frac{\partial N}{\partial s} = \text{const.}; \quad \frac{\partial N}{\partial h} = \text{const.} \quad \text{with} \quad k = -6.38\,\frac{\partial N}{\partial h}\,.$$

This model, suitable for flat terrain with the waves propagating at least 40 m above the ground, describes the path of the rays as a circular arc. Signal transmission techniques utilize

$$\frac{\partial N}{\partial h} = -41.5\ \text{km}^{-1}$$

which corresponds to case a) with $k = 0.26$.

b) Gladstone's exponential function for the refractive index can be used when there is a large elevation difference between the end points especially in alpine areas. This model is described as

$$\frac{\partial N}{\partial s} = 0; \quad N = N_0 \cdot \mathrm{e}^{-0.136\mathrm{H}}$$

and leads to the formula

$$\lg N = \lg N_1 - \frac{1}{16900}\left(\frac{H_2 - H_1}{2} - 0.00981\, D^2\right).$$

using H in metres, and D in kilometres.

For curvature reduction, k is used with a linear gradient for the refractive index:

$$\frac{\partial N}{\partial h} = \frac{N_2 - N_1}{H_2 - H_1}.$$

4.4.7 Electro-optical Distance Meters

4.4.7.1 Electro-optical Distance Meters for Short and Medium Ranges

Distances of less than 10 to 15 km are usually encountered for projects in cadastral and engineering surveys. More frequently, these distances are between 0.01 km and 1 km. For this range, measuring signals can be generated and demodulated with semiconductor elements, which are of small volume and weight (section 4.4.1.6). Their voltage requirements also cause no problems and can be handled with the aid of small and light components. Thus a special group of distance meters was developed for the above mentioned projects, namely the short range distance meters (short range EDM), and the medium range EDM instruments. Their functional principle is described in section 4.4.1.4.

These instruments have small dimensions, are light, and have low electric energy requirements. Newer instruments have a *largely automated measuring cycle*. Thus the handling of the instruments consists primarily in aiming and in activating the measuring process. The results of the measurement are displayed at a display unit. By activating a switch or depressing keys, it is possible to generate displays of different parameters for some of the EDM units, such as horizontal distance, elevation difference, coordinate difference between station and target. Elevation and coordinate differences are computed on the internal microprocessor (section 4.4.1.4). The necessary zenith angle is either measured by an internal zenith angle sensor or else entered manually on the key board. Short period interruptions of the rays do not influence the results for most instruments. The internal microprocessor of these instruments monitors the measuring procedure, such that the originally planned number of observations is obtained in a series of sequential measuring steps. Some instruments even apply atmospheric corrections (section 4.4.4) automatically after the observer has keyed in the necessary parameters.

The aforementioned distance meters are often coupled with a theodolite without interfering with its basic functions. Thus, they can be very effective when directions and distances have to be measured at the same station, and when the stations change frequently. Such measuring tasks occur, for instance, when traversing, for polar surveys, and lay-out and staking work.

The *range* of short range EDM instruments is about 2 to 5 km, while medium range instruments can be used for distances of up to 15 km.

Based on *accuracy*, one distinguishes between two groups of instruments: A large group of instruments has an accuracy of $\pm 5 \ldots 10$ mm $\pm 1 \ldots 6 \cdot 10^{-6} D$, where D represents the distance. These instruments are ideally suited for the everyday tasks of a surveying engineer, such as traversing, polar surveys and the staking of buildings. A smaller group reaches accuracies of $\pm 0.2 \ldots 2$ mm $\pm 1 \cdot 10^{-6} D$. These instruments are well suited for precision surveys, e.g., engineering surveys.

Based on their *external design* and possible applications in the field, distance meters can be categorized as:

- instruments, designed solely for distance measurements, which are mounted in a U-type support that permits rotation around a vertical and a horizontal axis,
- instruments which can be fastened by means of an adapter to the turning top part of a theodolite,
- instruments where theodolite and distance meter form one unit.

The following *additional devices* are important for the staking and searching of points: automatic computation and display of coordinate and elevation differences, tracking of measured or computed values (tracking function), and a telemetric system which transmits information to the carrier of the prism. The telemetric system, for instance, may consist of an internal microphone at the instrument and a loud speaker at the prism support (Fig. 4.4.25).

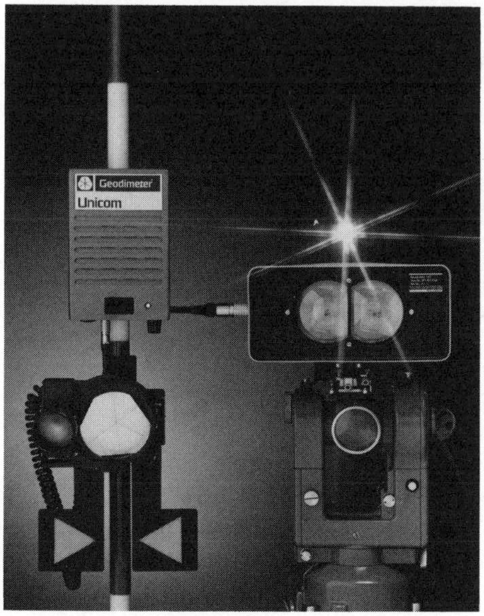

Fig. 4.4.25 Telemetric system consisting of microphone and loud speakers (*AGA* Geodimeter 122)

Another possibility is the transmission of the microprocessor data to a numerical display unit at the prism (Fig. 4.4.26). The Kern RD 10 System, for instance, transmits the following information: slope distance, horizontal distance, elevation difference, correction values in direction and normal to the line of sight, if a point has been approximately located. Furthermore, optical or acoustic signals are helpful to indicate whether the reflector is within the optical axis of the emitted bundle of rays.

Figs. 4.4.27 to 4.4.32 show different types of short and medium range EDM instru-

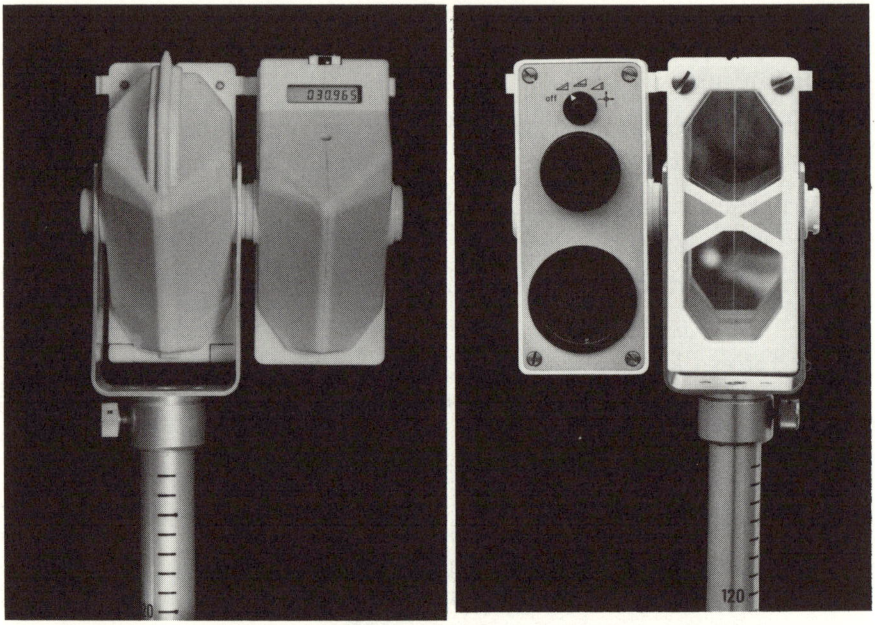

Fig. 4.4.26 Telemetric system with numerical display unit (*Kern* RD 10)

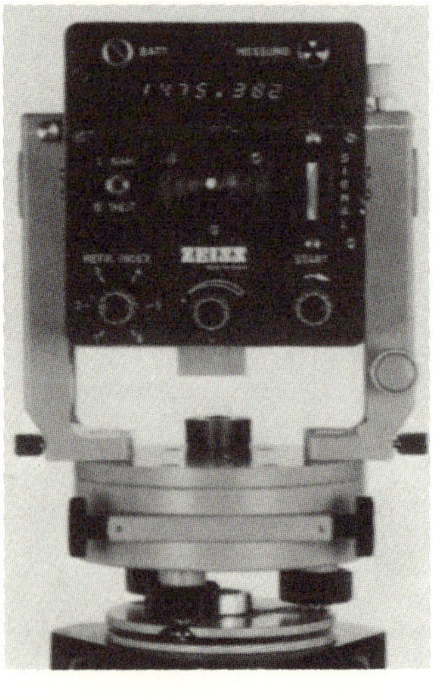

Fig. 4.4.27 Short range EDM as standalone unit: *Zeiss* Eldi 2 (also suitable as theodolite attachment)

ments. Table 4.4.1 provides a summary of the technical data of some of these distance meters. A more detailed description may be found in *Kahmen* (1978).

4.4.7.2 Long Range Electro-optical Distance Meters

The development of electro-optical distance meters for long ranges triggered electromagnetic distance measuring. The first instruments were the Geodimeter (geodetic distance meter), manufactured by the Swedish company AGA. The first design principle was given in 1943 by *E. Bergstrand*, of Stockholm, who developed it from instruments used for the determination of light velocity.

The operational principle of electro-optical distance meters is described in section 4.4.1.4. The measuring signal is generated by a laser which is indirectly modulated by a KDP modulator. The demodulation of the measuring signal is accomplished with the aid of a secondary electron multiplier. The voltage supply is accomplished with the aid of a secondary electron multiplier. The voltage supply is more difficult, because of the need to operate with high voltage. Thus the structural elements of long range EDM are much larger and heavier than for short and medium range EDM instruments. More complex circuitry is needed for signal processing, and even the optical components are larger and heavier. Thus long range EDM instruments have

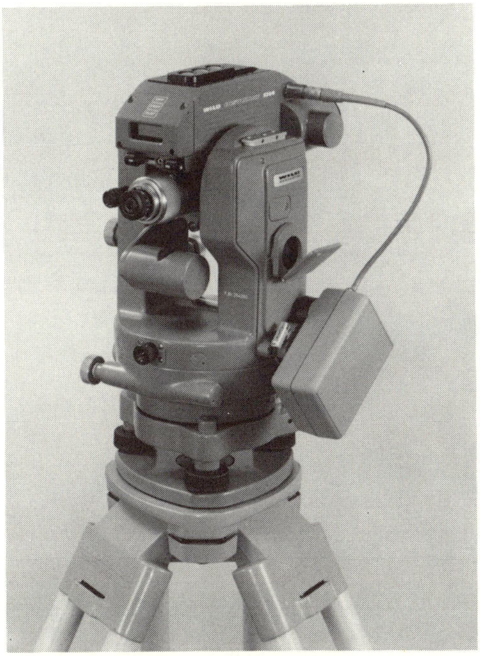

Fig. 4.4.28 Theodolite *Wild* T16 with distance meter *Wild* DI4 as attachment

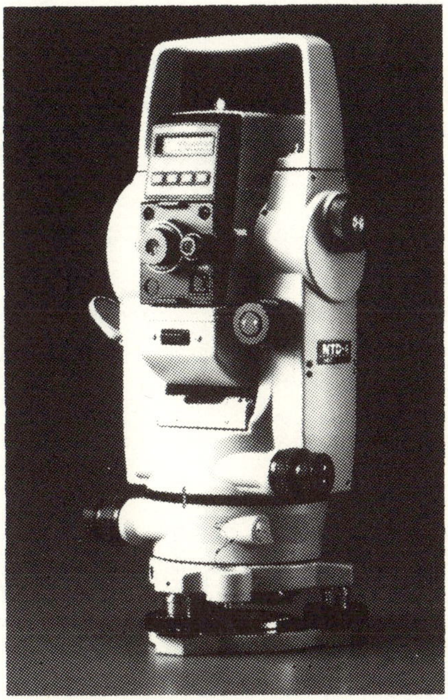

Fig. 4.4.29 Optical theodolite with integrated distance meter Nikon NTD 4

Table 4.4.1 Electro-optical distance meters for short and medium ranges

Manu- facturer	Type	External Design*	Range Reflectors [m]	Accuracy (Stand. dev.) [mm]	 [ppm]	Weight [kg]	Tracking [s]	V.-Angles electron. (Stand. dev.) [gon] [mgon]	 [°] [″]	Specialities
AGA Sweden	Geodimeter 110	b.	2100 3	±5	±3	2.8	2.0			
AGA Sweden	Geodimeter 112	b.	3600 3	±5	±3	2.8	0.4			
AGA Sweden	Geodimeter 116	b.	1400 3	±5	±5	2.8	0.4	±5 ±20	±5° ±1′	Tracklight Telemetry
AGA Sweden	Geodimeter 220	b.	2400 3	±5	±5	1.3	0.4	±20 ± 6	±20° ±20″	Tracklight Telemetry
AGA Sweden	Geodimeter 122	b.	3600 3	±5	±3	2.8	0.4	±20 ± 6	±20° ±20″	Tracklight Telemetry
AGA Sweden	Geodimeter 14 A	b.	8000 3	±5	±3	2.8				
Hewlett *Packard* USA	HP 3808 A	a.	6000 3	±5	±1	7.8	2			
Hewlett *Packard* USA	HP 3810 B	a.	5000 3	±5	±1	12	2	±23 ±15	±11 ± 6	Horiz. angles optic. ±1 mgon
Zeiss E. Germany	EOT 2000	c.	700 3	±10		11				Horiz. and vert. angles optic. ±0.2 mgon
Kern Switzerland	DM 102	b.	1700 3	±5	±5	1.7				

Table 4.4.1 Electro-optical distance meters for short and medium ranges (continued)

Manu- facturer	Type	External Design*	Range Reflectors [m]	Accuracy (Stand. dev.) [mm]	 [ppm]	Weight [kg]	Tracking [s]	V.-Angles electron. (Stand. dev.) [gon] [°] [mgon] [″]	Specialities
Kern Switzerland	DM 503	b.	00	±3	±2	1.6	2		
Keuffel *& Esser* USA	Autoranger A	b.	1600 3	±1 ... 2	±3	2.4			and coordin. diff.
Keuffel *& Esser* USA	Autoranger III	b.	4000 3	±5	±2	2.4			and coordin. diff.

* a. solely for distance measurements; b. instruments can be fastened by means of an adapter to the turning top part of a theodolite; c. theodolite and distance meter form one unit.

Table 4.4.1 Electro-optical distance meters for short and medium ranges (continued).

Manu-facturer	Type	External Design*	Range Reflectors [±m]	Accuracy (Stand. dev.) [mm]	[ppm]	Weight [kg]	Tracking [s]	V.-Angles electron (Stand. dev.) [gon]	Specialities
Nikon	ND 250	a.	2500	±5	±5	2.4	1.2		
Japan									
Sokkisha	RED	a.	500	±5	±5	0.9	1.3		
Japan	Mini	b.	3						
Nikon	RED 2	a.	3000	±5	±5	2	0.3		
Japan		b.	3						
Tellurometer	MA 100	a.	2000	±1,5 … 2	±2	17			
Inc. U.K.			7						
Topcon	DM – A 1	b.	1000	±5	±5	1.9	2		
Japan			3						setting out diff.
Topcon	DM – C 3	b.	2500	±5	±5	2.9			
Japan			3						
Topcon	DM – S 2	b.	2000	±5	±5	1.0	0.5		
Japan			3						setting out diff.
Wild	DI 1000	b.	500	±5	±5	0.6	0.3		
Switzerland			3						
Wild	DI 4	b.	1500	±5	±5	1.9	2.5		
Switzerland			3						
Wild	DI 4 L	b.	3500	±5	±5	1.9	2.5		
Switzerland			3						
Wild	DI 5	b.	3500	±3	±2	1.9	2.5		
Switzerland			3						ΔE ΔN H
									setting out diff.
Wild	DI 20	b.	7000	±5	±1	3.7	1		
Switzerland			3						

Manu-facturer	Type	External Design*	Range Reflectors [±m]	Accuracy (Stand. dev.) [mm]	[ppm]	Weight [kg]	Tracking [s]	V.-Angles electron (Stand. dev.) [gon]	Specialities
Wild u. Precision Internat. USA	Citation	b.	2300	±5	±5		1		
Zeiss W. Germany	Eldi 1	a.	4000 / 3 } II**)	±5…10	±2 } II	8.0	1		
		b.	3000 / 3 } I**)	±10	±2 } I				
Zeiss W. Germany	Eldi 2	a.	2000 / 3 } II	±5	±2 } II	4.2	1		
Zeiss			1600 / 3 } I	±10	±2 } I				
Zeiss W. Germany	SM 41	c.	1400 / 3 } II	±20	±2 } II	6.5			Horiz. and Vertic. angles optic ±1 mgon
			1000 / 3 } I	±10	±2 } I				
Com-Rad U.K. *Kern* Switzerland	Mekometer ME 3000	a.	2500 / 3	±0.2	±1	18.7			distance meter of highest accuracy
Com-Rad U.K.	Geomensor CR 204	a.	5000	±0.1	±0.5				distance meter of highest accuracy

** II: Range II, I: Range I

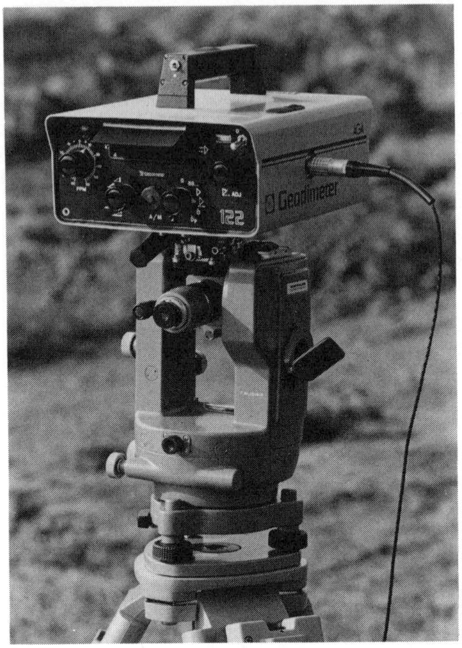

Fig. 4.4.30 Distance meter with internal zenith angle sensor and automatic distance reduction: *AGA* Geodimeter 122

larger weight and volume than the instruments discussed in section 4.4.7.1. Furthermore, the purchase price is significantly higher.

These instruments can be used for distances up to 60 km. However, the *effective range* is seldom longer than the sighting distance, which causes organizational difficulties for the efficient use of these instruments.

The manufacturers of these instruments usually claim an *accuracy* of $\pm(5 + 1 \times 10^{-6} D)$, with D denoting the distance. For long distances, this accuracy is primarily limited by errors in the determination of meteorological parameters. Under ideal external conditions (overcast and light wind for a well mixed atmosphere) an accuracy of 1 to 2×10^{-6} D, i.e., 1 to 2 ppm, can be achieved for the scale. For this, the temperature along the line needs to be measured to 1 to 2 °C, and the air pressure to 3 to 6 mbar.

The instruments operate with *analogue or digital phase detector systems*. When using analogue phase measurements, the process is not completely automatic. The result is obtained by stepwise changes and subsequent evaluation (section 4.4.1.2). Instruments with digital phase measurements operate fully automatically.

The measuring consists basically of sighting to the target, tuning the signal intensities, inputting scale and zero corrections, and initiating the measuring process. An automatic control makes sure that in case of a signal interruption, the planned number

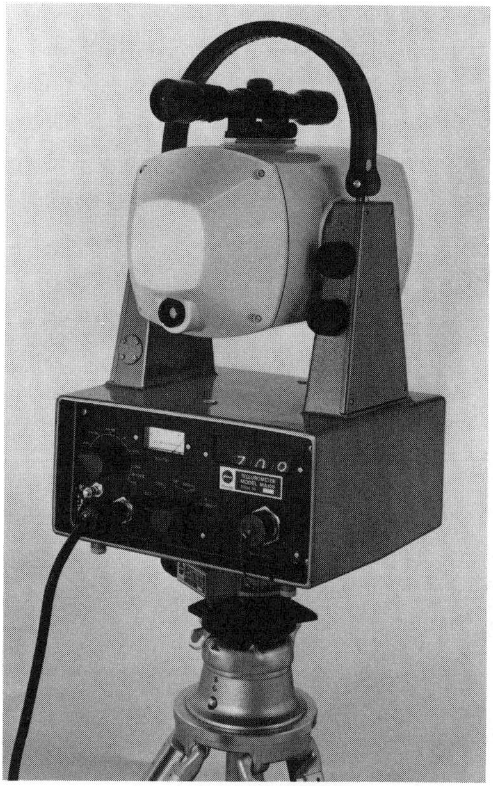

Fig. 4.4.31 High precision EDM: *Tellurometer* MA 100

Fig. 4.4.32 Medium range EDM *Wild* DI 20

of measurements is obtained by sequentially performing sets of measurements. This is especially useful on windy days when the carrier signal is diverted off and on beyond the range of the reflector. Distance meters with digital phase meters also have the capability of automatic data recording.

The AGA company in Sweden produced distance meters with analogue phase measurement. Fig. 4.4.33 illustrates an instrument of this type, having a range of 40 km.

Fig. 4.4.33 Long range distance meter, Geodimeter *AGA* 600

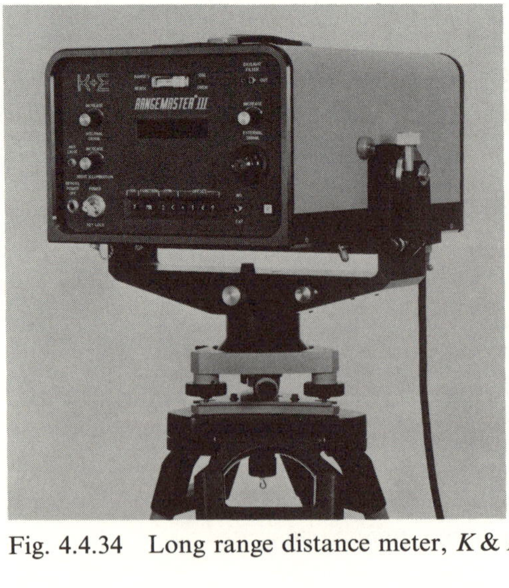

Fig. 4.4.34 Long range distance meter, *K & E* Rangemaster III

Distance meters of longer range using a digital phase meter are available through Keuffel & Esser (Fig. 4.4.34). The measuring process of these instruments is automatic. They also have an automatic data recording unit.

Other distance meters utilizing light waves as carrier are produced in the U.S.S.R. Table 4.4.2 provides an overview of the most frequently used instruments of this type.

4.4.7.3 Reflectors and Other Accessories

Reflectors are a necessary part of the measuring equipment. It is their task to reflect the light emitted from the transmitter and return it to the transmitter stations. A simple plane mirror is not suitable since it would have to be aligned extremely accurately. Therefore, so-called triple prisms are predominantly used as reflectors. They are glass bodies with three mutually orthogonal surfaces, like the cut-off edge of a glass cube. A light ray hitting a triple prism is reflected parallel to itself even though the axis of the prism may deviate from the direction of the light by 10 to 20 gon (up to 20°). For short distances (up to 1 km), a single prism is sufficient for most instruments. For longer distances, reflectors with 9 or more prisms are required.

The reflectors are usually set up on tripods with tribracks (Fig. 4.4.35). For detail surveys it is often sufficient to use a simple plumbing rod (Fig. 4.4.36) carrying a

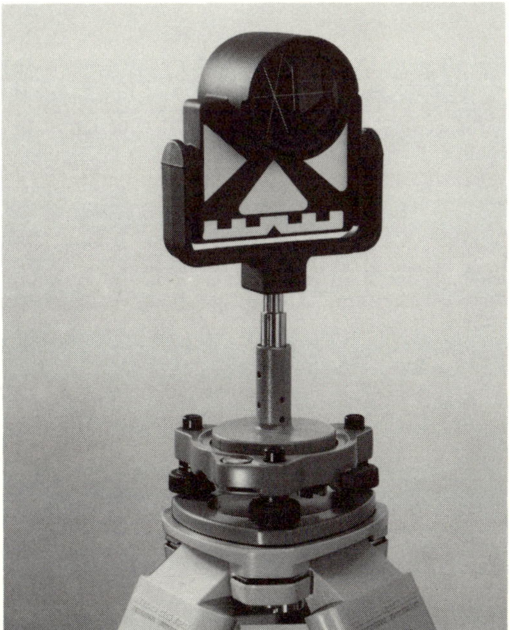

Fig. 4.4.35 Single prism on support in tribrack

Table 4.4.2 Long range electro-optical distance meters

Manufacturer	Type	Range [km]	Light source Carrier	Modulation	Fine Scale [m]	Phase measurement	Accuracy (Stand. der.) [mm]	[ppm]	Weight [kg]	Specialities
AGA Sweden	8	60	HeNe-Laser 1 mW 0.6328 μm	KDP Crystal	5	nullpoint meter	±5	±1	23	
AGA Sweden	6 BL	20–25	HeNe-Laser 1 mW 0.6328 μm	KDP Crystal	5	nullpoint meter	±5	±1	16	
AGA Sweden	600	40	HeNe-Laser 1 mW 0.6328 μm	KDP Crystal	5	nullpoint meter	±5	±1	15	
Spectra-Physics USA	Geodolit 3 Gr	65–80	HeNe-Laser 10 mW 0.6328 μm		1 ft	digital automatic	±1	±1	45	
USSR	Kvarts	30–50	HeNe-Laser 0.6328 μm		5	nullpoint meter	±10	±2	25	
Keuffel & *Esser*, USA	Ranger V-A	25	HeNe-Laser 0.6328 μm	KDP Crystal	10	digital automatic	±5	±2	16	Data storage (Printer)
Keuffel & *Esser*, USA	Range-master III	60	HeNe-Laser 0.6328 μm	KDP Crystal	10	digital automatic	±5	±1	20	Data storage (Printer)

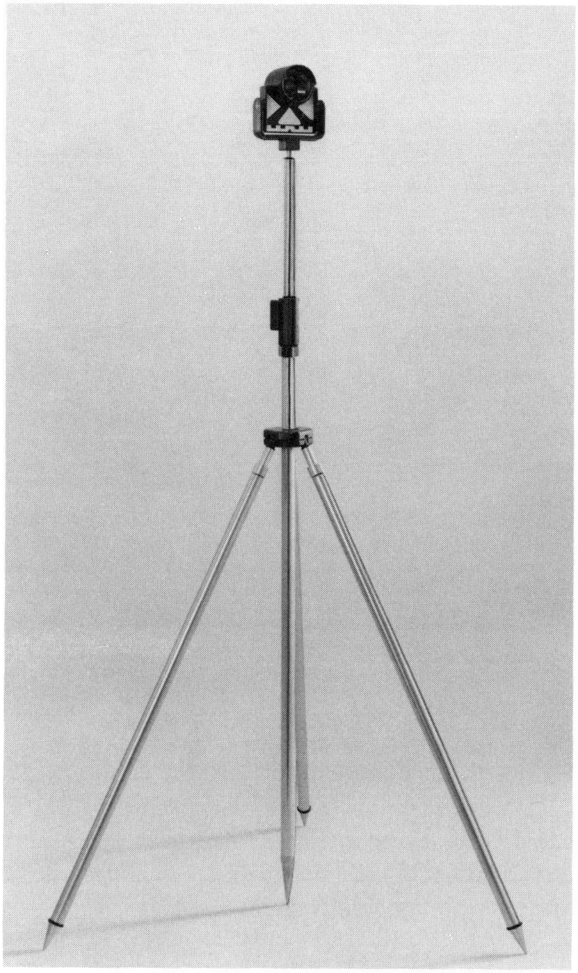

Fig. 4.4.36 Single prism with plumbing rod (*Wild*)

reflector and to hold it plumb with the aid of a bull's-eye level. Fig. 4.4.37 shows a reflector with 3 prisms, while Fig. 4.4.38 serves as an example for an arrangement of a prism reflector (Zeiss, Oberkochen).

Except for the reflectors, the various accessories are not essential, but help in reducing the measuring effort or providing new possibilities for applications. The companies offer a variety of collections. The following accessories ease the measuring effort: Transport containers, carrying frames, changeable batteries, battery chargers, additional tripods, even complete traversing equipment.

Additional applications are encouraged with devices for forced centring where EDM instruments can be exchanged for theodolites without changing the centring. Adapters are even more important. They are used to fasten an EDM or its sighting

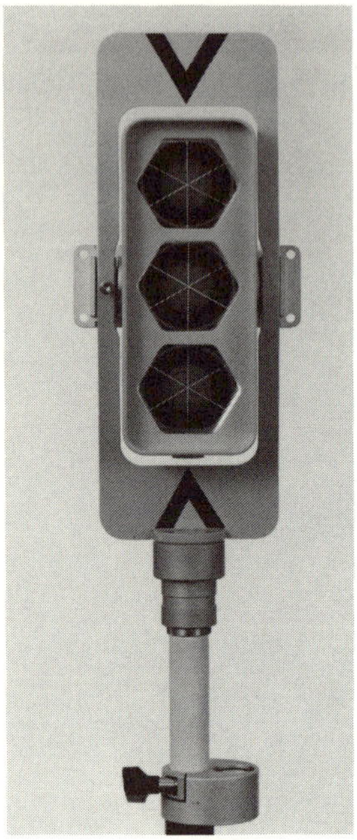

Fig. 4.4.37 Three prism reflector (*Zeiss*)

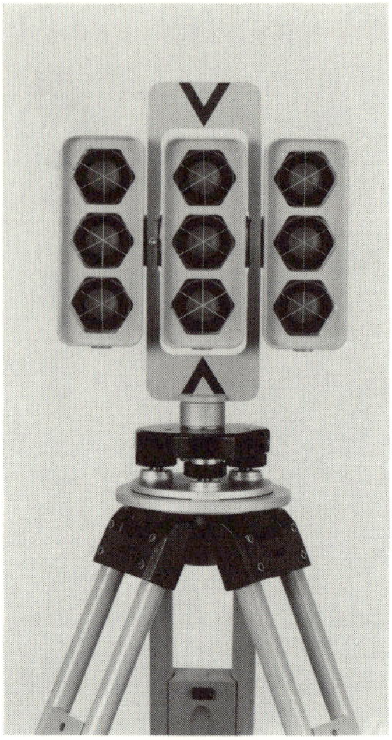

Fig. 4.4.38 Nine prism reflector (*Zeiss*)

head containing optical and electronic sighting and receiving components onto the telescope standards on the telescope itself such that when rotating it the EDM follows. Thus the observer has a quasi-tacheometer. Unfortunately, the optical and electronic axes are not usually identical, which means that they have to be set parallel using special procedures for determining zenith distances.

4.4.8 Microwave EDM

4.4.8.1 Range, Accuracy and Instrument Design

When the first microwave distance meters, the Tellurometer MA1/RA1, were produced in 1957, finally geodetic instruments were available which were capable of measuring long distances virtually independent of the weather. Instruments with ranges of up to 200 km were developed. However, the measuring accuracies are lower than for electro-optical distances meters. Errors in the determination of meteorological data are more critical, because of the influence of the refractive index. A relative accuracy of 3 to 4×10^{-6} can be reached under ideal conditions (overcast sky, well mixed atmosphere, light wind).

Since microwave carriers are less narrowly bundled, ground swing influences the measuring accuracy as well. The carriers of the first instruments had a wavelength of

10 cm and a frequency of 3.5 GHz. In order to reduce the unfavourable ground swing, higher carrier frequencies were utilized in later instruments (10 GHz or 3 cm, 25 GHz or 8 mm). Furthermore, fine readings are repeated with slightly modified carrier frequencies. Thus, a further reduction of the ground swing influence is achieved by averaging them. A similar effect is achieved in newer instruments by changing the modulation frequencies of 7.5 or 15 MHz (20 m or 10 m fine scale) to 75 or 150 MHz (2 m or 1 m fine scale). Therefore, a variation of carrier frequencies is not necessary for the latter.

Furthermore, ground swing needs to be considered when planning a trilateration network. Lines should avoid areas of strong reflectivity (water, highways, snow). In hilly and wooded terrain, normally little disturbing reflections are encountered.

The functional principle of microwave distance meters is described in section 4.4.1.5. While the original instruments generated the carrier with a clystron, newer models utilize a gun-diode. A clystron requires high voltage, thus the elements, including the voltage supply, are bulky and heavy. The voltage supply for gun-diodes, however, provides no problem. Thus instruments utilizing these can be built smaller and lighter. Older instruments require manual activating of each measuring sequence. Today, the whole procedure is automatic, monitored by a microprocessor.

Some newer instruments have separate measuring and antenna units, connected by cable of up to 30 m in length. Thus for poor sight connections, the antenna can be elevated on a mast, which is easier than erecting of measuring towers.

4.4.8.2 Selected Microwave Distance Meters

The first microwave instrument, the *Tellurometer* MA 1/RA 1 built by *T. L. Wadley* (1956) at the South African Research Institute for Electrical Engineers, utilized a microwave with $\lambda = 10$ cm ($F = 3$ GHz) as the carrier. The range of this instrument exceeded 100 km, which could be measured to within a few decimetres under suitable conditions.

Due to the more favourable characteristics of shorter microwaves, Tellurometer Inc. developed the MRA 3, with 3 cm carrier (10 GHz), MRA 4 with 8 mm carrier (35 GHz), MRA 5 with 3 cm carrier (10 GHz) and MRA 6 with 1.8 cm carrier (16 GHz). The MRA 5 has separate measuring and antenna units, with the latter fastened to a tripod or mast and connected with a 7 m or 25 m long cable. The measuring procedure is automatically monitored in models 5 and 6.

In some Tellurometer instruments, master and remote functions are combined in one instrument. Each can then selectively be switched to master or remote.

This idea has been abandoned for the inexpensive model CA 1000, first built in 1972. Master and slave stations are not identical. The instrument itself is of solid state design and thus small and light. The range is somewhat less than for the other Tellurometer models. With an extended antenna attachment, a range of up to 30 km can be reached. This can be extended to 100 km by use of parabolic mirrors according to the Casse-grain-principle. The modulation frequency is 10 GHz. The carrier frequency is automatically tuned at the slave. Therefore the ground swing has to be carefully monitored.

The *Electrotape* DM 20 by Cubic Corp. (U.S.A.) was the first microwave instrument equipped with a 3 cm carrier. It operates according to the same principle as the tellurometer, even though its technical realization is different. The range could be extended to 250 km with enlarged parabolic antenna, provided the meteorological conditions were suitable. A number of similar instruments by differt manufacturers followed.

The *SIAL MD 60* microwave distance meter by Siemens-Albis in Zürich (Fig. 4.4.39) represents a further step. The modulation frequency of the fine measurement is magnified 10 times to 150 MHz in this instrument. This reduces ground swing which eliminates the need for variation of the carrier frequency of 2 to 3 cm. After the instrument is directed, the measurement is fully automatic and takes 30 seconds. A range of 150 km with an accuracy of $\pm 1\,\text{cm} + 3 \cdot 10^{-6}\,D$ is claimed by the manufacturer.

There are two variants of the SIAL MD 60. The reflector is solidly built in the compact version SIAL D 60 C, while the reflector in the SIAL MD 60 S is in two parts. Measuring unit and antenna can be connected by a 50 m cable.

The Volna and Luc microwave distance meters are used predominately in the U.S.S.R.

Table 4.4.3 provides the technical data for some selected microwave distance meters.

Fig. 4.4.39 Distance meter SIAL MD 60

Table 4.4.3 Microwave distance meters

Manufacturer	Type	Range [km]	Carrier	Readout	Antenna Diameter [cm]	Antenna Beam width	Accuracy (Stand. dev.) [cm]	Accuracy (Stand. dev.) [ppm]	Weight [kg]
Tellurometer Ltd. U.K.	MRA 101	0.1–50	~3 cm	digital	33 cm	6°	±1.5	±3	16
	MRA 3	0.1–50	~3 cm	digital	33 cm	9°	±1.5	±3	17
	CA 1000	0.1–30	~3 cm	digital	Horn-Antenna		±1.5	±3	3.6
	MRA 5	0.1–50	~3 cm	digital		6°	±1.0	±5	12
	MRA 6	0.1–50	~1.8 cm	digital			±1.0	±3	7
Tellurometer Ltd. U.K.	MRB 201	0.1–30	~10 cm	digital	36 cm	6°	±1.5 m		30
Cubic Corp. USA	Electrotape DM 20	0.05–50	~3 cm	digital	33 cm	6°	±1.0	±3	11.5
Siemens Albis u. *Wild* Switzerland	Sial MD 60	0.02–150	~3 cm	digital	35 cm	6°	±1.0	±3	13.5
USSR	Luč	0.02–40	~3 cm	digital	38 cm	5°	±3.0	±3	14
USSR	Volna	15					±0.3	±3	9

5. Fundamentals of Plane Coordinate Computations

Geodetic surveys on a national basis require the determination of *geographic coordinates* as well as of *plane Cartesian coordinates* for selected points on the earth's surface. Practical detailed work, however, is nearly always carried out in plane coordinates because of the simpler computations. The plane coordinate systems are obtained by mathematically (geodetic ally) mapping of the ellipsoidal coordinates. For this, several approaches are possible (sections 7.2.2 and 7.2.3).

5.1 Rectangular and Polar Coordinates

A *plane geodetic coordinate system* is a *Cartesian x, y system* with the positive x-axis (abscissa) pointing east, and the positive y-axis (ordinate) pointing north. It should be noted here, that in central Europe the axes are reversed, i.e., x-axis pointing north, and y-axis pointing east.

The position of a point is determined by its rectangular coordinates x and y, or by its polar coordinates α and s (Fig. 5.1.1).

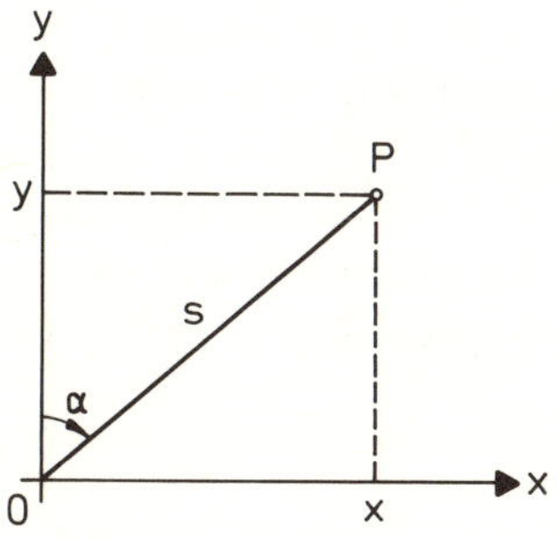

Fig. 5.1.1 Rectangular and polar coordinate systems

The position of the origin *0* depends on the selected geodetic projection.

The azimuth of a distance $P_1 P_2$ is the angle between a line parallel to the northing axis through P_1 and the line $P_1 P_2$. It is counted clockwise from the north branch of this parallel (Fig. 5.1.2) and denoted as $\alpha_{1,2}$.

Similarly, $\alpha_{2,1}$ is the azimuth of the line P_2P_1 at P_2. In a plane system, therefore:

$$\alpha_{2,1} = \alpha_{1,2} \pm 200\text{ gon} = \alpha_{1,2} \pm 180^\circ . \tag{5.1.1}$$

An azimuth may have any value between 0 and 400 gon (0° and 360°). The rules for trigonometric functions for angles larger than 100 gon (90°) are most easily remembered by using a unit circle.

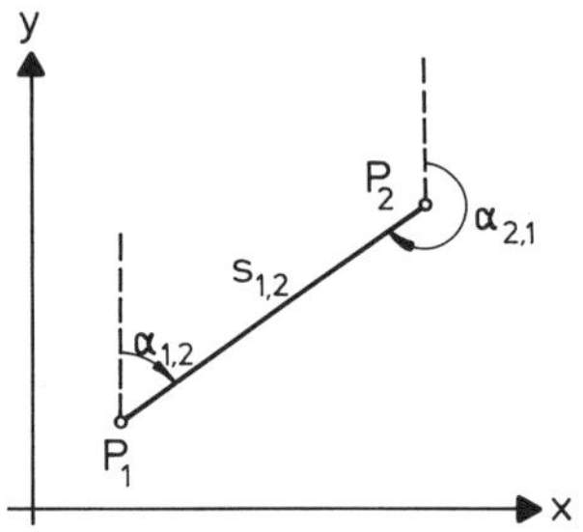

Fig. 5.1.2 Azimuths

a) The sine of an azimuth α is the projection onto the x-axis, while the cosine is the projection onto the y-axis. The algebraic signs of these functions are evident from the directions of the respective axis sections.

b) If φ represents the amount by which α exceeds full right angles, then the absolute amounts of $\sin\alpha$ and $\cos\alpha$ in the first and third quadrants are equal to the respective function of φ, while in the second and fourth quadrants they are equal to the cofunction. Therefore:

$\alpha =$	φ	$100\text{ gon} + \varphi$	$200\text{ gon} + \varphi$	$300\text{ gon} + \varphi$
$\sin\alpha =$	$+\sin\varphi$	$+\cos\varphi$	$-\sin\varphi$	$-\cos\varphi$
$\cos\alpha =$	$+\cos\varphi$	$-\sin\varphi$	$-\cos\varphi$	$+\sin\varphi$

c) The algebraic signs and functional values of tangent and cotangent are obtained from

$$\tan = \sin : \cos; \quad \cot = \cos : \sin .$$

Local coordinate systems are frequently selected for practical tasks (Fig. 5.1.3). In these systems the rectangular coordinates may be denoted as x', y' or ξ, η; while polar coordinates are often used as α', s' or ϑ, σ.

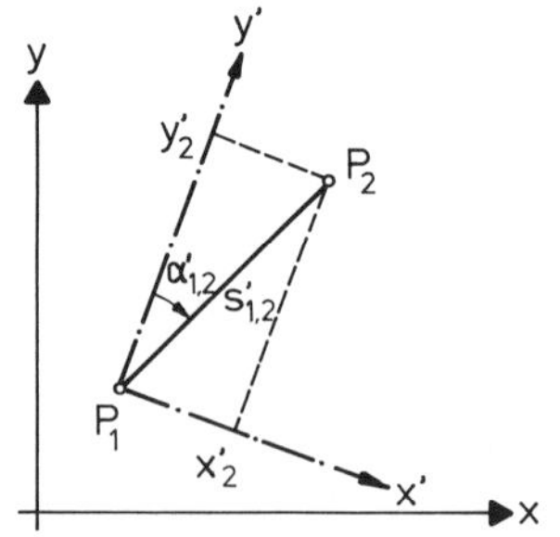

Fig. 5.1.3 Local coordinate system

5.1.1 Rectangular Coordinates Using Distance and Azimuth (First Geodetic Problem)

Given: $x_1, y_1, s, \alpha_{1,2}$; wanted: x_2, y_2.

With

$$\sin\alpha_{1,2} = \frac{x_2 - x_1}{s} = \frac{\Delta x}{s}$$

$$\cos\alpha_{1,2} = \frac{y_2 - y_1}{s} = \frac{\Delta y}{s}$$

one obtains

$$\begin{aligned} x_2 &= x_1 + s\sin\alpha_{1,2} \\ y_2 &= y_1 + s\cos\alpha_{1,2}. \end{aligned} \tag{5.1.2}$$

Numerical example:

Given: $x_1, y_1, s = 135.25\,\text{m}$; $\alpha_{1,2} = 29°40'05''$

$$\begin{aligned} \sin\alpha_{1,2} &= +\ \ 0.49497 \\ \Delta x = s\sin\alpha_{1,2} &= +\ \ 66.95\,\text{m} \\ x_1 &= \ \ 713.64\,\text{m} \\ \hline x_2 &= \ \ 780.59\,\text{m} \end{aligned}$$

$$\begin{aligned} \cos\alpha_{1,2} &= +\ \ 0.86891 \\ \Delta x = s\cos\alpha_{1,2} &= \ \ 117.52\,\text{m} \\ y_1 &= \ \ 496.72\,\text{m} \\ \hline y_2 &= \ \ 614.24\,\text{m} \end{aligned}$$

5.1.2 Distance and Azimuth From Rectangular Coordinates (Second Geodetic Problem)

Given: $x_1, y_1; x_2, y_2$; wanted $\alpha_{1,2}, s$.

According to Fig. 5.1.4 we get

$$s = \sqrt{(x_2 - x_1)^2 + (y_2 - y_1)^2} = \sqrt{\Delta x^2 + \Delta y^2} \tag{5.1.3}$$

$$\tan\alpha_{1,2} = \frac{x_2 - x_1}{y_2 - y_1} \tag{5.1.4}$$

$$\sin\alpha_{1,2} = \frac{x_2 - x_1}{s} = \frac{\Delta x}{s} \quad \text{or} \tag{5.1.5a}$$

$$\cos\alpha_{1,2} = \frac{y_2 - y_1}{s} = \frac{\Delta y}{s}. \tag{5.1.5b}$$

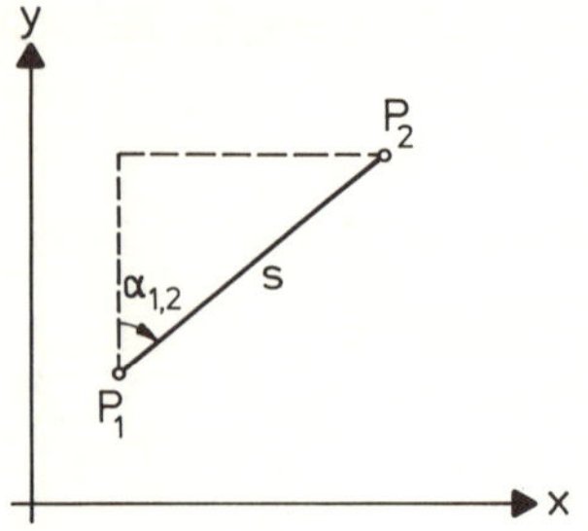

Fig. 5.1.4 Relations between polar and rectangular coordinates

It should be noted that the azimuth is in the third or fourth quadrant if $x < 0$.

Example:

$$
\begin{array}{ll}
x_2 = 795.17 & y_2 = 525.10 \\
x_1 = 528.15 & y_1 = 407.65 \\
\hline
\Delta x = 267.02 & \Delta y = 117.45
\end{array}
$$

$$s = \sqrt{267.02^2 + 117.45^2} = 291.71 \text{ m}$$

$$\tan\alpha_{1,2} = 2.27348 \qquad \alpha_{1,2} = 73.6194 \text{ gon}\,.$$

It should be noted that the more accurate value for $\alpha_{1,2}$ or s is obtained from equation (5.1.5a) for $\Delta x > \Delta y$, and equation (5.1.5b) for $\Delta y > \Delta x$.

Coordinate transfers (rectangular to polar and vice versa) can easily be done directly on small pocket calculators.

5.2 Coordinate Transformation

The problem arises frequently that points coordinated in one system have to be transformed into a neighbouring system. The following applications are common:

- Transformation of local coordinates into a net of higher order. Usually the local coordinates are the result of recent measurements (orthogonal or polar detail survey).
- Transformation of given coordinates into a local system. The local coordinates often serve as staking values.
- Transformation of coordinates of a higher order network into another higher order system. This case occurs frequently in overlap areas between regional coordinate systems.

Generally, the conditions for a *similarity transformation* are fulfilled: The two coordinate systems may have a different scale, however, the scale within a system is the same along both coordinate axes. If the latter condition is not fulfilled or if the coordinate axes intersect with different angles, then an *affine transformation* is to be used.

5.2.1 Similarity Transformation

The array of points (Fig. 5.2.1) that is processed through a similarity transformation undergoes the following coordinate changes:

1) shifts in both coordinate directions (x_0 and y_0)
2) rotation by an angle φ
3) multiplication of the scale of a factor q such that the scale becomes equal to the one of the other system.

To simplify the notation, the coordinates of the first system are denoted by ξ, η, while x, y is used for the second system.

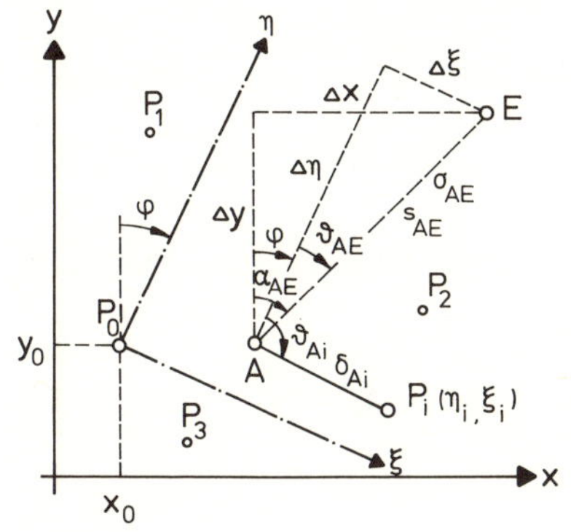

Fig. 5.2.1 Coordinate transformation

5.2.1.1 Unique Solution

The problem has a unique solution if two points are coordinated in both systems. Thus we have the following task: Given are two points A and E in the ξ, η and x, y systems and the points $P_1, P_2 \ldots P_i$ in the ξ, η system. Wanted are the x, y coordinates of points $P_1, P_2 \ldots P_i$. After checking the identity of points A and E, the transformation parameters are calculated. The scale factor becomes:

$$q = \frac{s_{AE}}{\sigma_{AE}} = \frac{\sqrt{\Delta x^2 + \Delta y^2}}{\sqrt{\Delta \xi^2 + \Delta \eta^2}} \tag{5.2.1}$$

where

$$\Delta \xi = \xi_E - \xi_A \qquad \Delta x = x_E - x_A$$
$$\Delta \eta = \eta_E - \eta_A \qquad \Delta y = y_E - y_A .$$

The rotational angle φ is obtained from the azimuth differences of the line AE in the two system:

$$\varphi = \alpha_{AE} - \vartheta_{AE} = \arcsin \frac{\Delta x}{s_{AE}} - \arcsin \frac{\Delta \xi}{\sigma_{AE}}$$

or (5.2.2)

$$\varphi = \arccos \frac{\Delta y}{s_{AE}} - \arccos \frac{\Delta \eta}{\sigma_{AE}} .$$

According to Fig. 5.2.1 we get

$$\left.\begin{aligned}\Delta x &= q \cdot \sigma_{AE} \cdot \sin(\vartheta_{AE} + \varphi) \\ \Delta y &= q \cdot \sigma_{AE} \cdot \cos(\vartheta_{AE} + \varphi)\end{aligned}\right\} . \tag{5.2.3}$$

Applying the addition theorem gives:

$$\Delta x = \sigma_{AE} \cos\vartheta_{AE} \cdot q \cdot \sin\varphi + \sigma_{AE} \sin\vartheta_{AE} \cdot q \cdot \cos\varphi$$
$$\Delta y = \sigma_{AE} \cos\vartheta_{AE} \cdot q \cdot \cos\varphi - \sigma_{AE} \sin\vartheta_{AE} \cdot q \cdot \sin\varphi$$

with

$$\left.\begin{aligned}\sigma_{AE} \cdot \sin\vartheta_{AE} &= \Delta\xi \qquad & q \cdot \sin\varphi &= o \\ \sigma_{AE} \cdot \cos\vartheta_{AE} &= \Delta\eta \qquad & q \cdot \cos\varphi &= a\end{aligned}\right\} \tag{5.2.4}$$

we get:

$$\left.\begin{aligned}\Delta x &= a \cdot \Delta\xi + o \cdot \Delta\eta \\ \Delta y &= a \cdot \Delta\eta - o \cdot \Delta\xi\end{aligned}\right\} . \tag{5.2.5}$$

Now the points $P_1, P_2 \ldots P_n$ can be transformed from the ξ, η system to the x, y system, starting with point A as the rotational centre:

a) If polar coordinates (σ_{Ai}, ϑ_{Ai}) are given:

$$\left.\begin{aligned}x_i &= x_A + q \cdot \sigma_{Ai} \sin(\vartheta_{Ai} + \varphi) \\ y_i &= y_A + q \cdot \sigma_{Ai} \cos(\vartheta_{Ai} + \varphi)\end{aligned}\right\} . \tag{5.2.6}$$

b) If rectangular coordinates (ξ_i, η_i) are given:

$$\left.\begin{aligned}x_i &= x_A + a(\xi_i - \xi_A) + o(\eta_i - \eta_A) \\ y_i &= y_A + a(\eta_i - \eta_A) - o(\xi_i - \xi_A)\end{aligned}\right\} . \tag{5.2.7}$$

If the rotational centre is placed at the origin P_0 of the local coordinate system, then we get (Fig. 5.2.2);

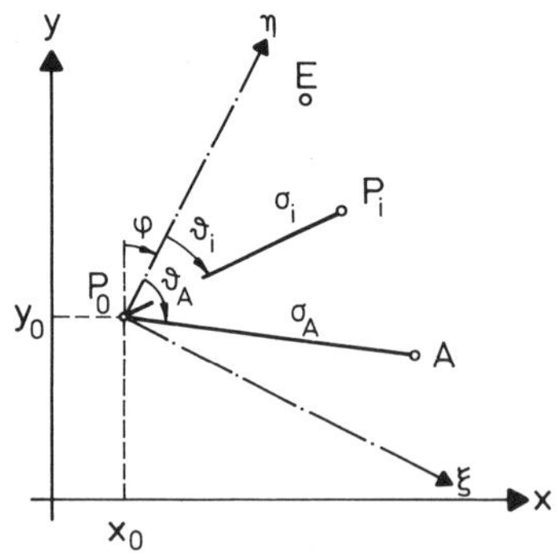

Fig. 5.2.2 Coordinate transformation using origin of ξ, η system as centre of rotation

$$\left.\begin{aligned} x_i &= x_0 + q \cdot \sigma_i \sin(\vartheta_i + \varphi) \\ y_i &= y_0 + q \cdot \sigma_i \cos(\vartheta_i + \varphi) \end{aligned}\right\} \tag{5.2.8a}$$

while P_0 has the coordinates

$$\left.\begin{aligned} x_0 &= x_A - q \cdot \sigma_A \cdot \sin(\vartheta_A + \varphi) \\ y_0 &= y_A - q \cdot \sigma_A \cdot \cos(\vartheta_A + \varphi) \end{aligned}\right\} . \tag{5.2.8b}$$

This replaces equation (5.2.6).

Equation (5.2.7) then becomes

$$\left.\begin{aligned} x_i &= x_0 + a \cdot \xi_i + o\eta_i \\ y_i &= y_0 + a \cdot \eta_i - o\xi_i \end{aligned}\right\} \tag{5.2.9a}$$

with coordinates of P_0 being

$$\left.\begin{aligned} x_0 &= x_A - a\xi_A - o\eta_A \\ y_0 &= y_A + \sigma\xi_A - a\eta_A \end{aligned}\right\} . \tag{5.2.9b}$$

Example (Transformation between two Gauss-Krueger systems (section 7.2.5)):

Given are the coordinates of A and E in both the x, y and ξ, η systems. The points P_1 and P_2, coordinated in the ξ, η system are to be transformed into the x, y system. After selecting a rotational centre in the ξ, η system (e.g. point A), q and φ or a and o are computed according to equations (5.2.1) and (5.2.2). Then equation (5.2.7) is used for the transformation.

Pt.	ξ, η System		x, y System		
No.	ξ_i	η_i	x_i	y_i	
A	97 319.35	08 802.06	96 935.27	08 922.55	A
E	98 858.81	09 717.54	98 511.77	09 772.69	E
$\Delta\eta, \Delta\xi$	+1 539.46	+915.48	+1 576.50	+850.14	$\Delta y, \Delta x$
θ, σ	65.8456	1 791.100	68.5154	1 791.114	t, s
o, a	+0.041 925	+0.999 129	2.6698	1.000 007 8	φ, q
P_1	98 338.99	08 586.69	97 944.99	08 664.62	P_1
P_2	97 918.31	09 538.01	97 564.56	09 632.75	P_2

5.2.1.2 Coordinate Transformation With Several Identical Points

(Helmert transformation)

A Helmert transformation is applied when the number r of identical points coordinated in both systems exceeds 2. Ideally, the identical points should be located at the edges of the areas to be transformed and should enclose these if possible. The task is then: Given are points $P_1, P_2 \ldots P_r$ in both the ξ, η and x, y systems plus the points $P_{A+1} \ldots P_n$ in the ξ, η system only. Wanted are the x, y coordinates of points $P_{r+1} \ldots P_n$.

For numerical reasons it is advisable to refer the coordinates to the centre of gravity of the identical points and then proceed with the computation of the transformation parameters.

The coordinates of this centre of gravity are:

$$x_s = \frac{\Sigma x}{n}, \quad y_s = \frac{\Sigma y}{n}, \quad \xi_s = \frac{\Sigma \xi}{n}, \quad \eta_s = \frac{\Sigma \eta}{n}.$$

According to *Wolf* (1975) the transformation parameters become:

$$\left.\begin{aligned} a &= q \cdot \cos\varphi = (\Sigma \Delta\eta \Delta y + \Sigma \Delta\xi \Delta x)/\Sigma(\Delta\xi^2 + \Delta\eta^2) \\ o &= q \cdot \sin\varphi = (\Sigma \Delta\eta \Delta x - \Sigma \Delta\xi \Delta y)/\Sigma(\Delta\xi^2 + \Delta\eta^2) \end{aligned}\right\} \tag{5.2.10}$$

with

$$\begin{aligned} \Delta x_i &= x_i - x_s & \Delta y_i &= y_i - y_s \\ \Delta \xi_i &= \xi_i - \xi_s & \Delta \eta_i &= \eta_i - \eta_s . \end{aligned}$$

The check values are

$$\Sigma \Delta x = \Sigma \Delta y = \Sigma \Delta \xi = \Sigma \Delta \eta = o .$$

Rotational angle φ and scale factor q are derived from equation (5.2.10) as:

$$q = \sqrt{a^2 + o^2} \tag{5.2.11}$$

$$\varphi = \arctan\frac{o}{a} = \arccos\frac{a}{q} = \arcsin\frac{o}{q} . \tag{5.2.12}$$

The points $P_{r+1} \ldots P_n$ are transformed using

$$\begin{aligned} x_i &= x_0 + a\xi_i + o\eta_i \\ y_i &= y_0 + a\eta_i - o\xi_i \end{aligned} \qquad (i = r+1, \ldots, n) . \tag{5.2.13}$$

The rotational centre P_0 (origin of the η, ξ system) has the coordinates:

$$\left.\begin{aligned} x_0 &= x_s - o\eta_s - a\xi_s \\ y_0 &= y_s - a\eta_s + o\xi_s \end{aligned}\right\} . \tag{5.2.14}$$

Tensions within the networks can be judged from linear residuals at the identical points.

$$\left.\begin{aligned} dx_i &= (x_s + a\Delta\xi_i + o\Delta\eta_i) - x_i \\ dy_i &= (y_s - o\Delta\xi_i + a\Delta\eta_i) - y_i \end{aligned}\right\} . \tag{5.2.15}$$

As a check the following holds:

$$\Sigma dx_i = \Sigma dy_i = o$$
$$\Sigma dx_i^2 + \Sigma dy_i^2 = \Sigma(\Delta\xi^2 + \Delta\eta^2)(a^2 + o^2) + \Sigma(\Delta x^2 + \Delta y^2)\,.$$

5.2.2 Five-parameter Transformation

The five-parameter transformation (affine transformation) permits different scales in the x and y directions. Thus, it is fundamentally different from the similarity transformation, where angular relationships are preserved.

Such a transformation is used, for instance, when the Soldner coordinates (section 7.2.2) are to be transformed to the Gauss-Krueger coordinates (section 7.2.3).

It is also successfully applied in photogrammetry and occasionally for numerical boundary delimitations.

5.3 Rectangular Coordinate Systems

Rectangular coordinate systems can be based on different design principles. These coordinates are not only used to define a fixed location of a point on the reference ellipsoid but also to provide the base for mapping onto a plane at 1 : 1 while retaining as much similarity to the original ellipsoidal image as possible. Mathematical congruence cannot be obtained because the ellipsoid cannot be mapped onto a plane without distortions. Thus, it is necessary to find means to approximate the desired conditions. Two rectangular systems proved to be rather suitable: the system of the Bavarian astronomer *Soldner*, and the system of the mathematician *C. F. Gauss*. The latter is nearly exclusively used at present all over the world (*United Nations*, 1976).

5.3.1 Soldner Coordinates

Soldner's system* is defined as follows: A principal meridian located in the centre of the surveying area is the abscissa axis (x-axis), while a selected point on this meridian serves as the origin. The ordinate of a point P is the elliptic normal from P to the x-axis, while the abscissa equals to the meridian arc from the origin to the foot of this normal and increases towards north, while y increases towards east.

As is obvious from Fig. 5.3.1, the ordinates converge with increasing distance from

* In some countries (e.g., in the U.K.) Soldner systems are also called Cassini projections.

the principal meridian. However, this convergence is so small in the vicinity of the principal meridian that two ordinate lines 1 km apart at the principal meridian are 5 cm closer at 64 km distance.

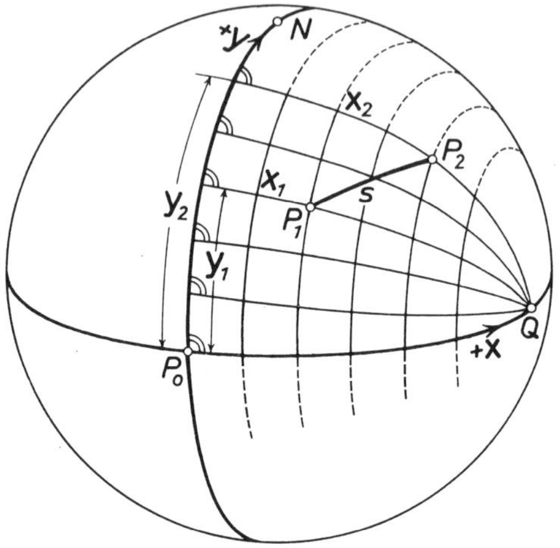

Fig. 5.3.1 *Soldner's* coordinate system on a sphere

This leads to the following mapping method: The ellipsoidal principal meridian is mapped onto the abscissa axis of a plane Cartesian system while the ellipsoidal ordinates are plotted as plane ordinates with their ellipsoidal length. When leaving the principal meridian, the ellipsoidal abscissa differences are elongated. To keep this distortion below ≈ 5 cm per km, the extension of such Soldner systems is limited to 64 km on both sides of the principal meridian. Surveying systems with larger extension may utilize several Soldner systems.

Most cadastral surveys in central Europe, carried out in the nineteenth century, were based on such Soldner systems. This has the advantage of simplicity in the computations. Due to the elongation near the edges of these systems, the directions of triangle sides of 1 km may be rotated by up to 5″ (1.5 mgon) which is a major disadvantage. Since this amount exceeds the angular observation accuracy, it has to be considered for first- to third-order triangulation. Furthermore, the narrow systems require numerous coordinate transformations in overlapping zones.

5.3.2 Gauss's Coordinates

To avoid the angular distortions, *C.F. Gauss* developed a method of conformal mapping, named after him, and utilized it in the geodetic survey of Hannover (1822–1847). Such a conformal map keeps a differential similarity between the original elliptic situation and the map image.

Since the derivation of mapping equations is well beyond the scope of this book, the difference between Gauss's and Soldner's system is illustrated with differential values in Fig. 5.3.2.

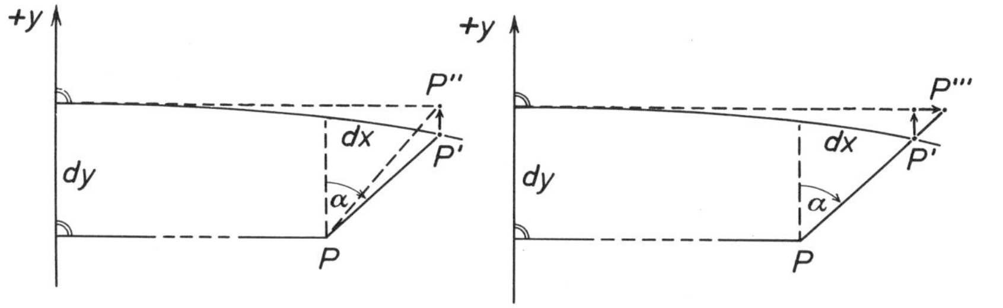

Fig. 5.3.2 Distortions of the ellipsoidal differential triangle for the *Soldner* and *Gauss* projection systems

Both systems utilize the principal meridian, which is the ellipsoidal abscissa axis, as the abscissa axis in the Cartesian system in true ellipsoidal length. Thus, utilizing the indices E, S, G, for ellipsoid, Soldner, and Gauss:

$$x_E = x_S = x_G \,. \tag{5.3.1}$$

This means that the abscissa values are the same for both mapping systems. Furthermore, Gauss also neglects the convergency of the ordinates $x =$ const. when imaging them in the plane system. The ordinates, however, obtain different values for Soldner and Gauss in the ellipsoidal differential triangle (Fig. 5.3.2) with azimuth α.

The left part of Fig. 5.3.2 shows the situation for the *Soldner method* where the point P' is mapped at P''. This means that the ellipsoidal ordinate difference dy, corresponding to the arc PP', is retained in its length. The abscissa difference, however, is elongated because the convergency of the ellipsoidal ordinates has been neglected, which leads to a distortion of the azimuth α. The scale factor for the abscissa is $(1 + y^2/2R^2)$ where y equals the ellipsoidal ordinate, and r the radius of curvature of a best fitting sphere to the reference ellipsoid at the location of the respective point.

The right part of Fig. 5.3.2 shows the *Gauss method* where the ordinate difference is elongated by the same factor as the abscissa difference, thus eliminating the angular distortion. Therefore point P' is mapped at P'', and the azimuth α of PP' remains unchanged. Since the rectangle remains unchanged, two angles in the differential triangle remain, which means that Gauss has achieved a similar image.

The elongation of the ordinate difference is repeated in subsequent differential triangles along the ordinate. Therefore, it is necessary to integrate $(1 + y^2/2R^2)\, dy$ in order to obtain the Gauss ordinate y_G from the ellipsoidal ordinate or the Soldner ordinate y_s which has the same length.

Neglecting terms of the magnitude $1/R^4$, the mapping equation for the Gauss ordinates becomes

$$y_G = y_s + \frac{y_s^3}{6R^2}\,. \tag{5.3.2}$$

The imaging equations (5.3.1) and (5.3.2) describe a conformal mapping system where similarity is retained. This was achieved by introducing other distortions which are mentioned in the following sections.

5.3.3 Reduction of Measured Quantities for the Gauss System

The ratio between an arc section on the map and its ellipsoidal original length is generally known as the magnification ratio. For Gauss's method, it is independent of the azimuth of the line, because both abscissa and ordinate difference are distorted by the same factor, namely by $(1 + y^2/2R^2)$.

Theoretically, such an arc section is differentially small. For practical geodetic work, however, a reasonably short line, such as a traverse leg or even a short traverse, can be taken in this context. If such a distance S is measured on the ellipsoid, its plane image s in Gauss's conformal projection is

$$s = \left(1 + \frac{y_m^2}{2R_m^2}\right) S \quad \text{or} \quad s - S = \frac{y_m^2}{2R_m^2}\, S \tag{5.3.3}$$

where y_m represents a mean ordinate value. The ellipsoidal distance S can be replaced by the plane length s on the right-hand side of the second equation without loss in accuracy. R_m is the radius of the best fitting sphere at the foot of the normal of x_m. The following table shows the numerical influence of equations (5.3.2) and (5.3.3).

Soldner Ordinate [km]	Ordinate Correction in [m]	$(s-1)$ per/km [mm]
10	0.00	1
20	0.03	5
30	0.11	11
40	0.26	20
50	0.51	31
60	0.88	44
70	1.40	60
80	2.10	79
90	2.98	99
100	4.09	123
150	13.81	276
200	32.75	491

This table provides round Soldner ordinate values and ordinate corrections by which these are transformed to Gauss ordinates. The third column provides elongation values for a kilometre distance with a mean ordinate as shown in the first column.

It is evident, that a distance expands more in the Gauss plane if it is farther away from the absicssa axis. Thus in a figure with a large east-west extension the farther parts are more enlarged than the closer ones, which means that for large areas the similarity is not retained. This means that for large areas not only the distances are distorted, but also the directions and thus the angles. Therefore observed distances and directions have to be reduced to the Gauss projection prior to introducing them into the computations.

The *directional distortion* is so small for the Gauss projection that the direction of a 1 km long line has a distortion of barely 0.2 mgon ($\approx 0.5''$) at a distance of 200 km from the principal meridian. This amount can almost always be neglected. The directional distortion increases for longer lines and then must be corrected for. For this, the reader is referred to geodetic literature.

The *distance distortion*, however, increases according to equation (5.3.3) with the square of the distance to the principal meridian and therefore cannot be neglected, even for certain cases in simple everyday surveying practice. The amount by which the measured distances have to be reduced prior to their input into computations within the Gauss projection is referred to as distance reduction. For 1 km long distances, it is practically identical with the expansion amount listed in the third column of the previous table. This, however, is only valid for distances at sea level. For higher elevations, an additional reduction is necessary, namely, the *height reduction*, because all terrain surface measurements are projected onto the surface of the reference ellipsoid. For horizontal surveys (not elevations!), this reference surface can be approximated by the mean sea level. Thus results of distance measurement and areas, determined at the physical earth surfaces at an elevation H_r above mean sea level, have to be reduced.

The projection S of a distance S^{Hr} measured at an elevation H_r is reduced according to Fig. 5.3.3 as

$$S = S^{Hr} \frac{R}{R + H_r} \approx S^{Hr}\left(1 - \frac{H_r}{R}\right) + \ldots$$

or

$$S = S^{Hr} - \frac{H_r}{R} S^{Hr} . \tag{5.3.4}$$

This means that the distance is shortened, which counteracts somewhat the expan-

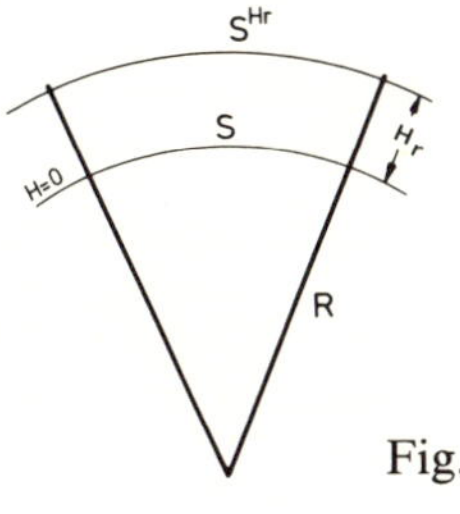

Fig. 5.3.3 Height reduction

sion due to the projection. It is therefore advisable to combine distance reduction and height reduction to obtain one reduction value in millimetres for a distance S_i in kilometres

$$\Delta_i = S_i\left(-\frac{H_r}{R_m} + \frac{y_m^2}{2R_m^2}\right)10^{-6}, \tag{5.3.5}$$

where H_r represents the average mean sea level elevation, and y_m the mean ordinate of the distance (i.e., the distance of the centre point from the principal meridian or its image). Table 5.3.1, which is based on values from *Jordan*, *Eggert*, and *Kneissl* et al. (1956), provides some numerical values.

Table 5.3.1 Reduction to the Gauss projection of a 1 km long distance measured at an average mean sea level elevation H_m at a distance y_m from the principal meridian.

Reduction Δ in [mm] for 1 km

H_r [m]	y_m [km]										
	0	10	20	30	40	50	60	70	80	90	100
0	0	1	5	11	20	31	44	60	79	100	123
50	− 8	− 7	− 3	3	12	23	36	52	71	92	115
100	− 16	− 14	− 11	− 5	4	15	28	44	63	83	107
200	31	− 30	− 26	− 20	− 12	− 1	29	29	47	68	92
300	− 47	− 46	− 42	− 36	− 27	− 16	− 3	− 13	32	52	76
400	− 63	− 61	− 58	− 52	− 43	− 32	− 18	− 2	16	37	60
500	− 78	− 77	− 73	− 67	− 59	− 48	− 34	−18	0	21	44
600	− 94	− 93	− 89	− 83	− 74	− 63	− 50	−34	−15	5	29
700	−110	−108	−105	− 99	− 90	− 79	− 65	−50	−31	− 10	13
800	−125	−124	−120	−114	−106	− 95	− 81	−66	−47	− 26	− 3
900	−141	−140	−136	−130	−121	−110	− 97	−81	−62	− 42	− 18
1000	−157	−156	−152	−146	−137	−126	−112	−97	−78	− 57	− 34

Notes:

1) For larger distances, the following corrections apply at sea level ($H = 0$)

y_m = 150 km	200 km	300 km
Δ = 276 mm	491 mm	1105 mm

2) A 10 km long distance in east-west direction at y_m = 100 km can be reduced to within 1 cm using Table 5.3.1. The following formula provides the exact reduction for a distance $P_1 P_2$ = S:

$$s - S = \frac{S}{12\,R_m^2}(y_1^2 + 4\,y_m^2 + y_2^2)\,. \tag{5.3.6}$$

Height and distance reductions are specifically required for the following cases:

a) Coordinate computation using distance measurements (trilateration), especially for arc sections and for combined methods. Each measured distance has to be reduced prior to entering it into the computations.

b) It is also advisable to reduce traverse legs, even though the closure error is distributed and thus automatically includes both measuring errors and projection distortions. However, both may be of the same magnitude, thus making it difficult to judge whether a certain traverse fulfils the given specifications or not.

c) When computing the distance between two coordinated points, the reduction has to be applied with the opposite sign. This is especially critical for precision staking and other engineering applications at larger distances to the principal meridian.

d) The same holds for area computation using Gaussian coordinates. With F being the field area, and f the projected area, the area reduction becomes

$$f - F = \frac{y_m^2}{R^2}\,f \tag{5.3.7}$$

and the field area

$$F = f - (f - F) \tag{5.3.8}$$

which is sufficiently accurate for legal surveys.

Since Gauss has not left a comprehensive treatment of his projection, Prof. *Krueger*, then director of the German Geodetic Institute in Potsdam, put it together at the beginning of this century. Therefore, this projection is often referred to as the *Gauss-Krueger projection*, especially in German speaking areas. A more generally used name is the *transverse Mercator projection.*

5.3.4 Gauss-Krueger Meridian Strip Systems (3° Transverse Mercator)

In order to avoid excessive distortions, *Gauss-Krueger systems* have been established with a principal meridian at every 3° of longitude, i.e., for Germany at 6°, 9°, 12° and 15° east of Greenwich. Each system extends on both sides to 1°40′ longitude (roughly 100 km) which provides an overlap of 20′ in longitude, i.e., an approximate 23 km wide strip in which all geodetic points carry two sets of coordinates (Fig. 5.3.4).

The system is arranged in Germany as follows: To avoid negative values, a value of 500 000 m is added to all Gauss ordinates, and an index digit, representing the longitude divided by 3, preceeds the coordinate. The ordinates are referred to as right value, while the abscissa is counted from the equator and referred to as high value.

Thus St. Michael's Church in Luneburg has the following *G*, *K* coordinates:

In the system of the 9° meridian:

$$H = 5\,902\,863.21 \text{ m}; \quad R = 3\,593\,571.20 \text{ m}$$

and in the system of the 12° meridian:

$$\text{H} = 5\,903\,137.40 \text{ m}; \qquad \text{R} = 4\,393\,360.64 \text{ m}\,.$$

Thus, in the first system, the foot of the ordinate is 5902863.21 m from the equator. The church is east of the 9° meridian with the ordinate being 93571.20 m.

In the system of the 12° meridian, the foot of the ordinate is 5903137.40 m north of the equator, while the church is west of the meridian with a Gauss ordinate of −106639.36 m. If there is no ambiguity, the first two digits are often omitted for practical calculations.

By limiting the system of meridian strips to roughly 100 km on both sides, the distance distortion (equation (5.3.3)) is so small (maximum 12 cm per 1 km) that it can be neglected for many practical cases. For precision surveys, however, it always has to be applied, and it should be kept in mind that distances computed from Gauss-Krueger coordinates are always longer than on the ellipsoid.

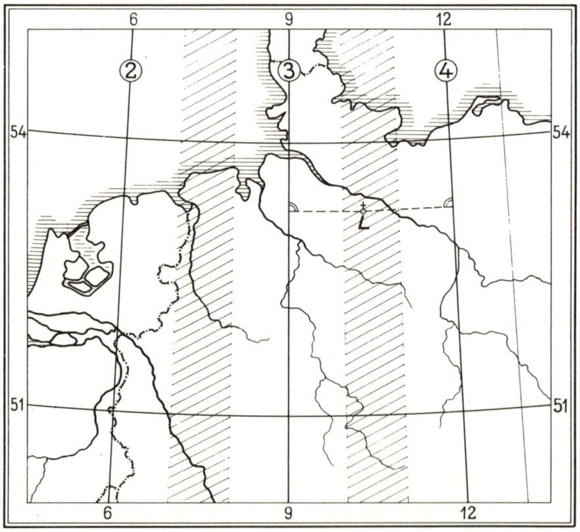

Fig. 5.3.4 *Gauss-Krueger* meridian strips in central Europe

5.3.5 The Universal Transverse Mercator System (UTM System)

The universal transverse *Mercator* grid system (UTM) in the U.S.A. is based on the International Earth Ellipsoid of 1924 as the reference surface and covers the whole earth with 60 strips, covering 6° of longitude each. Starting from 180° longitude, covering 180° to 174° N, the strips are numbered increasingly towards east with 1, 2, ..., 60. In order to counteract the increase of the enlargement ratio at the edges of the strips, the distances obtained from the net adjustment are reduced by a scale factor of 0.9996 before computing the coordinates. In latitude, the UTM system is limited by the 80° N or 80° S parallels. All NATO maps are based on this system.

The U.S.S.R. has a similar system of Gauss coordinates spanning the earth, which uses the *Krassowski* ellipsoid as the reference surface.

6. Determination of Plane Horizontal Coordinates

If the location of a number of points is defined in a rectangular coordinate system, then additional points can be coordinated using these points. The given points are often called *fixed* or *control points* to distinguish them from *new points*.

There are basically two type of measurements used for determining new points:

1) *geodetic measurements*, utilizing directions, angles, distances and distance differences. These may take place between points on the earth's surface or between points on the earth's surface and extra terrestrial points (satellites).

2) *photogrammetric measurements* utilizing distance or coordinate measurements on photographic or related imagery (remote sensing). This type of approach, while closely related to the first one, is covered in photogrammetry and thus beyond the scope of this book. Therefore, only geodetic measurements are treated in the following sections.

6.1 Types of Point Determinations

6.1.1 Numerical Point Determinations

Numerical approaches can be divided into three categories:

- *single point determination*, where each point is individually coordinated;
- *sequential point determination*, where a number of points are coordinated in one common linear approach;
- *networks*, where a number of points are coordinated in an area approach utilizing nets consisting of triangles, quadrilaterals, and similar geometric figures.

If the number of measurements is restricted such that it is sufficient to fulfill the geometric conditions to coordinate a point, a *unique solution* is obtained. If more measurements than necessary are taken, an *overdetermined solution* is achieved. Due to inevitable measuring errors, the superfluous measurements cause discrepancies which need to be treated by an "adjustment" process.

The following sections concentrate mainly on simple point determinations. Since smaller problems with overdetermination can today be quite comfortably solved on pocket calculators, examples are provided for the solution. The theory can be found in the adjustment literature (*Koch* 1980; *Vaníček* and *Krakiwsky* 1986).

6.1.2 Types of Technical Approaches

Different techniques are used for the determination of networks, namely triangulation (from Latin: angulus = angle), trilateration (from Latin: latus = side), combined methods and satellite techniques.

For *triangulation*, which has been used extensively since the seventeenth century, all angles in a triangle network are covered if possible. Furthermore, at least one side has to be measured in order to fix the scale of the network (Fig. 6.1.1). This side used to be determined by means of a base enlargement net, but today it is measured directly by electronic distance measurement. Triangulation is rather expensive and time consuming, especially for large point spacing and is thus only used under exceptional circumstances for new surveys.

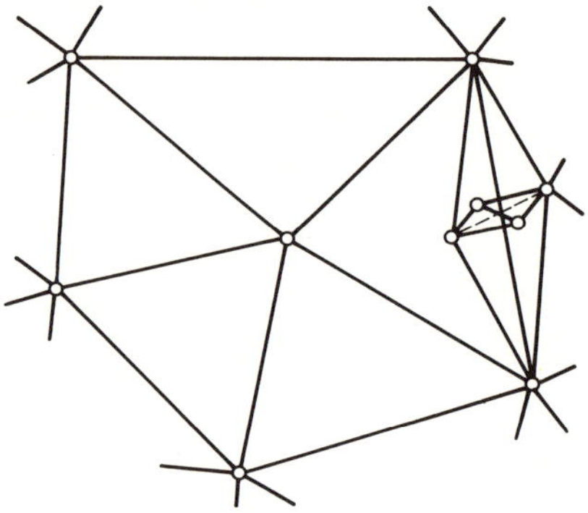

Fig. 6.1.1 Triangulation network with base enlargement net

Trilateration became economical after 1956 with the advent of electronic distance meters. In addition to the sides of the triangles, several diagonals are measured in order to obtain additional observations (Fig. 6.1.2).

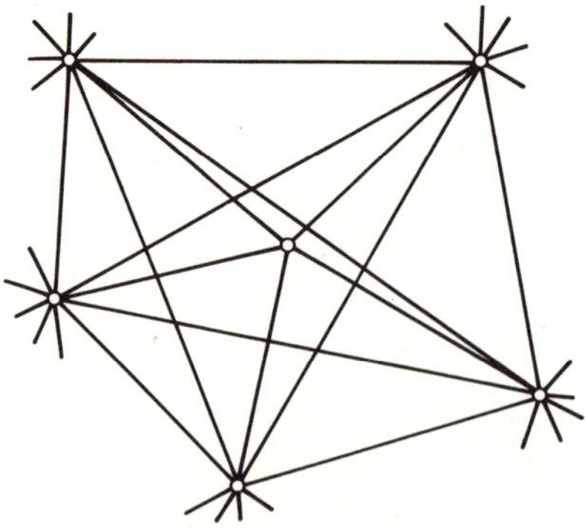

Fig. 6.1.2 Trilateration network

The *combined methods* utilize both angular and distance observations. Triangles, quadrilaterals with diagonals, and traverse nets form the basic geometric configuration.

Satellite techniques using the Transit navigation satellite system or the Global Positioning System (GPS) provide advantages over triangulation and trilateration since intervisibility between stations is not required. Control surveys done with terrestrial distance and direction measurements often require the erection of towers and favourable observing weather. Both requirements extend the time and expense of surveys. As well, without the constraint of intervisibility control points can be selected to better optimize network geometry.

With *inertial surveying systems* (ISS) and *very long baseline interferometry (VLBI)*, control points can also be determined without requiring intervisibility. These systems, however, use very costly equipment.

When establishing a surveying network, first a *base net* is established using homogeneous measurements and computations. In such a base net, normally point separations of 10 to 20 km are used. The net may cover the whole area (see Fig. 7.1.1), or it may consist of extended chains knotted together (see Fig. 7.1.4). *Additional fixed points* are coordinated within this base net by *individual, sequential or network approaches.*

The best known methods of point determination are (Figs. 6.1.3 and 6.1.4):

a) *Intersection*, by measuring the angles w_1 and w_2.
b) *Resection*, by measuring the angles w_1 and w_2.
c) *Arc section*, by measuring the distances D_1 and D_2.

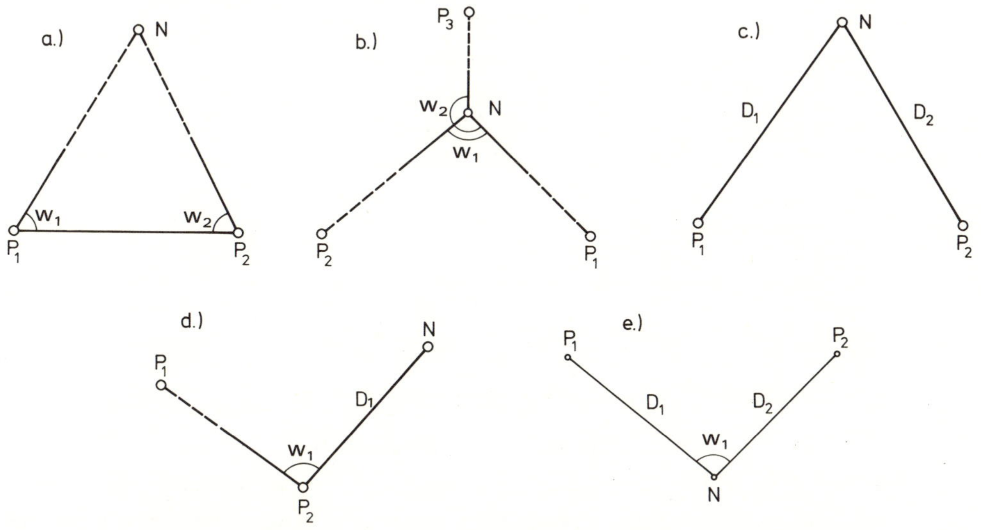

Fig. 6.1.3 Schematic illustration of the different methods used for determination of single points or groups of points (P_i = fixed points; N_i new points)

d) *Polar method*, by measuring angle w_1 and distance D_1.
e) *Combined angle and distance method*, by measuring angle w_1 and two distances D_1 and D_2.
f) *Traversing*, by measuring angles w_i and distances D_i.
g) *Traverse nets* (see Fig. 7.1.4).

In the near future there will be a competition between terrestrial and space techniques. Fig. 6.1.5 gives an impression of this competition, showing estimated one sigma uncertainties in baselines that can be achieved using a variety of positioning techniques.

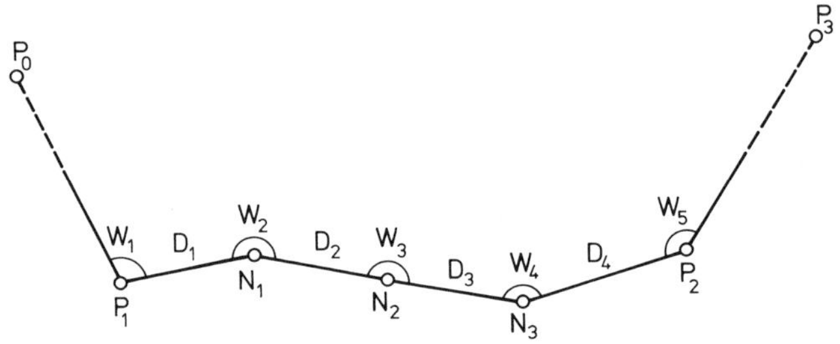

Fig. 6.1.4 Schematic illustration of a traverse

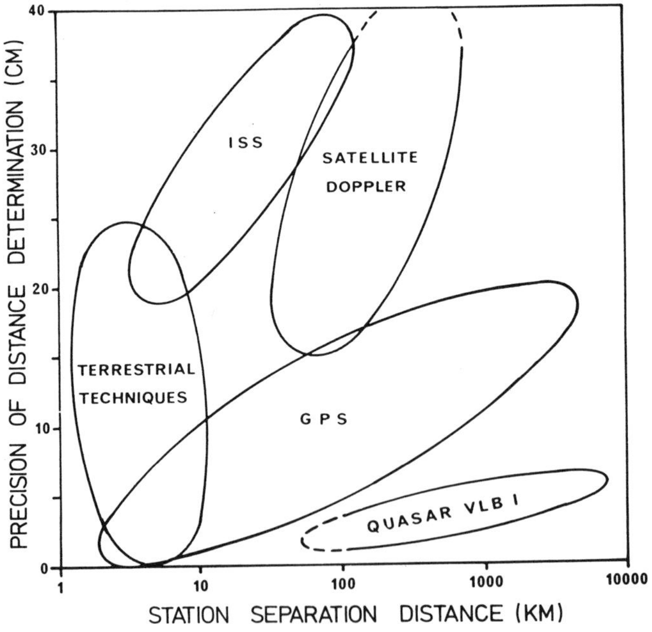

Fig. 6.1.5 Precision of distance determinations achieved or achievable using different techniques. Ellipses have been used to delineate the ranges of feasible station separations and ranges of precision. The delineations of separation and precision are at best fuzzy and should be interpreted accordingly (*Langley* et al. 1982)

6.2 Uncertainties in Determination and Definition of Horizontal Points

According to Helmert, the uncertainty of a planimetric point is expressed by its standard deviation.

$$\sigma_p = \sqrt{\sigma_a^2 + \sigma_b^2} \tag{6.2.1}$$

where σ_a or σ_b are the standard deviations of the point in two mutually orthogonal directions. If these directions coincide with the x- and y-coordinate axes of the point, then σ_x and σ_y are substituted for σ_a and σ_b.

A planimetric point can only be defined within a certain limited accuracy, because of

- small, unavoidable movements of the monumentation
- different interpretations of the centre of the monument
- centring errors.

As a rule, the local uncertainty in the point definition can be assumed as 10 mm in both coordinate directions. Thus the local definition of the point has an uncertainty of $\pm\lambda$ 15 mm.

The neighbouring accuracy of planimetric points within a suitable network configuration is basically the same as determined by the measuring method. Generally, the measuring methods are selected such that the uncertainty of coordination is in a well balanced ratio to the definition uncertainty.

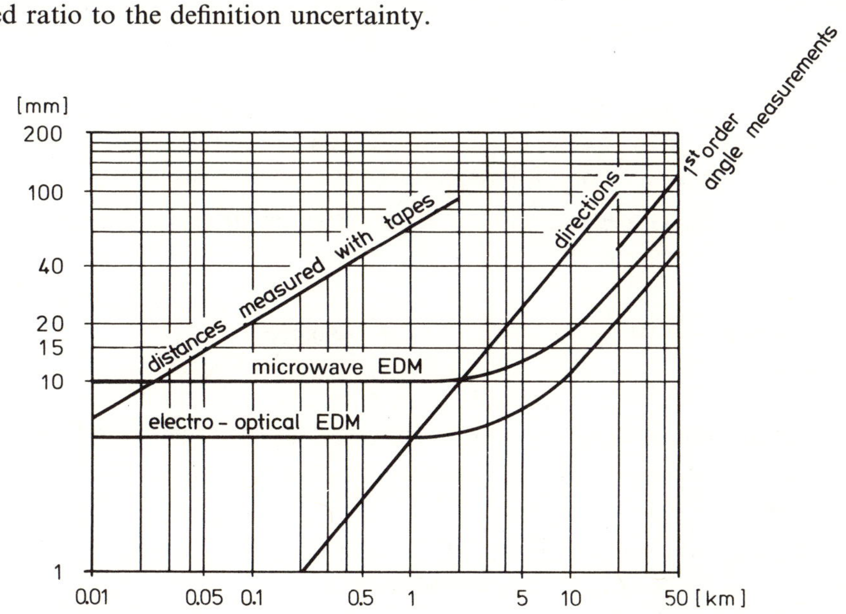

Fig. 6.2.1 Measuring accuracy and relative point accuracy as functions of the point separation

Planimetric points are determined by distance, direction and angle measuring methods. The uncertainty of the measuring method is a function of the distance with the exception of short range electronic distance measurements. Thus, depending on the method used, the relative point accuracy is also dependent or independent of the point separation. Fig. 6.2.1 shows this for terrestrial techniques.

When measuring distances with tapes, the measuring uncertainty already reaches the definition uncertainty at about a 50 m point separation. Methods where direction measurements are combined with electronic distance measurements are most suited for ranges up to 3 km. The uncertainties of electronic distance measurements alone exceed the definition uncertainty when the point separation is more than 15 km.

6.3 Preparatory Computations

Before starting the actual coordinate computations, one has to check the measurements because corrections and reductions may have to be applied. For distances these are: zero-correction, scale correction and geometric reductions.

If the angular measurements were performed in one telescope position only, then collimation and horizontal axis errors may affect the results. If reguired, their effect has to be mathematically compensated. Furthermore, direction corrections may have to be applied for large nets for specific projection systems.

These, however, are insignificant for the problems dealt with in the subsequent sections.

Often it is not possible to set up the theodolite or distance meter at the centre of the observation station or to erect targets centred above the target station. In such cases, preparatory centring computations are needed (section 6.3.1).

For certain tasks, it is necessary to orient the measured directions in a specific manner (section 6.3.2).

6.3.1 Centring of Observed Directions and Distances

The following cases may occur:

a) The centre of the station is accessible. However, the target can be sighted only from an eccentric station from which the centring elements (distance and direction to the centre) can readily be measured.

b) The centre of the station is inaccessible (i.e. steeple top, flag pole, etc.); the centring elements have to be indirectly determined.

c) The signal at the target station is set up eccentrically.

d) Eccentricities are avoided by using broken rays.

6.3.1.1 Station Eccentricity

Z denotes the centre, while B represents an eccentric observation station in Fig. 6.3.1. The directions $\alpha_1, \alpha_2, \ldots, \alpha_n$ and the direction α_z to the centre are observed at B. Furthermore, the distance $BZ = e$ is measured. The reduced directions $r_1\, r_2 \ldots r_n$ are wanted.

To solve this problem, the bundle of rays $\alpha_1\,\alpha_2 \ldots \alpha_z \ldots \alpha_n$ observed at B is assumed to be shifted parallel into Z such that the ray to P_1 is located as illustrated with a dotted line in Fig. 6.3.1. Since the angle δ_1 generated in this manner also occurs at P_1, r_1 can be computed as follows:

$$\varepsilon_1 = \alpha_1 - \alpha_2; \tag{6.3.1}$$

$$\sin\delta_1 = \frac{e}{S_1}\sin\varepsilon_1\,; \tag{6.3.2}$$

$$r_1 = \alpha_1 + \delta_1\,. \tag{6.3.3}$$

The directions $\alpha_2 \ldots \alpha_n$ are similarly centred. δ_i becomes negative for $\varepsilon_i > 200$ gon (180°).

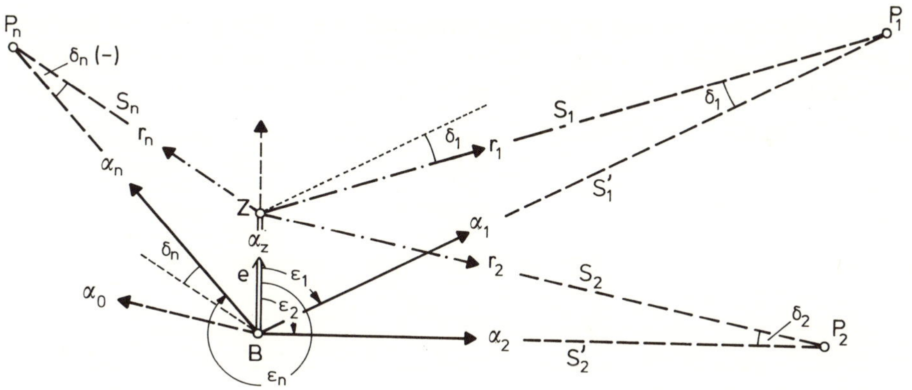

Fig. 6.3.1 Station eccentricity

If the distances S'_i were measured to point B and then reduced to horizontal distances, the centred distances S_i are computed as

$$S_i = \sqrt{(S'_i)^2 + e^2 - 2\,S'_i\,e\cos\varepsilon_i}\,. \tag{6.3.4}$$

In order to estimate the accuracy with which e and S have to be measured, equation (6.3.2) is differentiated, which leads to

$$\frac{d\delta}{\delta} = \frac{de}{e} + \frac{dS}{S}\,. \tag{6.3.5}$$

This means that e and S should be determined with the same relative accuracy. Thus for shorter distances, e is to be determined in millimetres, and S in decimetres, while for longer distances S in metres is sufficient.

If no distances were measured, then approximate values, computed from coordinates, can be used for equation (6.3.2).

Example: Centring of $n = 3$ directions (Fig. 6.3.1)

	α_i / α_z	$\varepsilon_i = \alpha_i - \alpha_z$	S_i / e	δ_i	$r_i = \alpha_i + \delta_i$
1	2	3	4	5	6
P_1	84°41′53″	52°58′17″	2353.2	+0°15′45″	84°57′39″
P_2	139°39′30″	107°55′53″	1346.8	+0°32′49″	140°12′19″
P_3	253°18′27″	221°34′51″	1592.9	−0°19′21″	252°59′06″
Z	31°43′36″		13.512		

6.3.1.2 Indirect Determination of Centring Elements

If the centring elements cannot be directly measured, e.g. if Z is the tip of a church steeple while B is located on the top platform, then e and α_z have to be determined indirectly. As illustrated in Fig. 6.3.2, this can be achieved with the aid of an auxiliary base SN at the foot of the tower.

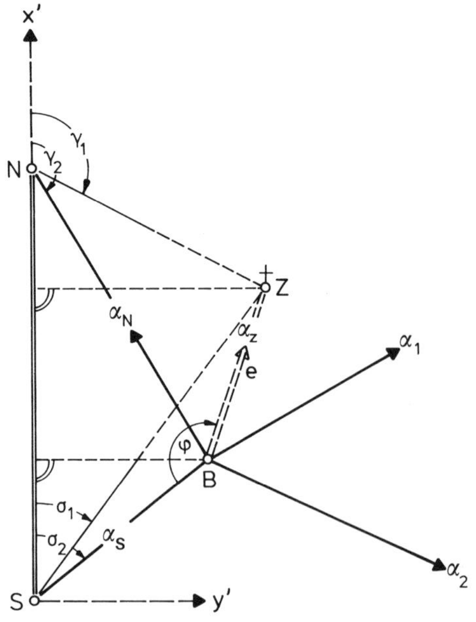

Fig. 6.3.2 Determination of centring elements

Since some of the sights are extremely steep, the vertical axis of the theodolite has to be perfectly plumb. The following angular observations are carried out

at B: the directions $\alpha_1 \alpha_2 \ldots \alpha_S, \alpha_N$
at S: the angles σ_1 and σ_2, and
at N: the angles γ_1 and γ_2.

In the triangle SNB the angles are added and the discrepancy to 200 gon (180°) is distributed evenly. The auxiliary base is measured with millimetre accuracy.

The easiest method to compute e and α_z utilizes the base SN as the abscissa axis of a local coordinate system for computing the following:

a) rectangular coordinates of B and Z with respect to SN using trigonometric relations in the triangles or intersection (section 6.4.1);

b) applying the second geodetic problem to obtain $BZ = e$ and α_{BZ};

c) $\varphi = \alpha_{BZ} - \alpha_{BS} = \alpha_{BZ} - \sigma_2 \pm 180°$;

d) $\alpha_z = \alpha_S + \varphi$.

Since local conditions vary, no fixed rules can be given. If, for instance, the station B is located on the ground, the auxiliary base can start at B which simplifies the situation. In any case, it is highly recommended to first sketch the situation and then select the most suitable approach. To check both computations and measurements, it is advisable to determine the centring elements independently using another auxiliary base.

If it is desirable to coordinate the eccentric point, the azimuth α_{ZB} can be determined using the direction observed at B to another known point. This azimuth plus e inserted into the first geodetic problem provides the coordinates of B from Z.

Example: The following was observed:

$$\alpha_1 = 0°00'00''; \ \alpha_S = 145°14'37''; \ \gamma_1 = 141°49'39''; \ \sigma_1 = 64°38'43''$$
$$SN = 94.376 \text{ m}; \ \alpha_N = 220°14'07''; \ \gamma_2 = 142°53'28''; \ \sigma_2 = 67°53'52''.$$

The computations yielded:

$$y_B = 22.181 \text{ m}; \ y_Z = 25.616 \text{ m}; \ \alpha_{BZ} = 350°41'31'';$$
$$x_B = 54.619 \text{ m}; \ x_Z = 54.056 \text{ m}; \ \alpha_Z = 284°02'15''; \ e = 3.481 \text{ m}.$$

6.3.1.3 Target Eccentricity

When observing at B, the direction r_t to the eccentric target T is measured instead of the desired direction r to point P (Fig. 6.3.3). For the reduction to the centre P, the eccentricity e and the angle ε are measured at T. Then

$$\sin\delta = \frac{e}{S}\sin\varepsilon \quad \text{and} \quad r = r_i + \delta\,. \tag{6.3.6}$$

It is quite common that an eccentricity is generated when the tip of an originally centred signal moves slightly. In this case, a graduated circle printed onto a sheet of paper of $20 \times 20\ \mathrm{cm}^2$ is placed onto a thin board and centred on the monument (Fig. 6.3.4). If the signal top is plumbed with a theodolite onto this board in two directions normal to each other, then the eccentricity e can be measured directly on the board. The direction to the observation station B is drawn onto the sheet such that $e_1 = e \sin \varepsilon$ can be directly read. The centring angle δ is then computed as $\delta = \varrho \cdot e_1 / S$. This method is often referred to as "plumbing".

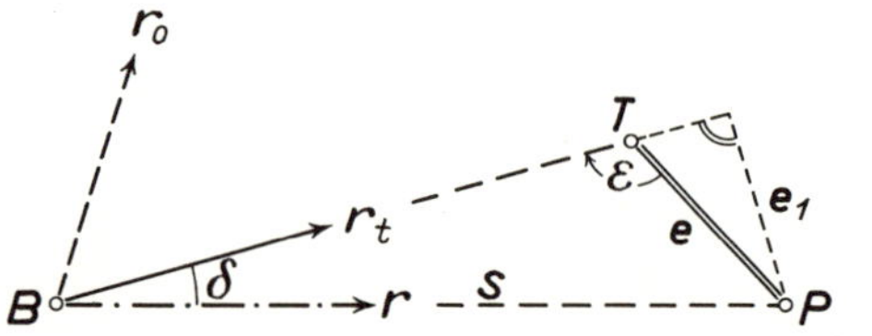

Fig. 6.3.3 Target eccentricity

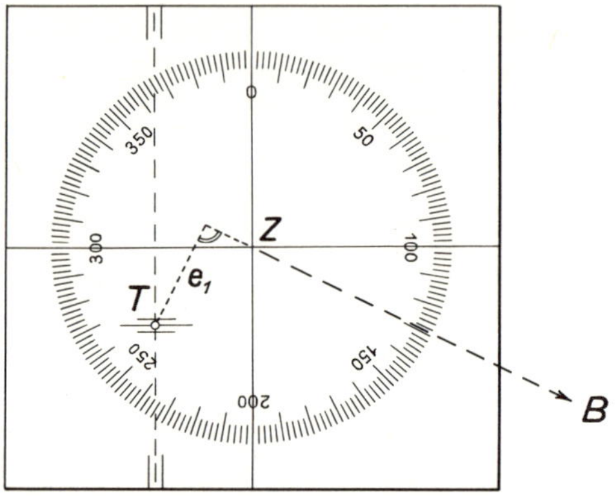

Fig. 6.3.4 Plumbing diagram

Often it is necessary to have eccentricities both at the station and the target. For practical work, it is not necessary to have special formulae for this case, as it is just as easy to perform the reductions consecutively, starting with the target eccentricities.

6.3.1.4 Broken Rays

If there is no intervisibility between points P_1 and P_2 (Fig. 6.3.5), at both points the direction to a nearby auxiliary point A is observed. These directions deviate by δ_1, or δ_2 from the desired directions. Furthermore, the angle $P_1 A P_2$ is measured, which provides the angle δ. As evident from Fig. 6.3.5, $\delta = \delta_1 + \delta_2$.

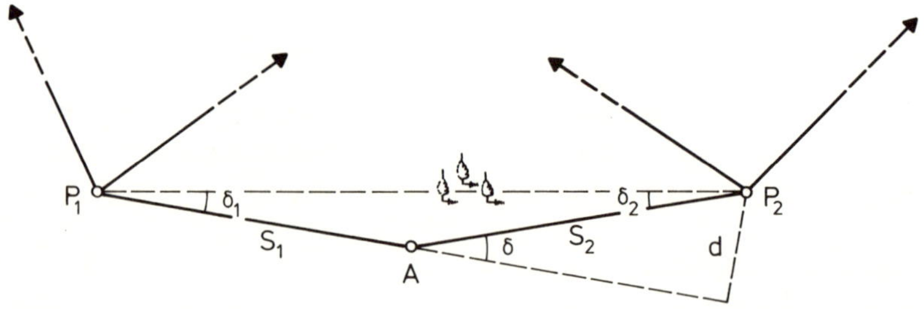

Fig. 6.3.5 Simple broken ray

If S_1 and S_2 are known, δ_1 and δ_2 are computed according to the tangent rule as

$$\tan\frac{1}{2}(\delta_2-\delta_1)=\frac{S_1-S_2}{S_1+S_2}\tan\frac{1}{2}\delta\,. \qquad (6.3.7)$$

To obtain δ_1 and δ_2 to within ± 0.5 mgon (1.5″), $\delta \leqq 2.2$ gon (2°) and $\sigma_\delta \leqq 0.7$ mgon (2″); S_1 and S_2 have to fulfil the condition $3/2 \geqq S_1/S_2 \geqq 2/3$, and their relative errors cannot exceed $\pm 2.5\cdot 10^{-4}$, i.e.10 cm per 400 m.

With approximate values for S_1 and S_2, and $\delta < 2$ cgon (1′) the following relations hold (Fig. 6.3.5):

$$\begin{aligned}&\delta_1(S_1+S_2)\approx(\delta_1+\delta_2)S_2=d\\&\delta_1=\delta\frac{S_2}{S_1+S_2};\qquad \delta_2=\delta\frac{S_1}{S_1+S_2}\,.\end{aligned} \qquad (6.3.8)$$

The point A can also be located beyond P_1P_2 (Fig. 6.3.6), where $\delta=\delta_2-\delta_1$ is measured.

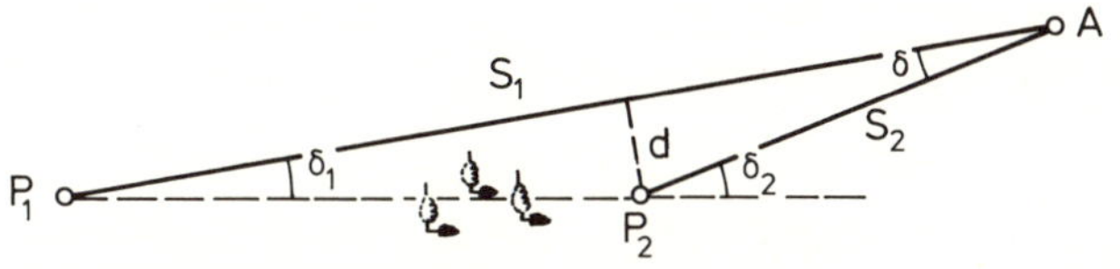

Fig. 6.3.6 Extended broken ray

$$\tan\frac{1}{2}(\delta_1+\delta_2)=\frac{S_1+S_2}{S_1-S_2}\tan\frac{1}{2}\delta \qquad (6.3.9)$$

and for $\delta < 2$ cgon (1′):

$$\begin{aligned}&\delta_1(S_1-S_2)\approx\delta S_2=d\\&\delta_1=\delta\frac{S_2}{S_1-S_2};\qquad \delta_2=\delta\frac{S_1}{S_1-S_2}\,.\end{aligned} \qquad (6.3.10)$$

If the ray is broken more than once (Fig. 6.3.7), it is computed as a free traverse with P_1A as the x-axis, which according to section 6.8 leads to

$$\delta_1=\alpha_{P,4}-\alpha_{P,P_2}\qquad \delta_2=\alpha_{P_2P_1}-\alpha_{P_2C}\,. \qquad (6.3.11)$$

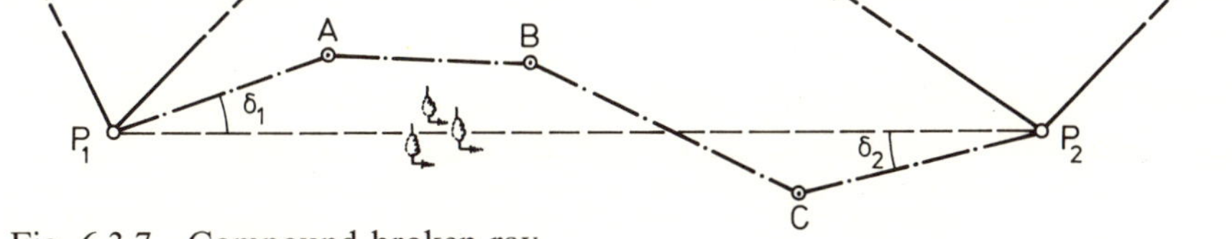

Fig. 6.3.7 Compound broken ray

In this case, theodolite and signals have to be extremely well centred (forced centring if at all possible).

The broken ray approach provides an excellent means for circumventing obstacles, which usually is faster than working with eccentricities.

6.3.2 Orientation of Observed Directions

Directions observed with a theodolite are initially oriented to a randomly selected direction. Once the zero direction is rotated such that it falls into the y-axis of the local coordinate system, then the bundle of directions r_i is considered to be oriented.

The zero direction may be the first direction of the reduced set, or the zero of the graduated circle. For both cases the same mathematical model is applied.

The following formulae provide the *orientation unknown* φ (rotational angle) *of a reduced bundle of directions,* if reduced directions r_i are used, while they provide the *orientation unknown* φ^T *of a graduated circle* if directly measured directions R_i are used. The solution for reduced directions is given.

6.3.2.1 Orientation with the Aid of a Reference Point

Given: Fixed points P_0 and P_1 and the directions r_1, r_N from P_0 to P_1 and N.

Wanted: The oriented direction r_N^0 pointing towards the new point N (Fig. 6.3.8):

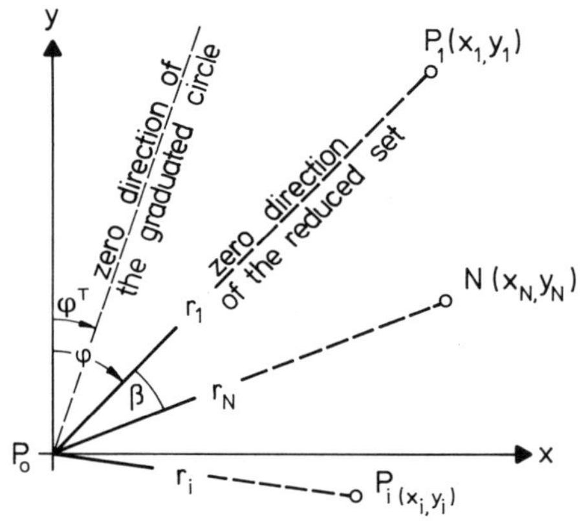

Fig. 6.3.8 Orientation of directions

$$r_N^0 = \alpha_1 + \beta_1 \tag{6.3.12}$$

α_1 is computed according to section 5.1.2, and

$$\beta = r_N - r_1 \cdot r_N^0 \tag{6.3.13}$$

is obtained from

$$r_N^0 = r_N + \varphi \tag{6.3.14}$$

with

$$\varphi = \alpha_1 - r_1 \; . \tag{6.3.15}$$

6.3.2.2 Orientation with the Aid of Several Reference Points

Given: Fixed points P_0, $P_1 \ldots P_n$ and the directions r_i observed at P_0.

Wanted: The oriented direction r_N^0 pointing toward the new point N (Fig. 6.3.8).

If observations are made to n fixed points, then the orientation unknown can be computed n times. Due to network tensions and observational errors, these values will differ. The arithmetic mean (equation (1.4.2)) provides a suitable final value:

$$\varphi_i = \alpha_i - r_i$$

$$\varphi = \frac{\Sigma(\alpha - r)}{n} \; . \tag{6.3.16}$$

Finally, the oriented direction pointing toward the new point becomes

$$r_N^0 = r_N + \varphi \, . \tag{6.3.17}$$

Check and error evaluation: The following discrepancies occur between oriented and measured directions:

$$v_i = \alpha_i - r_i^0 \; .$$

As a check we have

$$\Sigma v_i = 0 \; .$$

The accuracy of the measurements plus network tensions are evident from the standard deviation of the tan oriented direction

$$\sigma_{r^0} = \pm \sqrt{\frac{\Sigma v^2}{n(n-1)}} \tag{6.3.18}$$

and by lateral discrepancies of the directions

$$Q = v \cdot S$$

where S is the sighting distance.

Example: Orientation of directions

	azimuth α_i	meas. dir. R_i	$\alpha_i - R_i$ φ^T	orient. dir. $R_i + \varphi^T$ R_i^0	$\alpha_i - R_i^0$ v_i
1	41°50′33″	10°55′16″	30°55′17″	41°50′34″	1.0″
2	172°56′20″	142°01′05″	30°55′15″	172°56′23″	2.6″
N	(230°37′47″)	199°42′27″		230°37′45″	(1.9″)
3	292°10′19″	261°14′58″	30°55′21″	292°10′16″	3.2″
Σ			92°45′54″ 30°55′18″ = φ^T		

$s(R^0) = \sqrt{[vv] : (n(n-1))} = 1.7''$

6.3.2.3 Varying Sighting Distances

If the sighting distances vary significantly, one has to consider that coordinate errors caused by network tensions and centring errors caused by setting up the instrument and targets are independent of the distance. Therefore one must check whether these influences have to be considered by using a weight $p = \text{const.}\ S$. If so, the weighted mean has to be used instead of the arithmetic mean.

6.4 Trigonometric Point Determination

Basically there are two methods for trigonometric point determination, namely *intersection*, when the directions are measured at the fixed points, and *resection*, when the directions are measured at the new point.

Today trigonometric methods are used primarily when the targets are not easily accessible, while formerly they were almost exclusively used for network densification.

6.4.1 Intersection

In order to determine the new point N, the directions from two or more fixed points P_i to N are measured (Fig. 6.4.1). A unique solution is obtained if one angle each is measured at two fixed points (Fig. 6.4.2).

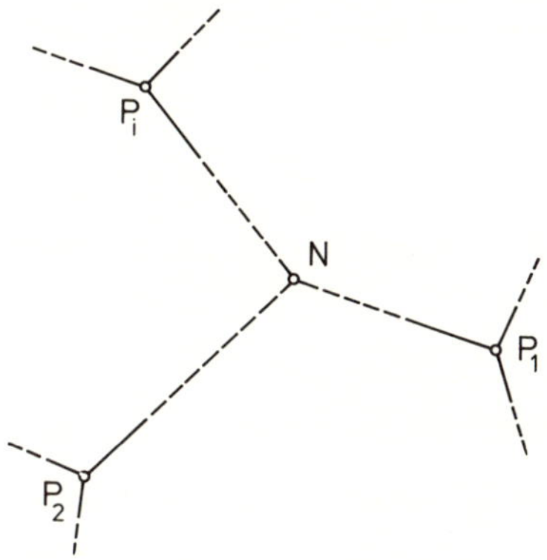

Fig. 6.4.1 Intersection (general case)

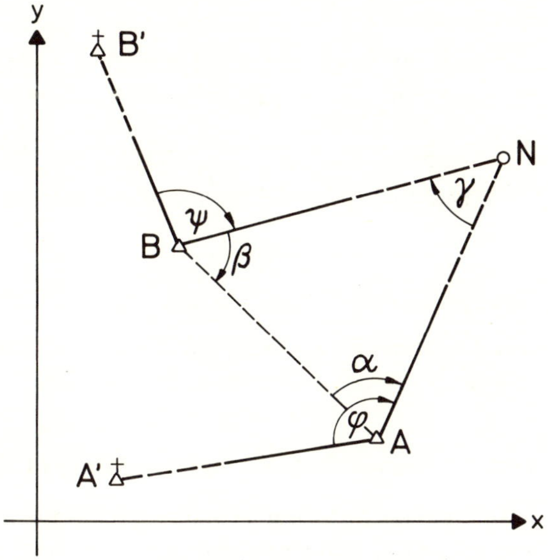

Fig. 6.4.2 Intersection with two angles

6.4.1.1 Unique Solution with First Geodetic Problem

Given: Coordinates of two stations A and B and two reference points A' and B'.

Measured: Angle φ at A and angle ψ at B.

Wanted: Coordinates of new point N (Fig. 6.4.2).

The solution is inherent in the first geodetic problem:

$$\begin{aligned} x &= x_A + \overline{AN}\sin\alpha_{AN} = x_B + \overline{BN}\sin\alpha_{BN} \\ y &= y_A + \overline{AN}\cos\alpha_{AN} = y_B + \overline{BN}\cos\alpha_{BN}\,. \end{aligned} \tag{6.4.1}$$

The azimuths α_{AN} and α_{BN} are computed according to section 6.3.2:

$$\alpha_{AN} = \alpha_{AA'} + \varphi; \quad \alpha_{BN} = \alpha_{BB'} + \psi\,. \tag{6.4.2}$$

The distances $\overline{AN}$ and $\overline{\mathrm{BN}}$ are obtained by applying the sine rule to triangle ABN:

$$\overline{AN} = \overline{AB}\,\frac{\sin\beta}{\sin\gamma}; \quad \overline{BN} = \overline{AB}\,\frac{\sin\alpha}{\sin\gamma} \tag{6.4.3}$$

where

$$\overline{AB} = \sqrt{(x_B - x_A)^2 + (y_B - y_A)^2} \tag{6.4.4}$$

$$\alpha = \alpha_{AN} - \alpha_{AB}; \quad \beta = \alpha_{BA} - \alpha_{BN}$$

$$\alpha_{AB} = \arcsin\frac{x_B - x_A}{\overline{AB}}; \quad \alpha_{BA} = \alpha_{AB} \pm 200\ \text{gon} = \alpha_{AB} \pm 180^\circ$$

$$\gamma = 200\ \text{gon} - (\alpha + \beta) = 180^\circ - (\alpha + \beta)\,.$$

Special cases:

a) If the sights from A to B and from B to A are possible, then points A' and B' are not necessary. In this case, equation (6.4.2) is replaced by

$$\alpha_{AN} = \alpha_{AB} + \alpha; \qquad \alpha_{BN} = \alpha_{BA} - \beta. \tag{6.4.2a}$$

b) If the angles φ and γ are measured instead of φ and ψ, then

$$\alpha_{AN} = \alpha_{AA'} + \varphi; \qquad \alpha_{BN} = \alpha_{AN} + \gamma. \tag{6.4.2b}$$

This case is called "side section".

c) If directions r_i to several fixed points are measured at A and B, then the oriented directions are computed as in section 6.3.2.

6.4.1.2 Unique Solution as Intersection of Straight Lines

A different set of formulae is obtained if one considers the new point as the point of intersection of two straight lines:

$$x - x_A = \tan\alpha_{AN}(y - y_A); \quad x - x_B = \tan\alpha_{BN}(y - y_B). \tag{6.4.5}$$

After computing α_{AN} and α_{BN} using equation (6.4.2) the solution becomes:

$$\left.\begin{aligned} y - y_A &= \frac{(x_B - x_A) - (y_B - y_A)\tan\alpha_{BN}}{\tan\alpha_{AN} - \tan\alpha_{BN}} \\ \text{and} \quad x - x_B &= (y - y_A)\tan\alpha_{AN}. \end{aligned}\right\} \tag{6.4.6}$$

Equations (6.4.7) provide a check:

$$\begin{aligned} y - y_B &= \frac{(x_B - x_A) - (y_B - y_A)\tan\alpha_{AN}}{\tan\alpha_{AN} - \tan\alpha_{BN}} \\ x - x_B &= (y - y_B)\tan\alpha_{BN}. \end{aligned} \tag{6.4.7}$$

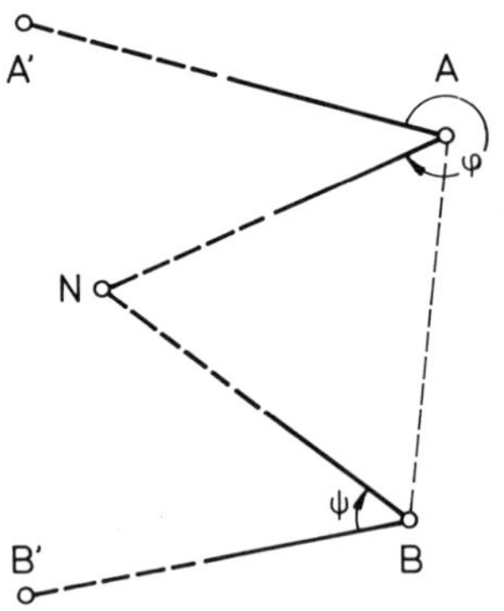

Fig. 6.4.3 Intersection with external angles

Example: Intersection using straight lines (Fig. 6.4.3)

	Pt. No.	Easting x	Northing y
Given	A	24681.92	90831.87
	A'	23231.58	91422.92
	B	24877.72	89251.09
	B'	22526.65	89150.52

from coord.'s:	$\alpha_{AA'}$	$= +292°10'19''$	$\alpha_{BB'}$	$= 267°33'02''$
Observed:	$+\varphi$	$= 289°27'20''$	$+\psi$	$= 54°40'33''$
	α_{AN}	$= 230°37'40''$	α_{BN}	$= 322°13'36''$
	$\tan\alpha_{AN}$	$= +1.218621$	$\tan\alpha_{BN}$	$= -0.774937$
	$\tan\alpha_{AN} - \tan\alpha_{BN}$	$= +1.993558$		
	$x - x_a$	$= -\ 629.13$	$y - y_a$	$= -\ 516.27$
	x	$= 24052.79$	y	$= 90315.60$

6.4.1.3 Intersection with Overdetermination and Least-Squares Adjustment*

Newly determined coordinates used to be checked by measuring and computational controls. Only additional observations can safeguard against blunders, such as sighting to a wrong target or extracting wrong initial data, e.g., copying the wrong coordinates. Therefore, at least one additional direction is observed, see section 6.4.1.2.

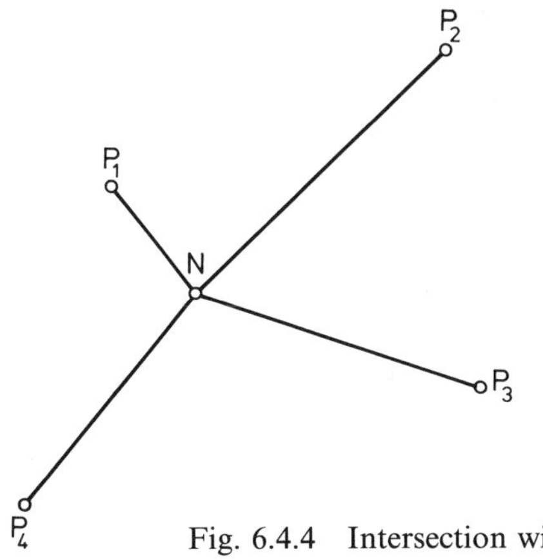

Fig. 6.4.4 Intersection with overdetermination

* Note: The following formulae can even be used if no additional measurements are taken. However, no accuracy evaluation is then possible! The approximate values can be scaled from a plan or map. If they are too far off, the method will not immediately provide the desired coordinates, but may need several iterations. By comparing new and previously computed coordinates, one can judge whether the accuracy is sufficient or not.

Thus, the coordinates of the new point are determined by least-squares adjustment (section 1.4.9).

The individual steps of the method are explained with an example. Using a given form, it can easily be handled with a modern pocket calculator. Since these now contain matrix operations, the example utilizes matrix algebra.

The coordination of the point N (Fig. 6.4.4) is used as the example. This point is to be coordinated from directions measured to four fixed points. The coordinates of the fixed points are given together with the oriented directions $(r_N^0)_i$ (also section 6.3.2)

Point No.	Easting x [m]	Northing y [m]	$(r_N^0)_i$
1	48 177.62	06 531.28	140° 37′ 20″
2	49 600.15	07 185.19	222° 34′ 46″
3	49 830.93	05 670.69	287° 03′ 20″
4	47 863.91	05 077.24	35° 32′ 31″
N	48 565.2 $= y_0$	06 059.0 $= x_0$	

Each oriented direction provides one observation equation of the form:

$$v_i = a_{i_1} dx + a_{i_2} dy - l_i$$

with

$$a_{i_1} = \frac{y_0 - y_i}{(s_i^0)^2} \cdot \frac{180°}{\pi}; \quad a_{i_2} = -\frac{x_0 - x_i}{(s_i^0)^2} \frac{180°}{\pi}$$

$$(s_i^0) = (x_0 - x_i)^2 + (y_0 - y_i)^2$$

$$l_i = (r_N^0)_i - \alpha_i^0$$

$$\alpha_i^0 = \arctan \frac{x_0 - x_i}{y_0 - y_i}.$$

First, the approximate coordinates of $N(X_0/Y_0)$ are computed (Table 6.4.1), and with them the coefficients a_{ij} and the absolute terms l_i of the observation equations.

Table 6.4.1

	$x_0 - x_i$ [m]	$y_0 - y_i$ [m]	s_i^0 [m]	α_i^0 [°]	α_{i_1}	α_{i_2}	l_i [°]
$P_1 N$	387.58	472.28	610.956	140.6257	−0.07250	−0.05949	−0.0035
$P_2 N$	−1034.95	−1126.19	1529.518	222.5825	−0.02759	0.02534	−0.0032
$P_3 N$	−1265.73	388.31	1323.955	287.0554	0.01269	0.04137	0.0003
$P_4 N$	701.29	981.76	1206.508	35.5389	0.03865	−0.02760	0.0032

Thus the observation equations $\mathbf{v} = \mathbf{A}\mathbf{x} - \mathbf{l}$ are given:

$$\mathbf{v} = \overset{\mathbf{A}}{\begin{bmatrix} -0.07250 & -0.05949 \\ -0.02759 & 0.02534 \\ 0.01269 & 0.04137 \\ 0.03865 & -0.02760 \end{bmatrix}} \overset{\mathbf{x}}{\begin{bmatrix} dx \\ dy \end{bmatrix}} - \overset{\mathbf{l}}{\begin{bmatrix} -0.0035 \\ -0.0032 \\ 0.0003 \\ 0.0032 \end{bmatrix}} .$$

Now the following is computed:

$$\mathbf{A}^T\mathbf{A} = \mathbf{N}; \qquad \mathbf{N}^{-1}; \qquad \mathbf{A}^T\mathbf{l} = \mathbf{n}; \qquad \mathbf{x} = \mathbf{N}^{-1}\mathbf{n} - \begin{bmatrix} +0.071 \\ -0.024 \end{bmatrix} .$$

After finding the corrections:

$$\mathbf{v} = \begin{bmatrix} -0.0002 \\ 0.0005 \\ -0.0004 \\ 0.0003 \end{bmatrix}$$

the computations are being checked:

$$\mathbf{v}^T\mathbf{v} = 0.000\,000\,5$$
$$-\mathbf{l}^T\mathbf{v} = 0.000\,000\,4 .$$

Then the empirical variance

$$s_0^2 = \frac{\mathbf{v}^T\mathbf{v}}{(m-n)} = \frac{0.000\,000\,540}{4-2} = 2.7 \cdot 10^{-7}\ (\text{grad}^2)$$

and the standard deviations of the unknowns are computed:

$$s_x = 0.007\ \text{m}$$
$$s_y = 0.007\ \text{m} .$$

The uncertainty of the point determination becomes

$$sp = \pm 0.0096\ \text{m}.$$

Finally, the adjusted coordinates of N are:

$$\hat{x} = x_0 + dx = 48\,565.271\ \text{m}$$
$$\hat{y} = y_0 + dy = 06058.976\ \text{m} .$$

Note: Because of the low degree of freedom, s_0^2 is statistically uncertain. If $(m-n) < 5$, s_0^2 should be replaced by practical experience values.

6.4.1.4 Accuracy of Intersection

The accuracy of an intersection depends on the measuring accuracy of the directions and on the geometry of the triangle ABP (Fig. 6.4.5). The standard deviation of the

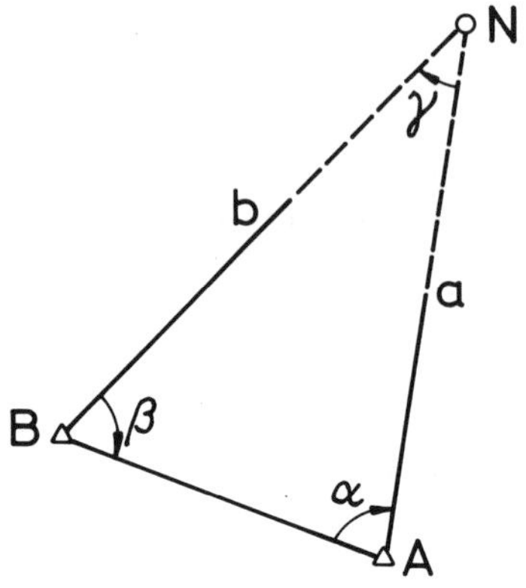

Fig. 6.4.5 Intersection triangle

position of the point is used to express the quality of its determination. It is estimated with the aid of the standard deviation σ_{r_0} of the oriented directions:

$$\sigma_N = \frac{1}{\sin\gamma}\sqrt{a^2+b^2}\cdot\sigma_{r_0}\,. \tag{6.4.8}$$

It is assumed in this case, that the standard deviation is the same for both oriented directions, and that the coordinates of the fixed points are free of error. σ_N assumes a minimum value for $\gamma = 121$ gon (109°). For $\gamma = 0$ or ≈ 200 gon (180°) an indeterminant problem is obtained.

6.4.2 Resection (Three Point Problem)

The directions r_i to the fixed points are measured at the new point N (Fig. 6.4.6). The new point is uniquely determined if two angles α and β between three fixed points M, A, B are observed (Fig. 6.4.7).

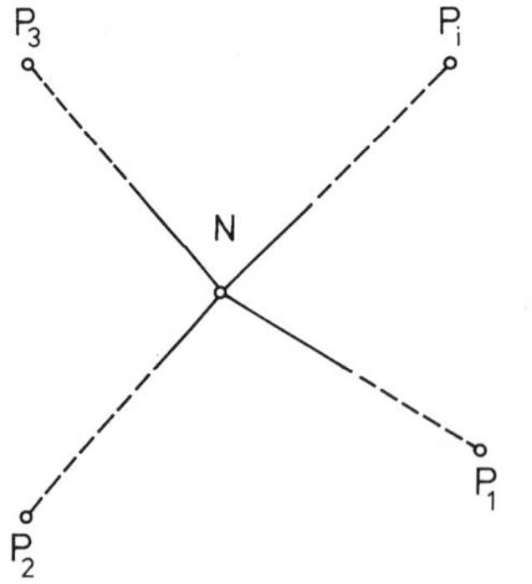

Fig. 6.4.6 Resection, general case

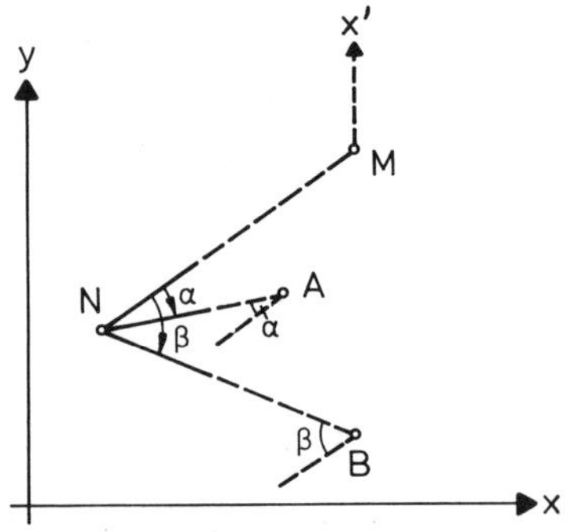

Fig. 6.4.7 Resection with two angles

6.4.2.1 Resection as Triple Arc Section

The three fixed points M, A and B are given (Fig. 6.4.7). The directions r_n, r_A and r_B are measured at N. Wanted are the coordinates of N. The solution is based on the intersection of three straight lines:

$$\left.\begin{aligned} x = x_M + (y - y_M)\tan\alpha_{M,N} &= x_A + (y - y_A)\tan\alpha_{A,N} \\ &= x_B + (y - y_B)\tan\alpha_{B,N} \end{aligned}\right\} . \tag{6.4.9}$$

To simplify this, relative directions and coordinates with respect to M are defined:

$$\begin{aligned} &r_A - r_M = r'_A = \alpha; \quad r_B - r_M = r'_B = \beta \\ &x_i - x_M = x'_i; \quad y_i - y_M = y'_i \\ &\alpha_{A,N} = \alpha_{M,N} + \alpha; \quad \alpha_{B,N} = \alpha_{M,N} + \beta . \end{aligned}$$

Instead of equation (6.4.9), the following can be written in the new system:

$$\begin{aligned} x' = y'\tan\alpha_{M,N} &= x'_A + (y' - y'_A)\tan(\alpha_{M,N} + \alpha) \\ &= x'_B + (y' - y'_B)\tan(\alpha_{M,N} + \beta). \end{aligned} \tag{6.4.10}$$

Since two of the equations (6.4.10) can be set equal, we get:

$$y' = \frac{x'_A - y'_A\tan(\alpha_{M,N} + \alpha)}{\tan\alpha_{M,N} - \tan(\alpha_{M,N} + \alpha)} \tag{6.4.11}$$

or

$$y' = \frac{x'_B - y'_B\tan(\alpha_{M,N} + \beta)}{\tan\alpha_{M,N} - \tan(\alpha_{M,N} + \beta)} . \tag{6.4.12}$$

Since

$$\tan(\alpha_{M,N} + \alpha) = \frac{\tan\alpha_{M,N} + \tan\alpha}{1 - \tan\alpha_{M,N}\tan\alpha} \tag{6.4.13}$$

and

$$\tan(\alpha_{M,N} + \beta) = \frac{\tan\alpha_{M,N} + \tan\beta}{1 - \tan\alpha_{M,N}\tan\beta} \tag{6.4.14}$$

we get after substituting equations (6.4.13) or (6.4.14) into (6.4.11) or (6.4.12):

$$y' = \frac{(x'_A + y'_A\cot\alpha)\tan\alpha_{M,N} + (y'_A - x'_A\cot\alpha)}{1 + \tan^2\alpha_{M,N}} \tag{6.4.15}$$

or

$$y' = \frac{(x'_B + y'_B\cot\beta)\tan\alpha_{M,N} + (y'_B - x'_B\cot\beta)}{1 + \tan^2\alpha_{M,N}} . \tag{6.4.16}$$

$\mathrm{Tan}\,\alpha_{M,N}$ is still unknown. It is obtained by equating the numerators of equations (6.4.15) and (6.4.16):

$$\tan\alpha_{M,N} = \frac{x'_A\cot\alpha - x'_B\cot\beta(y'_B - y'_B)}{y'_B\cot\alpha - y'_B\cot\beta - (x'_B - x'_A)}\,. \tag{6.4.17}$$

Finally, the coordinates of N are obtained:

$$x_N = x' + x_M;\quad y_N = y' + y_M\,.$$

Numerical Example for resection:

Pt. No.	Easting x	Northing y
	[m]	[m]
A	46867.94	05537.00
B	51293.86	06365.89
M	49666.56	04448.58

Given: $r_A = 60°07'50''$
$r_B = 265°18'22''$
$r_M = 326°33'59''$

Values reduced to M:

$r'_A = \alpha = 93°33'51''$; $x'_A = -2798.62$; $y'_A = 1088.42$
$r'_B = \beta = 298°44'23''$; $x'_B = 1627.30$; $y'_B = 1917.31$.

$$\tan(\alpha_{M,N}) = \frac{x'_A\cdot\cot\alpha - x'_B\cdot\cot\beta + (y'_B - y'_A)}{y'_A\cdot\cot\alpha - y'_B\cdot\cot\beta - (x'_B - x'_A)} = -0.5506757$$

$$y' = \frac{(x'_A + y'_A\cdot\cot\alpha)\cdot\tan(\alpha_{M,N}) + (y'_A - x'_A\cdot\cot\alpha)}{1+\tan^2(\alpha_{M,N})} = 1912.590$$

$$y' = \frac{(x'_B + y'_B\cdot\cot\beta)\cdot\tan(\alpha_{M,N}) + (y'_B - x'_B\cdot\cot\beta)}{\alpha_{M,N}1+\tan^2(\alpha_{M,N})} = 1912.590$$

$$x' = y'\cdot\tan(\alpha_{M,N}) = -1053.217$$

$$x_N = x' + x_M = 48613.34;\quad y_N = y' + y_M = 06361.17$$

6.4.2.2 Resection with Overdetermination and Least-Squares Adjustment *

The coordinates of the new point N are computed by least-squares adjustment if more than 3 directions are observed (section 6.4.2.1).

The individual steps of the method are explained with an example. As in section 6.4.1.3, the determination of the new point N (Fig. 6.4.4) serves as this example. The directions r_i are measured from N to the fixed points P_1, P_2, P_3, P_4 (Table 6.4.2). Furthermore, the coordinates for N have been computed.

Table 6.4.2 given coordinates for resection

Pt. No.	Easting x [m]	Northing y [m]	r_i
1	48 177.62	06 531.28	320° 37′ 19″
2	49 600.15	07 185.19	42° 34′ 49″
3	49 830.93	05 670.69	107° 03′ 17″
4	47 863.91	05 077.24	215° 32′ 34″
N	$48\,565.2 = x_0$	$06059.0 = y_0$	

Each measured direction provides an observation equation (section 1.4.9):

$$v_i = a_{i_1} dx + a_{i_2} dy - l_i \qquad \text{(weight: } p = 1\text{)}$$

with

$$a_{i_1} = \frac{y_0 - y_i}{(s_i^0)^2} \cdot \frac{180°}{\pi}; \quad a_{i_2} = \frac{x_0 - x_i}{(s_i^0)^2} \cdot \frac{180°}{\pi}$$

$$l_i = r_i - \alpha_i^0$$

$$a_i^0 = \arctan \frac{x_0 - x_i}{y_0 - y_i} \pm 180°$$

$$(s_i^0)^2 = (x_0 - x_i)^2 + (y_0 - y_i)^2 .$$

A fictitious equation (1.4.48) has to be introduced as well:

$$v_{n+1} = \sum_1^n a_{i_1} dx + \sum_1^n a_{i_2} dy - \Sigma l_i \qquad \left(\text{weight: } p_{n+1} = -\frac{1}{\sum_1^n p_i}\right).$$

First the approximate coordinates x_0, y_0 of N are computed, followed by the coefficients a_{ij} and the absolute terms l_i of the observation equations and of the fictitious equation.

---

* see also note on page 229

	$x_0 - x_i$	$y_0 - y_i$	s_i^0	r_i	α_i^0	a_{i_1}	a_a	l_i
	[m]	[m]	[m]	[°]	[°]			[°]
NP_1	387.58	− 472.28	610.956	320.6219	.2257	−0.07250	−0.05949	−0.0039
NP_2	−1034.95	−1126.19	1529.518	42.5803	.2825	−0.02759	0.02534	−0.0023
NP_3	−1265.73	388.31	1323.955	107.0547	.8554	0.01269	0.04137	−0.0006
NP_4	701.29	981.76	1206.508	215.5428	.4389	0.03865	−0.02760	0.0039
						−0.04874 $= [a_{i_1}]_1^4$	−0.02038 $= [a_{i_2}]_1^4$	−0.0029 $= [l_i]_1^4$

Thus the observation equations $\mathbf{v} = \mathbf{Ax} - \mathbf{l}$ are given:

$$\mathbf{v} = \overset{\mathbf{A}}{\begin{bmatrix} -0.07250 & -0.05949 \\ -0.02759 & 0.02534 \\ 0.01269 & 0.04137 \\ 0.03865 & -0.02760 \\ -0.04874 & -0.02038 \end{bmatrix}} \overset{\mathbf{x}}{\begin{bmatrix} dx \\ dy \end{bmatrix}} - \overset{\mathbf{l}}{\begin{bmatrix} -0.0039 \\ -0.0023 \\ -0.0006 \\ 0.0039 \\ -0.0029 \end{bmatrix}} ; \quad \mathbf{P} = \begin{bmatrix} 1 & & & & \\ & 1 & & & \\ & & 1 & & \\ & & & 1 & \\ & & & & -\frac{1}{4} \end{bmatrix}$$

Now the following is computed:

$$\mathbf{A}^T\mathbf{PA} = \mathbf{N}; \qquad \mathbf{N}^{-1}; \qquad \mathbf{A}^T\mathbf{Pl} = \mathbf{n}; \qquad \mathbf{x} = \mathbf{N}^{-1}\mathbf{n} = \begin{bmatrix} 0.074 \\ -0.028 \end{bmatrix}.$$

The check calculations can be performed once the corrections v are known:

$$\mathbf{v} = \begin{bmatrix} 0.0002 \\ -0.0005 \\ 0.0004 \\ -0.0002 \\ 0.0000 \end{bmatrix},$$

$$\mathbf{v}^T\mathbf{Pv} = 0.000\,000\,5$$
$$-\mathbf{l}^T\mathbf{Pv} = 0.000\,000\,7\,.$$

The empirical variance becomes:

$$s_0^2 = \frac{\mathbf{v}^T\mathbf{Pv}}{m-n} = \frac{0.000\,000\,490}{4-3} = 5 \cdot 10^{-7}$$

(variance of one direction in gon²).

This leads to the standard deviation of the unknown coordinates:

$$s_x = 0.009\ \text{m}$$
$$s_y = 0.010\ \text{m}$$

as well as to the combined point uncertainty: $s_p = \pm 0.014$ m.

The adjusted coordinates of the new point N are thus:

$$\hat{x} = x_0 + dx = 48\,565.274\ \text{m}$$
$$\hat{y} = y_0 + dy = 06\,058.972\ \text{m}$$

6.4.2.3 Accuracy of Resection

The new point N is geometrically determined as the intersection of two circles. One circle passes through the points B, M, N, the other through M, A, N (Fig. 6.4.8). If the new point is located on the circle passing through the fixed points A, M, B, then the solution is undetermined. In this case, both auxiliary circles coincide.

Therefore, the circle defined by M, A, B is called the "dangerous circle", because the solution becomes rather uncertain if point P is near the circle. To avoid this, choose a geometric arrangement similar to the ones illustrated in Fig. 6.4.9.

Again, the accuracy of the resection is expressed by the standard deviation of $P(\sigma_P)$ as in section 6.4.1.4. Without deriving the error formulae, only the result is presented.

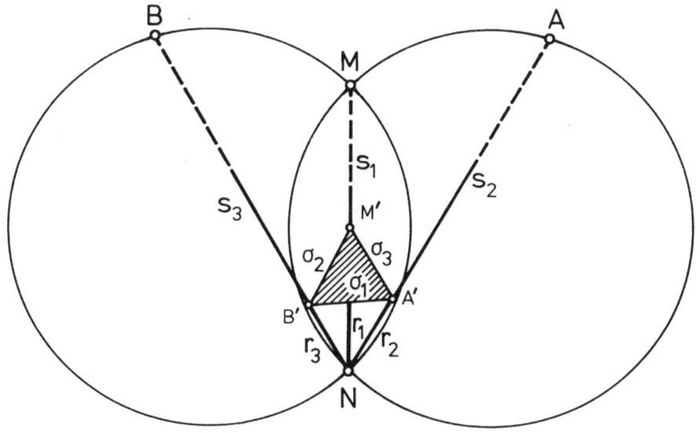

Fig. 6.4.8 Resection geometry and the reciprocal triangle

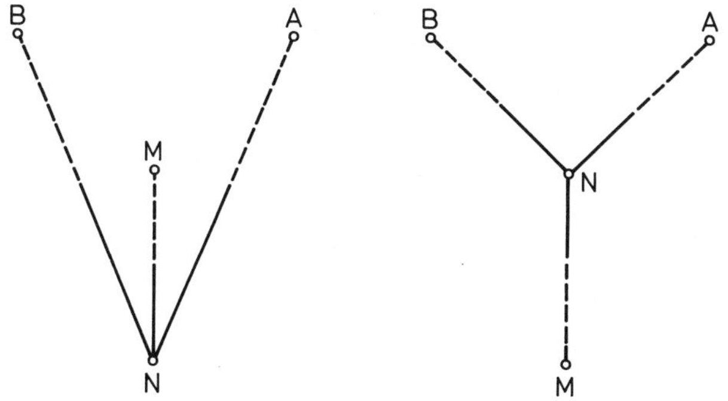

Fig. 6.4.9 Ideal configurations for resection

As illustrated in Fig. 6.4.8, the reciprocal lengths of the three determination rays $r_1 = 1/s_1; r_2 = 1/s_2; r_3 = 1/s_3$ are plotted at a selected scale into a scaled graph or map of the points. This produces the reciprocal auxiliary triangle $M'A'B'$. With the sides $\sigma_1, \sigma_2, \sigma_3$, and the area Δ. Assuming that directions rather than angles were observed in P with a standard deviation σ_r then the standard deviation of P becomes

$$\sigma_P^2 = \frac{\sigma_1^2 + \sigma_2^2 + \sigma_3^2}{4\Delta^2} \cdot \frac{\sigma_r^2}{\varrho^2} .$$

Thus the best case is where the σ_i are small, while Δ is as large as possible.

6.5 Point Determination Based on Distance Measurements

In order to coordinate the new point N, the distances s_i are measured to a number of fixed points P_i. N is then determined as the intersection of two circles with the centres P_1 and P_2, and the radii s_1 and s_2 (Fig. 6.5.1).

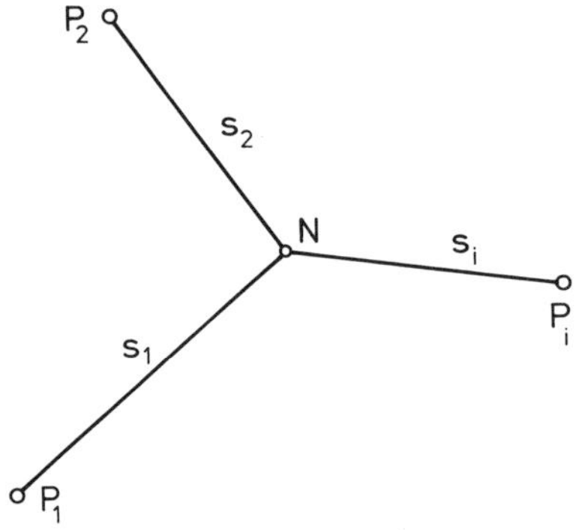

Fig. 6.5.1 Point determination via arc section

Thus a unique solution is obtained if two distances are measured. In such a case, however, it is not possible to check whether the scale of the measured distances coincides with the one for the network of fixed points. The scale factor g and the coordinates of N can be determined, if at least one distance is measured between fixed points (section 6.5.1), or if additional distances between N and fixed points are observed (section 6.5.3). The latter approach is preferable because it permits a more thorough test on the scale relationships of the surrounding fixed points.

The simple arc section is only of limited use, just like simple intersection or resection. Under certain geometric configurations it becomes rather weak or even undetermined.

6.5.1 Simple Arc Section

Given: The coordinates of two fixed points P_1 and P_2 plus the distances s_1 and s_2 which are reduced within the projection system (Fig. 6.5.2).

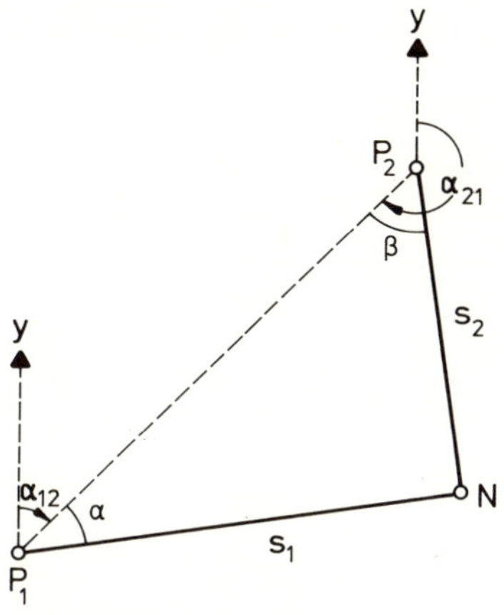

Fig. 6.5.2 Simple arc section

Wanted: The coordinates of the new point N.

The coordinates are again determined with the aid of the first geodetic problem:

$$\begin{aligned} x &= x_1 + s_1 \sin\alpha_{1,N} = x_2 + s_2 \sin\alpha_{2,N} \\ y &= y_1 + s_1 \cos\alpha_{1,N} = y_2 + s_2 \cos\alpha_{2,N} \,. \end{aligned} \tag{6.5.1}$$

This requires the distances $s_{1,2}$ and the azimuth $\alpha_{1,2}$ which is computed according to section 5.1.2:

$$s^*_{1,2} = \sqrt{(x_2 - x_1)^2 + (y_2 - y_1)^2}$$

$$\alpha_{1,2} = \arcsin\frac{x_2 - x_1}{s_{1,2}} = \arccos\frac{y_2 - y_1}{s^*_{1,2}} \,.$$

Furthermore:

$$\alpha_{1,N} = \alpha_{1,2} + \alpha$$

$$\alpha_{2,N} = \alpha_{2,1} - \beta \quad \text{where} \quad \alpha_{2,1} = \alpha_{1,2} \pm 200 \text{ gon } (180^\circ)$$

and

$$\alpha = \arccos(s_1^2 + s^{*\,2}_{1,2} - s_2^2)/2s_1 \cdot s^*_{1,2}$$

$$\beta = \arccos(s_2^2 + s^{*\,2}_{1,2} - s_1^2)/2s_2 \cdot s^*_{1,2} \,.$$

The signs of α and β are reversed if N is located on the opposite side of the line $P_1 P_2$. If, in addition, the distance $s^*_{1,2}$ was measured and reduced, then the scale factor q can be computed. This factor describes possible scale differences between the measuring unit of the distance meter and the given network:

$$q = \frac{s^*_{1,2}}{s_{1,2}} .$$

Using this, equations (6.5.1) become:

$$\begin{aligned} x &= x_1 + q \cdot s_1 \sin\alpha_{1,N} = x_2 + q \cdot s_2 \sin\alpha_{2,N} \\ y &= y_1 + q \cdot s_1 \cos\alpha_{1,N} = y_2 + q \cdot s_2 \cos\alpha_{2,N} . \end{aligned} \tag{6.5.2}$$

Example:

Given:

Pt. No.	Easting x [m]	Northing y [m]
P_1	328.76	1207.85
P_2	925.04	954.33

Measured:

$s_1 = 294.33$ m, $s_2 = 506.42$ m, $s_{1,2} = 648.08$ m

1) $s^*_{1,2} = \sqrt{(596.28)^2 + (-253.52)^2} = 647.937$ m; $\alpha_{1,2} = 113°02'01''$

$q = 647.937 : 648.08 = 0.999779$

2) $\alpha = \arccos\{(294.33^2 + 648.08^2 - 506.42^2) : (2 \cdot 294.33 \cdot 648.08)\}$

$= 49°01'18''$

$\beta = \arccos\{(506.42^2 + 648.08^2 - 294.33^2) : (2 \cdot 506.42 \cdot 648.08)\}$

$= 26°01'34''$

$\alpha_{1,N} = 113°02'01'' + 49°01'18'' = 162°03'19''$

$\alpha_{2,N} = 113°02'01'' - 26°01'34'' = 87°00'27''$;

3) $x = 328.76 + q \cdot s_1 \cdot \sin\alpha_{1,N} = 328.76 + 90.66 = 419.42$

$y = 1207.86 + q \cdot s_1 \cdot \cos\alpha_{1,N} = 1207.85 - 279.95 = 927.90$

4) Check: $y_2 = y + q \cdot s_2 \cdot \sin\alpha_{N,2} = 419.42 + 505.62 = 925.04$

$x_2 = y + q \cdot s_2 \cdot \cos\alpha_{N,2} = 927.90 + 26.43 = 954.33$.

6.5.2 Arc Section with Overdetermination and Least-Squares Adjustment*

The measurements are better checked and the accuracy of the coordinates is increased when additional differences are increased for the arc section. The degree of freedom is

* see also note on page 229

$(n - m)$ if n distances are measured from known fixed points P_i to the new point N, while $m = 3$ unknowns, i.e., two coordinates plus a scale factor are to be determined.

The weight 1 can be assigned to all distances for the least-squares adjustment, because the accuracy of electromagnetic distance measurements is largely independent of the distance. The computations are explained with an example.

The computation of the Gauss-Krueger coordinates of point N (Fig. 6.4.4) is to serve as this example. In an area, located approximately 100 m above MSL, distances are measured from four known points P_i towards N.

a) In preparation for the adjustment, the distances are corrected and geometrically reduced. Subsequently, the distances S_i are reduced to s_i in the Gauss-Krueger projection. Two of these distances are used to compute approximate coordinates of N as described in section 6.5.1.

The given values for the example in Fig. 6.4.4 are:

Pt. No.	Easting x [m]	Northing y [m]	s_i [m]
1	48 177.62	06 531.28	611.023
2	49 600.15	07 185.19	1 529.482
3	49 830.93	05 670.69	1 323.884
4	47 863.91	05 077.24	1 206.524
N	$48\,565.2 = x_0$	$06\,059.0 = y_0$	

b) *Determination of the most probable coordinates for N*: Each distance s_i provides an observation equation (section 1.4.9)

$$v_i = a_{i1}\,dx + a_{i2}\,dy + a_{i3}\,dq - l_i$$

with

$$a_{i1} = q_0 \frac{x_0 - x_i}{s_i^0}; \quad a_{i2} = q \frac{y_0 - y_i}{s_i^0}; \quad a_{i3} = s_i^0$$

$$q = q_0 + dq \qquad \text{(scale factor)}$$
$$q_0 = 1 \qquad \text{(approximate value)}$$
$$l_i = s_i - s_i^0$$
$$s_i^0 = \sqrt{(x_0 - x_i)^2 + (y_0 - y_i)^2}\ .$$

First the approximate coordinates x_0, y_0 are computed (section 6.5.1) and then the coefficients a_i and the absolute terms l_i of the observation equations.

	$x_0 - x$	$y_n - y_i$	s_i	s_i	a_{i1}	a_{i2}	a_{i3}^*	l_i
	[m]	[m]	[m]	[m]				[m]
NP_1	387.58	− 472.28	610.023	611.023	0.63438	−0.77302	0.610956	0.067
NP_2	−1034.95	−1126.19	1529.518	1529.482	−0.67665	−0.73630	1.529518	−0.036
NP_3	−1265.73	388.31	1323.955	1323.884	−0.95602	0.29330	1.323955	−0.071
NP_4	701.29	981.76	1206.508	1206.524	0.58126	0.81372	1.206508	−0.016

This provides the observations equations $\mathbf{v} = \mathbf{Ax} - \mathbf{l}$

$$\mathbf{v} = \overset{\mathbf{A}}{\begin{bmatrix} 0.63438 & -0.77302 & 0.610956 \\ -0.67665 & -0.73630 & 1.529518 \\ -0.95602 & 0.29330 & 1.323955 \\ 0.58126 & 0.81372 & 1.206508 \end{bmatrix}} \overset{\mathbf{x}}{\begin{bmatrix} dx \\ dy \\ dq \end{bmatrix}} - \overset{\mathbf{l}}{\begin{bmatrix} 0.067 \\ -0.036 \\ -0.071 \\ 0.016 \end{bmatrix}} .$$

Now the following is computed:

$$\mathbf{A}^T\mathbf{A} = \mathbf{N} = \begin{bmatrix} 2.11213 & 0.20041 & -1.21181 \\ & 1.88786 & -0.22839 \\ & & 5.92124 \end{bmatrix};$$

$$\mathbf{A}^T\mathbf{l} = \mathbf{n} = \begin{bmatrix} 0.14404 \\ -0.03309 \\ -0.08883 \end{bmatrix};$$

$$\mathbf{N}^{-1} = \begin{bmatrix} 0.54009 & -0.04417 & 0.10883 \\ & 0.53580 & 0.01163 \\ & & 0.19160 \end{bmatrix};$$

$$\mathbf{x} = \mathbf{N}^{-1}\mathbf{n} = \begin{bmatrix} 0.070 \\ -0.025 \\ -0.002 \end{bmatrix}.$$

The check calculations can be performed once the corrections $\mathbf{v}$ are known

$$\mathbf{v} = \begin{bmatrix} -0.004 \\ 0.004 \\ -0.006 \\ 0.002 \end{bmatrix} \qquad \begin{aligned} \mathbf{v}^T\mathbf{v} &= 0.00007 \\ -\mathbf{l}^T\mathbf{v} &= 0.00005 \,. \end{aligned}$$

The empirical variance become s:

---

* The coefficients $a_{i_3}$are multiplied by a factor $f = 0.001$ so that all elements of the matrix **A** are of the same magnitude. Therefore the normal equation matrix **N** becomes homogeneous and its inverse $\mathbf{N}^{-1}$ is then numerically stable. The unknown d_q and its standard deviation S_q have to be multiplied with $f = 0.001$ as well.

$$s_0^2 = \frac{\mathbf{v}^T\mathbf{v}}{n-u} = \frac{0.00007}{4-3} = 7.06 \cdot 10^{-5}*,$$

where n = number of observations; u = number of unknowns.

This leads to the standard deviations of the unknowns

$$s(x) = \sqrt{7.06 \cdot 10^{-5} \cdot 0.540} = \pm 0.006 \text{ m}$$
$$s(y) = \sqrt{7.06 \cdot 10^{-5} \cdot 0.535} = \pm 0.006 \text{ m}$$
$$s(q) = \sqrt{7.06 \cdot 10^{-5} \cdot 0.129} \cdot 10^{-3**} = 3 \cdot 10^{-6},$$

as well as to the combined point uncertainty:

$$\sigma_p = 0.009 \text{ m}.$$

The adjusted coordinates and the scale factor are then

$$x = x_0 + dx = 48\,565.270 \text{ m}; \quad y = y_0 + dy = 06058.975 \text{ m};$$
$$q = q_0 + dq = 0.999998**.$$

6.5.3 Accuracy of Arc Sections

The accuracy of the simple arc section depends on the standard deviation and on the geometry of the triangle $P_1 P_2 N$ (Fig. 6.5.3).

$$\sigma_N = \frac{1}{\sin\gamma} \sqrt{2} \cdot \sigma_s \tag{6.5.3}$$

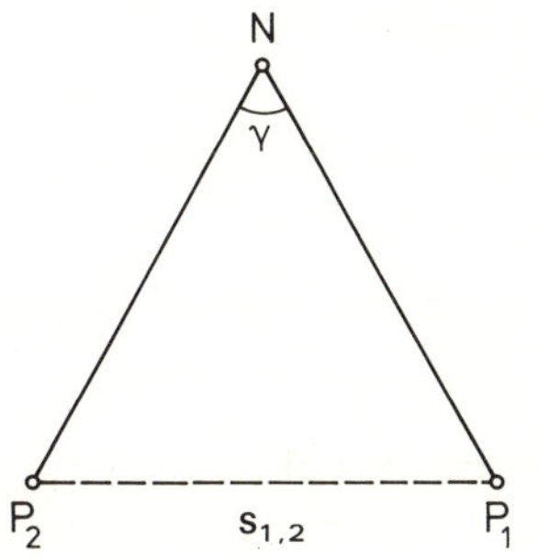

Fig. 6.5.3 Arc section triangle

---

* Because of the low degree of freedom, s_0^2 is statistically uncertain. If $(n-m) < 5$, s_0^2 should be replaced by practical experience values.

** see note on page 242

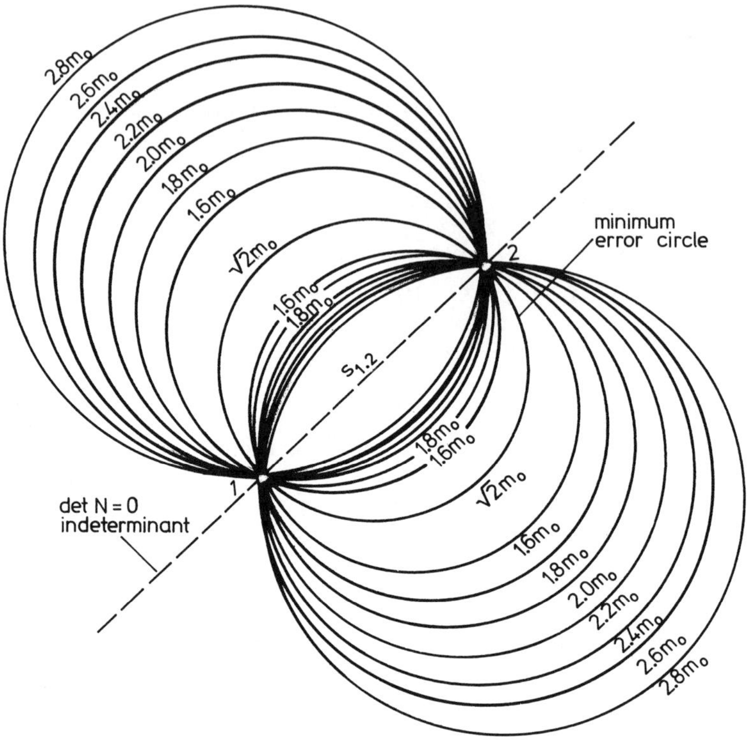

Fig. 6.5.4 Lines of equal point errors for simple arc section

where γ denotes the angle of intersection of the lines determining N. σ_s is considered as independent of the distance. The accuracy relationships are illustrated with lines of equal standard deviation of the new point ($\sigma_N = \text{const.}$), as shown in Fig. 6.5.4. This figure shows lines of equal σ_N based on the unit length $s_{1,2}$, representing the separation of the two fixed points.

This figure, which is based on $\sigma_0 = \sigma_s$ being independent of the distance, expresses the following:

- The lines of equal point errors are circles whose centres are located on the bisecting normal to $s_{1,2}$.
- The minimum error circle with $\sigma_N = \sqrt{2} \cdot \sigma_0$ is a circle with radius $R_{\min} = s_{1,2}/2$ with the centre halfway between P_1 and P_2. This is the well known Thales's circle, which represents the location of all points where the lines intersect at a right angle.
- The method is indeterminant, if the new point is located on the extension of the line $P_1 P_2$.

6.6 Point Determination Using a Combination of Angular and Distance Measurements

Combined angular and distance measurement methods are preferred when electronic distance meters are used in conjunction with optical or electronic theodolites. Such combined methods are far less dependent on the geometric configuration than the previously mentioned ones, which are based on one type of observation only. No matter what the geometry looks like, they will always provide an acceptable solution.

If the directions r_i and the distances s_i are measured from the new point N to several fixed points P_i (Fig. 6.6.1), then four unknowns have to be determined in order to coordinate N:

- the coordinates x_N, y_N
- the scale factor q which describes possible scale differences between the observed distances and the network, and
- the orientation unknown φ of the observed directions.

A unique solution is obtained, if distances and directions are measured to two fixed points.

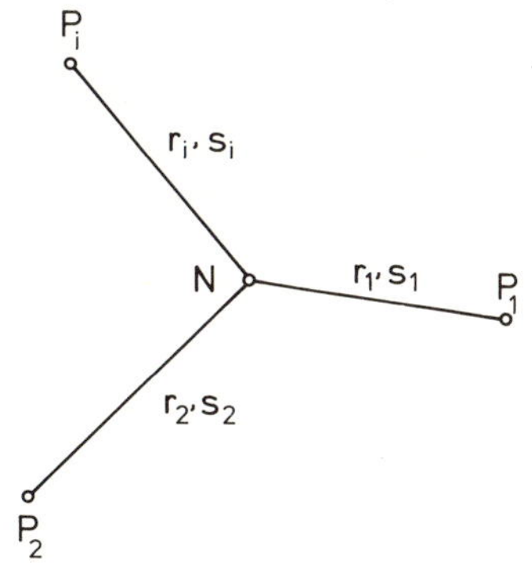

Fig. 6.6.1 Point determination by observing directions and distances at a new point

It is desirable to design the measurements such that directions and distances are measured with equal accuracy:

$$\sigma_0 = \sigma_r \cdot s = \sigma_s$$

where σ_s = standard deviation of distances s

σ_r = standard deviations of directions.

6.6.1 Unique Solution With Similarity Transformation

Given: Fixed points A and E, directions r_A and r_E, measured at the new point N, and distances s_A and s_E (Fig. 6.6.2).

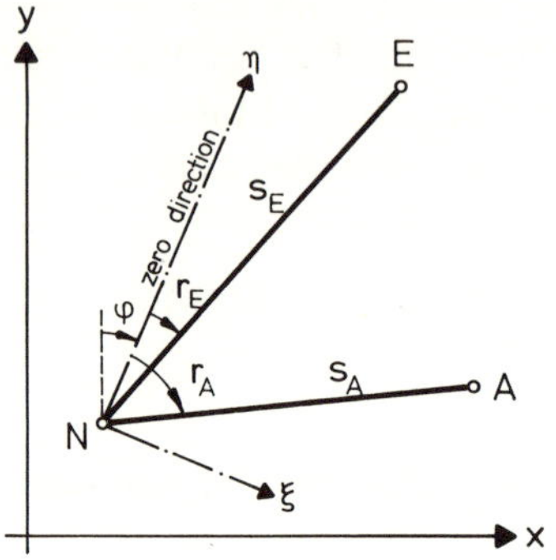

Fig. 6.6.2 Point determination with directions and distances

Wanted: Coordinates of the new point N.

The new point N is considered as the origin of a local ξ, η coordinate system, where the η-axis coincides with the zero direction of the measured directions. A and E are then defined by polar coordinates (r, s) in the ξ, η system, while the rectangular x, y coordinates are given. Thus, with two identical points known in both systems, the coordinates of N (origin of the local system) are easily computed by a similarity transformation.

According to Fig. 6.6.2 and equation (5.2.8.b) we have:

$$\left.\begin{aligned} x &= x_A - q \cdot s_A \sin(r_A + \varphi) = x_E - q \cdot s_E \sin(r_E + \varphi) \\ y &= y_A - q \cdot s_A \cos(r_A + \varphi) = y_E - q \cdot s_E \cos(r_E + \varphi) \end{aligned}\right\} \tag{6.6.1}$$

where $\vartheta_A = r_A$, $\vartheta_E = r_E$, $\sigma_A = s_A$, $\sigma_E = s_E$ and

$$\left.\begin{aligned} \xi_A &= s_A \sin r_A; \quad \eta_A = s_A \cos r_A \\ \xi_E &= s_E \sin r_E; \quad \eta_E = s_E \cos r_E \end{aligned}\right\}. \tag{6.6.2}$$

Example:

Pt. No.	r	s [m]	ξ [m]	η [m]	x [m]	y [m]
$1 = A$	0°25′07″	116.42	0.85	116.42	17520.66	06410.71
$3 = E$	171°23′20″	148.12	22.18	−146.45	17258.15	06435.37
$E - A$	170°58′13″	–	21.33	−262.87	−262.51	24.66

$\sigma_{AE} = 263.734$ m; $s_{AE} = 263.667$ m; $q = s_{AE} : \sigma_{AE} = 0.999\,746$

$\theta_{AE} = 175°21'40''$; $\alpha_{AE} = 275°21'59''$; $\varphi = 100°00'19''$

(θ_{AE} is azimuth in ξ, η system)

$y \quad = 17\,520.66 - q \cdot 116.42 \cdot \cos 111.582 = 17\,406.19$

$z \quad = 06\,410.71 - q \cdot 116.42 \cdot \sin 111.582 = 06\,431.77$

6.6.2 Point Determination With the Aid of a Helmert Transformation

The coordinates of a new point N are determined by least-squares adjustment, if the directions r_i and the distances s_i were measured from N to a number of fixed points $P_i (i > 2)$ (Fig. 6.6.3).

The Helmert transformation provides a solution that has the following advantages:

- no need for approximate coordinates
- no matrix operations
- no limit on the number of points used, even when operating with a small capacity computer.

It has the disadvantage that both direction and distance need be determined for all points. Thus measuring directions only, e.g., to inaccessible points, is no longer an option. Directions can be utilized when applying a general trigonometric solution with an observation equation for each observed quantity which is far more computer intensive. In the following, only the Helmert transformation is described, while the reader is referred to special geodetic literature for the general case.

As in the previous section, the new point N is again selected as the origin of a local ξ, η system, with the η-axis coinciding with the zero direction.

The points $P_1 \ldots P_r$ are coordinated in both the x, y and ξ, η systems, and thus represent identical points. Thus the coordinates of the new point N are determined with the aid of the transformation equations of the Helmert transformation.

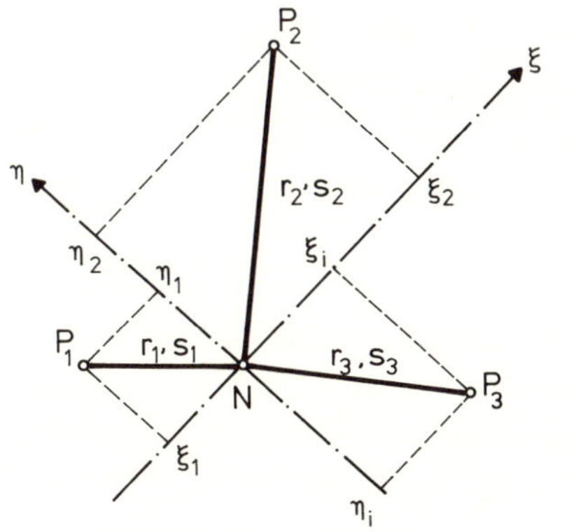

Fig. 6.6.3 Local coordinate system for point determination with Helmert transformation

ξ, η coordinates are computed for the polar points $P_1 \dots P_r$:

$$\xi_i = s_i \sin r_i; \quad \eta_i = s_i \cos r_i \quad (i = 1 \dots r)\,. \tag{6.6.3}$$

This leads to coordinates of the origin of the local system, which is the new point N:

$$\begin{aligned} x &= \frac{\Sigma x}{n} - a\frac{\Sigma\xi}{n} - o\frac{\Sigma\eta}{n} = x_s - a\xi_s - o\eta_s \\ y &= \frac{\Sigma y}{n} + o\frac{\Sigma\xi}{n} - a\frac{\Sigma\eta}{n} = y_s + o\xi_s - a\eta_s \end{aligned} \tag{6.6.4}$$

where the transformation parameters a and o are defined as in section 5.2.1.

Example:

Pt.	measured		local system		given coord's	
No.	r	s	ξ	η	x	y
		[m]	[m]	[m]	[m]	[m]
1	0°25′07″	116.42	0.85	116.52	17520.66	06410.71
2	74°30′36″	52.79	50.87	14.10	17411.28	06379.22
3	171°23′20″	148.12	22.18	−146.45	17258.15	06435.37
4	325°46′00″	209.65	−117.94	173.33	17597.31	06517.78
Σ	212°05′03″	526.98	− 44.04	157.40	69787.40	25743.08
	centre of gravity		− 11.01	39.35	17446.85	06435.77

	$\Delta\xi$	$\Delta\eta$	Δx	Δy	$[\Delta\eta^2 + \Delta\xi^2] = 75555.02$
1	11.86	77.07	73.81	−25.06	$[\Delta\xi \cdot \Delta x] = -23677.33$
2	61.88	− 25.25	− 35.57	−56.55	$[\Delta\xi \cdot \Delta y] = -12579.13$
3	33.19	−185.80	−188.70	− 0.40	$[\Delta\eta \cdot \Delta x] = 61805.77$
4	−106.93	133.98	150.46	82.01	$[\Delta\eta \cdot \Delta y] = 10558.53$

$(o = (61805.77 + 12579.13) : 75555.02 = 0.9845130)$
$a = (10558.53 + 23677.33) : 75555.02 = 0.1736324$
$(q = \sqrt{o^2 + a^2} = 0.999707; \quad \varphi = \arccos(a : q) = 100°00'07'')$
$x = 17446.85 - o \cdot 39.35 + a \cdot 11.01 = 17406.20$
$y = 06435.77 - a \cdot 39.35 - o \cdot 11.01 = 06431.76$

6.6.3 Accuracy of Points Determined with Directions and Distances

If the new point N is considered as being the origin of the local coordinate system, then the standard deviations of its coordinates equal the standard deviations of the translation parameters.

Assuming that $\sigma_0 = \sigma_r \cdot s = \sigma_s$, the standard deviation of the points is:

$$\sigma_p^2 = 2\sigma_0^2\left(\frac{1}{h} + \frac{x_N^2 + y_N^2}{\Sigma(x_i^2 + y_i^2)}\right) = 2\sigma_0^2\left(\frac{1}{n} + \frac{\xi_s^2 + \eta_s^2}{\Sigma(\Delta\xi^2 + \Delta\eta^2)}\right) \qquad (6.6.5)$$

where σ_r represents the standard deviation of the directions, σ_s the standard deviation of the distances, s denotes the distances, x_n, y_n the coordinates of the new point N, and x_i, y_i the coordinates of the fixed points.

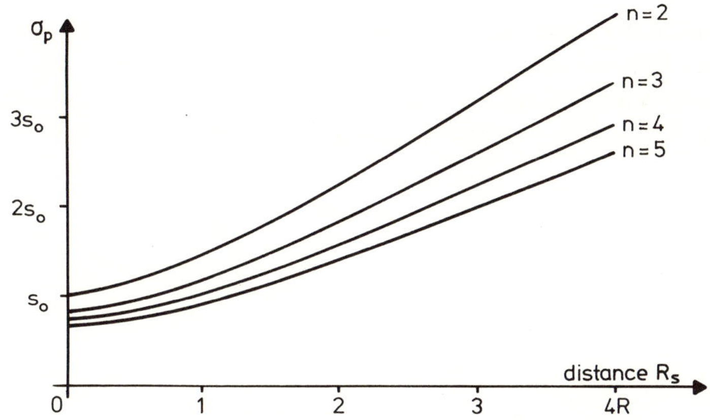

Fig. 6.6.4 Increase of the standard deviation of a new point, determined by directions and distances as a function of the radial distance

The accuracy of the point determination is visually represented by lines of equal point errors. These lines are circular arcs around the centre of gravity of the fixed points, which means that the minimum error $\sigma_{P_{\min}} = \sqrt{\frac{2}{n}} \cdot \sigma_0$ occurs at the centre of gravity of the fixed points. The accuracy only depends on the number of fixed points, not on their location. Fig. 6.6.4 illustrates this behaviour. In Fig. 6.6.5 several cases are illustrated with $n = 2, 3, 4, 5$ regularly spaced fixed points, all located on a circle with unit radius around the centre of gravity. Standard deviations of the new point as a

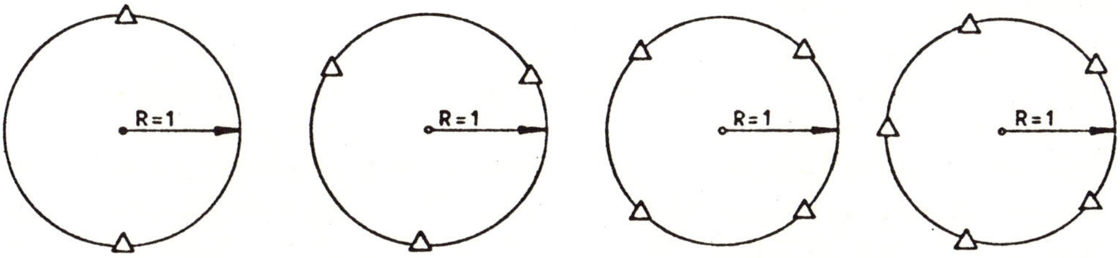

Fig. 6.6.5 Standard deviations of a new point, determined from $n = 2, 3, 4, 5$ fixed points

function of the distance R_s are obtained from Fig. 6.6.4 in units of the standard deviation σ_0 of unit weight.

As evident from Fig. 6.6.4, measurements to a fifth point do not significantly increase the accuracy.

6.7 Polar Surveys of Object Points

For detailed work, polar surveys are common (see also Chapter 8). As shown in Fig. 6.7.1, new points N_i are determined by measuring distances s_i and angles α_i with respect to a known reference direction. These polar coordinates are either used directly for plotting the survey plan, or else transformed into rectangular coordinates for further use.

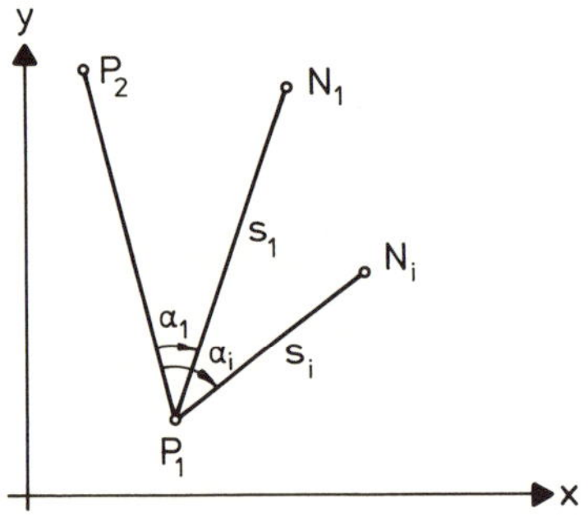

Fig. 6.7.1 Polar points

The measurements are performed with a tacheometer, usually consisting of an optical or electronic theodolite combined with an EDM instrument.The method is most suited for automation, provided all measurements are carried out with electronic instruments. Thus an *automatic data flow* can be generated from the field survey to the production of maps, plans and registers.

For economic reasons, directional *measurements are usually performed in one face only*. It is therefore necessary to determine collimation, horizontal axis, and height index errors prior to the survey, and apply mathematical corrections if needed (see section 10.3).

Observation stations are placed at 300 to 700 m intervals, if the terrain is reasonably open. The influence of refraction on the accuracy of the directions can be neglected for these distances. In more difficult terrain, i.e., dense vegetation or densely built-up areas, the stations are closer together. Fixed points may be selected as observation stations, but freely chosen unknown stations can be used as well. The latter case is referred to as "free stationing". It is far more flexible and less schematic than the orthogonal method (see section 8.3.1.1).

Although as a rule not all *polar points can be controlled in the field* by independent measurements, it is advisable to check important points in the field. Double measurements are only effective for directions, while it is better to check the distances by independent measurements, i.e. along the boundary, etc., rather than by repeating the EDM readings.

A better check is available by measuring points from two stations. Even then, an error will only show later when performing the computations.

The *polar method has the following advantages* over the methods described in section 8.3.1:

- there are no constraints of a rigid line net
- fixed points are needed every 300 to 500 m rather than 80 to 100 m as before
- points can be better secured, as they can be moved from high traffic areas to safer locations
- the automation proves to be highly economical
- the use of theodolite plus EDM instrument instead of rectangle prism and tape provides a higher accuracy.

6.7.1 Polar Survey From one Fixed Point

Given: Fixed points A and E plus the directions r_E and r_i, and the distance s_E and s_i (Fig. 6.7.2).

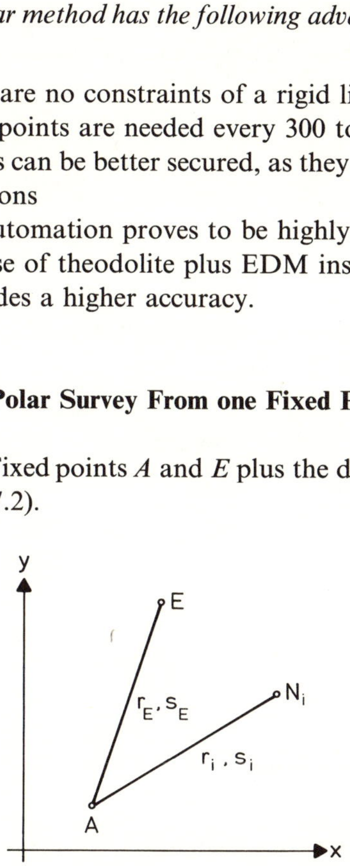

Fig. 6.7.2 Polar survey from one fixed point

The coordinates of the new points N are:

$$x_i = s_i \sin(r_i + \varphi)$$
$$y_i = s_i \cos(r_i + \varphi)\,. \tag{6.7.1}$$

The orientation unknown φ is obtained as

$$\varphi = \alpha_{A,E} - r_E\,. \tag{6.7.2}$$

For high precision work, it is necessary to check whether the scale of the measured distances matches the scale of the net of fixed points. Therefore, at least one distance between fixed points is to be measured, i.e., s_E, which provides the following scale factor:

$$q = \frac{s^*_{AE}}{s_E} \tag{6.7.3}$$

where

$$s^*_{AE} = \sqrt{(x_E - x_A)^2 + (y_E - y_A)^2}\,.$$

If q deviates significantly from one, then the coordinates of the new points are computed as

$$\begin{aligned} x_i &= q \cdot s_i \sin(r_i + \varphi) \\ y_i &= q \cdot s_i \cos(r_i + \varphi)\,. \end{aligned} \tag{6.7.4}$$

6.7.2 Polar Survey with Free Stationing

Given: Two fixed points A and E plus the directions r_A, r_E, r_i, and distances s_A, s_E, s_i as measured from a freely selected station S (Fig. 6.7.3).

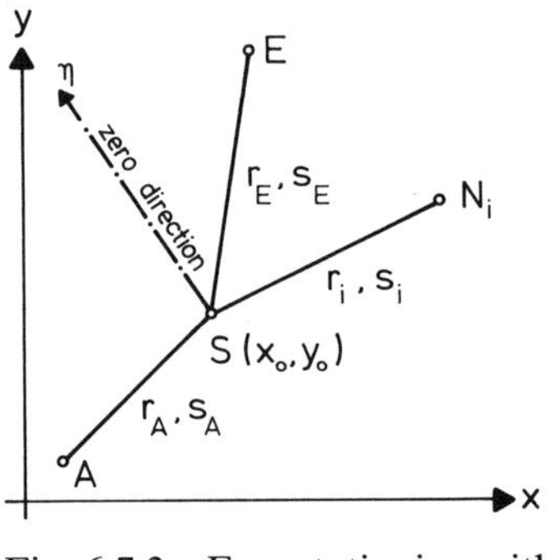

Fig. 6.7.3 Free stationing with two fixed points

The station is considered as the origin of a local ξ, η coordinate system whose η-axis corresponds to the measured zero direction. A and E are coordinated in the x, y system and also in the ξ, η system by their polar coordinates r, s, and thus represent identical points.

The new points N_i are then coordinated with the aid of a similarity transformation:

$$\begin{aligned} x_i &= x_0 + q \cdot s_i \sin(r_i + \varphi) \\ y_i &= y_0 + q \cdot s_i \cos(r_i + \varphi) \end{aligned} \tag{6.7.5}$$

with

$$\begin{aligned} x_0 &= x_A - q \cdot s_A \sin(r_A + \varphi) \\ y_0 &= y_A - q \cdot s_A \cos(r_A + \varphi)\,. \end{aligned} \tag{6.7.6}$$

Again the ξ, η coordinates are:

$$\left.\begin{aligned}\xi_A &= s_A \sin r_A; \quad \eta_A = s_A \cos r_A \\ \xi_E &= s_E \sin r_E; \quad \eta_E = s_E \cos r_E\end{aligned}\right\} \tag{6.7.7}$$

and q and φ are computed according to section 5.2.1.

6.7.3 Polar Survey with Free Stationing Using More than Two Fixed Points

Given: Fixed points $P_1 \ldots P_r$, directions $r_1 \ldots r_r$, and distances $s_1 \ldots s_r$ measured from a freely selected station S to the fixed points, and directions r_i and distances s_i from S to the new points N_i (Fig. 6.7.4).

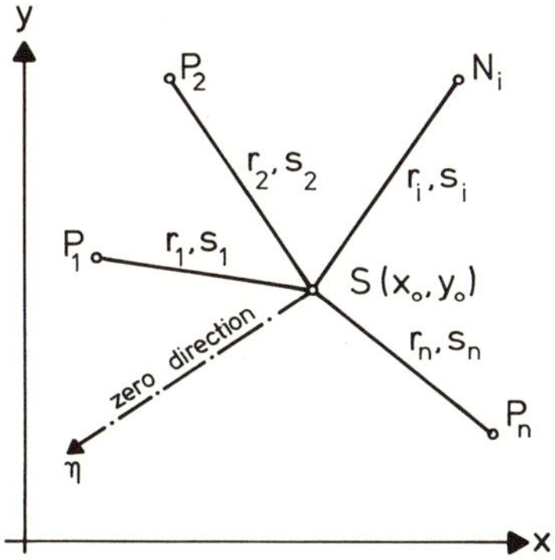

Fig. 6.7.4 Free stationing with more than two fixed points

Again, the station S is considered to be the origin of a local ξ, η coordinate system with η-axis coinciding with the measured zero direction. In preparation for the coordination computations, ξ, η coordinates are computed for the measured fixed points $P_i \ldots P_r$:

$$\xi_i = s_i \sin r_i; \quad \eta_i = s_i \cos r_i \quad (i = 1 \ \ldots \ n)\,. \tag{6.7.8}$$

These points also have known x, y coordinates and thus are identical points for the Helmert transformation which is used to compute the coordinates of the new points. As in section 5.2.1, the coordinates of the new points become:

$$\begin{aligned}x_i &= x_0 + a\xi_i + o\eta_i \quad (i = r + 1 \ \ldots \ n) \\ y_i &= y_0 - o\xi_i + a\eta_i\end{aligned} \tag{6.7.9}$$

where

$$\begin{aligned}x_0 &= \frac{\Sigma x_i}{n} - a\frac{\Sigma \xi_i}{n} - o\frac{\Sigma \eta_i}{n} \quad (i = 1 \ldots r) \\ y_0 &= \frac{\Sigma y_i}{n} + o\frac{\Sigma \xi_i}{n} - a\frac{\Sigma \eta_i}{n}\end{aligned} \tag{6.7.10}$$

and a, and o are defined as in equation (5.2.10).

Example: (Transformation parameters from example in section 6.6.2.)

Pt. No.	φ r	q s	r^0	sq	x	y
		[m]	[°]	[m]	[m]	[m]
S	$\varphi = 100°00'06''$	$q = 0.999\,707$	–	–	17406.20	06431.76
P_1	0°25′07″	116.42	100°25′13″	116.386	17520.67	06410.71
P_2	74°30′36″	52.79	174°30′42″	52.775	411.25	379.23
P_3	171°23′20″	148.12	271°23′26″	148.077	258.17	435.35
P_4	325°46′00″	209.65	65°46′06″	209.589	597.32	517.78
P_5	64°23′16″	83.15	164°23′22″	83.126	428.57	351.70
...	...	...	...	...	...	...
P_n	280°59′50″	114.61	20°59′56″	114.576	17447.26	06538.73

6.7.4 Accuracy of Polar Points

The point error σ_p of a polar point amounts to

$$\sigma_p = \sqrt{(s \cdot \sigma_{r^0})^2 + \sigma_s^2} \qquad (6.7.11)$$

where

σ_{r^0} is the standard deviation of oriented directions (see section 6.3.2),

σ_s is the standard deviation of distances,

$s \cdot \sigma_{r^0}$ describes the standard deviation of the new point normal to the determining ray, while σ_s describes the standard deviation along the determining ray. Normally, the surveys are designed such that both are equal, i.e.,

$$s \cdot \sigma_{r^0} = \sigma_s = \sigma_0 \,. \qquad (6.7.12)$$

Thus

$$\sigma_p = \sqrt{2}\,\sigma_0 \,. \qquad (6.7.13)$$

Under these conditions, the following holds (*Ruopp* 1971):

- The accuracy of a polar point is independent of the location of the terrain point; it depends primarily on the location of the polar points within the frame of fixed points.
- The lines of equal standard deviation of the coordinates are circles around the centre of gravity of the fixed points.
- The minimum error occurs at the centre of gravity.

The increase of the coordinate error with increasing radial distance is of interest (Fig. 6.7.5).

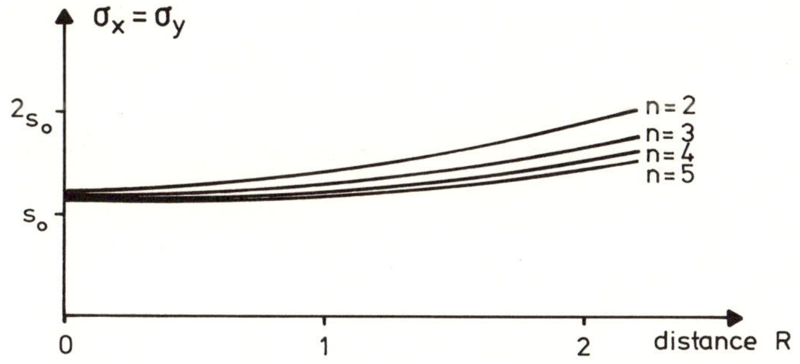

Fig. 6.7.5 Standard deviations σ_x and σ_y of the coordinates of polar points with error free fixed points and free stationing

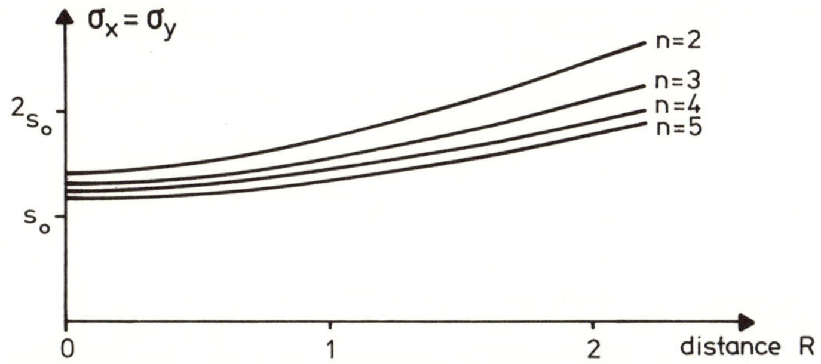

Fig. 6.7.6 Standard deviations σ_x and σ_y of the coordinates of polar points with free stationing, when the fixed points have a standard deviation $\sigma_x = \sigma_y = \sigma_0$.

The fixed points are arranged as shown in Fig. 6.6.5. For the first analysis, the fixed points are considered to be free of errors, while in the second analysis they are not. This leads to the results illustrated in Figs. 6.7.5 or 6.7.6.

From the figures it is apparent, that

- additional fixed points provide little improvement of accuracy;
- the polar points should be located within a circle with $R = 1$ around the centre of gravity of the fixed points.

6.8 Traversing

Until about 1960, traversing was considered to be inferior to triangulation, both in geometric strength and in accuracy, even though it has been used in surveying since about 1830. Thus the method served only for low order point densification, i.e., stations for detail surveys. In the meantime, accuracy relationships have reversed, because the angular accuracies when using forced centring, and distance accuracies when using EDM instruments have vastly improved. In addition, there are numerous other advantages, such as

- the net configuration can be arranged to suit local conditions
- it is not usually necessary to erect signals, which is costly and indispensable for triangulation
- work progresses faster.

Traversing is utilized for many types of work, e.g.,

- establishment of first-order network
- densification of second- and third-order trigonometric nets
- determination of fixed points (observation stations) for detail surveys
- determination of observation stations for topographic surveys of smaller areas
- staking.

The lengths of the sides may vary considerably. First-order base networks have legs of up to 50 km in length, while densification deals with 0.4 to 2.5 km long legs, and for the determination of observation stations they range from about 1 km to 50 m or less.

6.8.1 Layout and Measurement of Traverse Nets

6.8.1.1 Traverses, Ring Traverses, Traverse Nets

In traversing, the basic frame consists of an irregular polygon, with the angles being measured at each corner, and the distances between neighbouring points (Fig. 6.8.1). The measurements are arranged such that rectangular coordinates can be computed for all points. Depending on the survey situation, a ring traverse, a stretched traverse, or a traverse net is utilized.

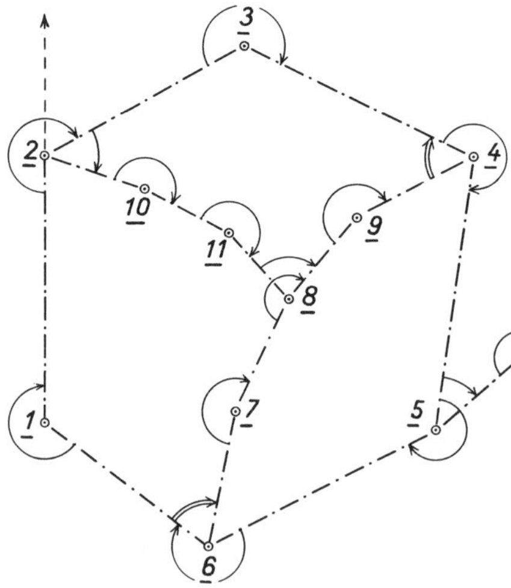

Fig. 6.8.1 Traverse net

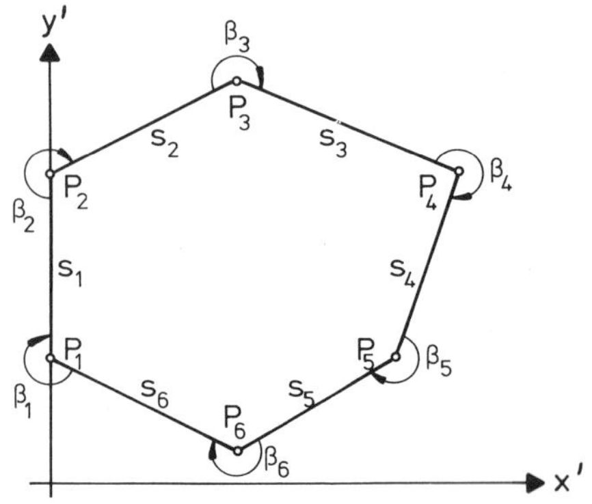

Fig. 6.8.2 Ring traverse

The following types can be distinguished.

Closed traverse, which begins and ends at previously coordinated points (i.e., traverse 6, 7, 8, 9, 4 in Fig. 6.8.1). In addition to the traverse angles and legs, reference angles are measured as indicated by double arrows in Fig. 6.8.1.

A *ring polygon* is a special case of a closed traverse, utilizing the same fixed point as the starting and terminal point (Fig. 6.8.2). It is primarily used to delineate closed boundaries and to provide a framework for detail surveys within an area.

An *open traverse* (traverse 5, 12, 13 in Fig. 6.8.1) is only connected at its starting point. Open traverses have an extremely unfavourable error propagation and should be avoided. This is, however, not always possible, e.g., for detail surveys in interior courts, dead end streets, or when staking a tunnel.

A *free traverse* is an end in itself and used, e.g., for staking long straight lines whose end points are given but not intervisible, and similar problems in highway surveys.

A *traverse net* consists of several traverses. Even the most difficult terrain can be made accessible for a coordinate survey by means of traverse nets.

The least-squares adjustment of traverse nets provides the most accurate results for the coordinates of traverse points (*Gründig* and *Joeckel* 1983).

6.8.1.2 Selection of Traverse Points

The selection of traverse points depends on the problem at hand. For the densification of third- and fourth-order trigonometric networks, the reconnaissance considers weather effect on range and accuracy of angular and distance measurements and tries to avoid the need for erecting costly signals. Thus is it most avantageous to work with 0.8 to 1 km long traverse legs (*Augath* 1976). In this case, not all traverse points become coordinated monuments; some are only temporarily used for the measurements and subsequent computations (Fig. 7.2.2). The points are placed such that a rather uniform distribution and a good connection to neighbouring points is obtained.

Traverses often follow transportation lines, water courses, or property boundaries. They are usually established such that as much information can be referenced as possible. Traverse points are selected at locations where the instrument can be set up without too many problems, and where clear visibility of the neighbouring stations exists.

The length of the legs depends on the density of the detail as well as on topographic conditions.

All points are monumented, e.g., by pins, pipes or bolts, and referenced geometrically to neighbouring house corners, power poles, etc.

6.8.1.3 Measuring Distances and Angles

Traverse legs are nowadays best measured with electronic distance meters (section 4.4.7). Tapes or the subtence bar (section 4.3.2) are used only in exceptional cases.

Depending on the projects, theodolites of low, medium or high accuracy (section 3.3) are used for angular measurements. Often a tacheometer is used, i.e., theodolite and distance meter are one unit (section 12.2).

Depending on the project, different accuracies are required. Densification of third- and fourth-order networks in densely populated areas require, for instance,

- a standard deviation σ_s of 5 mm for distances
- a standard deviation σ_r of 0.5 mgon ($1 \div 2''$) for angles.

While this can be relaxed for detail surveys to

$$\sigma_s = 1 \ldots 2 \text{ cm}$$
$$\sigma_r = 1 \ldots 2 \text{ mgon} \qquad (3'' \ldots 5'')$$

If at all possible, directional and distance accuracies should match each other, i.e., the lateral error caused by angular measurement errors should be equal to the longitudinal one, caused by distance measurement errors. Therefore

$$\sigma_0 = \sigma_s = s \cdot \sigma_r \, .$$

The accuracy required for precision traversing is primarily achieved by accurate centring. With the naked eye, a centring error of 0.5 cm is easily possible for centring instruments and/or targets. If normal to the traverse leg, such an error causes a rotation of 2.1 mgon ($\approx 7''$) for a 150 m leg. Thus three centring errors of this magnitude may cause an angular error of 8.4 mgon ($\approx 27''$) as illustrated in Fig. 6.8.3.

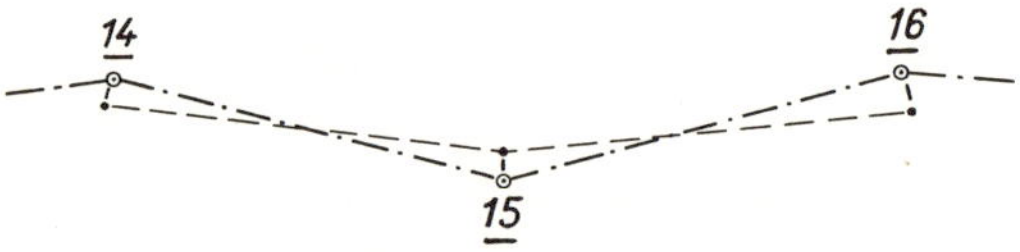

Fig. 6.8.3 Centring errors

Thus the effects of centring errors exceed significantly the reading errors at the theodolite.

This can be markedly improved by using instruments with centring rods and levels for the range poles. Only forced centring (section 3.5.4), however, overcomes this problem.

6.8.2 Traverse Computations

First the oriented directions r_i^0 and distances of the traverse legs s_i are determined. These are then used to coordinate the new points by consecutively applying the first geodetic problem (section 5.2.1), starting at a fixed point.

6.8.2.1 Closed Traverses

The traverse, illustrated in Fig. 6.8.4, is used to demonstrate the computation.

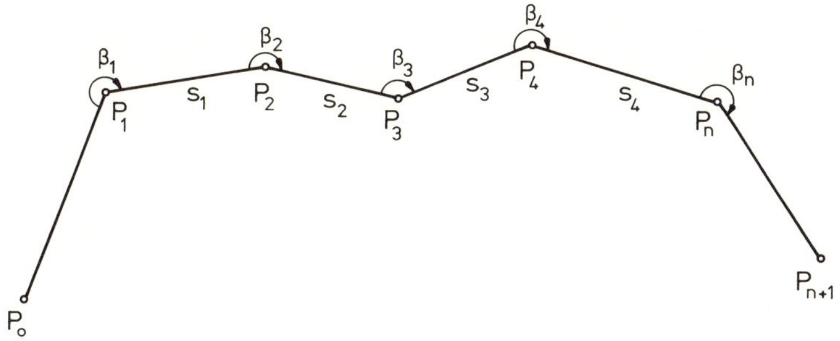

Fig. 6.8.4 Closed traverse

The new points P_2, P_3 and P_4 are uniquely determined by means of the angles β_1, β_2, β_3 and the distances s_1, s_2, s_3 provided the measurements are free of errors. Three superfluous measurements are given with s_4, β_4 and β_n. This overdetermination and the unavoidable measuring errors cause discrepancies which can be eliminated in two different ways.

a) The traverse is computed approximately as a free traverse (section 6.8.1) starting at one end and afterwards fitted in between both fixed points with the aid of a similarity transformation.

b) The angular closure is distributed evenly among the measured traverse angles, then the traverse is approximately computed.

Afterwards the remaining coordinate misclosure is distributed among the coordinate differences in relation to the individual leg lengths. Then the final coordinates are computed.

The first method does not place any demand on the geometry of the traverse, as long as angular and distance measurements are matched in accuracy. The traverses can be curved and thus fit optimally to the requirements of a detail survey. This method is also preferred when the newly measured traverse has a higher internal accuracy than the given network of fixed points. Thus it is ensured that the shape of the traverse is not changed by discrepancies in the coordinates of the closure points. Unfortunately, the angle β_n is only used to check the traverse angels β_i, but does not enter the computations.

The second method should be restricted to reasonably stretched traverses. This method was preferred in the past, because distances measured with tapes were less accurate than angles. The solution as described corresponds to a rigid least-squares adjustment.

Prior to the traverse computations, the azimuths α_0 and α_n are computed according to section 5.1.2 (Fig. 6.8.5).

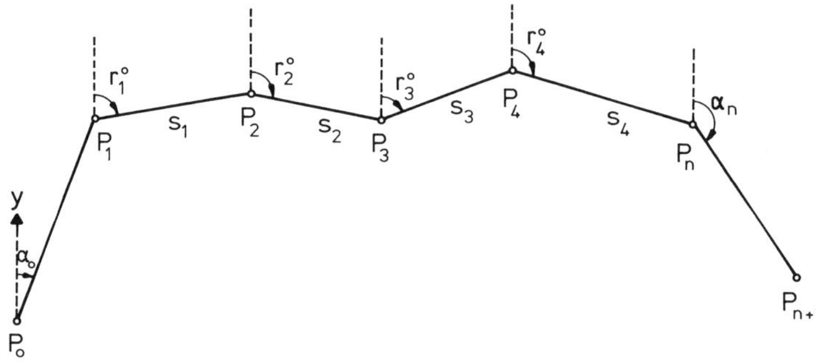

Fig. 6.8.5 Polar elements, determining a traverse

1st step: Checking of the measured angles β_i.

The sum $\alpha_0 + \Sigma\beta_i$ should be equal to β_n, except for a multiple of a 200 gon (180°). The angular misclosure due to measuring uncertainties becomes

$$w_\beta = (\alpha_n - \alpha_0) - \{\Sigma\beta + n \cdot 180° - a \cdot 360°\} . \tag{6.8.1}$$

If it is less than the given error limit, then the traverse computation continues.

2nd step: The traverse is now computed in a fictitious x', y' system starting with $x' = y' = 0$. The directions of the traverse legs are oriented by consecutively adding $\beta_i + 180°$ to the computed azimuth α_0:

$$\begin{aligned} r_1^0 &= \alpha_0 + \beta_1 + 180° \\ r_2^0 &= r_1^0 + \beta_2 + 180° \\ &\vdots \end{aligned} \tag{6.8.2}$$

Check:

$$\alpha_n = r_4^0 + \beta_n + 180° .$$

The oriented directions r_i^0 and the distances s_i are used to consecutively compute the local coordinates:

$$\begin{aligned} x_2' &= x_1' + s_1 \sin r_1^0; \quad y_2' = y_1' + s_1 \cos r_1^0 \\ x_3' &= x_2' + s_2 \sin r_2^0; \quad y_3' = y_2' + s_2 \cos r_2^0 \\ &\vdots \\ x_n' &= x_4' + s_A \sin r_4^0; \quad y_n' = y_4' + s_4 \cos r_4^0 \end{aligned} \tag{6.8.3}$$

P_n and P'_n would coincide if the measurements were perfect. Instead, lateral and longitudinal residuals C and L are obtained (Fig. 6.8.6):

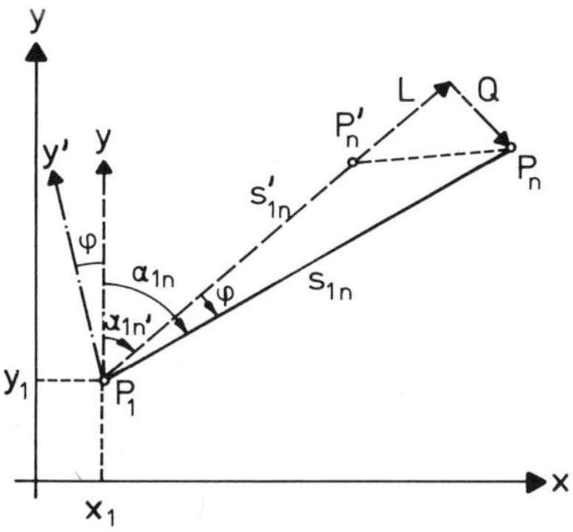

Fig. 6.8.6 Lateral and longitudinal residuals in a traverse

$$L \approx s_{1,n} - s'_{1,n}$$
$$C \approx \frac{\varphi \cdot \pi}{180^\circ} \cdot s_{1,n} \tag{6.8.4}$$

with

$$s_{1,n} = \sqrt{(x_n - x_1)^2 + (y_n - y_1)^2}; \quad s'_{1,n} = \sqrt{x_n'^2 + y_n'^2} \tag{6.8.5}$$

$$\varphi = \alpha_{1,n} - \alpha_{1,n'} = \arcsin \frac{x_n - x_1}{s_{1,n}} - \arcsin \frac{x'_n}{s'_{1,n}}. \tag{6.8.6}$$

If L and C fit within the given error limits, then the final coordinates are computed.

3rd step: The closure points P_1, P_n or P'_n can be considered as identical points of two coordinate systems, which are rotated by the angle φ, shifted by x, and y, and have different scales.

The final coordinates of the new points are then obtained by similarity transformation. With P_1 as the rotational centre we get:

$$x_2 = x_1 + ax'_1 + oy'_1; \quad y_2 = y_1 - ox'_1 + ay'_1$$
$$x_3 = x_1 + ax'_2 + oy'_2; \quad y_3 = y_1 - ox'_2 + ay'_2 \tag{6.8.7}$$

where

$$a = q \cdot \sin\varphi; \quad o = q \cdot \cos\varphi$$

$$q = \frac{s_{1,n}}{s'_{1,n}},$$

and φ as above in equation (6.8.6).

Example: See Table 6.8.1

2nd Method:

1st Step: The angular closure w_β is computed as in the first method. If it is less than a given error limit, then it is evenly distributed among the traverse angles $\beta_1 \ldots \beta_n$.

Table 6.8.1

	α	s [m]	Δx x [m]	Δy y [m]	Pt. No.
			35 84 099.32	55 36 105.52	P_0
α_0	13° 28′ 33″				
			84 292.22	36 910.51	P_1
$\alpha_{1,n}, s_{1,n}$	250° 43′ 20″	1 120.010			
			−1 057.21 83 235.01	−369.77 36 540.74	P_n
α_n	116° 43′ 39″				
$\alpha_n - \alpha_0$	103° 15′ 06″		35 84 099.32	55 36 105.52	P_{n+1}

	r^0 β	s	x' x	y' y	Pt. No.
α_0	13° 28′ 33″	–	0.000	0.000	
			84 292.22	*36 910.51*	P_1
β_1	36° 55′ 03″				
r_1^0, s_1	230° 23′ 36″	234.995	−181.049	−149.813	
			84 111.19	36 760.73	2
	205° 17′ 30″				
	255° 41′ 06″	242.206	−415.735	−209.699	
	146° 10′ 39″		83 876.55	36 700.87	3
	221° 51′ 45″	248.266	−581.414	−394.594	
	225° 29′ 02″		83 710.88	36 516.02	4
	267° 20′ 47″	269.017	−850.143	−407.049	
			83 442.20	36 503.58	5
	192° 49′ 07″				
r_{n-1}, s_{n-1}	280° 09′ 54″	210.534	−1 057.372	−369.893	
β_n	16° 34′ 08″	–	*83 235.01*	*36 540.74*	$P_6 = P_n$
{ }*	103° 15′ 30″				
w_β	−0° 00′ 24″		$C = +0.062$	$L = -0.194$	C, L
$\alpha_{1,n'}, s'_{1,n}$	250° 43′ 09″	1 120.204			
φ, q	−0° 00′ 11″	0.999 826 8	0.000 055 0	0.999 826 8	o, a

* $\{[\Sigma\beta] + n \cdot 180° - a \cdot 360°\}$

2nd Step: Now approximate coordinate differences between the points are computed. For this, the oriented azimuths are calculated by consecutively adding the corrected traverse angle to the starting azimuth α_0.

$$r_1^0 = \alpha_0 + \beta_1 + \frac{w_\beta}{n} + 180^\circ$$
$$r_2^0 = r_1^0 + \beta_2 + \frac{w_\beta}{n} + 180^\circ . \tag{6.8.8}$$

Check:

$$\alpha_n = r_4^0 + \beta_4 + \frac{w_\beta}{n} + 180^\circ.$$

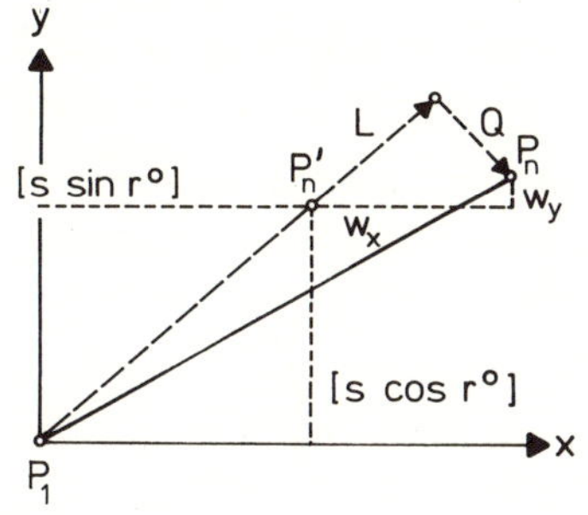

Fig. 6.8.7 Longitudinal and lateral errors

Now the coordinate differences are computed consecutively, using the oriented directions r_i^0 and the distances s_i.

$$\begin{aligned} \Delta x_1 &= s_1 \sin r_1^0; & \Delta y_1 &= s_1 \cos r_1^0 \\ \Delta x_2 &= s_2 \sin r_2^0; & \Delta y_2 &= s_2 \cos r_2^0 \\ &\vdots \\ \Delta x_4 &= s_4 \sin r_4^0; & \Delta y_4 &= s_4 \cos r_4^0 . \end{aligned} \tag{6.8.9}$$

This loads the following closure errors:

$$\begin{aligned} w_x &= (x_n - x_1) - \Sigma s \cdot \sin r^0 \\ w_y &= (y_n - y_1) - \Sigma s \cdot \cos r^0 \end{aligned} \tag{6.8.10}$$

which provide the lateral and longitudinal errors (Fig. 6.8.7).

$$L = \frac{w_x \Sigma s \cdot \sin r^0 + w_y \Sigma s \cdot \cos r^0}{\sqrt{(\Sigma s \cdot \sin r^0)^2 + (\Sigma s \cdot \cos r^0)^2}}$$
$$C = \frac{w_x \Sigma s \cdot \cos r^0 - w_y \Sigma s \cdot \sin r^0}{\sqrt{(\Sigma s \cdot \sin r^0)^2 + (\Sigma s \cdot \cos r^0)^2}} . \tag{6.8.11}$$

If L and C fall within the given error limits, then the coordinate closure errors are distributed proportionally to the distances s_i to Δx_i and Δy_i.

3rd Step: The coordinates of the traverse points are obtained by consecutively adding the corrected coordinate differences to the previous coordinates, starting at x_1 and y_1. At the end, the values x_n and y_n have to be obtained:

$$x_2 = x_1 + s_1 \sin r_1^0 + \frac{w_x}{\Sigma s} \cdot s_1; \quad y_2 = y_1 + s_1 \cos r_1^0 + \frac{w_y}{\Sigma s} \cdot s_1$$
$$x_3 = x_2 + s_2 \sin r_2^0 + \frac{w_x}{\Sigma s} \cdot s_2; \quad y_3 = y_2 + s_2 \cos r_2^0 + \frac{w_y}{\Sigma s} \cdot s_2 . \tag{6.8.12}$$

Example: Table 6.8.2

6.8.2.2 Ring Traverse

It is assumed that the traverse 1, 2, 3, 4, 5, 6 in Fig. 6.8.2 has no connections to a fixed coordinate system. However, all angles and distances are measured.

A local coordinate system is introduced by, for instance, selecting the side 1, 2 as the y-axis. To avoid negative values, the origin is moved, e.g., to $y_1 = 500$ m.

When plotting a ring traverse with ruler and protractor it becomes apparent that the angles at points 1 and 6 and the distance 6–1 are not required for a unique solution. As a result of the three redundant measurements, three closure discrepancies will appear when computing the traverse.

This is equivalent to the closed traverse. The computations are equivalent to the second method.

1st step: The sum of the exterior angles of a polygon with n sides equals $(n - 2) \cdot 180°$. The sum of the observed angles β_i should give the same amount:

$$w_\beta = (n + 2) \cdot 180° - \Sigma \beta - a \cdot 360° . \tag{6.8.13}$$

If w_β does not exceed the easily obtainable limit of $1' \sqrt{n} \, (1' \sqrt{n})$, then w_β is equally distributed to all traverse angles.

2nd step: The azimuths are obtained by sequentially adding the corrected traverse angles $\beta_i \pm 180°$ (200 gon) to the starting azimuth $r_1^0 = 0°$ as follows:

$$r_2^0 = r_1^0 + \beta_2 + \frac{w_\beta}{n} + 180° . \tag{6.8.14}$$

$$\vdots$$

Check:

$$r_1^0 = r_6^0 + \beta_1 + \frac{w_\beta}{n} + 180° .$$

Generally:

$$r_i^0 = r_{i-1}^0 + \beta_i + \frac{w_\beta}{n} + 180° .$$

Then:

$$\Delta x_i' = s_i \sin r_i^0; \quad \Delta y_i' = s_i \cos r_i^0$$

and

Table 6.8.2 Computation of a closed traverse

Pt.	β r^0	s	x $\Delta x = s \sin r^0$	y $\Delta y = s \cos r^0$
		[m]	[m]	[m]
1	2	3	4	5
P_0			0.00	500.00
	129°25′17″ 4″			
P_1	64°02′16″		197.20	352.69
	13°27′29″ 4″	*148.11*	−3 34.47	−1 144.04
P_2	196°12′40″		213.64	496.72
	29°40′05″ 4″	*135.25*	−2 66.95	117.52
P_3	190°22′46″		280.57	614.24
	40°02′47″ 4″	*121.17*	−2 77.96	92.76
P_4	191°05′55″		358.51	707.00
	51°08′38″ 4″	*138.28*	−2 107.68	−1 86.76
P_5	65°48′07″		466.17	793.75
	296°56′42″			
P_6			223.81	916.95
comp. from coords.	167°31′44″ 167°31′25″		287.06 286.97	441.08 441.06
w_β =	19″		$-0.09 = w_x$	$-0.02 = w_y$

$$w_{x'} = 0 - \Sigma s_i \sin r_i^0; \quad w_{y'} = 0 - \Sigma s_i \cos r_i^0 \tag{6.8.15}$$
$$w_s = \sqrt{w_{x'}^2 + w_{y'}^2} .$$

If $w_{x'}$, $w_{y'}$ and w_s fall within acceptable limits, then the final coordinates are computed:

$$x_i' = x_{i-1}' + s_{i-1} \sin r_{i-1}^0 + \frac{w_{x'}}{\Sigma s} \cdot s_{i-1}$$
$$y_i' = y_{i-1}' + s_{i-1} \cos r_{i-1}^0 + \frac{w_{y'}}{\Sigma s} \cdot s_{i-1} . \tag{6.8.16}$$

Example: Table 6.8.3

6.8.2.3 Traverses with Incomplete Closure, Open Traverses

An *open traverse*, such as traverse 5, 12, 13 in Fig. 6.8.1, has a directional connection at the starting point only. With it, the traverse can be coordinated like a closed traverse, except that there are no closure errors. This eliminates the check on both computation and measurement. Therefore, if such a traverse cannot be avoided, extreme care has to be taken both for measurements and centring as well as the computations.

A *traverse with absolutely no connection* is computed like an open traverse by arbitrarily selecting a coordinate system (i.e. x- or y-axis along the first leg).

A *traverse with coordinate closure at both ends* but no angular closure is computed like a free traverse and then transformed onto the coordinate system using a coordinate transformation (section 5.3).

6.8.2.4 Blunder Search

Angular blunders manifest themselves in the angular closure, and distance blunders in the coordinate closure, provided the traverse is properly closed. In both cases, it is often possible to localize the blunder.

To find an *angular blunder*, one computes the traverse first forward from A and then backward from E without distributing the angular errors. The point of intersection between the foreward and backward computation represents the location where the angular blunder has occurred, provided only one blunder was made. The same procedure can be performed graphically with ruler and protractor (Fig. 6.8.8).

A *distance blunder* causes a shift of the traverse section following the incorrect leg in direction of this leg by a distance error f_s. First, one checks whether f_s is near a round number (i.e., 5 m or 20 m). Then one checks whether the azimuth of f_s is similar to the azimuth of one of the traverse legs. If this is the case, the blunder likely occurred when measuring this leg and can be corrected by remeasurement. This method does not work if several legs have nearly the same azimuth.

Table 6.8.3 Computation of a ring traverse

Pt.	β r^0	s	x $\Delta x = s \sin r^0$	y $\Delta y = s \cos r^0$
	[°]	[m]	[m]	[m]
1	2	3	4	5
1			0.00	500.00
	0.0000 +1.3″	*294.13*	+4 0.00	−3 294.13
2	241°13′27″		0.04	794.10
	61°13′28″ +1.3″	*255.25*	+4 223.73	−3 122.88
3	235°43′12″		223.81	916.95
	116°56′42″ +1.3″	*271.84*	+4 242.32	−3 −123.17
4	250°20′12″		466.17	793.75
	187°16′55″ +1.3″	*308.14*	+5 −39.07	−4 −305.64
5	234°05′34″		427.15	488.07
	241°22′31″ +1.3″	*282.51*	+4 −247.99	−3 −135.35
6	248°02′45″		179.20	352.69
	309°25′17″ +1.3″	*232.03*	−4 −179.24	−3 147.34
1	230°34′42″		0.00	500.00
2	0°00′00″		−0.25 0.00	0.19 0.00
w_β	359°59′52″ 8″	$w_s = 0.31$	$w_x = 0.25$	$w_y = -0.19$

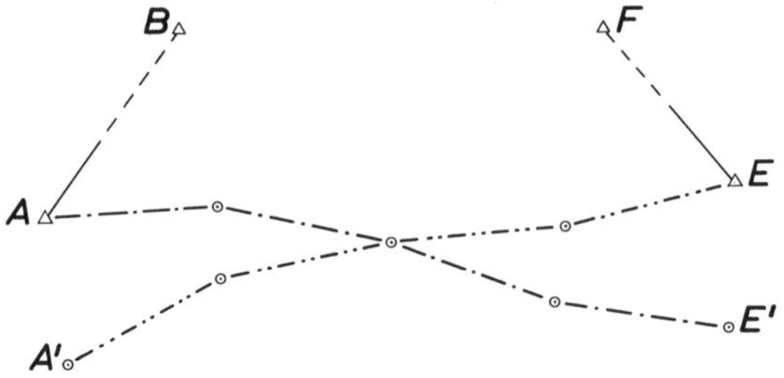

Fig. 6.8.8 Angular blunder

6.8.3 Accuracy of Traversing

The following treatment is based primarily on fully closed traverses.

6.8.3.1 Error Theory of a Stretched Traverse

The effects of measuring uncertainties in traverses are difficult to predict, because many different measuring elements are used with only three redundant observations. When looking at stretched traverses, distance errors affect primarily the longitudinal direction, and angular errors the lateral direction.

As a base for the evaluation, the (true) longitudinal error L and the true lateral error C are either graphically determined or derived using equation (6.8.11).

For a graphical solution, $\Sigma s \sin\alpha$ and $\Sigma s \cos\alpha$ are plotted on a map (i.e., 1 : 5000) from A which gives E'. From E, W_x and W_y are plotted at 1 : 5 or 1 : 10. The thus obtained point E is projected onto the line AE' or its extension. Then L and Q are read at enlarged scale.

The *longitudinal error* primarily shows the systematic and random error components of the distance measurements. The systematic components are proportional to the lengths of the legs and thus have to be distributed proportionally. The random components increase with the square root of the distance. If, however, the end points are fixed, these random components are also to be distributed proportional to the distance as a rigid least-squares adjustment will prove. Therefore, the distribution of both systematic and random error components follows the same principle which makes a separation unnecessary. Thus, the distribution of the coordinate closure errors, as performed in section 6.8.2, conforms with the theory provided that all distances were measured with the same instruments.

The *lateral error* primarily shows inaccuracies of angular measurements. To evaluate this, an open traverse is first considered. An angular error rotates the remaining part of the traverse. Therefore observation errors at the beginning of the traverse are especially critical, while towards the end the influence is less significant. In fully closed traverses – the usual case – the angular closure error is evenly distributed on the observed angles. The traverse is then computed as if only angular closure but no coordinate closure were available. Based on error theoretical investigations, a standard lateral deviation σ_q is to be expected at the end of a stretched traverse with equal legs as

$$\sigma_p = \Sigma s \cdot \frac{\sigma}{\varrho} \sqrt{\frac{n(n+1)}{12(n-1)}} \tag{6.8.17}$$

where σ is the standard deviation of each of the n traverse angles. This expression provides the basis for establishing error specifications.

By distributing the coordinate closure errors according to section 6.8.2, the lateral error is eliminated together with the longitudinal one. The lateral uncertainty in the middle of the traverse amounts approximately to

$$\sigma_{qm} = \Sigma s \cdot \frac{\sigma}{\varrho} \sqrt{\frac{n}{192}} \,. \tag{6.8.18}$$

Numerically, this represents about one quarter of the amount obtained with equation (6.8.16).

The simple method of distribution given in section 6.8.2 is therefore in reasonably good agreement with the error theory. The results, however, become worse for larger closing errors and for traverses that differ from the ideal case of a stretched traverse with equal sides.

6.8.3.2 Specifications

Survey specifications should represent the technical capabilities and the legal requirements. When setting them, practical experience of the surveying institutions is also considered. Basically, however, they represent three times the expected standard deviation, for reasons given in section 1.4. Therefore, one should normally not exceed one third of their value.

In Germany, the following specifications have been valid since 1970:

I. *Basic Considerations*

a) The formerly applied distinction between favourable, average and difficult conditions for distance measurements has been dropped together with the division into primary and secondary traverses.

b) For certain areas (i.e., urban areas) the specifications may be reduced to $^2/_3$ or $^1/_2$ of the following values under II below.

c) Under certain special circumstances differences up to 1.5 times the amount may be acceptable.

II. *Specific Error Limits for Traverses of up to 2 km Length*

a) Error limit E_α for angular closure

$$E_\alpha[°] = 0.5' + 80'/\Sigma s \cdot (n-1)\sqrt{n}$$

where Σs is the sum of all traverse legs, and n is the number of traverse points. 0.5′ in the first term provides the lower limit, while 80′ in the second one is the upper limit. These differences are acceptable because of different conditions in the various states.

b) Error limit E_s for distances

$$E_s[\text{m}] = 0.06 + 0.00015\,S - 0.04\sqrt{s}$$

where S is the total distance between the terminal points of the traverse in metres.

c) Error limit E_q for lateral error

$$E_q[\text{m}] = 0.06 + 0.00007\,S + 0.007\,n\sqrt{n}$$

with S as in II b and n as in II a).

d) The largest acceptable difference for two measurements of a traverse leg

$$D_s[\text{m}] = 0.002 + 0.006\sqrt{n}\ .$$

III. *The Largest Acceptable Difference for Two Measurements of any Traverse Leg*

$$D[\text{m}] = 0.05 + 0.003\,s + 0.08\sqrt{S}$$

Considerations I b and I c apply here as well with 1.5 being replaced by 2.0 for I c.

6.8.4 Special Cases in Traversing

6.8.4.1 Closure to Inaccessible Points

The starting point P_1 is inaccessible, which means that the closure angle $P_0\,P_1\,2$ and the side s_1 cannot be observed directly (Fig. 6.8.9).

By transferring the coordinates of P_1 to point 2, this problem can be solved. First, an

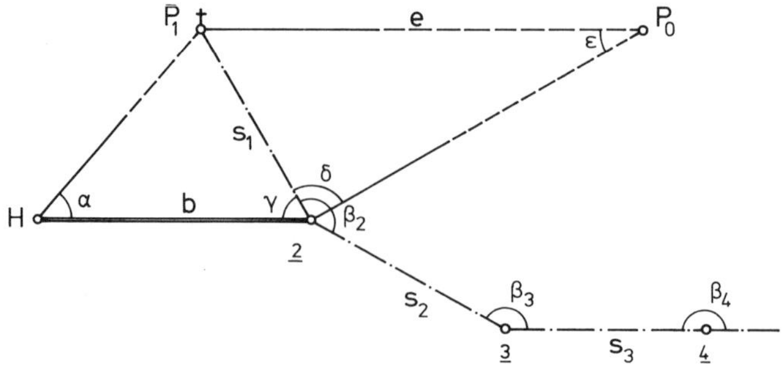

Fig. 6.8.9 In accessible closure point

auxiliary point H is selected. Then the base b is measured as well as the angle α at H and the directions to H, P_1, 0 and 3 at point 2. These observations are sufficient to derive the angles β_2, γ and δ. Then the distance s_1 is computed as

$$s_1 = b \frac{\sin \alpha}{\sin(\alpha + \gamma)} . \tag{6.8.19}$$

After computing the azimuth $\alpha_{0,1} = \alpha_0$ and the distance e from the coordinates of P_0 and P_1 according to section 6.8.2, the angle ε at P_0 is obtained as

$$\sin \varepsilon = \frac{s_1}{e} \sin \delta . \tag{6.8.20}$$

The azimuth r_1^0, is then

$$r_1^0 = \alpha_0 - (\delta + \varepsilon) + 180° . \tag{6.8.21}$$

As a check, the coordinates of point 2 can be determined using another auxiliary triangle.

6.8.4.2 Elimination of Short Legs

Short legs have a very unfavourable influence on the angular measurements because the influence of dangerous centring errors increases with decreasing length of the legs.

To eliminate the short leg 2–3 in Fig. 6.8.10, one observes at points 2 and 5 the directions to points 5 or 2, in addition to the directions to the neighbouring points. In the office the length of the auxiliary leg 2–5 is computed using the section 2, 3, 4, 5 of the traverse. Then the traverse is computed as 1, 2, 5, 6 ... with the section 2, 3, 4, 5 being subsequently connected.

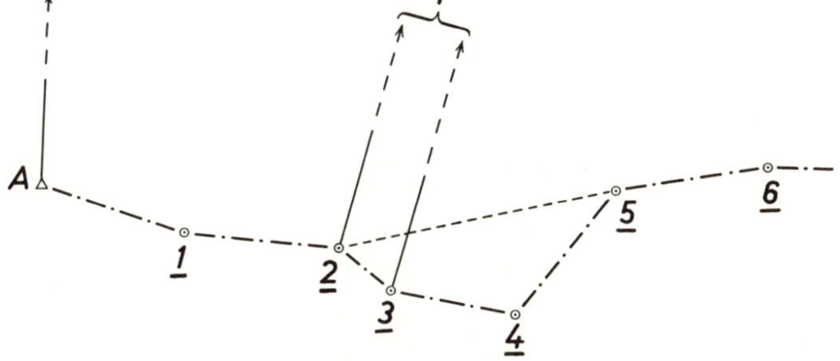

Fig. 6.8.10 Elimination of short legs

Another means of avoiding rotations of the traverse is intermediate orientation. This is achieved in Fig. 6.8.10 by sighting from 2 and 3 to a far distant coordinated point F.

In the office the traverse is first calculated as usual along 1, 2, 3, 4, 5, 6 .... This leads to preliminary coordinates for points 2 and 3, which are then used to compute the practically correct azimuths $\alpha_{2,F}$ and $\alpha_{3,F}$. Using the angles $F\,2\,1$ and $F\,3\,4$ preliminary azimuths for the adjacent sides are obtained which are then used for distributing the angular closure over the observed traverse angles. This also gives corrections to the azimuths of the adjacent sides, which, as a rule, are half the amount of the other traverse angles.

If the angles are observed with forced centring, then the above mentioned aids are not used.

7. Fundamentals of Horizontal Geodetic Networks

If a larger segment of the earth's surface is to be surveyed, it is first covered by a network of fixed control points. These points are either monumented or else coincide with specific man-made points (e.g. steeples, towers, etc.).

In conjunction with a vertical network, these geodetic control points provide the basic framework for topographic mapping as well as for legal surveys. The design of such networks as well as the technical means for coordinating them have been undergoing changes in time. As a rule, a large horizontal network is laid out in several steps, corresponding to different "orders".

7.1 Older Horizontal Networks

Older geodetic networks have been established by *triangulation*, usually in four steps. The first-order net consists of triangles with 30 to 50 km sides and first-order geodetic control points at the corners. These points are usually located on top of mountains or hills, if possible, to avoid erecting large observation towers. Church steeples or similar buildings may be used as well; they require, however, eccentric reference points on the ground as a safeguard against movements of the building. The first-order net is densified by a second-order net (10 to 20 km sides), a third-order net (3 to 10 km sides), and, if needed, particularly in difficult terrain, by a fourth-order net with 1 to 3 km sides. Traverses are fit into the third- and fourth-order networks. Fig. 7.1.1 shows an example for a first-order network generated by triangulation. It is the German first-order net, established between 1870 and 1950 by connecting individual net sections. Its reference surface is the ellipsoid, computed in 1841 by Bessel from measurements of 10 arc sections ($a = 6\,377\,397$ m; $f = 1:299.15$). Its fundamental point is the point Rauenberg with an assumed zero deflection of the vertical. The net is oriented by the astronomic azimuth of the triangle side Rauenberg – Berlin (St. Mary's Church). The scale is determined by five baselines, enlarged by base enlargement nets (*Torge* 1980).

The *angles within a triangulation net* were measured with precision theodolites – often equipped with photographic recording – to fractions of a second of arc. Concrete pillars are used at the observation stations. Often it was necessary to erect signals or observation towers of steel of up to 30 m height or more above the ground point (Fig. 7.1.2).

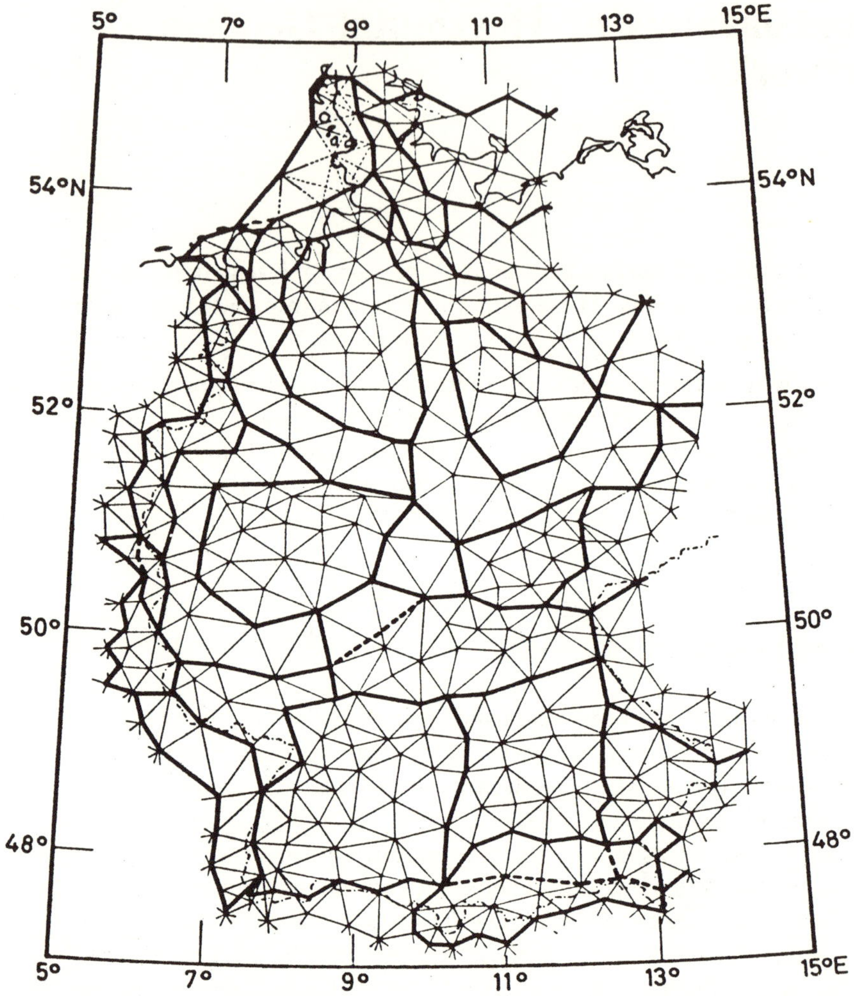

Fig. 7.1.1 German first-order horizontal network. (Courtesy: Lower Saxonian Ministry of the Interior).

These towers consisted of two separate elements inside each other, namely the very stable interior tower for the instrument, and the lighter exterior one which carried the observer and targets and light targets for sighting from other points. For first-order measurements, only the illuminated points of the neighbouring towers were used as targets, or else the so-called helitropes which directed the sunlight toward the observer at the next station by means of a mirror system. The angles were observed in all combinations or else according to the sector method (section 3.7.5.3). Disturbing refraction influences were counteracted by spreading the observations of each angle over several days with different weather conditions as well as mixing these with night measurements. Because of limited sighting distances, observations at one station

Fig. 7.1.2 Steel tower

required an average of 6 weeks (*Augath* 1976). All observations were subjected to a station adjustment, providing the most probable value for each direction.

Until a few decades ago, the *sides* were measured at 100 to 200 km separation using base measurements which required months of work each. For this, a 5 to 8 km long base was measured several times with invar wires (section 4.3.1) to reach a relative accuracy of 1 : 1 million. This base was then transferred to the triangle side using a base magnification net (Fig. 7.1.3) which had to be measured with highest precision such that an accuracy of 1 : 100 000 was reached for the triangle side using angular measurements only.

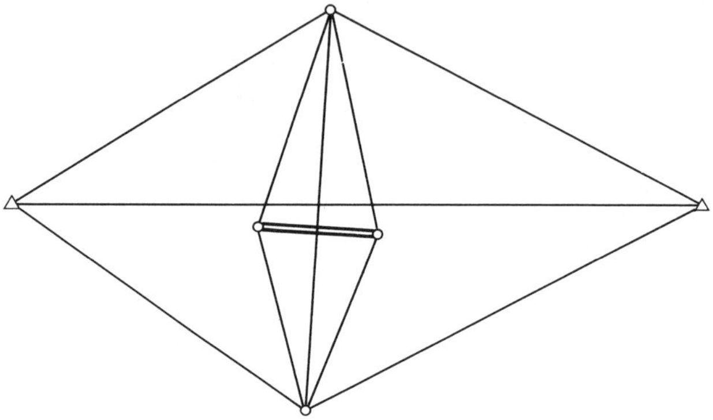

Fig. 7.1.3 Base magnification net

In order to *orient such a horizontal network* on the reference ellipsoid, the geographic latitude and longitude as well as the direction of the meridian had to be determined at selected points by astronomic means.

The *densification of the first-order net* to second-, third-, or fourth-order should follow the same approach as the first-order net. In practice, however, this was not the case, as the method which provided the results with the least effort was always selected, e.g., combined intersection – resection. Generally, traverses were used in the final densification step.

The *computation within the triangles* usually started with the least-squares adjustment of the measuring residuals within the main triangles. Starting from astronomically determined points, the adjusted observations were utilized to determine lengths and directions of the triangle sides on the reference ellipsoid. The latter were then used for the determination of geographic coordinates of the first-order points. Subsequently, rectangular coordinates were derived from these geographic coordinates. In the rectangular coordinate system, the second-, third- and fourth-order points were then coordinated utilizing combined intersection – resection approaches. With the advent of electronic distance meters, distances or combined direction-distance approaches are now mainly applied.

Triangulation chains along meridians and circles of latitude were established in the U.S.A., in Russia, and other extended countries. Laplace azimuths for directional control and baselines for scale were measured at the junctions of such chains. The basic network in the U.S.A. has a separation of about 500 km, while it is 200 km to 400 km in the U.S.S.R. The accuracy of these networks is a few parts per million. Depending on the area, different methods were applied for densification: e.g., traverses or aerotriangulation.

The Australian base net consists primarily of *long traverses* (Fig. 7.1.4). The junction points are several hundred kilometres apart. Laplace azimuths were observed at nearly all observation stations. This fundamental network was densified by triangulation and traverse nets in the coastal areas. For virtually uninhabited area, aerodist was used to determine the fixed points.

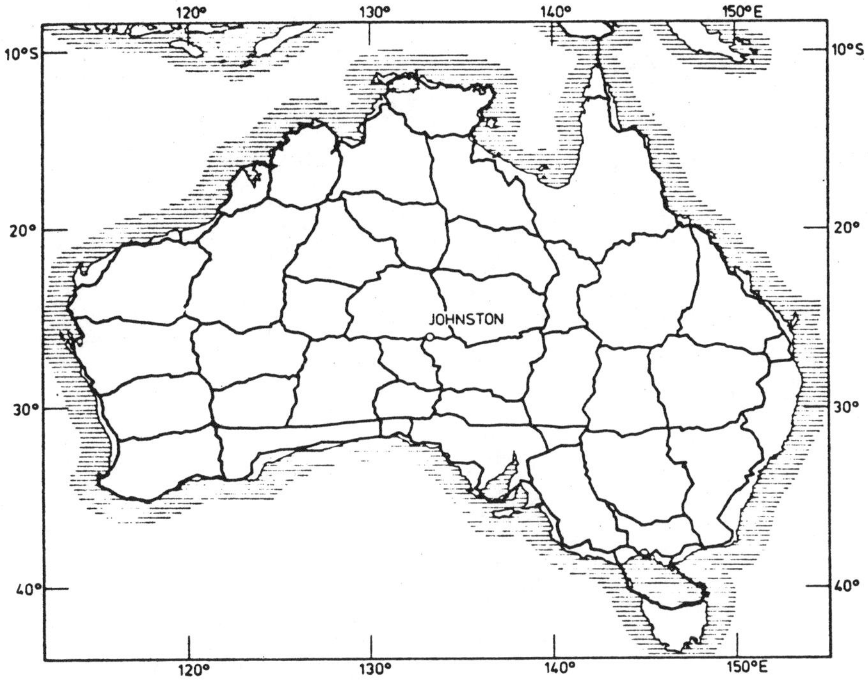

Fig. 7.1.4 Australian traverse network (simplified illustration; from "National Report for Australia", IUGG XVI General Assembly 1975)

7.2 Design and Observation of a Modern Horizontal Network

New developments in land registration demand horizontal networks of high accuracy, especially in highly developed and densely settled areas. Advances in electronic distance measuring and data handling techniques, together with a modified network design, make accuracy increases possible. The design of a horizontal net-

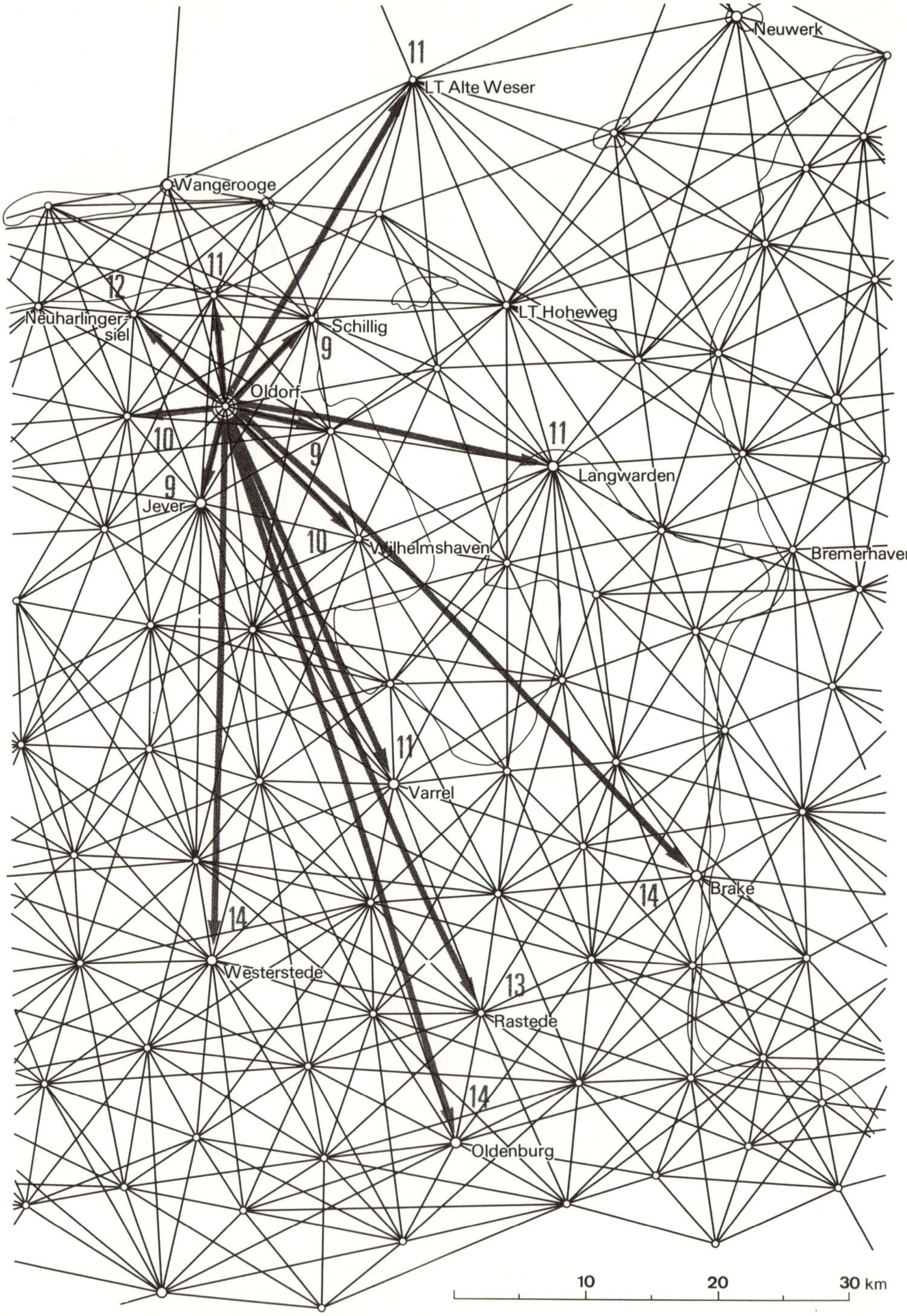

Fig. 7.2.1 Base net of first- and second-order. Network design and accuracy (*Augath* 1976) (⊙ starting point for the computation of standard deviations $s(s)$ of adjusted distances →; $s(s)$ in mm; ⊙——⊙ measured distance)

work determined after 1970 with high accuracy requirements is described below with the aid of an example.

Electronic distance meters and theodolites could be used most effectively and economically, if the network was laid out in two steps*. The first step consists of an area net of first and second order. The second step represents the densification to third and fourth order. The base net was measured by trilateration, using microwave instruments (Fig. 7.2.1).

The densification net is determined with combined direction and distance observations as traverse network (Fig. 7.2.2).

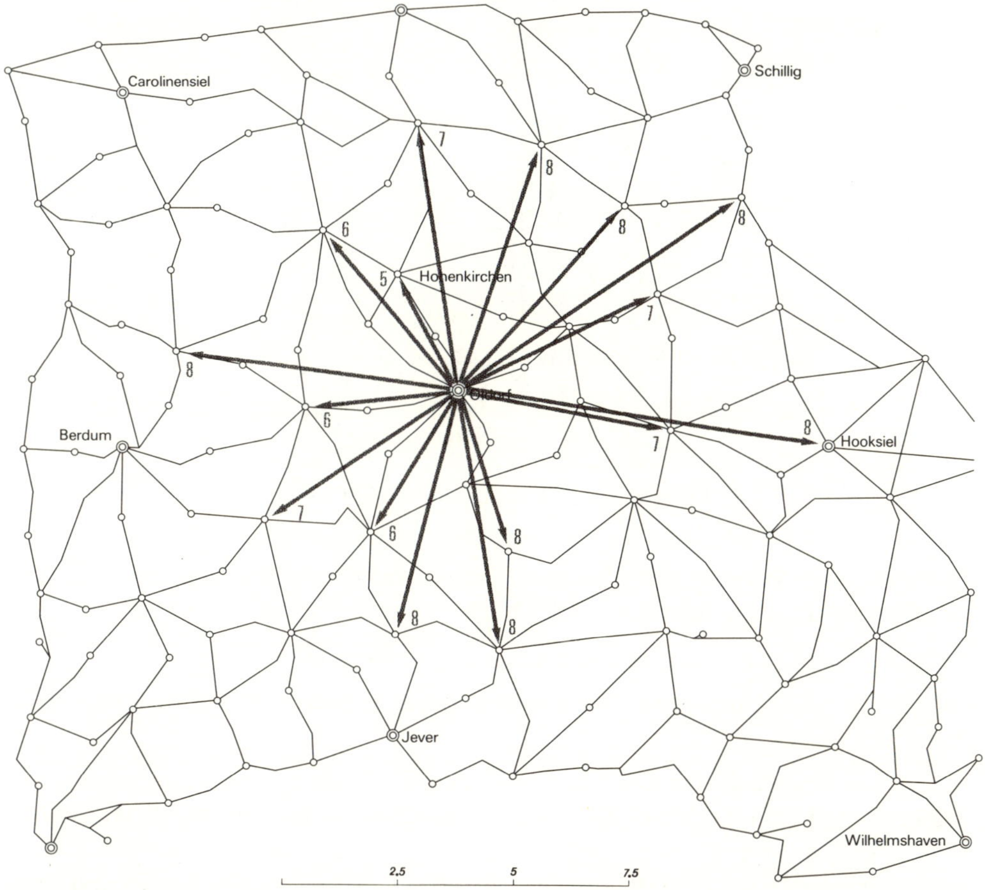

Fig. 7.2.2 Densification network, design and accuracy (*Augath* 1976) (⊙ starting point for the computation of standard deviations $s(s)$ of adjusted distances →; $s(s)$ in mm; measured distance)

---

* This approach was used in Lower Saxony in Germany.

The trilateration and traversing methods were preferred, because they were very economical, even for high accuracy requirements. Large distances were bridged with the aid of mobile observation ladders (Fig. 7.2.3), which had cost only a fraction of the observation towers used for triangulation. Furthermore, these methods were largely independent of the weather. Several trilateration stations could be observed on a single day.

Fig. 7.2.3 Observation ladder (base and platform)

In remote areas, either long traverses or photogrammetric aerotriangulation are used for network densification.

Different approaches may be applied for the orientation of the nets:

a) If a national datum exists, the new network is transformed into it by means of identical points.

b) As a rule, additional satellite measurements are utilized to improve orientation and geometry in a simultaneous adjustment. Three-dimensional points, determined by satellite measurements, can be connected to the terrestrial measurements if ellipsoidal heights are available. The latter can be derived, e.g., from levelling results and geoid determinations. If the geoid information is insufficient for the area in quation, normally the combination is carried out in planimetry only.

c) If a national datum has not been established, the satellite datum may well be used as the datum for the network. The results of terrestrial measurements can be fitted in with a transformation by means of identical points.

With the Transit navigation satellite system (section 12.4.1), points can be determined with a relative accuracy of a few decimetres. By selecting a suitable point separation, a scale accuracy of about $2 \cdot 10^{-6}$ can be achieved. The Canadian horizontal network, for instance, has been improved and densified in recent years by covering large areas with Doppler control points, at 100 to 300 km density.

By now it is possible to determine points with entimetre accuracy using differential methods (12.3.2) with the NAVSTAR Global Positioning System (GPS) (section 12.4.2). It becomes economical to determine networks with point separation of up to 1 ... 10 km with this method (*Beutler* et al. 1984). Classical geodetic nets lose then their importance.

7.3 General Remarks on Geodetic Network Design

The points within a geodetic network are connected by direction measurements, distance measurements or by combined methods. The design of a network depends on its purpose, e.g., for engineering projects a certain measuring accuracy may be given, however, certain financial limits should not be exceeded. Furthermore, the results must be highly reliable.

Thus there are three requirements for network design: measuring *accuracy*, minimizing *cost*, and striving for high *reliability*.

Grafarend (1974) defined the design of zero to third order and dealt with certain optimization models. The zero-order design deals with the determination of the geometry of the net within a coordinate system, e.g., by holding a certain amount of point coordinates fixed. First-order design deals with optimization of the planimetric position. It defines the optimal position of the points based on an observation plan and expected measuring accuracy. The second-order design optimizes the weights. The point locations are fixed at that stage, while the number of repetitions of the observations is variable. The third-order design deals with the improvement of existing networks by additional points and observations, as well as the optimization of the weights. A detailed treatment of the optimization equations can be found in the literature, i.e., *Grafarend* et al. (1979), and *Schmitt* (1979).

Only generally applicable basic network design criteria are given here.

a) A *homogeneous and high point accuracy* is achieved if the measurements are evenly spread over the horizon, and a suitable measuring method (section 6.2) is selected. Large networks are determined by trilateration, while combined distance and direction networks are both suitable for horizontal nets with shorter point separations.

b) *Neighbouring accuracy* has to be maintained by connecting neighbouring points, especially when direction measurements are performed.

c) A *high reliability* is achieved when additional observations are taken and comprehensive direction and distance measurements are performed using the neighbouring points. Reliability provides an indication of how much the results are influenced by random and systematic errors.

The treatment of rigorous least-squares adjustments of networks goes beyond the scope of this book. There is a wealth of information available in the geodetic literature. However, an approximate method suitable for smaller combined networks for many purposes is provided.

7.4 Approximate Method for the Computation of Small Combined Networks

Rigid net adjustments are usually performed in one step, which means that a large computer is needed, even for a small number of points. However, it is possible to use small computers (pocket calculators) if directions and distances are measured between the points, and if an approximate solution is sufficient.

Tacheometers (section 12.2) can be utilized most economically, if the observation station can be freely selected. Section 6.7 contains the coordination of a new point

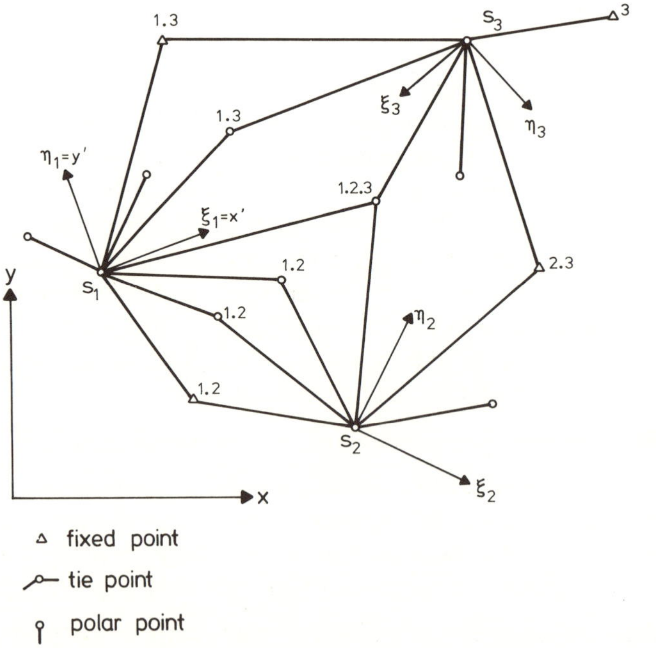

Fig. 7.4.1 Linked Helmert transformation: Transformation of several station systems ξ, η into one x', y'-system which is subsequent, fitted into the x, y system of the fixed points

which was observed by the polar method from a freely selected station. The balance of this section provides an approximate computation method for a tacheometer network where distances and directions were measured from several stations to further points. The method is based on linked Helmert transformations. The individual local polar systems are linked together by means of common linkage points (Fig. 7.4.1). The observation units are first referred to a freely chosen zero-direction as polar coordinates with the station as the origin. Not all of these polar systems contain fixed points, because normally there are only a few of them available at the edge or within the area of interest.

This method contains three steps:

1) The polar coordinates (r_i, s_i) of the local system are transformed into rectangular coordinates (ξ, η) (Fig. 7.4.2).

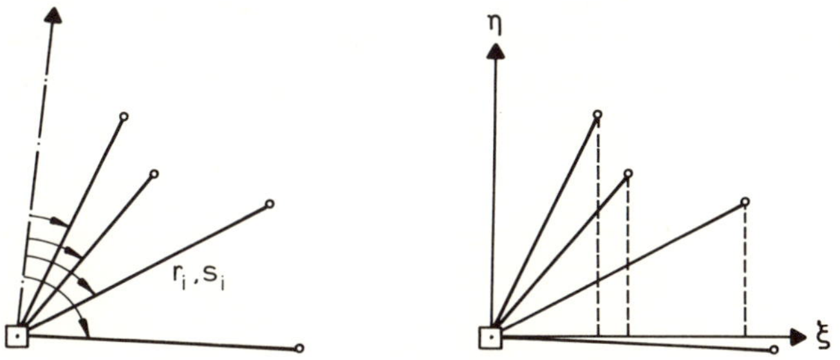

Fig. 7.4.2 Local system of polar coordinates (r_i, s_i) transformed into rectangular coordinates (ξ, η)

2) The individual elementary systems are then combined to a unique x', y' system in steps by coordinate transformation (section 5.2). The common x', y' system can be randomly selected. Usually the system of one station is held fixed.

3) Finally, the x', y' coordinates are to be transformed into the x, y system of the network of fixed points (section 5.2). The accuracy of the method depends primarily on the location and number of linkage points. For practical applications, the following should be considered:

- the separation of linkage points should be selected such that it is as large as possible;
- the centre of gravity of the linkage points should be in the middle of the common edge of two systems;
- additional linkage points near the centre of gravity provide no significant improvements: It is better to increase the number of external linkeage points;
- longitudinal or lateral linkage as well as a combination of the two can lead to equally good result.

The simultaneous determination of linked Helmert transformations is a modification of the above method. The transformations of the 2nd and 3rd steps are then carried out simultaneously. This approach is better for automation, however, it cannot be accomplished with a small calculator any more. Thus the first method has the advantage that the computation of the transformation parameters and the transformations themselves are performed in separate steps. Thus only a small amount of storage is required. The simultaneous adjustment has a much higher number of unknowns: 4 transformation parameters per station system plus two coordinates per new point. However, only new points, which are also linkage points, contribute to the adjustment. Ordinary polar points do not increase the system of normal equations.

The solutions described are approximate because all quantities are entered with unit weight under the assumption that they are not correlated.

Numerous studies have shown that these approximate methods provide a sufficient accuracy for many practical purposes, i.e., cadastral surveys (*Krack* 1977). If a rigorous least-squares adjustment is needed for higher accuracy requirements, the methods mentioned provide an easy means of calculating approximate coordinates.

8. Field Surveys with Simple Instrumentation and their Evaluation

8.1 Marking of Points and Straight Lines

8.1.1 Demarcation of Points in the Field

The main purpose of a cadastral survey is the demarcation and representation of boundary lines. Therefore, if possible, the corner points of a boundary are locally marked by boundary monuments, which – depending on the location – may be of different design. In addition to boundary points, fixed control points, which form the framework of the detail surveys, are monumented.

For the duration of the survey, the necessary points have to be visible for a certain distance. As mentioned earlier, all points which lie in one plumb line are projected onto the same point of the reference surface. It is therefore advisable to make the plumb line of the points visible.

If this is not already the case, e.g., the corners of a house, it is achieved with the aid of range poles. Range poles are usually 2.8 cm thick and 2 m long poles of ash or pine, which are alternately painted red and white at every 50 cm, and which are equipped with a steel tip to ram them into the ground. They are set plumb with the aid of a plumb bob, or a bull's-eye level which can be either clamped on or held against the pole (Fig. 8.1.1). On hard ground (e.g., pavement), metal range pole tripods are used (Fig. 8.1.2).

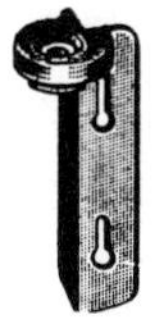

Fig. 8.1.1 Range pole level

8.1.2 Staking of Straight Lines

To stake a straight line *AB* in the field, first the two end points are staked with range poles. Additional range poles are lined in at 30 to 50 m distances by working from the farthest to the nearest one. Once all intermediate points are lined in, the straight line is "staked". An experienced observer can line in with the naked eye to about 200 m,

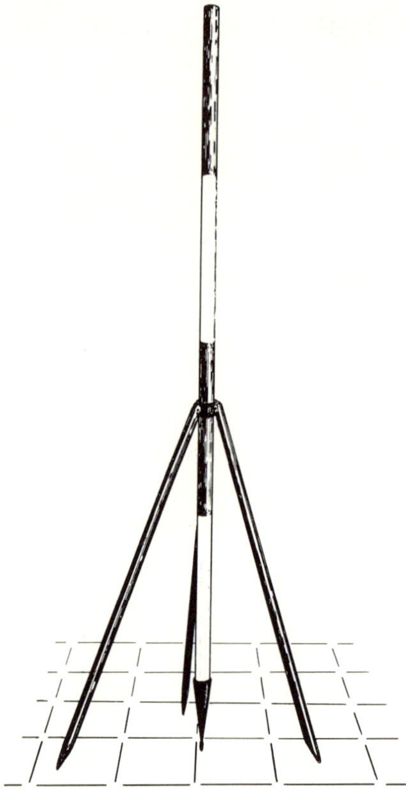

Fig. 8.1.2 Range pole tripod

while binoculars are required for longer distances. If one is careful, the intermediate points can be staked within about 2 range pole thicknesses in flat terrain. For higher accuracy or in mountainous terrain, a theodolite is required.

8.1.3 Straight Lines with Obstacles

8.1.3.1 Mutual Lining in

If points A and E are not accessible (e.g., house corners) or not intervisible due to the terrain, mutual lining in with 2 men may be applicable (Fig. 8.1.3).

As illustrated in Fig. 8.1.3, the two men are placed with their range poles at two points, 1 and 2, somewhere between A and E, such that they are intervisible and that one end point can be seen from each. # 1 lines up # 2 in the line 1 A, then # 2 lines up # 1 to position # 3 on line 2 E, then # 3 lines up # 2 to position # 4 on line 3 A, etc. Once there is no more deviation at the plumb range poles, both points are on the straight line. The accuracy can be improved by iterating from the other side of the straight line as well.

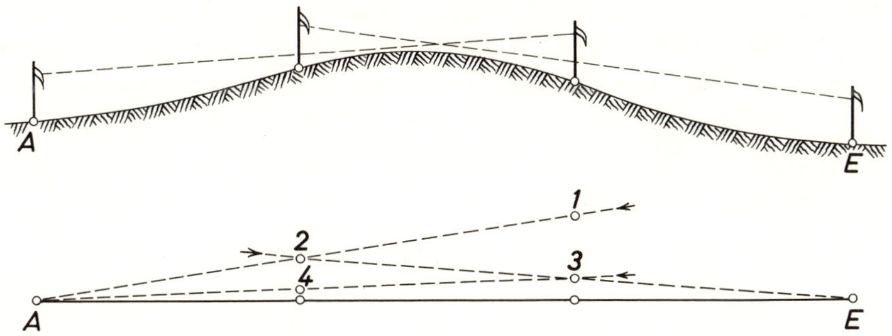

Fig. 8.1.3 Mutual lining in

8.1.3.2 Measuring Line with Obstacle

In such a case it is often advantageous to stake a line parallel to the original one (Fig. 8.1.4). The double arcs symbolize right angles in this figure.

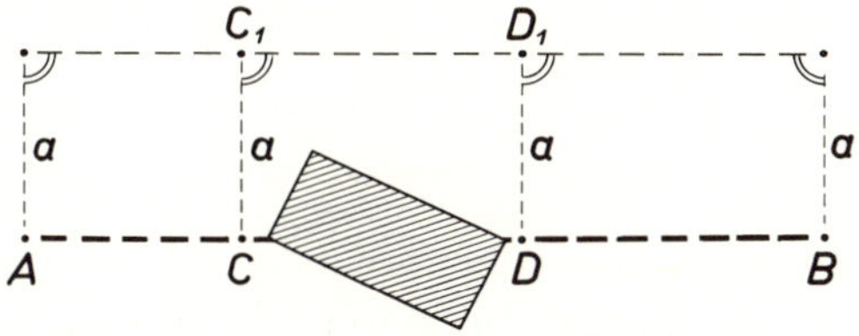

Fig. 8.1.4 Measuring line with obstacle

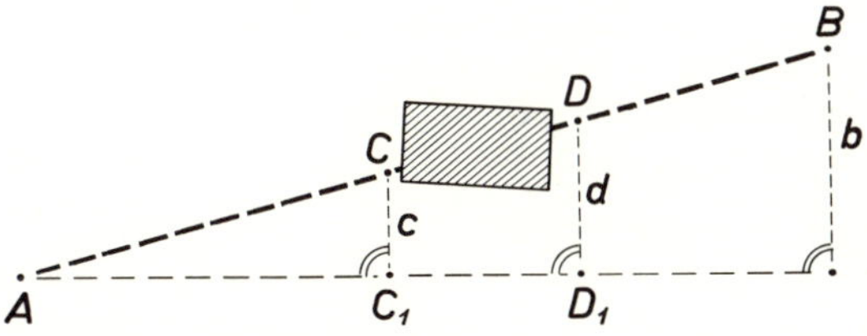

Fig. 8.1.5 Use of auxiliary base

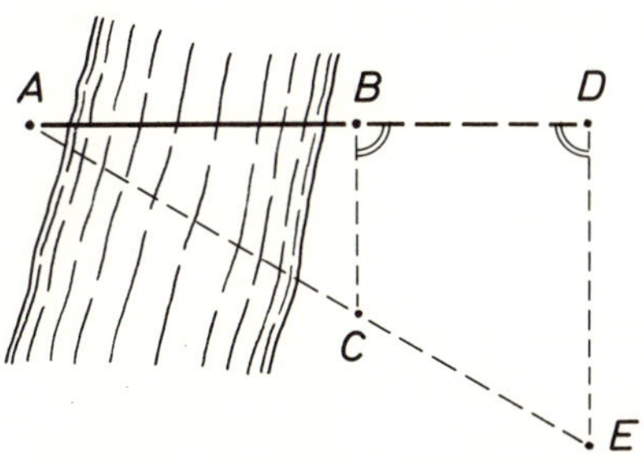

Fig. 8.1.6 Crossing of small water body.

The case illustrated in Fig. 8.1.5 utilizes an auxiliary base through A with a rectangle offset for B. After b is measured, the ordinates at C_1 and D_1 are computed by simple proportionality and then staked in the field. The section with the obstacle has the length

$$(CD)^2 = (C_1 D_1)^2 + (d - c)^2$$

as obtained by the Pythagoras formula.

Small bodies of water as shown in Fig. 8.1.6 can be crossed by using the ratio

$$AB = BC \frac{BD}{DE - BC}.$$

8.2 Staking of Fixed Angles

For the staking of right angles or for orthogonally coordinating points with respect to a measuring line, simple instruments like diopters, angle mirrors, or prisms are used.

8.2.1 The Cross Staff

This is a diopter instrument, where two lines of sight normal to each other are defined by slots in a cone or cylinder. It is still frequently used today in hilly terrain because the cross staff permits inclined sights of up to about ± 35 gon ($\pm 30°$). As a rule, a cross staff is nowadays also equipped with a double pentagon prism (Fig. 8.2.1 and

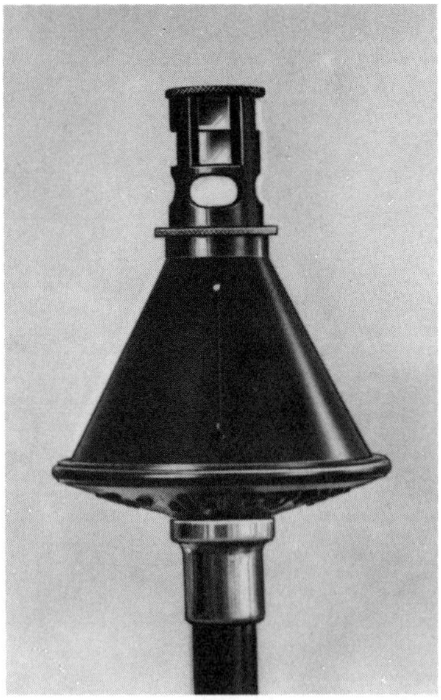

Fig. 8.2.1 Cross staff

section 8.2.4). The standard deviation of a staked right angle is about ± 5 cgon ($\pm 2'$), which represents a lateral uncertainty of ± 2 cm for 30 m.

8.2.2 Angle Mirror and Sextant

The angle mirror consists of two plane mirrors arranged at 50 gon (45°) to each other (Fig. 8.2.2).

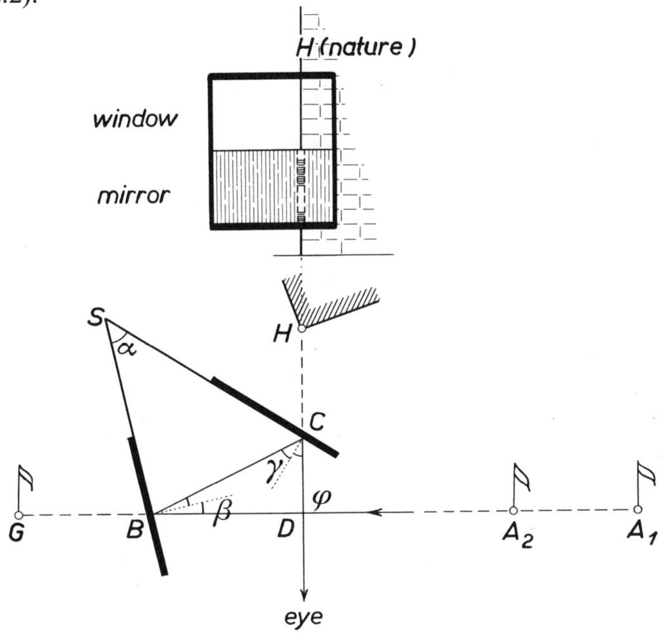

Fig. 8.2.2 Operating principle of the angle mirror

A ray coming from A_1 and A_2 is reflected at the mirrors B and C and thus deflected by the angle φ from its original direction.

Since

$$\varphi = 2\beta + 2\gamma \quad \text{and} \quad \alpha = \beta + \gamma$$

$$\varphi = 2\alpha = 2 \cdot 45° = 90° .$$

The angle mirror is rarely used today, because the mirrors have to be frequently adjusted and they are subject to positional change, and also become "blind" after some time. Both these shortcomings do not occur with right angle prims (section 8.2.3).

The principle of the angle mirror is included because it also applies to the sextant, which is still widely used in hydrographic surveys and thus merits mention.

Again, two mirrors are used to determine the angle α which is twice the measured angle β (Fig. 8.2.3).

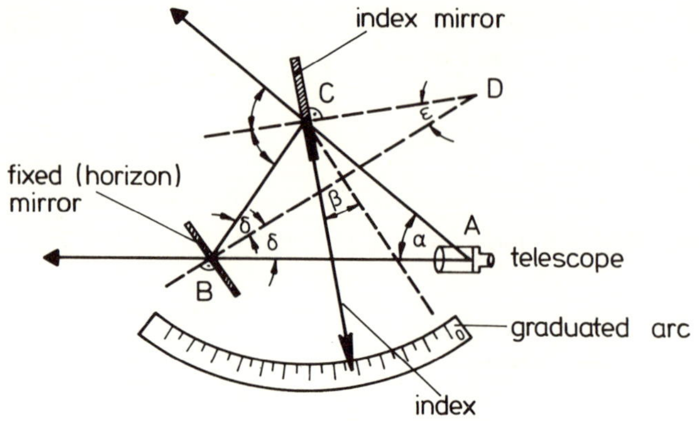

Fig. 8.2.3 Operating principle of the sextant

The sum of angles in triangle *ABC* is:

$$\alpha + 25 + (180^\circ - 2\gamma) = 180^\circ \quad \text{or} \quad \alpha = 2(\gamma - \delta)\,.$$

The sum of angles in triangle *BCD* is:

$$\delta + \varepsilon + (180^\circ - \gamma) = 180^\circ \quad \text{or} \quad \varepsilon = \gamma - \delta\,.$$

Since the sides forming the angles β and ε are mutually perpendicular, $\beta = \varepsilon$ and

$$\alpha = 2\beta\,.$$

The sextant is handheld and measures space angles. While it is possible to measure horizontal angles, the sextant is mainly used for measuring vertical angles in navigation (e.g., vertical angle of a star above the horizon).

8.2.3 Right Angle Prisms

Right angle prisms are ground glass bodies with parallel edges. The optical train can easily be derived using the laws of reflection and total reflection respectively.

8.2.3.1 Gowlier's Pentagon Prism

This prism (Fig. 8.2.4) was originally designed as a quadrilateral prism, with the fifth side being obtained when cutting off one edge. The two longer edges are mirror coated. The optical train is illustrated in Fig. 8.2.5.

Since the configuration is similar to the angle mirror (section 8.2.2), a deflection of 100 gon (90°) is achieved because the mirror coated sides include on angle of 50 gon (45°).

The field of view is much larger than for the triangle prism, and the images are brighter due to less loss of light within the prism. The vertex of the right angle falls inside the prism. If the upper and lower prism surfaces are mirror coated, inclined sights are possible. Because of these advantages, the pentagon prism has virtually replaced the Bauernfeind prism for practical work.

The prism housing sits on top of a handle and can be centred with the aid of a plumb bob.

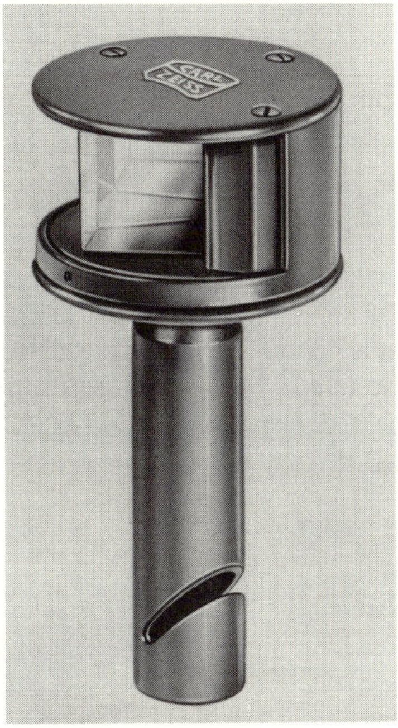

Fig. 8.2.4 Pentagon prism

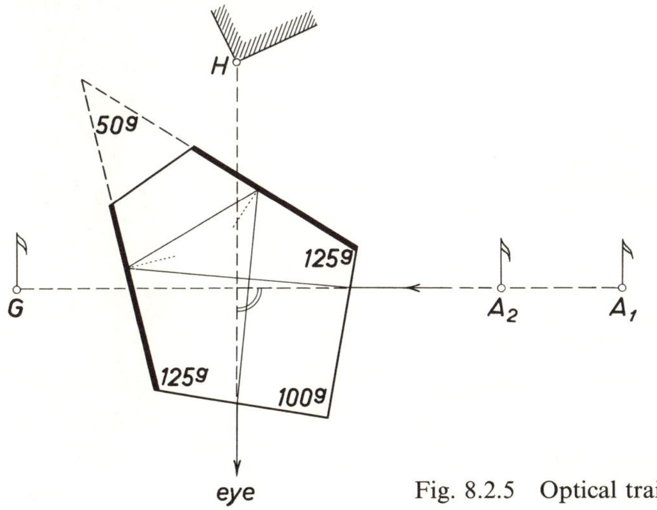

Fig. 8.2.5 Optical train of pentagon prism

In order to measure the house corner H in Fig. 8.2.5 normal to the line AG, the pentagon prism is shifted sideways until the images of A_1 and A_2 appear to be coming from H. At this position the plumb bob is dropped to mark the foot of the normal.

If a line normal to AG is to be staked at D, a helper with a range pole is sent into the approximate direction. The observer positions himself at D and directs the helper until his range pole appears in the same direction as the images of A_1 and A_2.

8.2.3.2 The Wollaston Prism

This prism (Fig. 8.2.6) is a quadrilateral with the right angle edge being cut off. It has no mirror coated surfaces but relies on total reflection at points C and D. Since $\gamma + \delta = 150$ gon, BC has to be normal to DE, thus ray $A_1 A_2 BF$ is normal to EFH.

8.2.3.3 Accuracy of Right Angle Prisms

In addition to grinding errors which are less than 2 cgon (1′), a sighting error of ±3 to 4 cgon (2′) is to be expected for freehand use. Coupled with a centring error of ±2 cm, a staking accuracy of ±2 to 3 cm can be expected with a pentagon for a 30 m long orthogonal line. When using a plumbing rod (Fig. 8.2.7) this uncertainty can be reduced by about 1 cm.

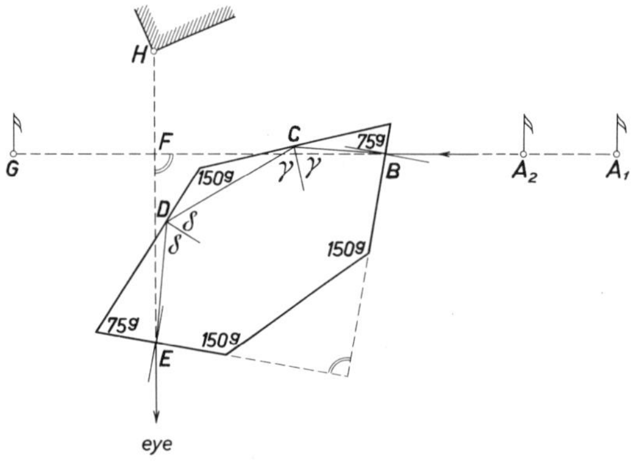

Fig. 8.2.6 Optical train of Wollaston prism

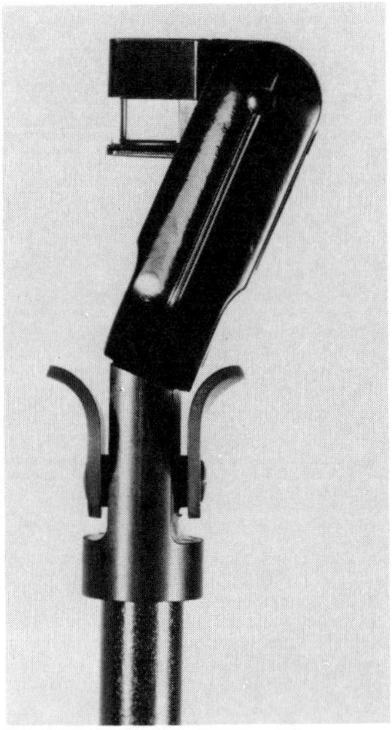

Fig. 8.2.7 Double pentagon with plumbing rod attached to it

8.2.4 Prism Crosses

Prism crosses consist of two right angle prisms above each other, and can be used for staking straight lines and right angles. The double pentagon (Figs. 8.2.8 and 8.2.9) is most widely used. The Zeiss P9 prism (Fig. 8.2.9) has been newly designed according to user suggestions. It contains a parallel plane plate between the prisms for straight sights and has mirror coated upper and lower surfaces for inclined sights.

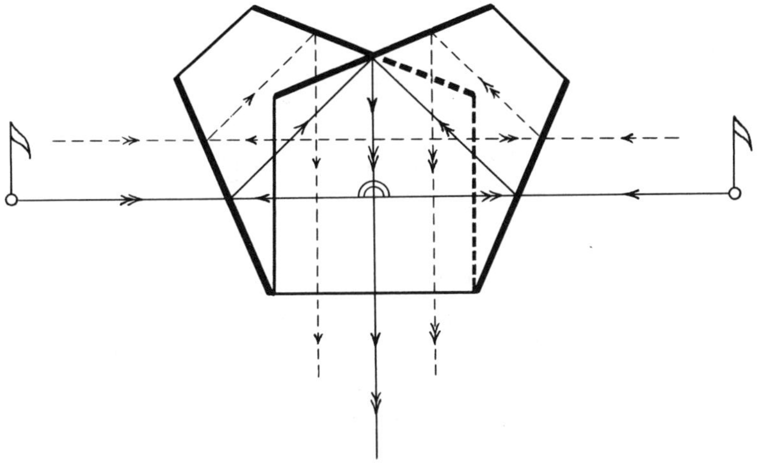

Fig. 8.2.8 Optical train of double pentagon prism

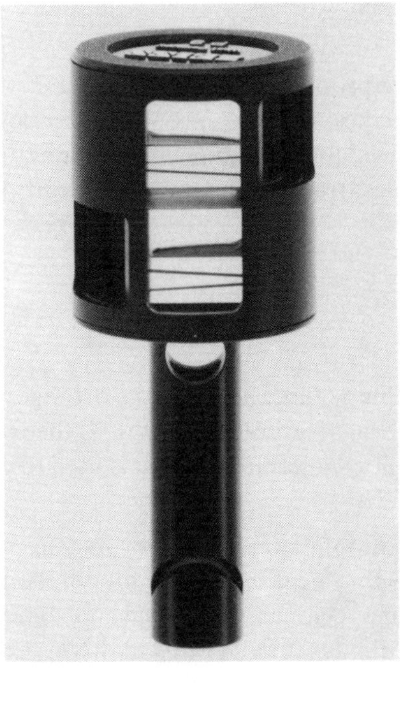

Fig. 8.2.9 Double pentagon

With such a prism cross, an observer can place himself on a straight line defined by two range poles by moving back and forth until the two images appear in one vertical line.

The accuracy is the same as for right angle prisms. With increasing distance from the end points, the accuracy of lining in decreases rapidly. It is therefore advisable to repeat the procedure after turning around.

8.3 Detail Surveys

Although detail surveys can be performed either by *aerial photogrammetry* or by *ground survey methods*. Only the latter are covered in this context. Photogrammetric methods are covered in the pertinent photogrammetric literature, and the interested reader is advised to refer to it.

Detail surveys are based on a horizontal control network (Chapter 7).

An area is properly surveyed if sufficient measurements (angles and distances) are taken to permit the plotting of all necessary points onto a scaled plan or map.

For precision work, larger areas, and for subsequent automatic handling of the data, it is most advantageous to use a theodolite in conjunction with an electronic distance meter. Small areas can be surveyed with tape and rectangle devices.

Basically two methods are used, namely the *orthogonal method* and the *polar method*. Occasionally the *tie-in approach* is utilized when working with the orthogonal method.

Items to be included in a detail survey are property boundaries, lot and vegetation lines, with associated walls, fences and ditches; highways, roads, water courses and railways with all associated structures and dams; buildings of all types; and topographic items such as hydrants, lamp poles, traffic signs, utilities, hydro-lines etc.

8.3.1 Orthogonal and Tie-in Method

8.3.1.1 Field Surveys

The orthogonal method uses a surveying line as the x-axis and the orthogonal offsets to boundary points, building corners and other features of interest as the y-ordinates (Fig. 8.3.1). A right angle prism, cross staff or similar instrument is used to determine the foot of each normal.

Afterwards all ordinates as well as the foot points on the survey line are measured, thus providing rectangular coordinates in this local system for all points. If the end points of the survey line are coordinated in a regional or national coordinate system, a simple transformation will provide coordinates for all points in this system as well.

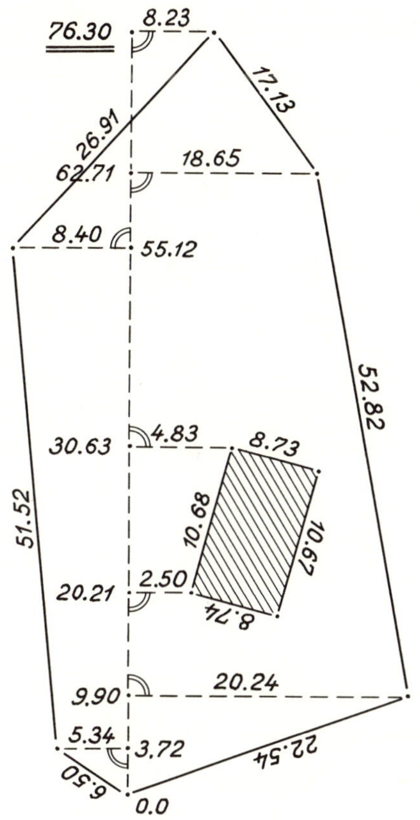

Fig. 8.3.1 Orthogonal survey

The tie-in approach (Fig. 8.3.2) is virtually never used as a separate method, but rather to supplement the orthogonal method.

This method utilizes distance measurements only. The area in question is covered by a framework of survey and/or boundary lines with all other features being tied into this frame. Thus other boundary lines, the sides of buildings, etc., are prolonged on both sides until they intersect with one of the frame lines. The points of intersection are measured along the line. Interior features are usually tied into secondary lines, which in turn are tied into the framework.

For larger surveys, both methods are often combined. The tie-in approach is of advantage with regular boundary lines. It also has the advantage that the areas can be directly derived from the field data. Often both types of parcels are next to each other. Then it is advisable to put a framework of boundary and measuring lines around the perimeter of the area to be surveyed. This frame is densified by measuring lines inside into which straight boundaries and sides of buildings are tied. Irregular boundaries are measured by offsets from selected measuring lines. The details are to be measured

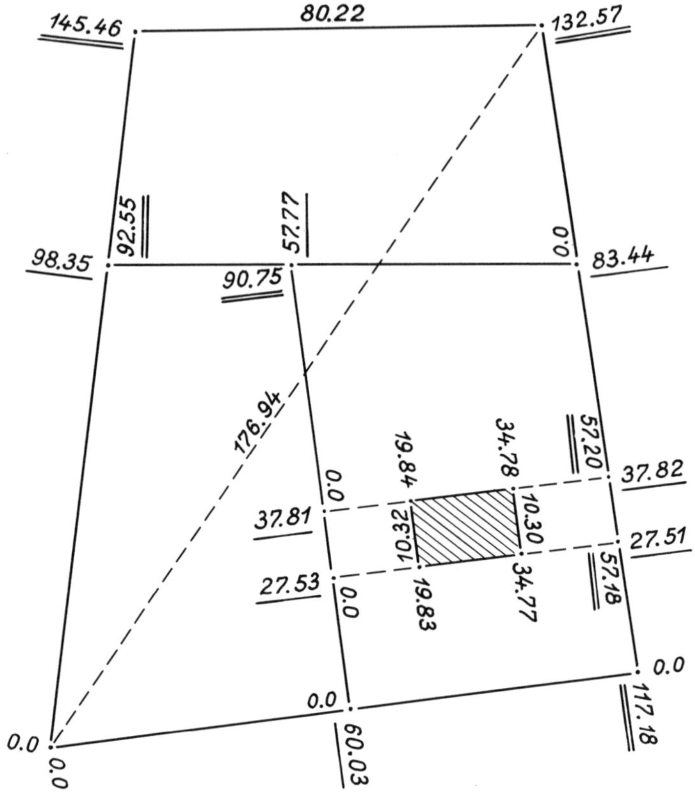

Fig. 8.3.2 Tie-in survey

with short offsets and extensions which are near normal to the lines, such that few numerical values are needed. It is also necessary to include effective checks.

These check measurements serve to detect blunders, especially at the plotting stage. Therefore, distances between neighbouring boundary markers, along buildings, and occasionally some diagonals are helpful in checking offsets.

All boundary points – and principal project points in technical surveys – should be controlled by check measurements. Fig. 8.3.3 illustrates an example of a detail survey with proper control measurements, using both tie-in and orthogonal methods.

8.3.1.2 Simple Computations

Field surveys are evaluated to provide survey plans at certain scales together with area values. Both are based on rectangular local coordinates, which normally are tied into a regional or higher order coordinate system. Usually the measuring line is referred to as the local x-axis.

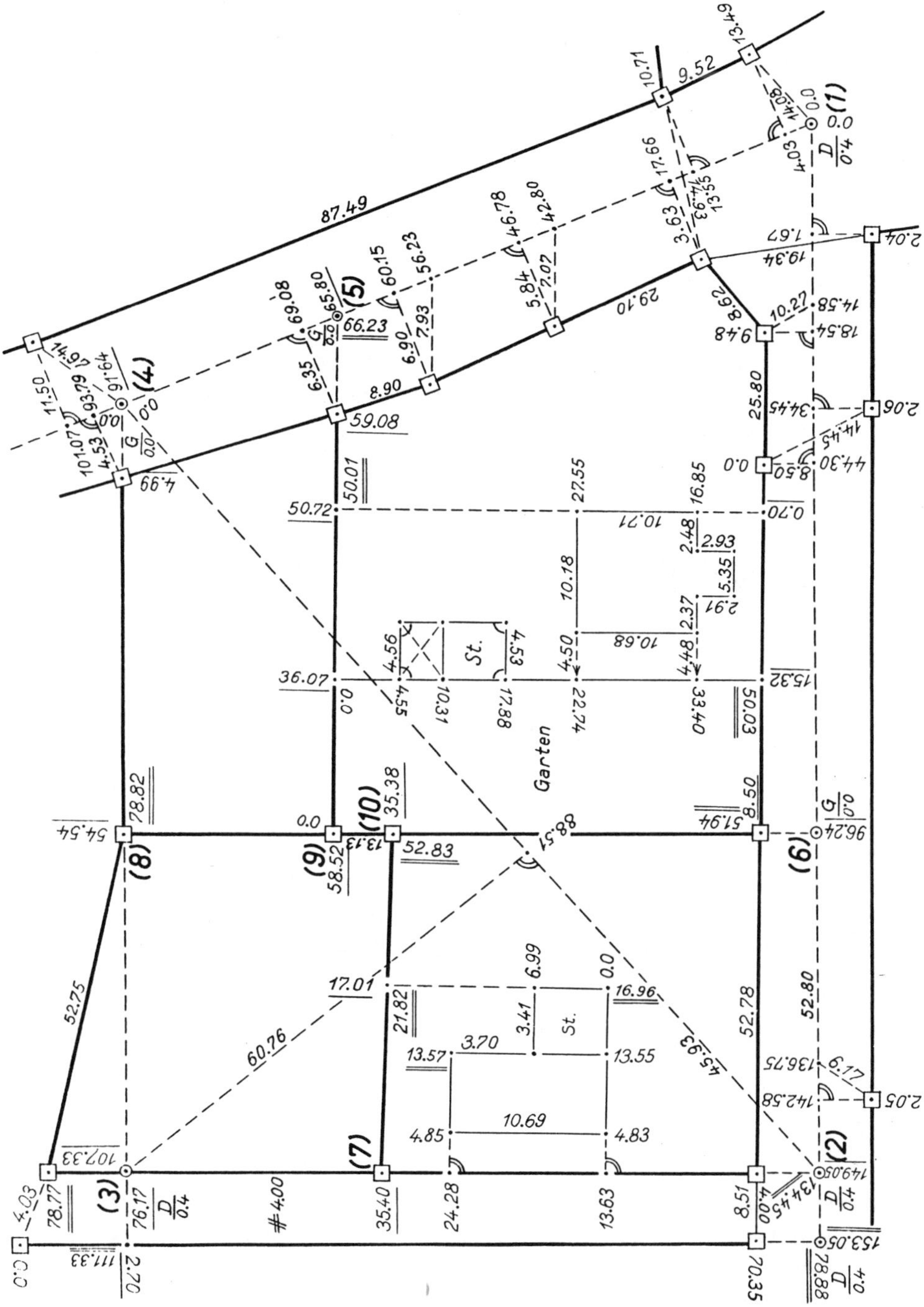

Fig. 8.3.3 Detail survey using tie-in and orthogonal methods

(1) *Computation of the height of a triangle*

If the three sides of a triangle are known (Fig. 8.3.4) then the following relation holds

$$b^2 - p^2 = c^2 - q^2 = h^2 \tag{8.3.1}$$

which leads to

$$\frac{p+q}{2} = \frac{a}{2} = \frac{a^2}{2a} \tag{8.3.2}$$

$$\frac{p-q}{2} = \frac{(b+c)(b-c)}{2a} = \frac{b^2-c^2}{2a}\,. \tag{8.3.3}$$

Addition provides p, subtraction 2, while h can be computed twice (check!) from equation (8.3.1). If the latter falls outside the triangle, the arithmetic sign has to be taken into account.

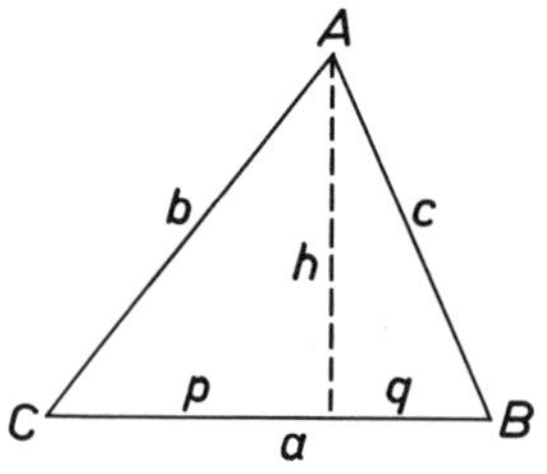

Fig. 8.3.4 Height of triangle

Example:

In Fig. 8.3.3, the triangle (1), (2), (4) has the sides $a = 149.05$ m, $b = 134.45$ m, $c = 91.64$ m. The coordinates of point (4) with respect to the line (1), (2) are to be computed. With the above formulae we obtain $p = 107.02$ m; $q = 42.04$ m, and $h = 81.42$. If point (1) is taken as the origin, we obtain

$$x_1 = 0.00 \quad x_4 = 42.04 \quad x_2 = 149.05$$
$$y_1 = 0.00 \quad y_4 = 81.42 \quad y_2 = 0.00\,.$$

(2) *Coordination of points on a line*

The coordinates of points (1) and (2) located on a previously coordinated line AE (Fig. 8.3.5) are to be computed.

The measurements provided $\Delta s_1, \Delta s_2 \ldots \Delta s_c$ for the total line $s = \Sigma \Delta s_i$. In order to check s as well as the coordinates of A and E, the theoretical distance s^* is computed as

$$s^* = \sqrt{(x_e - x_a)^2 + (y_e - y_a)^2} \tag{8.3.4}$$

and compared against s. The difference $d = s^* - s$, caused by measuring inaccuracies is to be within the prescribed tolerances (section 4.1.3.2). In order to distribute the

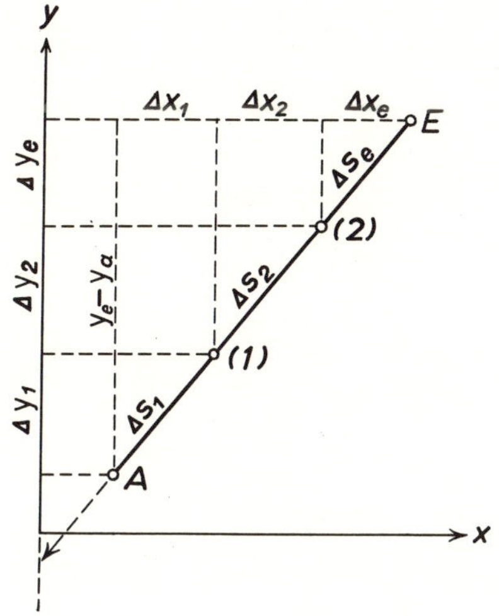

Fig. 8.3.5 Points on a line

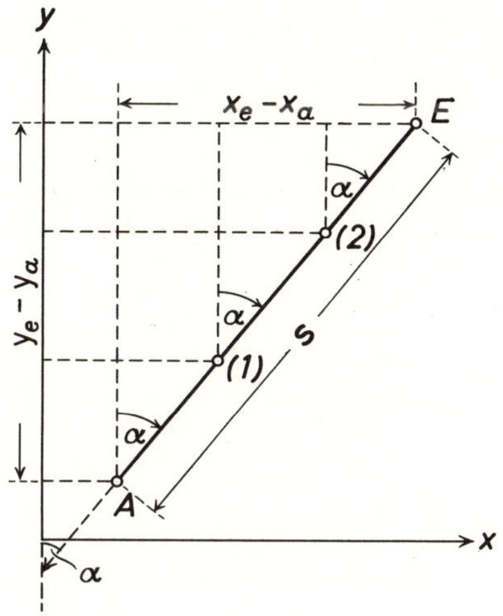

Fig. 8.3.6 Azimuth along line

discrepancy, s and Δs_i are transformed to s^* and Δs_i^* by applying a scale factor s^*/s. Now the coordinate differences are obtained as (Fig. 8.3.6)

$$\Delta x_i = \frac{x_e - x_a}{s^*} \Delta s_i^*; \quad \Delta y_i = \frac{y_e - y_a}{s^*} \Delta s_i^*$$

or with $s^* = s \cdot s^*/s$ and $\Delta s_i^* = s_i \cdot s^*/s$

$$\Delta x_i = \frac{x_e - x_a}{s} \Delta s_i; \qquad \Delta y_i = \frac{y_e - y_a}{s} \Delta s_i . \tag{8.3.5}$$

With

$$a = \frac{y_e - y_a}{s}; \quad o = \frac{x_e - x_a}{s} \tag{8.3.6}$$

the following coordinates are obtained

$$\begin{aligned} x_1 &= x_a + o\Delta s_1 & y_1 &= y_a + a\Delta s_1 \\ x_2 &= x_1 + o\Delta s_2 & y_2 &= y_1 + a\Delta s_2 \\ x_e &= x_2 + o\Delta s_e & y_e &= y_2 + a\Delta s_e \end{aligned} \tag{8.3.7}$$

where the last line serves as a check.

The azimuth of the line is easily derived from equation (8.3.6):

$$a = \cos\alpha; \quad o = \sin\alpha . \tag{8.3.8}$$

Example:

Coordination of point (8) in Fig. 8.3.3 using points (3) and (4):

$$s^* = \sqrt{(x_4 - x_3)^2 + (y_4 - y_3)^2} = 107.38 \text{ m}$$

s (measured) $= 107.33$ m; $\quad d = s^* - s = 0.05$ m

(max. permissible: 0.16 m)

$$o = \frac{x_4 - x_3}{s} = -0.0488\,; \quad a = \frac{y_4 - y_3}{s} = 0.9993\,.$$

Pt.	Distances Δs_i	x_i / $\Delta x_i = o \cdot \Delta s_i$	y_i / $\Delta y_i = a \cdot \Delta s_i$
(4)	54.54	81.42	42.04
		− 2.66	54.50
(8)	52.79	78.76	96.54
		− 2.58	52.75
(3)	[107.33]	76.18	149.29

(3) *Offset Computation*

In Fig. 8.3.7, the point P is coordinated by the normal to AE at the foot F. Thus, the following can be derived from the figure, assuming that the scale factor S/s is also applied to h:

$$\begin{aligned} x_p - x_f &= \frac{y_e - y_a}{s} \cdot h = +a \cdot h \\ y_p - y_f &= -\frac{x_e - x_a}{s} \cdot h = -o \cdot h\,. \end{aligned} \tag{8.3.9}$$

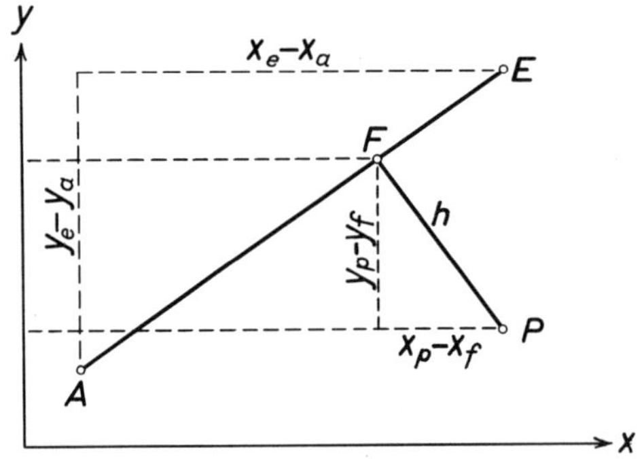

Fig. 8.3.7 Offset point

Example:

Computation of the coordinates of point (3) with respect to the line (4), (2) in Fig. 8.3.3:

$$s^* = \sqrt{(x_2 - x_4)^2 + (y_2 - y_4)^2} = 134.46\,m$$

s (measured) = 134.45 m; $d = s^* - s = 0.01$ m

(max. permissible: 0.18 m)

$$o = \frac{x_e - x_a}{s} = -0.6056; \quad a = \frac{y_e - y_a}{s} = 0.7959 .$$

Pt.	Distances Δs_i	h	x_i Δx_i	y_i Δy_i	Remarks	Formulae
(4)			81.42	42.04	$\equiv A$	$\Delta x_i = o \cdot \Delta s_i$
	88.51		−53.60	70.45		$\Delta y_i = a \cdot \Delta s_i$
			27.82	112.49	$\equiv F$	$\Delta x_i = a \cdot h$
		60.76	48.36	36.80		$\Delta y_i = -o \cdot h$
(3)			76.18	149.29	$\equiv P$	$\Delta x_i = a \cdot h$
		−60.76	−48.36	−36.80		$\Delta y_i = -o \cdot h$
			27.82	112.49	$\equiv F$	$\Delta x_i = o \cdot \Delta s_i$
	45.94		−27.82	36.56		$\Delta y_i = a \cdot \Delta s_i$
(2)	[134.45]		0.00	149.05	$\equiv E$	

Thus, to obtain the coordinates of P, first F is coordinated and then the coordinate differences from F to P are applied. h is to be taken as positive or negative depending on whether P is to the left or right of the line.

Equations (8.3.7) and (8.3.9) are special cases of the similarity transformation (equation (5.2.5)).

(4) *Intersection of two lines*

A straight line is commonly defined by either two points or by one point and its azimuth. The direction coefficient m is simply the ratio $o : a$.

$$\tan\alpha = \frac{x_e - x_a}{y_e - y_a} = m . \tag{8.3.10}$$

Thus with $P_1(x_1, y_1)$ and m, a straight line is uniquely defined and has the following equation:

$$x - x_1 = m_1(y - y_1) . \tag{8.3.11}$$

Similarly, a second straight line has the equation

$$x - x_2 = m_2(y - y_2) . \tag{8.3.12}$$

The point of intersection $P(x/y)$ has to fulfill both equations:

$$y - y_1 = \frac{(x_2 - x_1) - m_2(y_2 - y_1)}{m_1 - m_2} . \tag{8.3.13}$$

If one line is given by the points $P_1(x_1, y_1)$ and $P_3(x_3, y_3)$ and a second one by $P_2(x_2, y_2)$ and $P_4(x_4, y_4)$ (Figs. 8.3.8 and 8.3.9), then the coordinates of the point

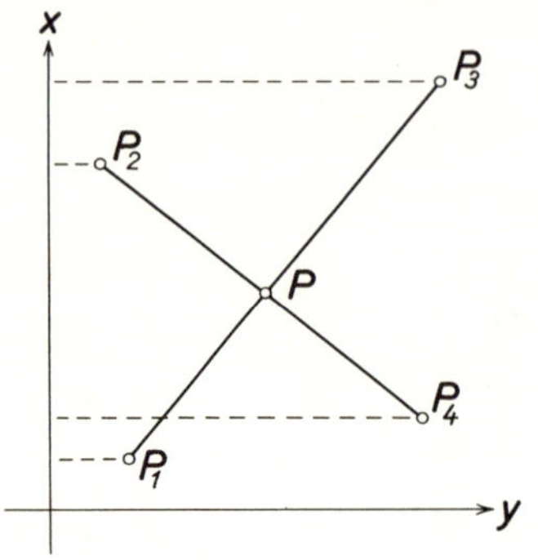

Fig. 8.3.8 Arc section (case I)

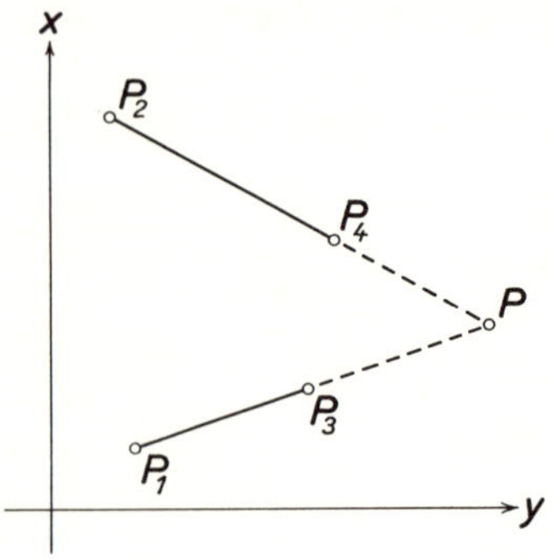

Fig. 8.3.9 Arc section (case II)

of intersection are obtained as

1. with equation (8.3.10): $m_1 = \dfrac{x_3 - x_1}{y_3 - y_1}, \quad m_2 = \dfrac{x_4 - x_2}{y_4 - y_2}$
2. with equation (8.3.13): $(y - y_1)$, with equation (8.3.11) $(x - x_1)$;
3. $y = y_1 + (y - y_1); \quad x = x_1 + (x - x_1)$.

Check: $m_1 = \dfrac{x_3 - x}{y_3 - y}; \quad m_2 = \dfrac{x_4 - x}{y_4 - y}$.

(5) *Checking of a line net*

The previously mentioned computations provide an opportunity to check the lines by computing various distances and comparing them with the measured values. This provides an independent check on the measurements and leads to the detection of virtually any blunder in the measurements provided the line net is properly established.

Example:

Checking of the survey lines in Fig. 8.3.3.

Distance	from coords.	measured	difference	acceptable
(6)–(8)	78.76	78.82	−0.06	0.14
(7)–(10)	52.79	52.83	−0.04	0.12
(5)–(9)	66.26	66.23	0.03	0.13

8.3.2 Polar Method

The polar method is based on measurements of direction based on a fixed reference direction plus the distance from a known station to the object point. This provides a unique definition of each point (Fig. 8.3.10).

Usually, the directions are measured with a theodolite, and the distances with an EDM instrument (section 4.4). For larger areas, the use of electronic tacheometers is most suitable. It is thus possible to generate an automatic data flow of the survey of the object points to the production of maps, plans or registers (section 12.2.2).

The station is either a coordinated point or else freely selected. The computations are described in section 6.7.

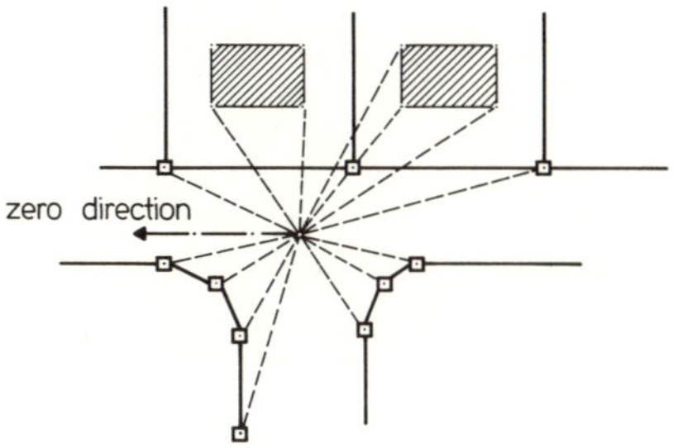

Fig. 8.3.10 Polar survey

8.4 Plotting of the Survey Plan

8.4.1 Scale, Materials and Manual Plotting

The purpose of a particular plan usually influences the selection of scale and material. Smaller plans are plotted at 1 : 250, 1 : 500, 1 : 1000. Transparent paper is the commonly used material for construction plans, etc., because of easy reproduction (ozalid copies). If higher demands are placed on scale and accuracy, transparent foils on a PVC base (e.g., Astralan) or polycarbon base (e.g., Pokalon), or polyester base (e.g., Hostaphan, Stabilene, or Safir PE) are used, which sometimes require special ink. The best accuracy is obtained with quality heavy paper with an aluminum base. Occasionally scribing onto specially coded glass or foil material is utilized.

A steel ruler and triangles are the basic tools, although small plotting instruments (e.g., Aristo Minicoordinatograph by Dennert and Pape) have become quite popular. This instrument has a range of 20 cm by 20 cm and scales for 1 : 500, 1 : 1000, 1 : 2000 and 1 : 2500, and provides a plotting accuracy of 0.2 mm.

Larger coordinate plans require the plotting of a grid (Fig. 8.4.1). With a long steel ruler (1 m long), a diagonal cross with equal sides is plotted. The end points are connected and subdivided into decimetres. As a check, the ruler is placed diagonally across the grid squares whose end points have to fall exactly onto a straight line.

If such nets have to be plotted frequently, a metal template with fine drill holes in a grid arrangement is often utilized.

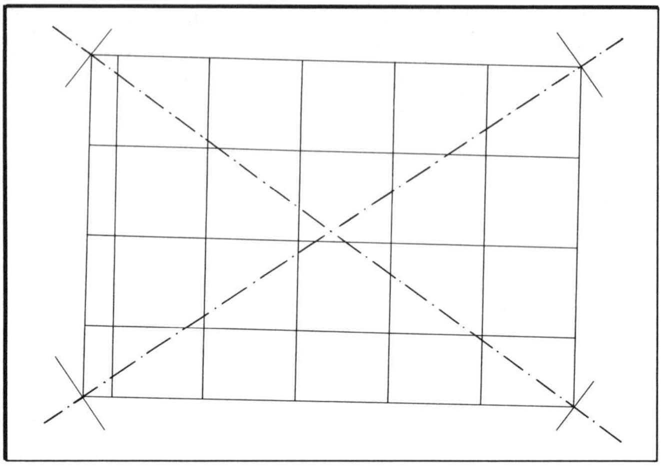

Fig. 8.4.1 Grid plot

Once the framework is established, manual plotting starts by positioning the survey lines using the coordinates of their end points. Details are then plotted with respect to these, and check measurements are measured with the ruler on the plan as a control.

8.4.2 Interactive Plotting and Data Bank Systems

(1) *Purpose and Requirements*

Large jobs concerning the evaluation, administration and plotting of surveying data are nowadays often handled with automated systems. Such a system should have the following characteristics:

a) Designed as a total system, capable of measuring, recording, computing, administering and plotting of surveying data.

b) Modular structure of both hardware and software.

c) Continuous data flow from the survey (e.g., with an electronic tacheometer) to plotted map (e.g., digitally driven plotter).

d) Simple operation with software applying dialogue techniques.

e) Manipulation of the data with geodetic evaluation and plotting programmes.

f) Installation of a "geodetic coordinate data bank" and of a "cartographic data bank".

g) Use of point members, containing point identification, classification, and status, as well as code for monitoring the data flow.

h) Use of a powerful desk computer with variable mass storage.

i) Use of an intelligent plotting table for precision and speedy plotting.

j) Support in data handling and preparation of plotting data by means of a graphics terminal.

(2) *Structure of the System*

Fig. 8.4.2 illustrates the structure of an automatic surveying and plotting system.

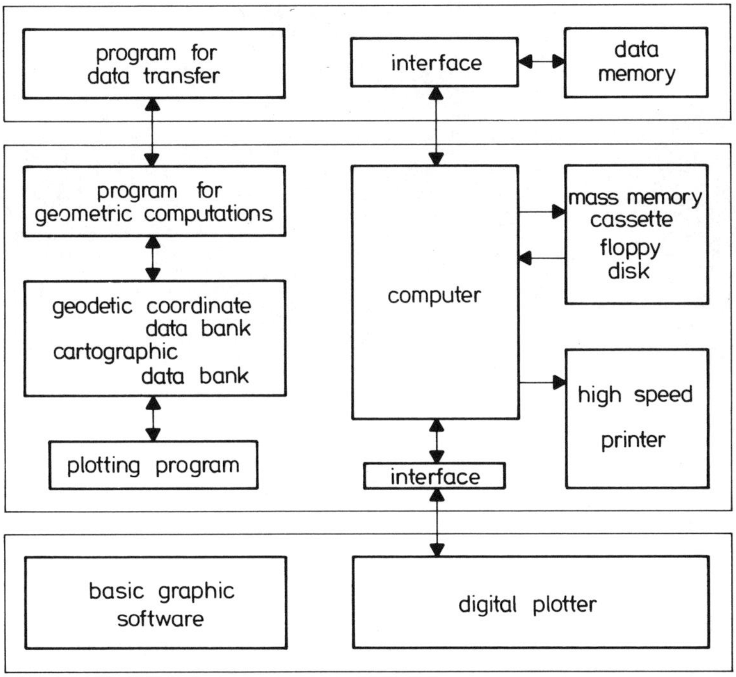

Fig. 8.4.2 Block diagram of an automatic surveying and plotting system

Field data are automatically fed by means of an interface into the computer from an electronic data device (e.g., an electronic tacheometer (section 12.2)). A special program triggers this data transfer. In addition to field data, other data, e.g., graphical information, can be transferred to the computer by means of a digitizing table or photogrammetric plotter (Fig. 8.4.3).

If a *digitizing table* is used, then by driving the cursor to the points of interest or along lines, the graphic plan is transformed into a digital map with the data stored in a data bank. For checking purposes, the plan is also displayed on the screen of the graphics terminal. Non-graphic information (i.e., text) can now be added to the graphic elements from the keyboard. The digitized points of the graphic plan are transformed into the reference coordinate system prior to storage.

The *computer* has usually a high speed execution and an extended working storage.

Fig. 8.4.3 Interactive Graphics Surveying System (*Wild*)

Furthermore, it should have well designed dialogue techniques for data handling and for graphic representation at the terminal. It also has to be capable of monitoring tasks in real time and of interrupt structures.

The *software* consists normally of programs

- for two-way data traffic between the measuring units (or their storage devices) and the computer;
- for interactive solution of geometric tasks (coordination of points, network adjustment, etc.);
- for the generation of data banks;
- for plotting.

The final link of the automation chain is a *digital plotting table*. Its task is to present in graphical form on paper or scribing material the map stored digitally in the data bank. A microprocessor supports the computer. A variety of drafting or scribing tasks can be utilized in the plotting head. The maximum plotting speed in the axis direction can be 30 mm/s with a maximum acceleration of 5 m/s^2. The graphical resolution of such systems is up to 0.02 mm.

The generation of the digital map can be monitored on the display terminal. Furthermore, texts, lists and commands to the operator can be displayed on a second (alphanumeric) terminal.

As a rule, the systems are designed such that sections of plans and maps can be enlarged or reduced at will. These are usually defined by a movable cross on the display screen.

8.5 Reproduction and Scale Changes

8.5.1 Reproduction of Plans

The easiest way of reproducing transparent plans is by the ozalid method, i.e., the plans are exposed onto light sensitive paper which is developed by ammonia vapour. Opaque originals are copied with common office copiers (e.g., Xerox). Photographic reproduction for larger plans might be required. This is accomplished by contact copying using a photographic negative. With office copiers or photographic reproduction systems small plans can also be enlarged or reduced.

8.5.2 Scale Changes

If simple figures are to be enlarged or reduced, a reduction compass can be used. This instrument consists of two rods held together by a screw with a pencil at each end. By changing the lengths of the rods from the pivot point, a certain magnification ratio can be obtained. A better instrument is the pantograph whose working principle is evident from Fig. 8.5.1.

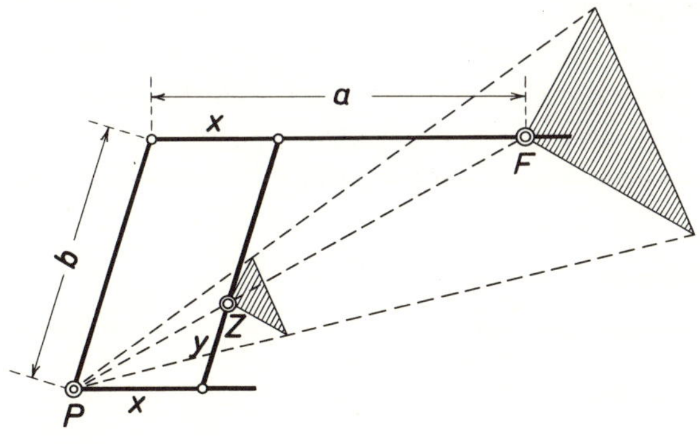

Fig. 8.5.1 Pantograph

If, for instance a plan is to be reduced from 1 : 400 to 1 : 1000, then with

$$\frac{PZ}{PF} = \frac{400}{1000} = \frac{2}{5} = \frac{x}{a} = \frac{y}{b}$$

x and y is computed, and the joints are set accordingly. Then the original plan is followed with the needle F, while the pen Z draws the reduced figure. For enlargements, needle and pen are interchanged (Fig. 8.5.2).

Of course it is far easier to use optical magnification or redution, which may be available on an office copying machine. The most precise way is photographically by using a reproduction camera. Its cost, however, restricts the use to larger companies or government agencies.

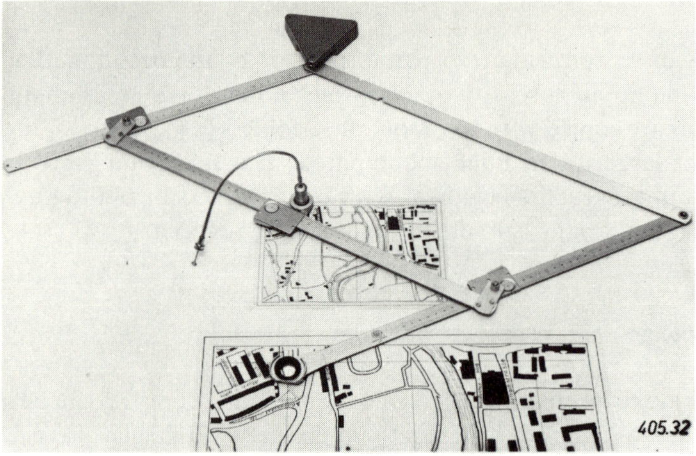

Fig. 8.5.2 Precision pantograph

8.6 Area Determination

If a closed figure, bounded by straight lines, is surveyed according to the procedures given in the previous sections, its area can be computed. The method applied depends on the type of survey as well as on the required accuracy. The following basic procedures are available:

1) Numerical area determination

2) Semigraphical area determination

3) Graphical area determination

8.6.1 Numerical Area Determination

8.6.1.1 Area Determination Using Field Values

This approach is suitable if the area was surveyed in such a manner that it can be easily divided into triangles and trapeziums. The area of these figures is equal to half the base line times the height. As shown in Fig. 8.6.1, a reversed trapezium follows the same rule, as the area is equal to the difference of the two triangles:

$$A = {}^1/_2 a(h_1 - h_2)$$

where a is the base line, h_1 is the height inside the area, while h_2 is the height outside the area.

Rather than carrying the factor ${}^1/_2$, usually double the area is computed and divided by 2 after completion of the computation. Often the accuracy requirements are such that the factors can be rounded to decimetres.

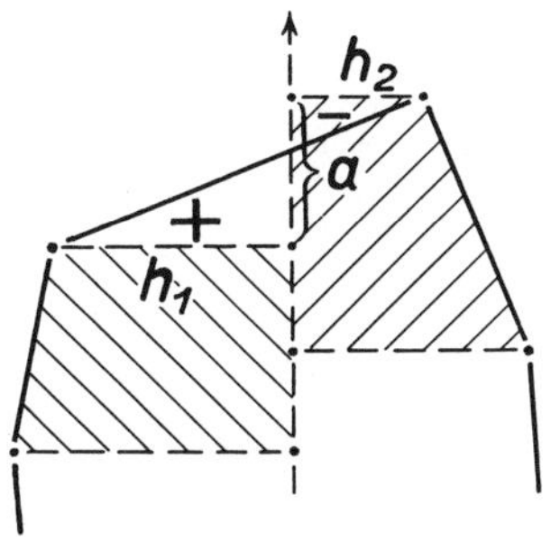

Fig. 8.6.1 Reversed trapezium

Example: Area determination for the lot shown in Fig. 8.3.1

a) with centimetre accuracy		b) rounded to decimetre	
(3.72 − 0.00) · (5.34 + 0.00) = 3.72 · 5.34 =	19.86	3.7 · 5.3 =	19.6
(55.12 − 3.72) · (8.40 + 5.34) = 51.40 · 13.74 =	706.24	51.4 · 13.7 =	704.2
(76.30 − 55.12) · (8.40 − 8.23) = 21.18 · 0.17 =	3.60	21.2 · 0.2 =	4.2
(76.30 − 62.71) · (8.23 + 18.65) = 13.59 · 26.88 =	365.30	13.6 · 26.9 =	365.8
(62.71 − 9.90) · (18.65 + 20.24) = 52.81 · 38.89 =	2053.78	52.8 · 38.9 =	2053.9
(9.90 − 0.00) · (20.24 + 0.00) = 9.90 · 20.24 =	200.38	9.9 · 20.2 =	200.0
	3349.16 : 2		3347.7 : 2
	A = 1674.58 m²		A = 1673.8 m²

8.6.1.2 Area Determination Using Coordinates

This approach is easily explained using the practical situation given in Fig. 8.3.1, however, the origin is mentally shifted to the left such that it falls outside the figure (Fig. 8.6.2). Then the area is obtained by subtracting the area 1′ − 1 − 6 − 5 − 4 − 4′ from 1′ − 1 − 2 − 3 − 4 − 4′:

$$\begin{aligned} 2A = {} & (x_1 - x_2)(y_1 + y_2) + (x_2 - x_3)(y_2 + y_3) + (x_3 - x_4)(y_3 + y_4) \\ & + (x_4 - x_5)(y_4 + y_5) + (x_5 - x_6)(y_5 + y_6) + (x_6 - x_1)(y_6 - y_1) \\ \text{or:} \quad & 2F = \Sigma (x_i - x_{i+1})(y_i + y_{i+1}) \,. \end{aligned}$$

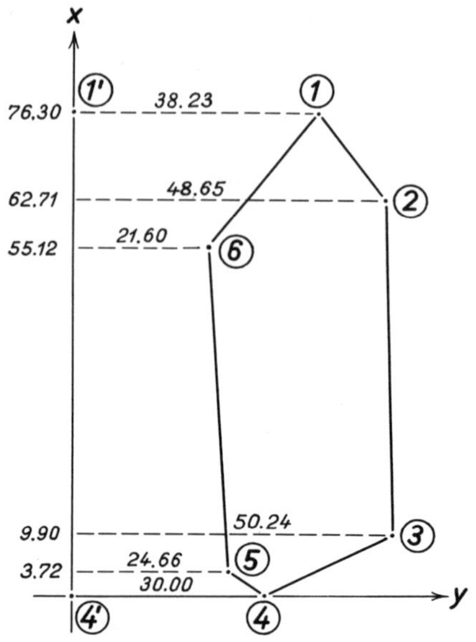

Fig. 8.6.2 Gauss's area formula

Similarly, by exchanging the x- and y-axes:

$$2A = \Sigma(y_{i+1} - y_i)(x_i + x_{i+1}) .$$

These formulae are called the "*Gauss* trapezium formulae". After multiplication and rearrangement the "Gauss triangle formulae" are obtained:

$$2A = \Sigma x_i(y_{i+1} - y_{i-1}) = \Sigma y_i(x_{i-1} - x_{i+1}) .$$

Example: Compute the area of the lot shown in Fig. 8.6.2 using the first triangle formula:

Point	y_i	x_i	$y_{i+1} - y_{i-1}$	$x_i \Delta y_i$
1	38.23			
2	48.65	62.71	12.01	753.15
3	50.24	9.90	−18.65	−184.64
4	30.00	0.00	−25.58	
5	24.66	3.72	−8.40	−31.25
6	21.60	55.12	13.57	747.98
1	38.23	76.30	27.05	2063.92
2	48.65			3349.16 : 2
				$A = 1674.58 \text{ m}^2$

8.6.2 Semigraphical Area Determination

The semigraphical area determination (Fig. 8.6.3) is based on the following consideration: Measuring errors $\mathrm{d}a$ and $\mathrm{d}h$ are present in a triangle with a short base a and a long height h. The area error is then $\mathrm{d}A = {}^1/_2(a\mathrm{d}h + h\mathrm{d}a)$. In order to keep $\mathrm{d}A$ small, the error $\mathrm{d}a$ of the short base a is to be kept small as it is multiplied by the large value h.

A semigraphical approach is suitable if the area is divided into narrow triangles with short base lines, measured accurately in the field. The long heights can then be extracted from an scaled plan with lower accuracy.

Example: The area of the lot shown in Fig. 8.6.3 is

$$2A = 19.95 \cdot 42.0 + 17.48 \cdot 40.2 + 19.52 \cdot 81.6 + 17.98 \cdot 66.6 + \\ + 18.91 \cdot 33.0 = 4955; \quad A = 2478\ \mathrm{m}^2.$$

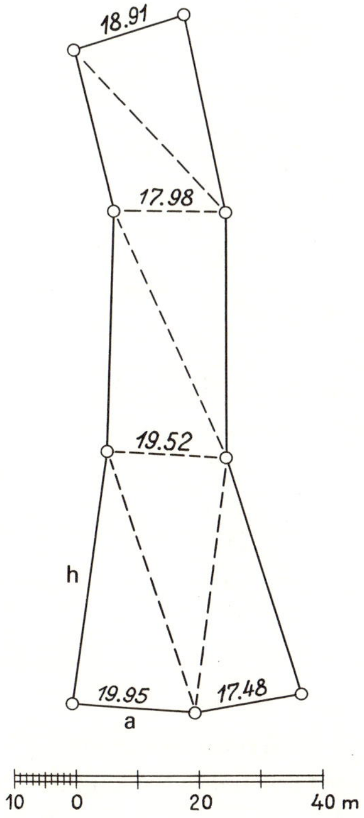

Fig. 8.6.3 Semigraphical area determination

8.6.3 Graphical Area Determination

8.6.3.1 Graphical Area Determination Using Simple Tools

Graphical methods are used when numerical methods are not feasible (e.g., for natural boundaries), as well as for independently checking both the computation and accuracy of the plotting of the plan.

Regular figures are divided into triangles or trapeziums whose base lengths and heights are measured in the plan with a ruler.

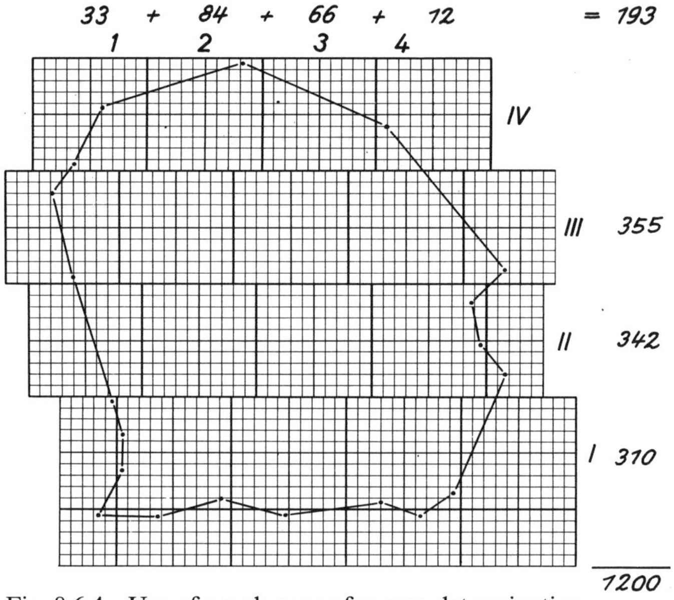

Fig. 8.6.4 Use of graph paper for area determination

Irregular figures are covered with millimetre graph paper (Fig. 8.6.4). The area is then obtained by adding the number of squares and multiplication with the scale factor.

Example:

		mm^2
Strip I (left to right)	=	310
Strip II (right to left)	=	342
Strip III (left to right)	=	355
IV/1 + IV/2 + IV/3 + IV/4 = 33 + 84 + 66 + 10	=	193
Total		1200

For *long narrow figures*, a parallel line graph (Fig. 8.6.5) is more suitable. A transparent sheet with parallel lines separated by h is placed normal to the length of the figure dividing it into a number of small trapeziums. With m_1, m_2 ... m_n being the median lines of these trapeziums, the area becomes

$$A = hm_1 + hm_2 + \dots hm_n = h\Sigma m .$$

The individual medians are mechanically added with a compass. h equals 10 in the reduced Fig. 8.6.5. Σm amounted to 247.5 leading to $A = 2475\ \mathrm{m}^2$ as compared to 2478 m^2 obtained with the semigraphical approach. It should be noted that irregular end figures have to be adjusted as shown in Fig. 8.6.5.

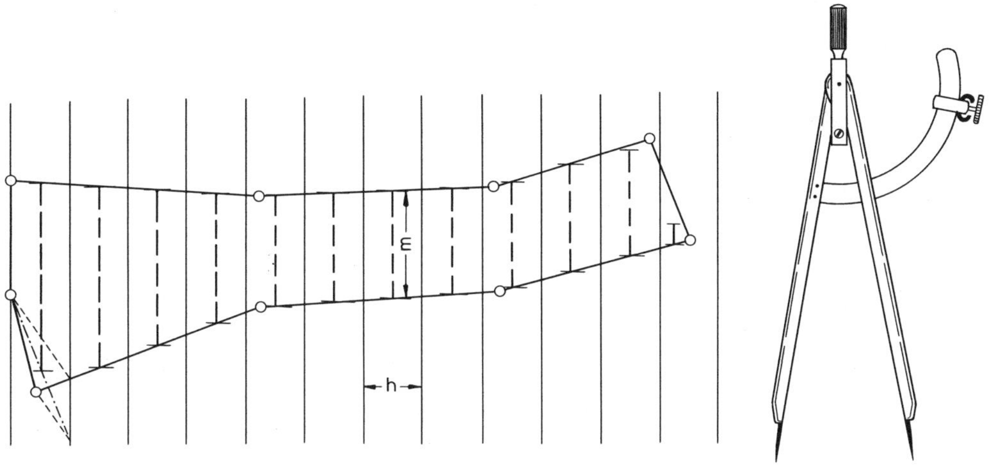

Fig. 8.6.5 Parallel line graph

8.6.3.2 Graphical Area Determination Using the Planimeter

a) *Description and operating principle*

The polar planimeter (Fig. 8.6.6) is a mechanical integration instrument which permits the determination of an area by following its boundary lines.

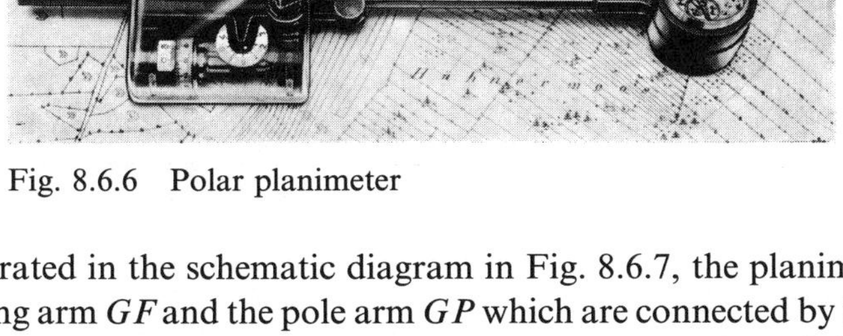

Fig. 8.6.6 Polar planimeter

As illustrated in the schematic diagram in Fig. 8.6.7, the planimeter consists of the measuring arm GF and the pole arm GP which are connected by the ball joint G. The pole arm ends at the pole P which is fixed onto the plan by weight or with a pin. The measuring arm supports at its free end the measuring point F, and on its other end,

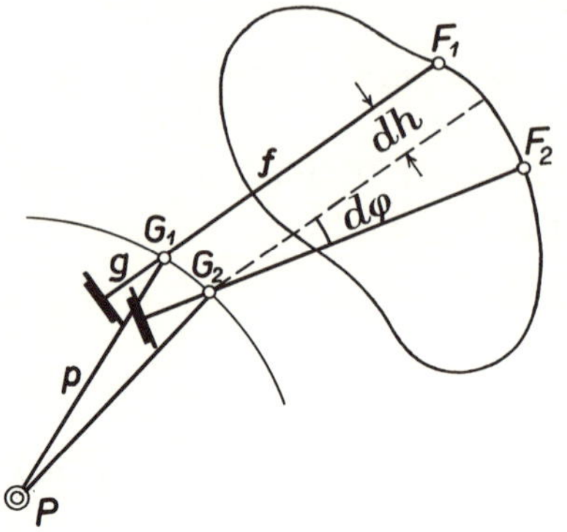

Fig. 8.6.7 Pole located outside the area

extending beyond G, the integration roll, which is held by a shaft parallel to the measuring arm. To simplify the explanation, the roll is drawn in Fig. 8.6.7 as if it were directly on the measuring arm at a distance g.

When the measuring point is moved along the boundary line of a given figure, the joint G moves along a circular arc with radius p around the pole. A differential movement of the measuring arm can be separated into a parallel shift and a rotation. A parallel shift $\mathrm{d}h$ by the measuring arm results in an area coverage of $f\mathrm{d}h$. When rotating by $\mathrm{d}\varphi$, a small sector of a circle with the area $^1/_2 f^2 \mathrm{d}\varphi$ is covered while the roll rotates backwards by $-g\mathrm{d}\varphi$.

Thus the following relationships hold for a differential area $\mathrm{d}\Delta$ covered by parallel shift and rotation with $\mathrm{d}a$ representing the roll movement:

	area element	roll reading
parallel shift	$f\mathrm{d}h$	$\mathrm{d}h$
rotation	$^1/_2 f^2 \mathrm{d}\varphi$	$-\mathrm{d}g\varphi$
	$\mathrm{d}A = f\mathrm{d}h + {}^1/_2 f^2 \mathrm{d}\varphi$	$\mathrm{d}a = \mathrm{d}h - g\mathrm{d}\varphi$ or $\mathrm{d}h = \mathrm{d}a + g\mathrm{d}\varphi$

By substituting $\mathrm{d}h$ into the left equation we obtain:

$$\mathrm{d}A = \mathrm{d}fa - (fg + {}^1/_2 f^2)\mathrm{d}\varphi\,.$$

By integration over the whole area we get

$$A = fa + (fg + {}^1/_2 f^2) \int \mathrm{d}\varphi + C$$

where C is an integration constant.

Two cases have to be distingushed for further treatment:

1st case: The pole is located outside the area (Fig. 8.6.7). The measuring arm moves forward and backward in this case and eventually ends up at its original position, which means $\int d\varphi = 0$. The area covered by the measuring arm is equal to the area traced by the measuring point. With r being the circumference of the roll and h the number of revolutions we get

$$A = fa = frn = kn \ ,$$

where the constant $K = fr$, and n is read at the roll.

2nd case: The pole is located inside the area. Since this case does not render sufficient accuracy for most practical purposes, it is recommended that the area be divided into several smaller sections which are then measured individually with the pole outside.

b) *Determination of the planimeter constant*

Although many planimeters permit the adjustment of the length of the measuring arm to obtain a round constant, it is usually easier to just determine the constant and multiply the roll readings.

The calibration can be carried out with the aid of a calibration ruler supplied with the instrument which provides a circle with a known area (e.g. 100 cm^2).

It is, however, better to perform a relative calibration using a grid square on the plan because differential scale changes due to paper shrinkage, etc., are then covered as well.

c) *Rules for the use of the planimeter*

- The most dangerous error is caused by an inclined roll, i.e., the axes of roll and the measuring arm are not exactly parallel. This systematic error can be eliminated by measuring the area in both roll positions with fixed pole and then taking the mean (Fig. 8.6.8). All modern planimeters are designed such that the measuring arm can be moved beneath the pole arm.

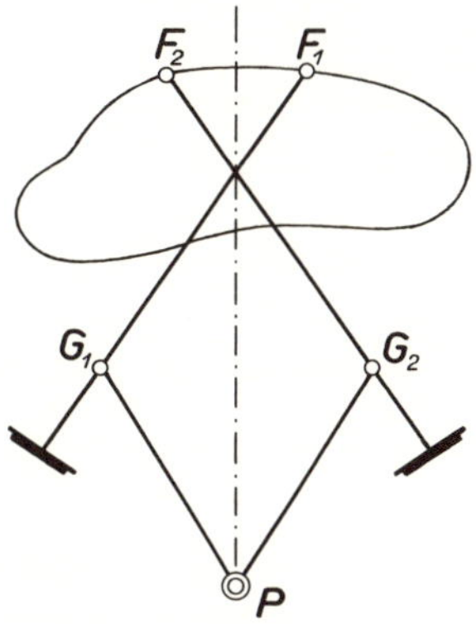

Fig. 8.6.8 Measuring in both roll positions

- When moving normal to the direction of the measuring arm, the roll is turning, while it slides when moving in its direction. When changing from sliding to turning, an inertial point has to be overcome, which represents a second error source, especially when working on rough paper. To eliminate sliding as much as possible, the planimeter should be placed outside the figure in such a manner that the basic circle passes through the centre of gravity of the figure and bisects it along its largest extension.

- If the measuring arm has a varying length, the vernier units should be selected such that the measuring arm is rather long, as extreme situations can be better avoided and the summation of random errors is smaller than for a short measuring arm.
- Each area should be measured at least once in each roll position.
- Figures with straight line boundaries are to be measured free hand. The use of a ruler can cause large systematic errors.

Examples:

1) Determination of the length of the measuring arm for the vernier unit $k = 10\ m^2$.

In a cadastral plan (1 : 1000) a 1 dm^2 grid was measured five times in both roll positions, using an arm length setting of $f_1 = 327.2$. This area represents 10000 m^2.

Pos. I	n	Pos. II	n
0443.5		0741.5	
	997.0		998.5
1440.5		1740.0	
	995.0		997.5
2435.5		2737.5	
	997.5		996.5
3433.0		3734.0	
	996.5		998.0
4429.5		4732.0	
	996.0		998.0
5425.5		5730.0	
Mean	996.4		997.7; $n_1 = 997.05$

Thus:

$$n_0 = \frac{F_0}{k} = \frac{10000\ m^2}{10\ m^2} = 1000\,,$$

$$\Delta f = f_1 \frac{n_1 - n_0}{n_0} = 327.2 \frac{997.05 - 1000}{1000} \approx -0.96\,.$$

The length of the measuring arm of some planimeters can be changed. If this is the case, then a new setting

$$f_0 = f_1 + \Delta f = 327.2 - 0.96 \approx 326.2\,.$$

is desirable.

As a check, the measurements were repeated with the new setting. This resulted in $n_1 = 1000.1$ and $\Delta f = +0.03$, which is too small to provide a further correction of the measuring arm length.

2) Determination of an area.

Using the above mentioned planimeter with the setting $f_0 = 326.2$ and the vernier unit $k = 10\ \text{m}^2$, a parcel was measured four times on a 1 : 1000 cadastral plan.

Pos. I	n	Pos. II	n
0787.5		0434.5	
	237.5		238.5
1025.0		0673.0	
	239.0		239.9
1264.0		0912.0	
	238.0		240.0
1502.0		1152.0	
	239.0		239.5
1741.0		1391.5	
Mean:	238.4		239.2; $n = 238.8$.

Thus:

$$A = nk = 238.8 \cdot 10\ \text{m}^2 = 2388\ \text{m}^2.$$

8.6.4 Accuracy of Area Determination

8.6.4.1 Checks

Since computational errors can easily occur when determining an area, it is advisable to determine an area using two different methods. If both methods are equivalent, e.g., graphical, the mean of both determinations is used; if one method is more accurate (i.e. numerical) the other is solely used as a check.

If a number of adjoining parcels is computed, the total area is determined to check the sum of the individual areas. If both methods are graphical, small discrepancies are divided proportional to the individual area. In this case corrections for paper shrinkage and map projection can be applied.

8.6.4.2 Evaluation of Different Methods

In order to be able to select the most suitable method for each case, and to judge possible discrepancies, it is necessary to know the peculiarities of the methods and their error sources.

a) The *numerical area determination* using coordinates of the boundary points provides the most accurate results. Wrongly identified points and input errors are the most common error sources.

b) The *area determination using field values* depends primarily on the accuracy with which the small distances are determined. If in a triangle the base A and height h have errors of $\mathrm{d}a$ and $\mathrm{d}h$, respectively, the area is erroneous by

$$\mathrm{d}A = \tfrac{1}{2}(a\mathrm{d}h + h\mathrm{d}a).$$

Similarly in a trapezium with sides a and b:

$$\mathrm{d}A = \tfrac{1}{2}(a+b)\mathrm{d}h + h(\mathrm{d}a + \mathrm{d}b)\,.$$

Thus the errors in the small distances are multiplied by larger values. If, however, the area computation is the result of a large sum of partial areas, some of these errors cancel, so that the error of the final result only increases by the square root of the number of parts.

c) The *semigraphical area determination* is practically as accurate as when using field values, if the figures are narrow and the short sides are measured in the field.

d) The *accuracy of graphical area determination* depends on the scale and stability of the plan, the tools used, and the skill of the operator.

The *area scale* is the square of the linear scale. By changing from 1 : 500 to 1 : 1000, all paper lengths are reduced to one half and all areas to one quarter. Thus an error in the paper area will increase with the square of the scale reduction.

The *stability of the plan* depends on the drafting material. Paper especially reacts to changes in humidity, with at times different amounts in different directions. It is advisable to determine shrinkage or expansion in two directions using a grid and to express it as a percent. If a and b are the sides of a rectangle measured in a plan with p% or g% paper shrinkage, then

$$A = ab\left(1 + \frac{p}{100}\right)\left(1 + \frac{q}{100}\right) = ab\left(1 + \frac{p}{100} + \frac{q}{100} + \frac{pq}{10000}\right)$$

or neglecting the small term

$$A = ab + ab\,\frac{p+q}{100}\,.$$

This means that the area is incorrect by $(p+q)$%. For normal drafting paper, up to 3% linear changes can be expected.

The accuracy of planimeter work also depends on the skill and calm hand of the operator. Generally, the planimeter provides better values if the circumference of a figure is small with respect to its area. Thus it is not suited for long narrow areas, such as roads, watercourses, etc.

8.6.4.3 Error Limits

In Germany, the surveying administrations agreed on error limits for areas, based on many years of experience. These specify by how much the values of two independent determinations of the same figure may differ. The actual limit is three times the standard deviation, so that the expected uncertainty should amount to about a third of the given values.

The admissible differences are for determination using

field values	$\Delta_A = 0.2\sqrt{A} - 0.0003\,A$
field values + plans 1 : 500	$\Delta_B = 0.2\sqrt{A} - 0.0003\,A$
field values + plans 1 : 1000	$\Delta_C = 0.4\sqrt{A} - 0.0003\,A$
field values + plans 1 : 2000	$\Delta_D = 0.6\sqrt{A} - 0.0003\,A$
maps 1 : 2000	$\Delta_E = 1.2\sqrt{A} - 0.0003\,A$

Using these values, the following table was computed:

A [ha]:	0.01	0.05	0.20	1.00	10.0	50.0
Δ_A [m^2]	2	5	10	23	93	291
Δ_B [m^2]	3	7	14	33	125	362
Δ_C [m^2]	4	9	18	43	156	433
Δ_D [m^2]	6	14	27	63	220	574
Δ_E [m^2]	12	27	54	123	409	999

9. Differential Levelling

9.1 Levelling Instruments

9.1.1 Simple Levelling Instruments

Fig. 9.1.1 illustrates the basic idea of differential levelling. The elevation difference between points A and B is determined by measuring their vertical distance from a horizontal line of sight. This requires a levelling instrument capable of generating the horizontal line of sight plus one or, better, two levelling rods on which the vertical distances are read. The elevation difference between points A and B in Fig. 9.1.1 is $1.96 - 1.12 = 0.84$.

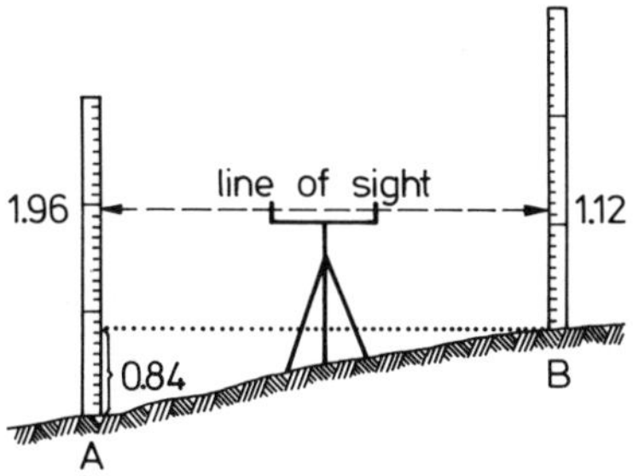

Fig. 9.1.1 The concept of differential levelling

9.1.1.1 The Simple Water Level

The simplest levelling instrument consists of two glass cylinders, connected with a pipe and filled with a coloured liquid. Fig. 9.1.2 illustrates such a simple water level,

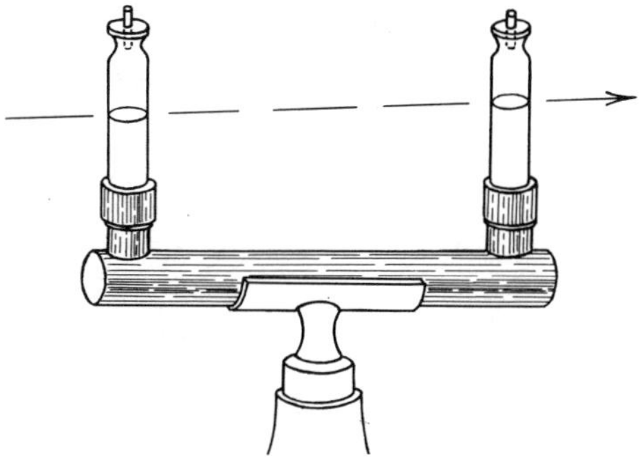

Fig. 9.1.2 The simple water level

where a line connecting the two water surfaces represents the horizontal line of sight. Its accuracy, however, is so poor that this instrument is not used anymore in this form. The accuracy improves somewhat if floating targets are placed onto the liquid surfaces in the two cylinders.

9.1.1.2 Hydrostatic Level

If the connecting pipe is replaced by a 20 to 30 m long hose, and if the two cylinders are graduated in millimetres, a hydrostatic level is obtained. This instrument is frequently used on construction sites, especially in difficult terrain.

The hose is moved around to get rid of possible air bubbles, then the two measuring cylinders are held side by side in order to determine a possible difference in the water levels, the so-called zero correction. Then the two cylinders are moved to their measuring positions, and after a few minutes pause for letting the water settle, both ends are read several times. The results are averaged and the zero correction is applied. If handled properly, a standard deviation of 1 to 2 mm can be expected for a single elevation difference. Precision hydrostatic levels are used in mining and yield a significantly better accuracy.

9.1.1.3 Horizontal Rod

This instrument, which is used for approximate elevation determination, consists of a 3 m long rod with spirit level, and a graduated staff with a bull's-eye bubble. Fig. 9.1.3 illustrates the measuring principle.

The horizontal rod provides correct results only when its bottom edge is parallel to the bubble axis. It is tested by taking the mean of two readings taken with opposite

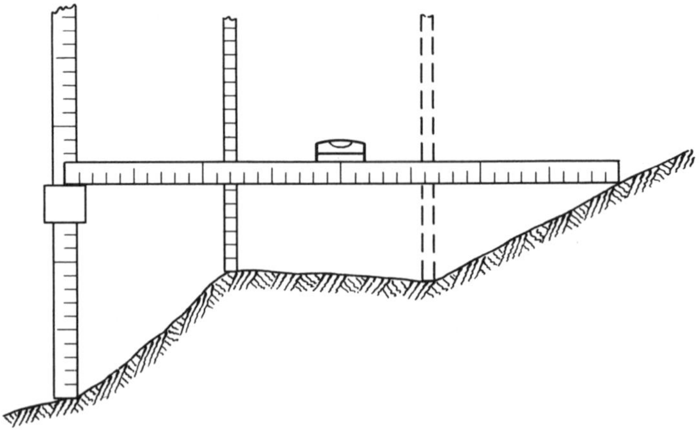

Fig. 9.1.3 Horizontal rod

orientation of the level. If these readings differ significantly, then the level bubble has to be adjusted for the mean reading. For careful work, a standard deviation of 0.5 cm per set up is expected, which leads to $\pm 0.5\sqrt{n}$ cm for n set-ups.

9.1.2 Spirit Levels

9.1.2.1 Mechanical Structures of Spirit Levels

A spirit level consists of a measuring telescope, its support which permits rotation around a vertical axis, and the telescope bubble which is used to bring the collimation axis into a horizontal position. If, as illustrated in Fig. 9.1.1, levelling rods are placed at A and B, the elevation difference Δh is obtained as the difference between "back-sight" to A and "foresight" to B as

$$\Delta h = B - F$$

which represents the fundamental equation of differential levelling.

Nearly all modern levels have telescopes with internal focussing lens and devices like mirrors or prisms which permit the observer to see the level bubble without moving, at times within the field of view of the telescope. The differences between instruments are found primarily in the supports of telescope and bubble, the axis arrangement,

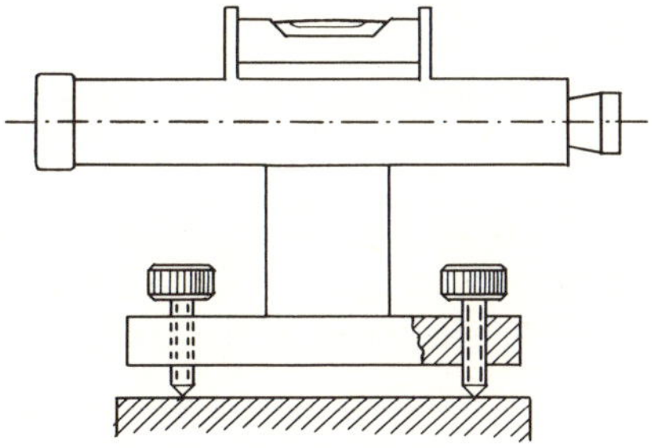

Fig. 9.1.4 Dumpy level

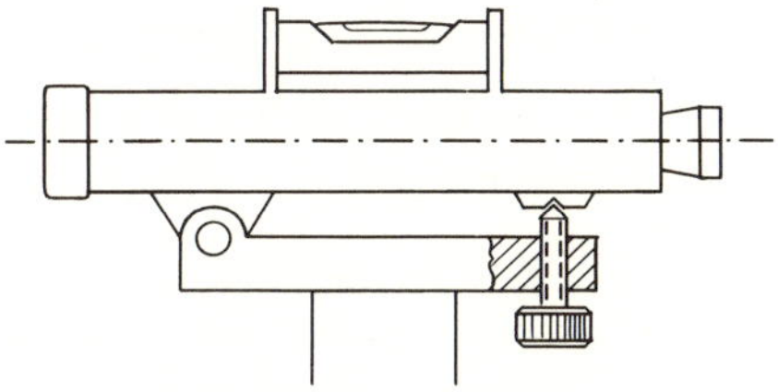

Fig. 9.1.5 Tilting level

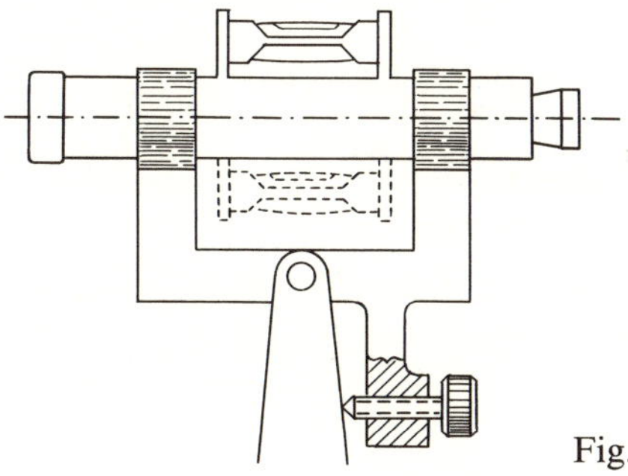

Fig. 9.1.6 Level with ring telescope

and the optical quality of the telescope. Concerning the supports, the following distinctions can be made:

a) Levels with a rigid telescope where there is a solid connection between the bubble and telescope and between the telescope and support (i.e., dumpy level, Fig. 9.1.4).

b) Levels with a rigid connection between the telescope and bubble, which can be tilted with respect to the vertical axis (i.e., tilting level, Fig. 9.1.5).

c) Levels with tilting screw and ring telescope, which can be rotated around its horizontal axis. This levels are equipped with a reversible bubble (Fig. 9.1.6).

9.1.2.2 Peg Test and Level Adjustment

The above mentioned differences are not always apparent in modern instruments because of their compact exterior design. However, they have to be considered when the level is to be adjusted.

(1) *Dumpy Level*

In this case, the telescope can be moved up and down with the foot screws of the tribrach only. Thus the vertical axis has to be exactly plumb for obtaining a horizontal line of sight. This, however, is only possible if the following two conditions are fulfilled:

1) Bubble axis and vertical axis have to be normal to each other. If this is not the case, the bubble has to be adjusted as in section 2.1.3.2.

2) The collimation axis has to be parallel to the bubble axis. Two steps are required to fulfill this condition:

1st step (Fig. 9.1.7, lower portion): The level is placed within a few decimetres in the middle between two levelling rods A and B, separated by $2s = 40$ to 160 m. After the bubble has been centred, the readings, a and b, are taken. If the collimation axis is

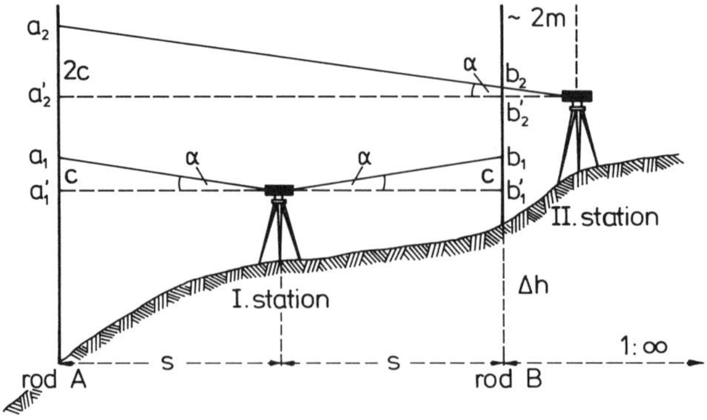

Fig. 9.1.7 Peg test

inclined by the angle α from the bubble axis, both rod readings are in error by the same amount $c = s \cdot \alpha / \varrho$. When the elevation difference is found, this error c cancels out. Thus, if a' and b' represent the correct readings we get

$$\Delta h_{A,B} = a_1' - b_1' = a_1 - b_1 \,. \tag{9.1.1}$$

This shows that the correct elevation difference is obtained, even with a misadjusted instrument when levelling from exactly in the centre.

2nd step: Now the instrument is moved as closely as possible (~ 2 m) towards rod B (Fig. 9.1.7, upper portion). After the bubble has been centred, the reading b_2 is obtained which equals the theoretical value b_2' unless the instrument is grossly out of adjustment.

There are different ways of obtaining b_2. Modern instruments have a shortest focussing distance of about 2 m. If the instrument requires an adjustment of $2c = 20$ mm at 40 m distance, the amount $(b_2' - b_2)$ which was neglected is 1 mm at 2 m, and thus barely acceptable. It means that the whole adjustment has to be repeated. To avoid this, the following methods have been suggested:

- A needle is aligned on the levelling rod by sighting through the telescope in reverse, i.e., by looking into the objective lens.
- The instrument is set up in such a manner that the objective lens touches the rod. Then its upper and lower edges are read at the rod, with b_2' being the mean value.

Both methods contain some inaccuracies which have led to the proposal of a modified peg test by Kukkamäki (section 9.1.2.3).

After b_2' is obtained, the rod at A is read. Because of the collimation error α the reading is not a_2' but

$$a_2 = \Delta h + b_2 + 2c = b_2 + a_1 - b_1 + 2c$$

which leads to

$$2c = (a_2 - b_2) - (a_1 - b_1) \,. \tag{9.1.2}$$

Thus the correct reading at A can be computed as

$$a_2' = a_2 - 2c = b_2 + (a_1 - b_1) = b_2 + \Delta h \,. \tag{9.1.3}$$

The instrument is adjusted by moving the cross hair up or down until the value $(b_2 + \Delta h)$ is reached at rod A.

Example:

Set Up I (centre)		Set Up II (near B)	
reading at A:	$a_1 = 2.261$	reading at B:	$b_2 = 1.532$
reading at B:	$b_1 = 1.342$	reading at A:	$a_2 = 2.465$
	$\Delta h = 0.919$	theoretical $= b_2 - \Delta h = 2.451$	

Check, according to equations (9.1.2) and (9.1.3):

$$2c = 0.933 - 0.919 = 0.014$$
$$a'_2 = 2.465 - 0.014 = 2.451 \, .$$

After the cross hair had been adjusted to the theoretical value at A, the following check readings were taken:

Set Up I:	$a_1 = 2.343$	Set Up II:	$a_2 = 2.569$
	$b_1 = 1.425$		$b_2 = 1.650$
	$\Delta h = 0.918$		$\Delta h = 0.919$

Thus the two conditions are fulfilled, and the level is now in proper adjustment and provides horizontal sights.

(2) *Tilting Level*

In this case, the telescope can be tilted with respect to the vertical axis within a limited range. Therefore the vertical axis is only approximately plumb as set with the aid of a bull's-eye bubble. Afterwards the telescope bubble is centred with the aid of the tilting screw. The only condition that is required for horizontal sight is that the bubble axis and collimation axis are parallel.

This is again checked with the aid of a peg test (Fig. 9.1.7). Although the instrument can be adjusted by moving the cross hair up or down, this is not practical. Rather the reading is set with the aid of the tilting screw, and then the bubble is centred using its adjustment screws.

When using a tilting level, a horizontal line of sight is obtained independent of the orientation of the vertical axis, provided that the telescope bubble is centred. This requires centring of the bubble prior to each reading which has to be kept in mind, especially when taking several readings from one station.

(3) *Level with Reversible Bubble and Ring Telescope*

These instruments, which are rarely used today, do not require an adjustment. The mean of the two rod readings obtained in both bubble positions and with rotated telescope represents a horizontal line of sight. This leads to a simple adjustment procedure. Once the correct value has been established as mentioned, this reading is set with the aid of the tilting screw and then the bubble is adjusted. Thus the instrument can be adjusted in one set-up which can be rather useful in mountainous terrain or on construction sites, where local conditions dictate different back- and foresight distances. The Wild $N2$ is an example of such an instrument.

This advantage is not without drawbacks. Additional error sources, such as convergency of bubble axes, or between bubble and collimation axes, less reliable telescope support, vibrations, etc., may degenerate the accuracy beyond the apparent gain.

9.1.2.3 Kukkamäki's Modified Method of Level Adjustment

In addition to the inaccuracies mentioned when obtaining the b_2' value for the peg test, even larger errors have to be expected when the telescope has to be focussed to extremely short distances. To eliminate these problems, Kukkamäki selected the location of the second instrument set-up beyond AB at a distance $2s$. With $s = 10$ m, the situation shown in Fig. 9.1.8 is obtained for a collimation error α: As in the ordinary peg test, the elevation difference $\Delta h_{AB} = a_1 - b_1$ is determined by setting up between A and B. Since α is the same in both portions of Fig. 9.1.7, its effects are evident from the upper portion: At B the reading b_2 is shifted by $2c$, while the reading a_2 at A is shifted by $4c$. This value of a_2 is obtained from the readings at B as:

$$a_2 = \Delta h_{A,B} + b_2 + 2c = a_1 - b_1 + b_2 + 2c$$

or

$$2c = (a_2 - b_2) - (a_1 - b_1)\,. \tag{9.1.4}$$

Therefore the cross hair is to be shifted such that the following reading is obtained

$$a_2' = a_2 - 4c; \quad b_2' = b_2 - 2c\,. \tag{9.1.5}$$

The computation of this value is checked by

$$a_2' - b_2' = a_1 - b_1 = \Delta h_{A,B}\,. \tag{9.1.6}$$

When using this approach for the adjustment for precision levels, the unbalanced effect of earth curvature and refraction has to be considered.

According to Kukkamäki, it amounts to

$$p = -1.68 \cdot 10^{-4}\, s^2 \tag{9.1.7}$$

where s is to be taken in metres. For the common sighting distances this becomes:

$s = 10$	20	30	40	50	60 m
$p = -0.02$	-0.07	-0.15	-0.27	-0.42	-0.60 mm

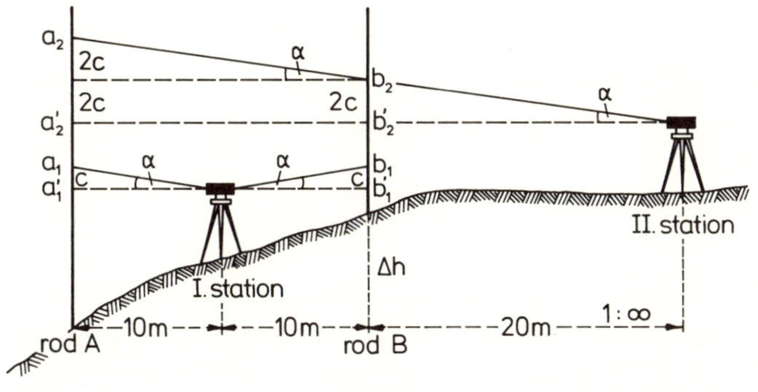

Fig. 9.1.8 Kukkamäki's modified method of level adjustment

Examples:

a) Without considering the curvature of the earth and refraction:

Rod [m]	Stn. I [m]	Stn. II [m]	theor. value according to equation (9.1.5) [m]
A	$a_1 =$ 1.499	$b_2 =$ 1.685	$a_2' =$ 1.679
B	$b_1 =$ 1.722	$b_2 =$ 1.905	$b_2' =$ 1.902
	$a_1 - b_1 = -0.223$	$a_2 - b_2 = -0.220$	$a_2' - b_2' = -0.223$

$2c = -0.220 + 0.223 = 0.003$ [m]; $4c = 0.006$ [m]

At station II the cross hairs are shifted until the theoretical value is obtained.

b) With the curvature of the earth and refraction consideration:

$$2c = [(a_2 + p_{40}) - (b_2 + p_{20})] - (a_1 - b_1)\,, \tag{9.1.4a}$$

$$a_2' = (a_2 + p_{40}) - 4c; \quad b_2 = (b_2 + p_{20}) - 2c\,, \tag{9.1.5a}$$

(note that mm/10 are carried in the computations)

Rod	Stn. I Reading [m]	Stn. II Reading [m]	s [m]	p [mm]	Stn. II Reading + p [m]
A	$a_1 =$ 1.4986	$a_2 =$ 1.6853	40	−0.27	1.6850
B	$b_1 =$ 1.7215	$b_2 =$ 1.9050	20	−0.07	1.9049
	$a_1 - b_1 = -0.2229$	$a_2 - b_2 = -0.2197$		−0.20	−0.2199

From equation (9.1.4a): $2c = -0.2199 + 0.2229 = 0.0030$ [m]; $4c = 0.0060$ [m]

From equation (9.1.5a):

$$a_2' = 1.6850 - 0.0060 = 1.6790 \text{ [m]}$$

$$b_2' = 1.9049 - 0.0030 = 1.9019 \text{ [m]}$$

$$a_2' - b_2' = -0.2229 \text{ [m] (debit} = a_1 - b_1)\,.$$

Considering p, the following value is to be set at station II:

$$(a_2) = 1.6790 + p_{40} = 1.6793 \text{ [m]}; \quad (b_2) = 1.9019 + p_{20} = 1.9020 \text{ [m]}\,.$$

When comparing the results of a) and b) the following is evident: For area levelling and geometric levelling with readings of 0.1 mm the earth curvature and refraction effects can be neglected.

9.1.2.4 Adjustment Using a Collimator

A collimator is a telescope with illuminated cross hairs focussed at infinity. Its collimation axis can be precisely levelled with the aid of a spirit bubble. A well adjusted precision level with reasonably long focal length can be used as collimator if needed.

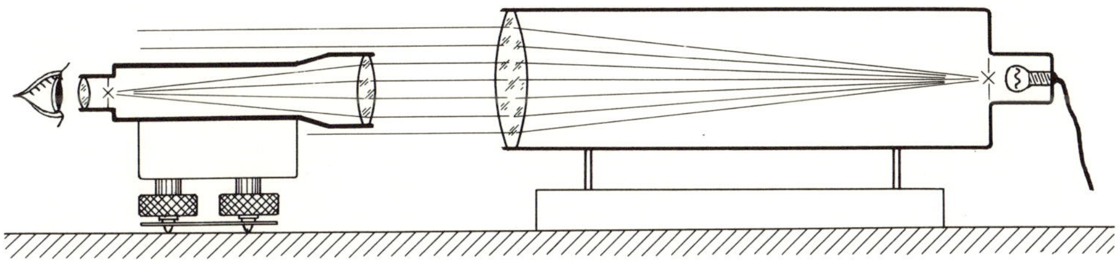

Fig. 9.1.9 Level and collimator

As illustrated in Fig. 9.1.9, the level which is to be adjusted is also focussed to infinity and lined up with the collimator.

Then the cross hair of the level is coincided with the one of the collimator by using the tilting screw and adjusting the telescope bubble, or by moving the cross hair with its adjustment screws.

9.1.2.5 Construction, Engineer's and Precision Levels

Based on their main uses, levels are classified as construction, engineer's, or precision levels.

Although this classification refers to the use of the instruments, another one, based on accuracy alone, is also widely used. *Deumlich* (1982) provides the following classes, based on the standard deviation for 1 km double levelling:

i) levels of low accuracy > 10 mm/km

ii) levels of medium accuracy $\leqq 10$ mm/km

iii) levels of high accuracy $\leqq 3$ mm/km

iv) levels of very high accuracy $\leqq 1$ mm/km

v) levels of highest accuracy $\leqq 0.5$ mm/km.

The central class (iii) represents the engineers' levels, while the lower two represent construction levels, and the higher two are commonly referred to as precision levels, which shows that the two types of classification are quite compatible.

a) *Construction levels* are designed for simple elevation measurements on construction sites, area levelling and short levelling loops, where a standard deviation of ± 5 to ± 10 mm per km is sufficient. Such instruments are either dumpy or tilting levels, have an objective lens diameter of 25 to 30 mm, 18 to 25 times telescope magnification, and bubbles with 30″ to 60″ sensitivity per 2 mm. Usually the level bubble is viewed by the observer by means of a mirror. Often a horizontal circle is incorporated in the instrument. Table 9.1.1a contains a selection of these instruments while Fig. 9.1.10 illustrates one.

b) *Engineer's levels* are designed for engineering tasks such as rail, bridge and tunnel

Table 9.1.1 Spirit levels (selection)

Manufacturer	Model		Telescope		Bubble		$\sigma_{1\,km}$**	Specialities
	without	with	magnif.	opening	sensitivity	accuracy		H = levelling tripod
	horizontal circle				2 mm			R = friction clamp
				[mm]	[″]	[″]	[mm]	K = tilting screw
a) Construction levels (with bubble mirror)								
Breithaupt, Kassel	Nakum	–	25	30	30		8	
Breithaupt, Kassel	–	Nifix	20	20	45		7	H, R, K
Ertel, Munich	–	BNL	20	25	60		6	H, R, K
Fennel, Ismaning	FN 01	FNK 01	25	30	30		5	
Fennel, Ismaning	FA 1	FNK 1	25	30	30		5	K
Kern, Aarau	GKO	GKO-C	18	24	45	±3	5	H, R, K
Wild, Heerbrugg	–	NK 01	19	25	60	±10	10	R
Zeiss, Oberkochen	–	Ni 52	20	25	60		5	
b) Engineer's levels (with tilting screw and coincidence bubble)								
Breithaupt, Kassel	Naguk	Nakre	30	40	20	±0.5	2	
Fennel, Ismaning	FN 10	FNK 10	25	30	30	±1	3	
Fennel, Ismaning	FN 2	FNK 2	30	40	20	±0.3	2	
Zeiss, Jena		NI 021 A	31	37	20		2; 0.7*	
Kern, Aarau		GK 1	22.5	30	45	±1		H, R, K
Kern, Aarau		GK 23	30	45	18	±0.4	2; 0.5*	H, R, K
Sokkisha, Japan		TTL 6	25	40	40	±2	2	K
Wild, Heerbrugg	N 2	NK 2	28	40	30	±0.75	2	reversible bubble R, K
c) Precision levels (with tilting secrew, coincidence bubble and reading micrometer)								
Breithaupt, Kassel	Nabon	–	42	50	8	±0.2	0.25	1′cross bubble
Wild, Heerbrugg	N 3	–	42	50	10	±0.2	0.2	

** Standard deviation for 1 km double levelling

* Obtained with a plan-parallel plate and micrometer; all other values are obtained without these

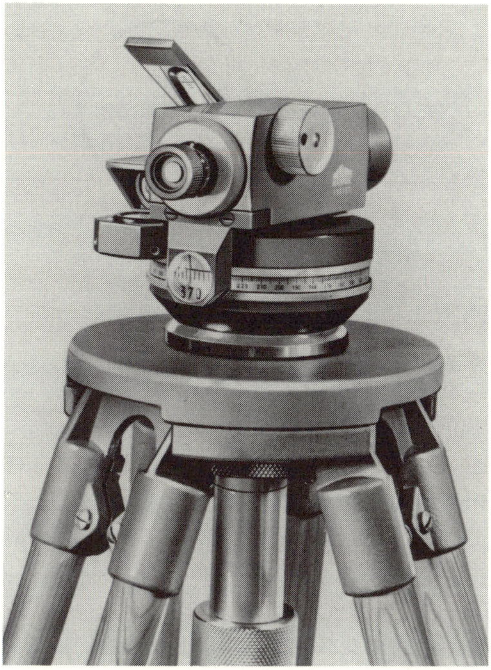

Fig. 9.1.10 Construction level Nifix (*Breithaupt*)

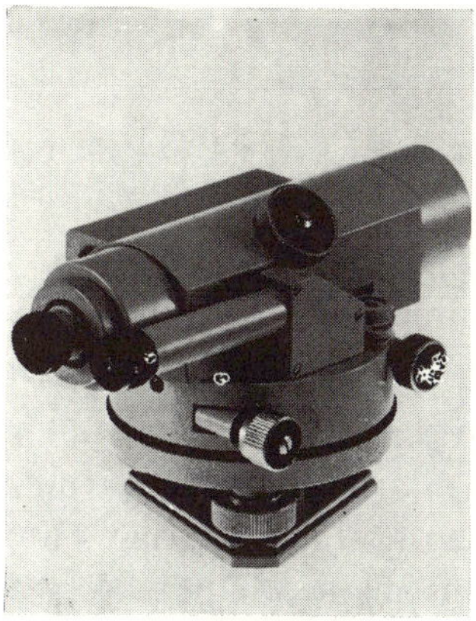

Fig. 9.1.11 Engineer's level FNK 2 (*Fennel*)

staking, as well as for levelling projects where a standard deviation of 2 to 3 mm per km is required. All modern spirit levels of this category are equipped with a tilting screw. They have an objective lens diameter of 30 to 40 mm, a 25 to 30 times telescope magnification, and coincidence bubbles (Fig. 2.1.7) with 20″ to 30″ sensitivity per 2 mm. Most of them are equipped with a horizontal circle.

In order to improve the reading accuracy, for some of them a parallel plate micrometer attachment is available, similar to the ones discussed in the following section. Table 9.1.1b contains a selection of these instruments, while Fig. 9.1.11 shows an example.

c) *Precision levels* are designed for highest accuracy (standard deviation: 0.3 to 1 mm per km). They are designed similarly to the engineer's levels, but have larger parts, heavier tripods and more precise screws. The telescope magnification is at least 40 times, the objective lens diameter about 50 mm, and the level sensitivity is 5″ to 10″.

Older precision levels are based on levelling with bubble readings. They have specially designed cross hairs which can be very precisely set onto the lines or sections of the rods and 5″ bubbles with continuous graduation permitting readings with an accuracy of $^1/_{10}$ pars (also section 9.2.2.4).

Modern precision levels are exclusively equipped with devices designed by H. Wild which permit levelling with centred bubble. Among these are the coincidence bubble

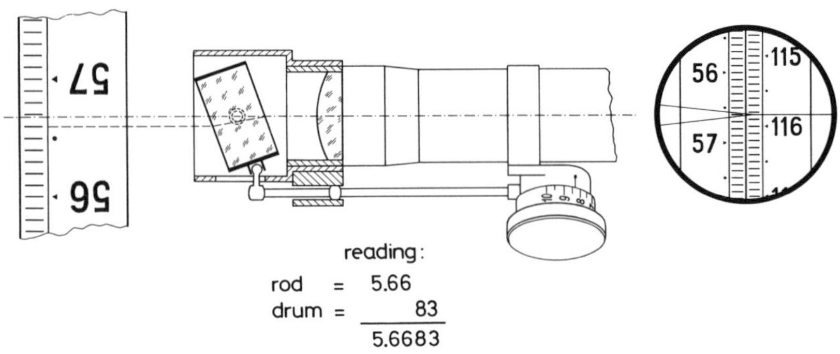

Fig. 9.1.12 Plan-parallel plate micrometer

(Fig. 2.1.7), whose sensitivity is approximately 10″ because of the high centring accuracy. Furthermore, a plan-parallel plate with thickness d which can be rotated around a horizontal axis is attached in front of the objective lens (Fig. 9.1.12).

When the plate is rotated by an angle α, the collimation axis is shifted by the amount $v \approx d(1 - 1/n) \tan\alpha$, where n is the refractive index of the plate. Thus the observer can set the cross hair onto a centimetre line of the rod and read the amount v in millimetres on the drum which is used to rotate the plate. Finally, in order to make the accuracy of setting the cross hair independent of the sighting distance, one half of the horizontal cross hair line is replaced by two lines forming a flat wedge. Thus the graduation line on the rod can be centred as illustrated in the right portion of Fig. 9.1.12.

Table 9.1.1c contains a selection of precision levels of this type, while Fig. 9.1.13 shows an example.

Fig. 9.1.13 Precision level *Wild* N3

9.1.2.6 Level Tacheometer

Construction and engineer's levels are often equipped with simple horizontal circles which can be used for rough angle measurements in flat terrain. The circles are graduated in half degrees or in gon and permit estimation of 1′ to 2′ or 1 cgon with a simple line index or a vernier (Fig. 9.1.14, on the right). These instruments are also equipped with stadia hairs with which the distance to the rod is determined to a few decimetres by reading the rod difference between upper and lower hair and multiplying it by 100. Modern instruments do not require the addition constant, which for older instruments may amount to 1.5 times the telescope length. Thus the level has become a level tacheometer.

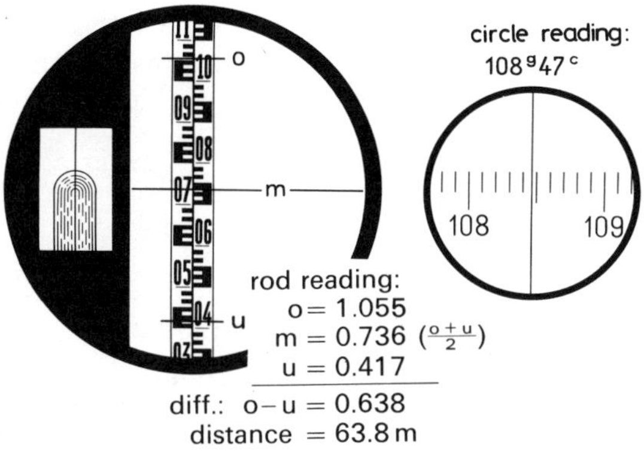

Fig. 9.1.14 Reading of a level tacheometer

9.1.3 Automatic Compensators for Levels

Preliminary Remarks: As mentioned in section 2.1.6, it is possible to obtain a horizontal line of sight or a horizontal height index without using a spirit level. The pendulum balance, known for about two centuries (Fig. 9.1.15), illustrates how a line

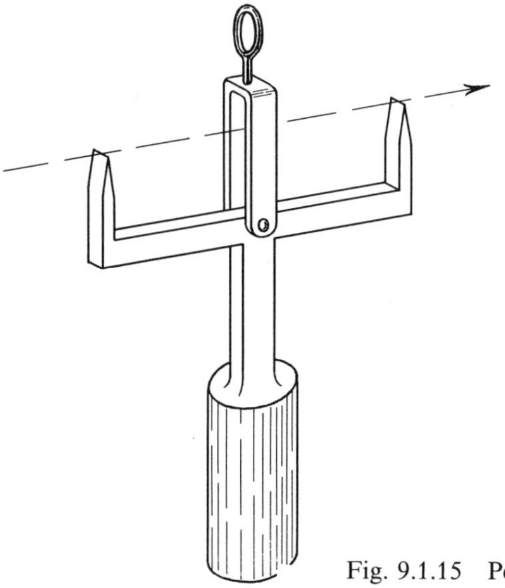

Fig. 9.1.15 Pendulum balance

of sight can automatically become horizontal under the influence of gravity. Based on this idea, Zeiss Oberkochen developed in 1959 the first "self levelling" instrument, where the disturbing effects of a slight inclination of the measuring telescope are automatically eliminated with the aid of an optical-mechanical component, namely the compensator.

This instrument, called Ni 2, has not only provided a reduction in measuring time by 30% to 40%, but also a significant increase in accuracy. As a consequence of the success of the Ni 2, nearly all manufacturers of surveying instruments produce nowadays levels with self levelling line of sight, usually called "automatic levels".

9.1.3.1 The Optical-Mechanical Foundations

The optical-mechanical foundations are explained using Fig. 9.1.16, which was provided by Zeiss when the Ni 2 was first introduced. The collimation axis of a level without spirit bubble will generally be inclined by a small angle α even though the bull's-eye level has been centred. A distant object point in the instrument horizon which is imaged at the focal plane of the objective lens, is, however, shifted upwards or down by $f \tan\alpha \approx f \cdot \frac{\alpha}{\varrho}$ (f = focal length). Thus, a horizontal ray entering the telescope has to be deflected at K with the aid of a gravity driven element (the compensator) such that it hits the horizontal line of the cross hair. If K is located at a distance s from the reticule, the deflection angle β is obtained from the following relationship:

$$f\alpha = s\beta \quad \text{with} \quad f : s = n \quad \text{as} \quad \beta = \alpha n \,. \tag{9.1.8}$$

To obtain a simple compensator, one selects $n = 2$ or $s = 1/2 f$ and suspends a mirror on a freely swinging pendulum between objective lens and reticule. When the pendulum has come to rest, the mirror surface has to be horizontal. If a horizontal ray enters a telescope that is inclined by the angle α, then the mirror will deflect the ray by $\beta = 2\alpha$ from its original direction. This fulfills the condition given in equation (9.1.8), which means that a pendulum mirror represents the simplest form of a compensator.

This solution, however, is only of theoretical value, as it is practically useless because the horizontal mirror would need to be extremely large for flat incident rays. A more practical solution is obtained if two 45° mirrors (or reflective prisms) facing each other

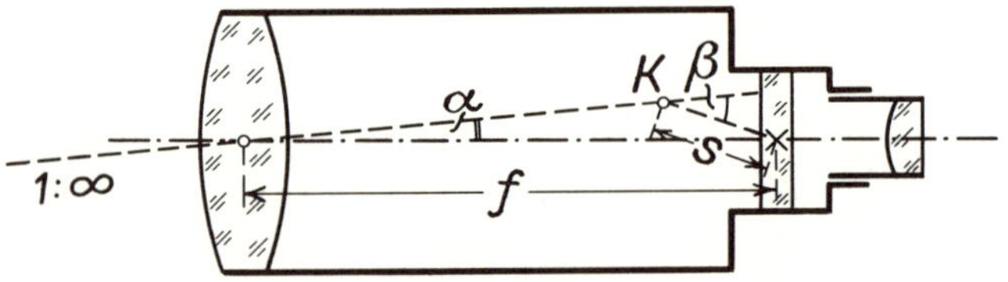

Fig. 9.1.16 Self-levelling line of sight

are inserted into the optical train such that they are parallel for horizontal sights. One of these is rigidly connected to the telescope, while the other one is kept at a constant inclination towards the horizontal with the aid of a pendulum (Figs. 9.1.17 and 9.1.18). If the telescope and with it one mirror is tilted by the angle α, the planes of the two mirrors will intersect with the angle α, and the ray which is reflected at both mirrors is then deflected as in an angle mirror by $\beta = 2\alpha$ from its original direction towards the horizontal cross hair line.

In order to automatically compensate for residual telescope tilts (of up to $\pm {}^1/_2{}^\circ$), generally three constructive prerequisites have to be met, namely an element rigidly connected to the telescope (e.g., in the case of Figs. 9.1.17 and 9.1.18, a reflecting prism), a movable element subjected to the influence of gravity (e.g., a pendulum mirror), and a dampening device, which quickly reduces the oscillations for speedy observations (here an air or magnetic dampening). Together, these three devices form the compensator. This definition is, however, not universal, and sometimes the movable element alone is referred to as a compensator.

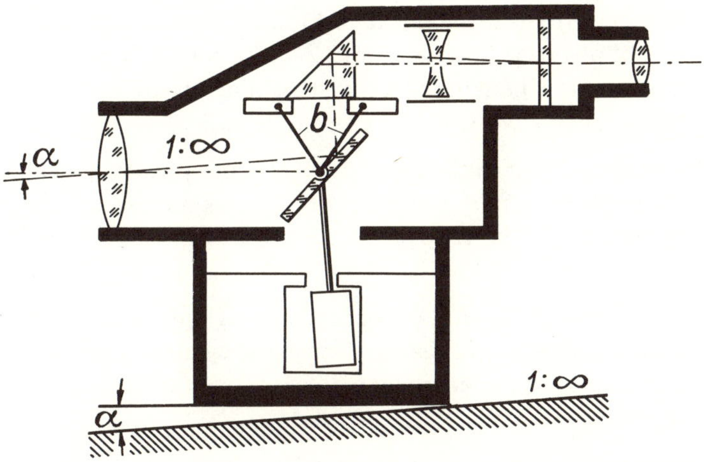

Fig. 9.1.17 *Askania* Na level

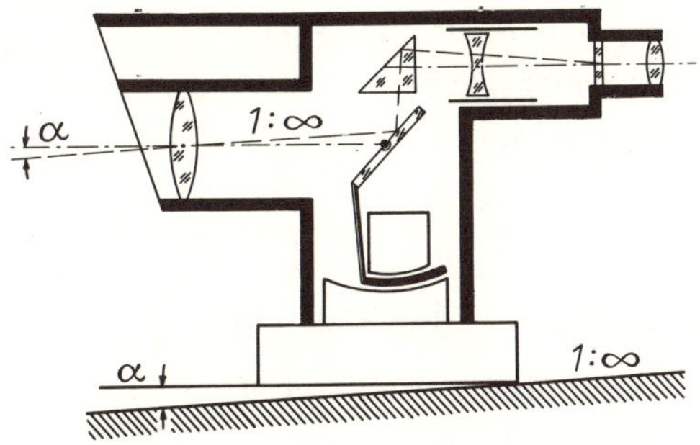

Fig. 9.1.18 Compensator of the *Breithaupt* Autom.

Compensators with $\beta = 2\alpha$ require a fair amount of space, thus hindering the use of an internal focussing lens. Therefore it is desirable to get $n > 2$. This can be realized, for instance, if the ray is reflected three times, which provides $\beta = 4$, and $s = {}^1/_4 f$. There are many other possibilities. Generally, compensators can be classified according to three major criteria.

The *first criterion* concerns the type of compensation. The compensation is either based on the (mechanical) principle of a gravity pendulum or it is a liquid compensator. The latter principle has been applied at the Ni 4 by Zeiss Oberkochen. This level, however, is not being produced anymore. Presently, liquid compensators are only used for stabilizing the index for reading the vertical circle. Thus there is no need to consider them here.

The mechanical compensators are classified according to the type of support, i.e. wire or ribbon elastic spring or spring joint, or else according to the axis support (edge, point, ball bearing). Pendula, supported by virtually friction free axes of this type, deflect the outcoming ray by the angle α with respect to the collimation axis, if the telescope is tilted by α. Thus the angular magnification equals one for these pendula. Angular magnification $n > 1$ is mechanically obtained with the aid of quadrilateral joints of wires or ribbons in a V or X shape, or with standing pendula of elastic springs, or with cross joints. The same effect can be obtained with optical means using mirrors and reflective prisms with one or more reflections.

The *second criterion* concerns the influence on the rays of optical elements within the optical train. The compensator monitors intermediate imaging systems, plan-parallel glass plates, glass wedges, or the reticule, in addition to the mirrors and reflective prisms.

The *third criterion* is concerned with the dampening of the compensator, which is needed for fast measurements. Air dampening with cylinder and piston is most prominent, while some use permanent magnets with swords which appear to be better than current dampening.

By combining the elements mentioned in these criteria, a practically unending variety of compensators can be developed. Table 9.1.2 provides an overview of the models most commonly used in central Europe. In addition to this table, the following text contains some explanations. We distinguish between primarily optical and primarily mechanical compensators, and state which elements are rigidly connected to the telescope and which are movable (within limits).

9.1.3.2 Compensators with Optical Angular Magnification

The compensators discussed in this section and in section 9.1.3.3 are based on the general principle: After the instrument has been roughly levelled with the aid of a bull's-eye level, its telescope axis is inclined by a rather small angle α. This angle α is to

Table 9.1.2 Automatic Levels (selection)

Manufacturer	Model	Telescope Magnif.	Telescope Opening [mm]	Compensator	Dampening	Accuracy [″]	$\sigma_{1\,km}$** [mm]	Specialities H: Levelling tripod
a) Construction levels								
Ertel, Munich	KNA	18	25	standing pendulum	magn.	1	5	H
Zeiss, Jena	NI 040 A	20	30			1	4	H
Zeiss, Oberkochen	Ni 42	22	30	pend. in ball bearing	magn.		5	H
b) Engineer's levels								
Askania	Na	25	36	ribbon pendulum	air		1.5	
Breithaupt, Kassel	Autom	27	33	cross spring joint	magn.		2.5	H
Ertel, Munich	BNA	24	30	standing pendulum	magn.	1	3	H
Zeiss, Jena	Ni 020 A	31	37			0.5	2; 0.7*	
Kern, Aarau	GK 1-A	25	45	magnetic supported pendulum	magn.	0.5	1.5	H
Nikon, Japan	AP-5	26	30	ribbon pendulum	air		2	
Topkon, Japan	AT-F 3	30	40	jointed quadrilateral	magn.	0.3	1.5	
Wild, Heerbrugg	NA 1	24	36	cross ribbon pend.	magn.	0.5	1.5; 1.0*	
c) Precision levels								
Zeiss, Jena	005 A	35	46			0.3	0.5	
Kern, Aarau	GK 2 A	32	45	intermediate imaging	magn.	0.3	0.7; 0.3*	
Sokkisha, Japan	B 2 C	32	40		magn.		1	
Zeiss, Oberkochen	Ni 3	24	40	jointed quadrilateral	air		1	
Zeiss, Oberkochen	Ni 2	32	40	jointed quadrilateral	air		0.7; 0.3*	
Zeiss, Jena	NI 007	32	40	ribbon pendulum	air	0.15	0.5*	
Zeiss, Jena	NI 002	40	55	ribbon pendulum	air	0.05	0.2*	
MOM, Budapest	Ni-A 3	30/50	67	ribbon pendulum	air	0.2	0.2*	
Salmoiraghi, Milano	Ni 1	30	45	wire supported pend.	air	0.2	0.2*	
Zeiss, Oberkochen	Ni 1	30/50	50	cross ribbon pend.	air	0.1	0.2*	

** Standard deviation for 1 km double levelling   * obtained with micrometer be magnified, if possible, by a factor n when the rays pass through the compensator, i.e., the effective inclination of the telescope axis becomes $n\alpha$ with $n > 1$.

a) *A reversable mirror, swinging on a ribbon*

The movable element of the compensator of the NI 002 precision level by Zeiss, Jena is a reversable mirror, swinging on a ribbon such that its mirror surfaces come to rest in a vertical position because of gravity. Fig. 9.1.19 illustrates this compensator with the incident bundle of rays indicated by the outside rays.

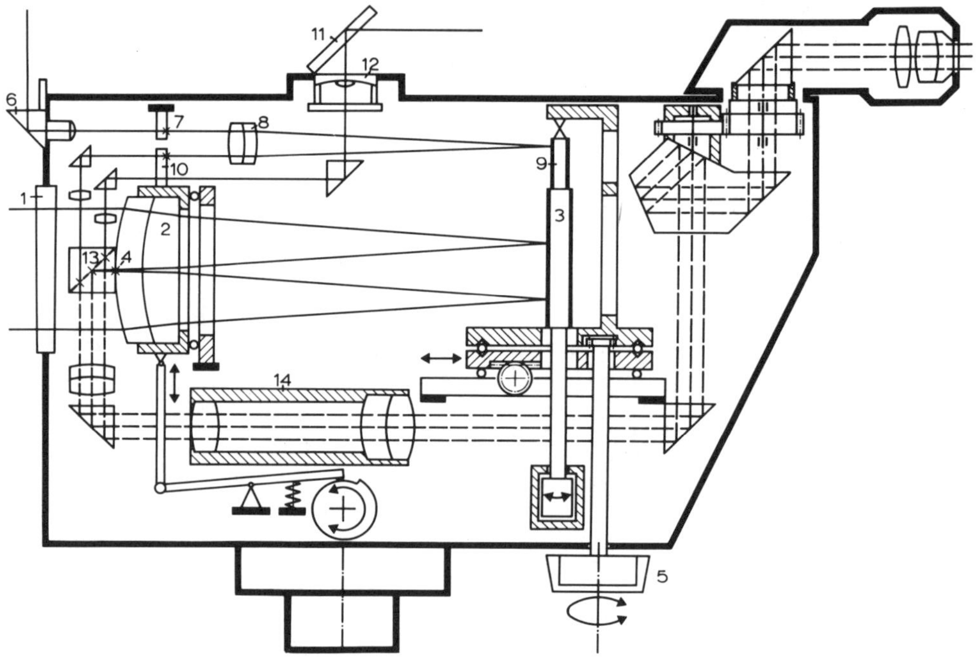

Fig. 9.1.19 Section through the *Zeiss*, Jena NI 002

The reversible mirror is located at $f/2$ from the objective lens. In order to eliminate possible changes in the collimation axis caused by intermediate lenses, this mirror is also used for focussing purposes. As illustrated in the figure, the measuring system consists of the objective lens (2) which carries the cross hair (4) in a fixed position and of the pendulum mirror (3). The latter reflects the incoming bundle of rays which is inclined by α, which causes a magnification of the angle of inclination for the principal ray by a factor of two ($n = 2$).

The imaging optics has the following arrangement: The cross hair (4) is imaged into the image plane (13) of the eye piece by means of an intermediate imaging system (14) and several prisms. Instead of the commonly used parallel plate micrometer, an objective lens micrometer is used whose reading is transferred onto the reticule together with the image of the bull's-eye bubble by means of the optical elements (6) to (10). The swivelling eye pieces permit reading of fore- and backsights without requiring the observer to move around the instrument. By activating the rotational knob (5), the pendulum mirror can be rotated by 180°, thus providing for an observation in a second comparator position. Residual errors of the instrument adjustment are eliminated by observing in two positions, and the mean of the two readings is referred to the "absolute" horizon. Therefore, it is not necessary to correct for adjustment errors, and foresights and backsights may be of different lengths. The compensator is dampened using air.

b) *A swinging roof prism acting as a mirror*

The movable element of the compensator of the GK 1–A engineer's level by Kern is a roof prism, acting as a mirror and swinging on a soft iron axis, held in place by a magnet (Fig. 9.1.20).

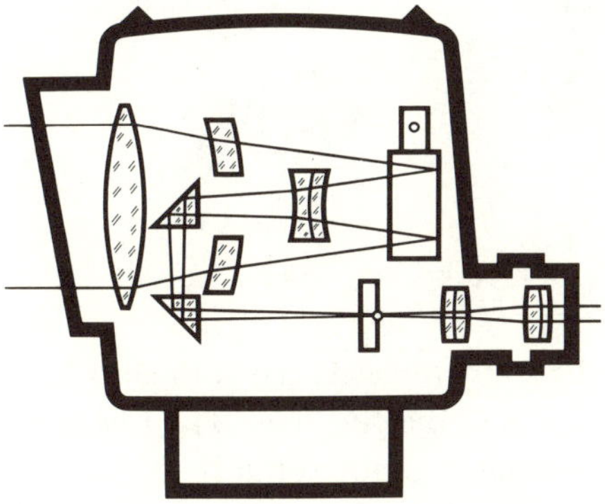

Fig. 9.1.20 Optical train of the *Kern* GKI-A

The axis is suspended virtually without friction by a yoke type permanent magnet at a distance of $f/2$ from the objective lens. Two mirror prisms, with 90° deflection each, are built into the objective lens system and represent the fixed parts. The angular magnification equals $n = 2$. The pendulum is dampened with two pistons using air.

The rays, which are reflected at the roof prism towards the objective lens, penetrate the focussing lens and are then directed towards the reticule and eyepiece by the two mirror prisms.

c) *A horizontal mirror swinging as a rigid pendulum*

The compensator of the Ni 42 construction level by Zeiss, Oberkochen, is a horizontal mirror swinging as a rigid pendulum. At a distance of $f/2$ from the objective lens, it is supported by an axis which rotates in shock protected precision ball bearings. Again, the magnification equals $n = 2$. The pendulum is dampened magnetically with the aid of an inductive brake (Fig. 9.1.21).

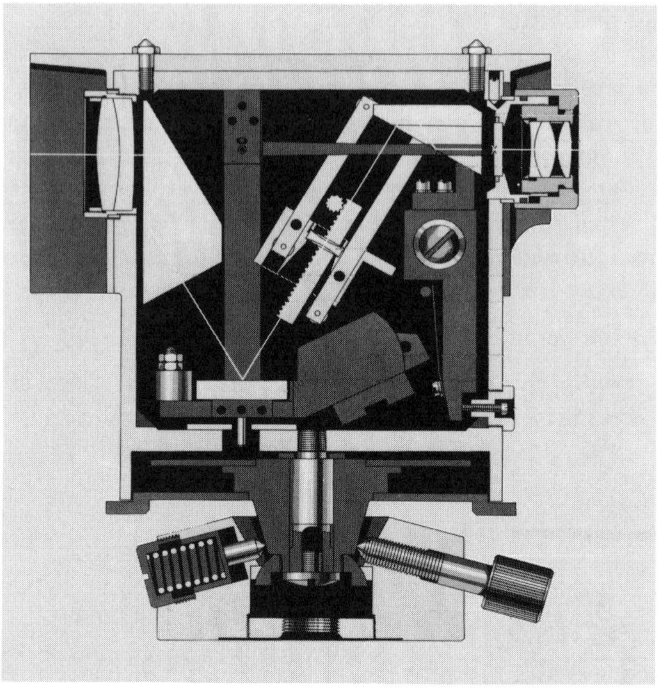

Fig. 9.1.21 Section through *Zeiss*, Oberkochen Ni 42

An incident horizontal ray is deflected by 60° towards the compensator mirror by means of a double prism located behind the objective lens. The ray penetrates the focussing lens after the reflection and is deflected horizontally towards the reticule and eyepiece by a double prism with roof edge. An index within the image plane of the compensator together with two limiting lines on the reticule mark the operating range of $\pm 1°$ of the compensator. The observer can thus at any time check the rough levelling and the proper functioning of the compensator.

d) *A mirror, inclined by 45° for horizontal sights*

The movable element of the compensators of the previously mentioned Askania Na (Fig. 9.1.17) and Breithaupt Autom (Fig. 9.1.18) levels is a mirror, inclined by 50 gon (45°) for horizontal sights. On the Na the mirror is mounted on an axis *b*, suspended by two fine ribbons, while on the Autom, it is fastened to a torsion ribbon stretched across the sighting direction. The magnification equals 2 in both cases ($n = 2$). The Na utilizes a piston for dampening in a cylinder with air pockets. The Autom has a magnet which surrounds the lower part of the pendulum, causing inductive currents which counteract the pendulum motion.

e) *A pendulum hanging from a spring joint*

The movable element of the NI 025 by Zeiss, Jena, is a pendulum hanging from a spring joint, with two rectangle prisms mounted at the top (Fig. 9.1.22, (8) with (3)

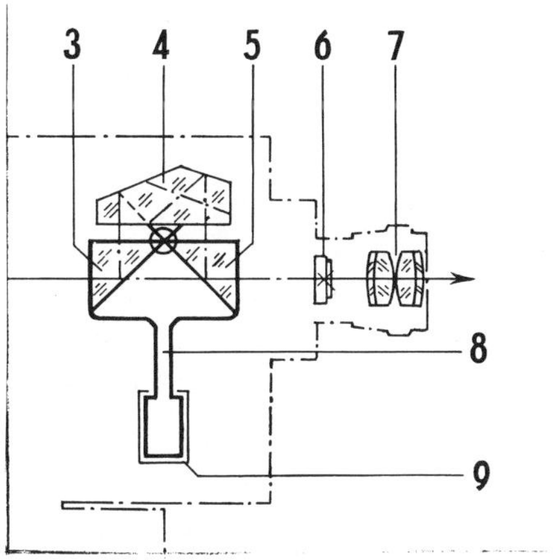

Fig. 9.1.22 Compensator of the *Zeiss*, Jena NI025

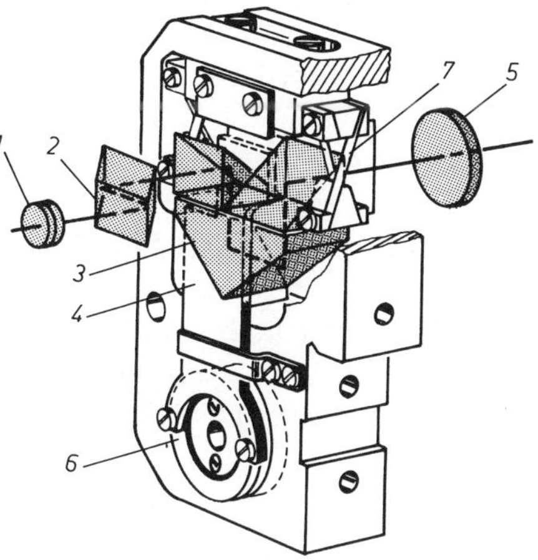

Fig. 9.1.23 Compensator of the *Fennel* FNA 1

and (5)). The roof prism (4) above the compensator acts as a fixed element and also generates an upright image. If the instrument is inclined by the angle α, the direction of the principal ray is deflected by 2α because of the reflection at the first rectangle prism. This angle is doubled by reflection at the second rectangle prism, which leads to $n = 4$. Thus the compensator can be placed at $s = f/4$ from the image plane (6) from where the rays continue to the eyepiece (7). The pendulum is air dampened as indicated by (9).

f) *A Porro prism system*

The characteristic part of the compensator of the Fennel FNA1 construction level (Fig. 9.1.23) is a Porro prism system of the second type, whose movable part swings on two ribbons. The two prisms shown in the upper part of the figure are mounted on the compensator pendulum (4) which swings on the two steel ribbons (7). The central prism (3) whose right angle points downwards, is solidly connected to the telescope body.

Once the pendulum (4) has come to rest in the direction of gravity, a ray that enters the objective lens inclined by α falls onto the right pendulum prism after passing through the focussing lens (5). This ray is reflected there towards the fixed prism (3), propagates from its right edge to the left one, and from there to the left pendulum prism. From there it passes to the reticule and eyepiece by means of the rhomb prism (2) which cancels the parallel shift caused by the Porro system. The angle of inclination α is magnified to 4α because of the double reflection, thus $n = 4$. The air piston system (6) dampens the pendulum. The FNA2 level by Fennel utilizes the same compensator slightly modified.

g) Compensator with intermediate imaging

Kern has developed a compensator with intermediate imaging for the GKO–A construction level. While all previously systems utilize mirrors or mirror prisms as optical elements, Kern utilizes the intermediate imaging system of the telescope itself for the GKO–A, which means that the compensator does not require any additional optical elements (Fig. 9.1.24). A rigid horizontal balance beam is the movable part of the compensator. Supported by two precision ball bearings, it swings around its horizontal axis (3) and carries the reverting lens (1) at the end facing the eyepiece. The counter weight (2) at the end facing the objective lens extends into the field of a magnet (4) which is used for dampening.

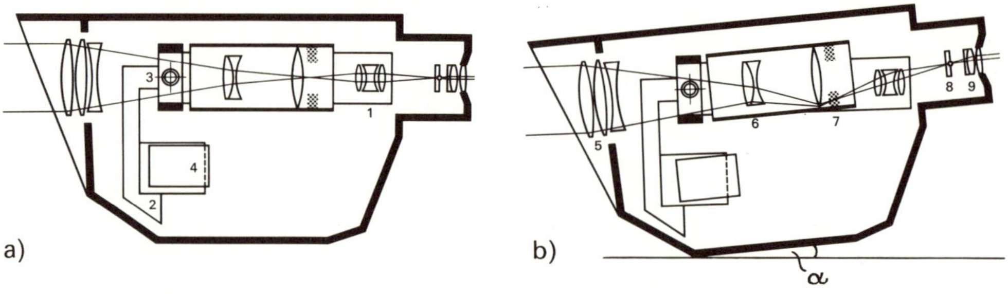

Fig. 9.1.24 Optical train of the *Kern* GKO-A
a) for levelled instrument; b) for inclined instrument

Objective lens and focussing lens generate an intermediate image in the first image plane (7) which remains fixed with respect to the telescope. This image is imaged upright onto the plane of the reticule by means of the reverting lens (1). The mechanical angular magnification is $n = 1$, while the total magnification is governed by the separation of the reverting lens from the centre of rotation. The dotted rectangles in (7) indicate a red warning diaphragm which appears in the field of view as soon as the telescope inclination exceeds the $\pm {}^1/_2°$ range of the compensator.

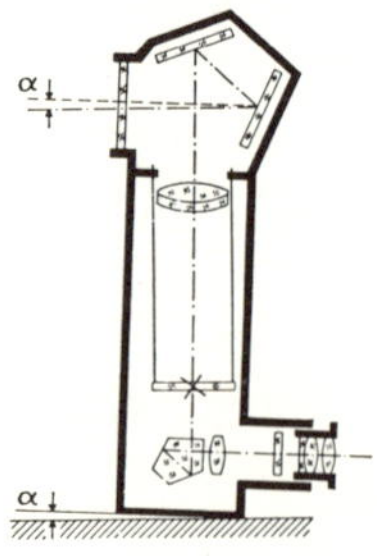

Fig. 9.1.25 Optical train of the *Salmoiraghi* 5190

h) *Optical pendulum in vertical levels*

The use of a pendulum consisting of optical elements in levels with vertical housing lead to a rather simple design. For the 5190 level by Filotecnica Salmoiraghi, Milano (Fig. 9.1.25) the reticule acts as a pendulum and is suspended by 3 steel wires whose length is equal to the focal length of the objective lens, i.e., $n = 1$. Thus the collimation axis and plumbline coincide, and the collimation axis outside the instrument is horizontal because of the 90° deflections by means of mirrors and prisms. The upper pentagon mirror can be shifted up and down, thus serving as a micrometer.

The NI 007 precision level by Zeiss, Jena (Fig. 9.1.26) has a 180° prism (i.e., a 90° prism with two reflections) at a pendulum with a length of $s = f/2$. A horizontal ray entering the telescope is deflected by 90° at the pentagon prism, then twice reflected at the pendulum prism, and falls onto the horizontal line of the crosshair after reflection on a triangle prism. This compensator does not deflect the ray but rather deflection of the compensator prism by x generates a parallel shift of $2x$. The pentagon prism at the upper part of the instrument can be tilted, and thus used as a micrometer as well.

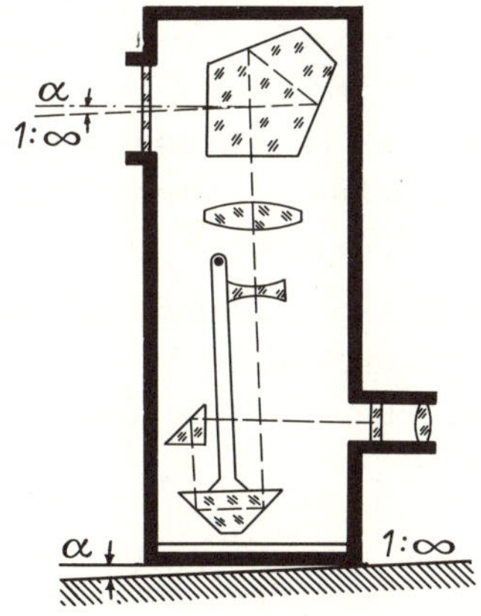

Fig. 9.1.26 Optical train of the *Zeiss*, Jena NI 007

9.1.3.3 Compensators with Mechanical Angular Magnification

Mechanical means for angular magnification are utilized in the following compensators:

a) *Jointed quadrilateral*

The optical element of the compensator of the Ni 2 engineer's level by Zeiss, Oberkochen is monitored by a jointed quadrilateral, acting like a mechanical level transmission. Fig. 9.1.27 illustrates this level with the jointed quadrilateral and the three characteristic prisms at the eyepiece end. The jointed quadrilateral is dimensioned such that if the upper base, which is rigidly connected to the telescope, is inclined by α,

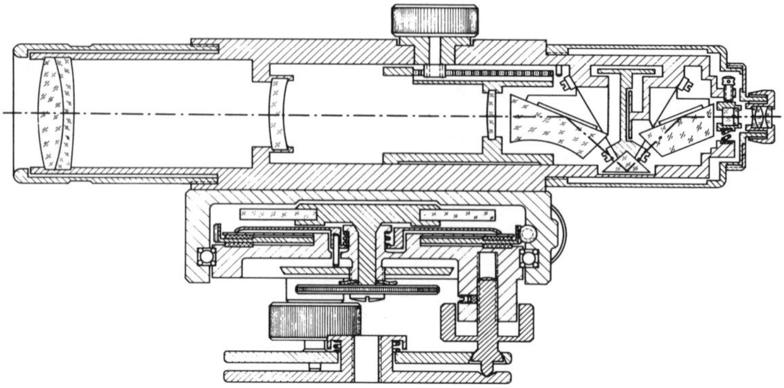

Fig. 9.1.27 Section through the *Zeiss*, Oberkochen Ni2

then the lower, movable base, settles at a tilt of 3.7 α (Fig. 9.1.28). The lower base is connected with the upper one by two wires each, which means that the joints are without play and friction. The optical element of the compensator – a 90° mirror prism – is mounted onto the lower base and doubles the tilt angle, which leads to a total magnification of $n = 7.4$. The pendulum is air dampened. After passing through objective and focussing lenses, the principal ray is deflected by 45° at the mirror prism connected to the telescope. It is then reflected by the movable 90° compensator prism and deflected into cross hair and eyepiece by the second mirror prism with roof edge. In conjunction with the roof edge, these three reflections generate an upright image.

The Ni1 precision level by Zeiss, Oberkochen also uses a jointed quadrilateral to generate a mechanical angular magnification in the compensator (Fig. 9.1.29). However, crossed ribbons are used instead of wires. The compensator dimensions are such that an angular magnification of $n = 16$ is achieved.

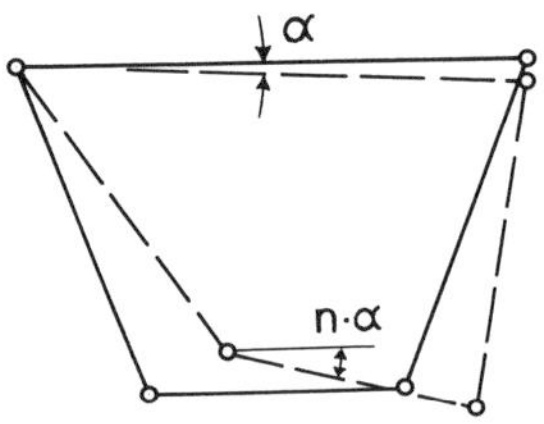

Fig. 9.1.28 Jointed quadrilateral of the Ni2

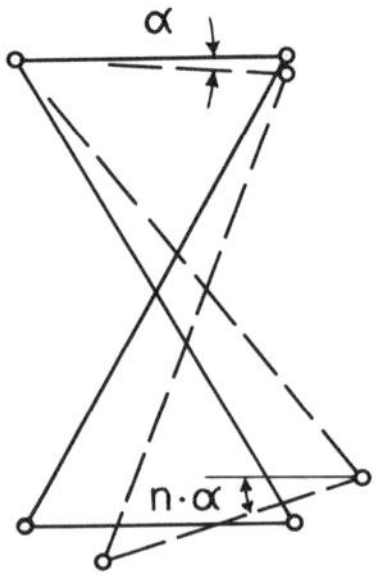

Fig. 9.1.29 Jointed quadrilateral of the Ni1

The eigen frequency of the swinging parts is so low that it represents a pendulum length of 1.12 m. The company thus refers to it as a long compensator with a reduced pendulum length of 1.12 m. Because of the large, reduced pendulum length, this

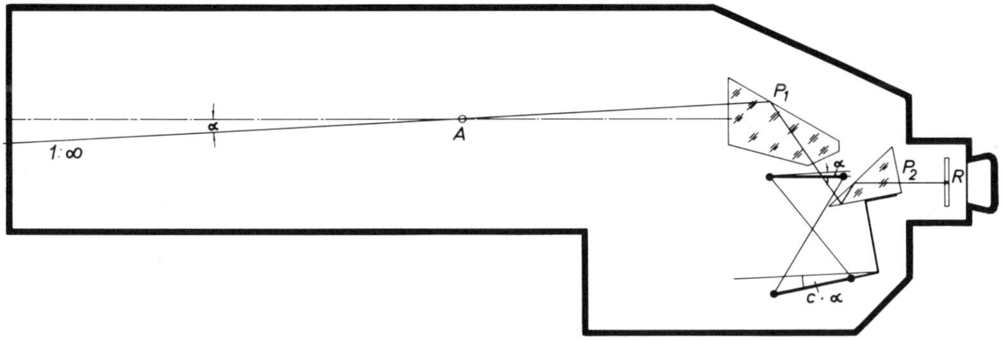

Fig. 9.1.30 Compensator of the *Zeiss*, Oberkochen Ni 1

compensator is hardly disturbed by external vibrations, e.g., those caused by traffic. Fig. 9.1.30 is a schematic illustration of the instrument. P_1 denotes the reversing prism which is equipped with a roof edge for mirror reversal. P_2 denotes a doubly reflecting prism which constitutes the movable and optically effective element of the compensator. The prism does not deflect the ray by a fixed angle, but causes a parallel shift if the telescope is inclined. The length of the lever arm of the prism from the rotational centre of the jointed quadrilateral is $^1/_{16}$ of the focal length, thus the shift of the ray is not subjected to an optical magnification. Contrary to the Ni 2 compensator, the position of the rotational axis of the jointed quadrilateral within the telescope is not dependent on the focal length. This holds for all compensators whose optical element causes a parallel shift. The triple reflection produces an upright image as a side benefit.

b) *An astatic pendulum*

The movable element of the compensators of the BNA and INA levels by Ertel is an astatic pendulum (Fig. 9.1.31). Its mechanical part is a standing elastic spring which is vertical when in rest. It is loaded on the top by a reversing prism, and is connected to

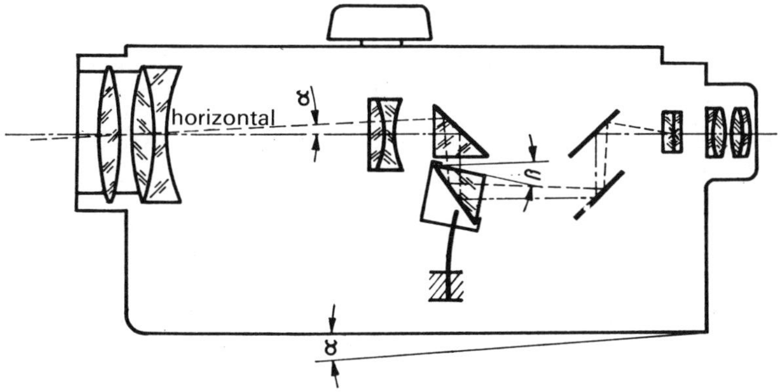

Fig. 9.1.31 Section through the *Ertel* BNA

the base by a thin piece of sheet metal. The optical part consists of a reversing prism, solidly mounted above the pendulum. It directs the incident rays onto the pendulum prism, from where they pass through two fixed mirror prisms towards the reticule.

If the instrument is tilted by α, then the upper part of the pendulum tilts by $n\alpha$ because of the unbalanced load. The angular magnification n is a function of the elasticity of the spring, the weight of the reversing prism and of its housing, and of the tilt α of the telescope with respect to the horizon. Induced currents are used to dampen the pendulum.

c) *A crossed spring joint*

The mechanical part of the compensator of the NA2 engineer's level by Wild is a crossed spring joint, which acts similar to an astatic pendulum. As illustrated in the right part of Fig. 9.1.32, the pendulum consists of four prestressed ribbons arranged as a cross and carries a reversing prism which is the optical element.

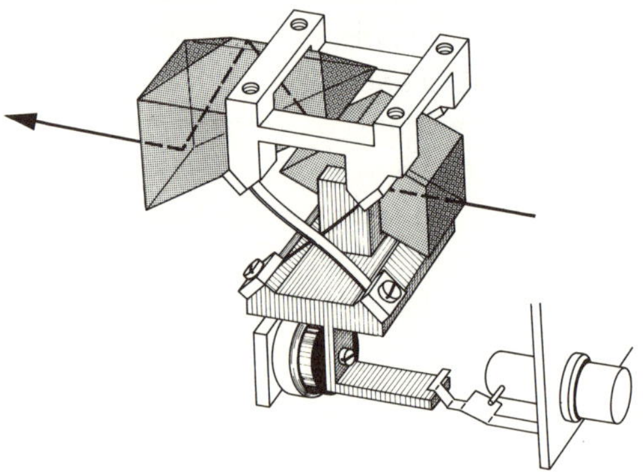

Fig. 9.1.32 Compensator of the *Wild* NA2

A ray, entering through the objective lens, is subjected to the mechanically generated angular magnification $n = 5$ when passing through the reversing prism. This amount is doubled in the level prism through reflection which gives a total magnification of $n = 10$. Therefore, the effective horizontal axis of the movable element could be placed at $f/10$ from the reticule normal to the optical axis of the telescope.

By depressing a control knob, shown on the bottom right in the figure, possible inertia of the pendulum can be eliminated.

The compensators of the NAK 0 and NAK 1 Wild levels operate under the same principle.

9.1.4 Automatic Levels

9.1.4.1 Construction, Engineer's, and Precision Levels

Like the spirit levels, automatic levels are categorized as construction, engineer's, and precision levels, based on their intended use. The external shape, dictated primarily by the operation principle of the compensator, deviates frequently from the commonly used stereotype of the spirit level. Of course, automatic levels do not need a tilting screw. Figs. 9.1.33 to 9.1.39 and Table 9.1.2 describe some of the common automatic levels.

Fig. 9.1.33 *Ertel* BNA

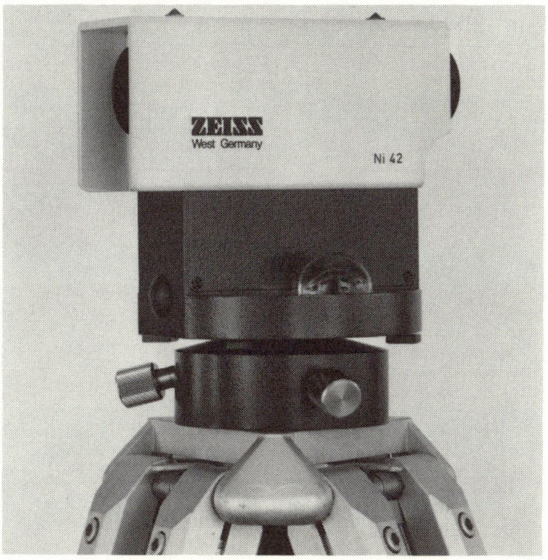

Fig. 9.1.34 *Zeiss*, Oberkochen Ni42

Fig. 9.1.35 *Sokkisha* B2C

Fig. 9.1.36 *Kern* GKI-A

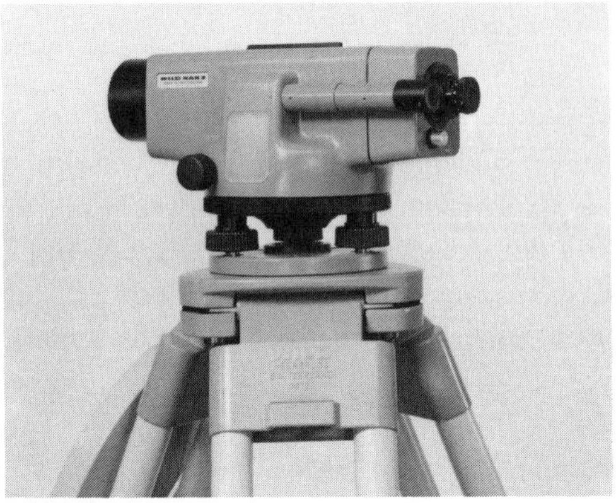

Fig. 9.1.37 *Wild* NA1

Fig. 9.1.38 *Zeiss*, Jena NI 002

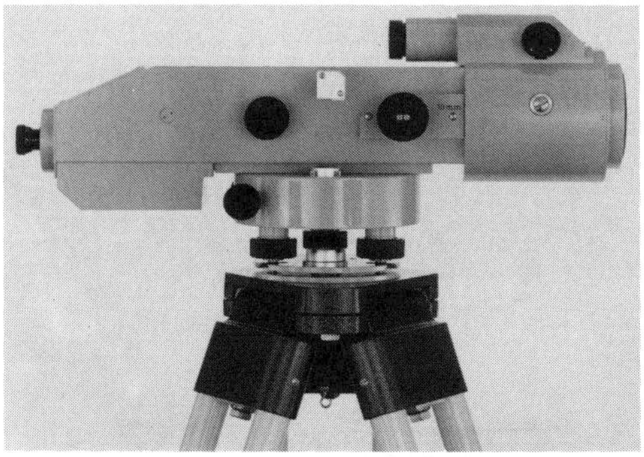

Fig. 9.1.39 *Zeiss*, Oberkochen Ni 1

9.1.4.2 Rules for the Use of Automatic Levels

An automatic level has to fulfill two conditions:

a) The tangent plane in the centre of the bull's-eye level has to be normal to the vertical axis of the instrument.

b) In a level with correctly functioning compensator, a ray passing through the principal point H of the objective lens has to fall onto the point S of the cross hair at the reticule (Fig. 9.1.40).

(1) *Testing and adjustment*. If condition a) is not fulfilled, then the adjustment is carried out as in section 2.1.1.

If condition b) is not fulfilled, then the level is out of adjustment – i.e. the collimation axis is inclined by the angle δ with respect to the horizon – then the ray hits the reticule at P instead of at the crosshair in S (Fig. 9.1.40).

The distance SP, however, is obviously a function of the sighting distance. For equal

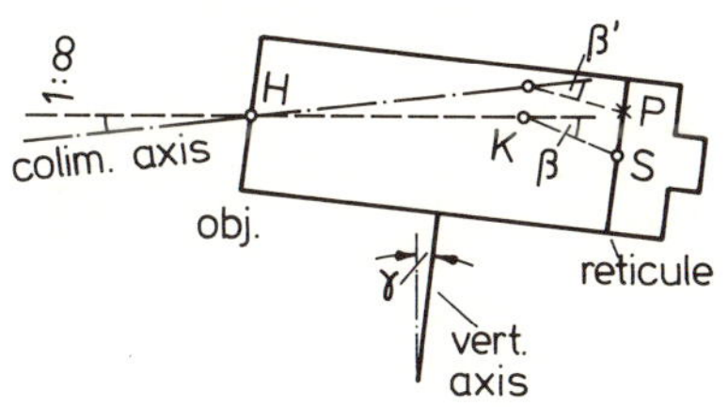

Fig. 9.1.40 Maladjustment

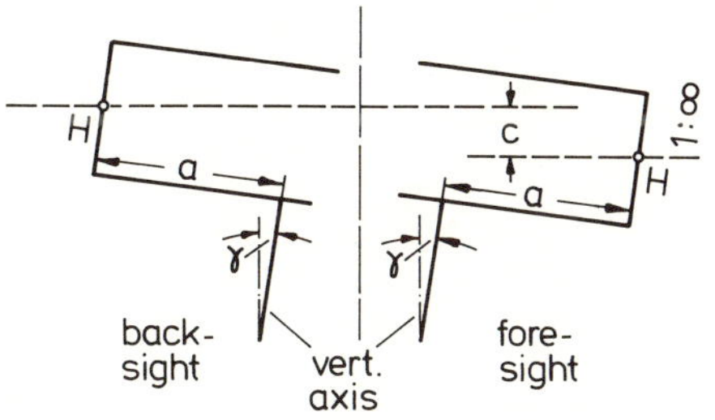

Fig. 9.1.41 Height shift

back-and foresights, such a maladjustment is practically without influence. Thus even automatic levels are suited for obtaining the correct elevation difference when levelling from the centre in between the points. Obviously, it is also possible to utilize a collimator as discussed in section 9.1.2.3. Depending on the design of the instrument, the position of the collimation axis is determined either by shifting the cross hair or else by rotating the slightly wedge-shaped glass in front of the objective lens.

(2) *Prelevelling with bull's-eye bubble.* In order to protect the compensator from damage during transportation, the range of the pendulum part is limited. Thus the bull's-eye level has to be carefully adjusted and centred to guarantee free movement of the pendulum part. Even then it can happen that the pendulum sticks to the restricting edge. It is therefore advisable to lightly tap on the tripod head prior to taking the reading, thereby making sure that the pendulum swings properly. The same effect is accomplished for the Wild NA 2 when depressing the control knob, shown in Fig. 9.1.32

(3) *Height shift of the objective lens and inclined horizon.* Both errors are unknown when working with spirit levels. When using automatic levels, however, they can occur because of residual errors in the adjustment of the bull's-eye bubble and insufficient prelevelling at the observation stations. Even for properly operating compensators an error c is obtained because of the inclination of the vertical axis by an angle γ, which – as illustrated in Fig. 9.1.41 – amounts to $c = 2a\gamma/\varrho$. If the object principal point is located $a = 150$ mm from the vertical axis, and γ amounts to one half of the common sensitivity of a bull's-eye bubble, i.e., $\gamma = 4'$, then a height shift of $c = 0.4$ mm is obtained. This cannot be tolerated for precision levelling. Thus the earlier mentioned requirement, that the bull'e-eye bubble has to be carefully centred, obtains here additional justification.

The so-called "inclined horizon" is another error caused by a tilted vertical axis. Often the compensators do not operate perfectly, but tend to over or under compensate. Possible causes are inaccuracies of the position of the compensator within the optical train, friction at the points, and similar influences which vary from instrument to instrument. For instance, for over compensation (Fig. 9.1.42 a) and b)) the actual compensation angle β' exceeds the theoretical value $\beta = n\alpha$. For a tilt of the vertical

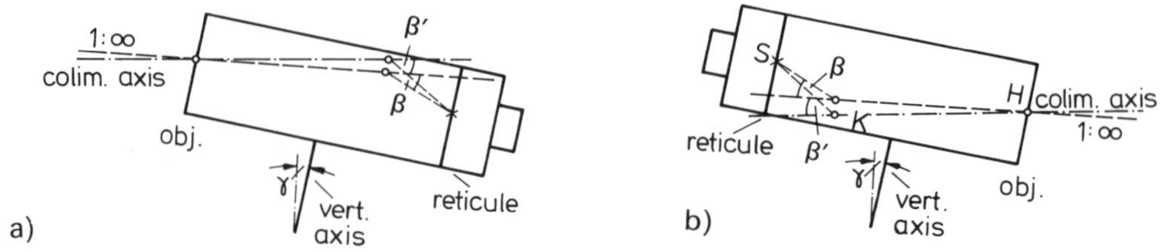

Fig. 9.1.42 Inclined horizon; a) backsight, b) foresight

axis as assumed in the figures, the collimation axis is inclined upward for the foresight, and downward for the backsight. Thus

$$B' < B \quad \text{and} \quad F' > F$$

where B and B' represent the readings with correct and inclined horizon, respectively, for the backsight, while F and F' denote foresight readings. Instead of $\Delta h = B - F$, an elevation difference $\Delta h' = B' - F'$ which is too small is obtained. The magnitude of this error depends primarily on the tilt γ of the vertical axis and on the sighting distance, and can reach a few tenths of a millimetre. Again, precise centring of the bull's-eye bubble will virtually eliminate this error.

Height shift of the objective lens and inclined horizon are not critical as long as the bull's-eye bubble is properly adjusted with respect to the vertical axis. By carefully centring the bubble, these errors can be controlled, especially since they occur as random errors with alternating sign and varying amounts for longer level lines.

A residual adjustment error of the bull's-eye level can never be completely avoided. Then, however, these errors take on a systematic character. This case occurs when the observer centres the bull's-eye level at each station habitually with the telescope pointing in the same direction, e.g., backsight. Although the mean of a forward and return level line is free of the error influence, the individual results may differ which is often rather disturbing.

An alternating measuring method helps. The bull's-eye bubble is centred at one station with the telescope pointing backward, while it is centred at the next station with the telescope pointing toward the forsight. This method is continued for the whole level line. If two rod men are utilized, the telescope is always first pointed toward the same rod, i.e., the same rod man, which has to be recognized. As a joke, someone suggested to equip him with a pair of red pants. Thus this method is often referred to as the "method of the red pants".

(4) *Periodic vibrations*. Traffic, construction work, high winds, etc., affect a sensitive compensator far more than a rather inert spirit bubble. Under such circumstances, the use of a spirit level is preferable. If, in spite of this, an automatic level is to be used, the tripod has to be placed solidly into the ground by stepping heavily on its legs, thus creating some stress because the disturbances cause vibrations of the tripod. Fur-

thermore, tripod vibrations can be significantly dampened by using an old trick of experienced instrument men, namely either laying a hand lightly onto the tripod head, or else by lightly touching two tripod legs when taking the readings.

Compensators are far less sensitive than spirit bubbles to temperature changes. Therefore, with automatic levels, an umbrella is only needed for precision work.

(5) *Influence of the earth's magnetic field.* Practically all level compensators are based on a pendulum operating within the earth's gravity field. Thus in self-levelling instruments the line of sight is being automatically levelled because of the influence of gravity. However, in addition to gravity, the magnetic field of the earth acts upon the pendulum as well. Possible distortions of the pendulum arrangement cause errors in the line of sight (Fig. 9.1.43). Assuming that the horizontal component of the magnetic field remains constant within a given working area at a constant latitude, then

$$\delta = \delta_0 \cos \mu \tag{9.1.9}$$

where μ denotes the magnetic azimuth (i.e., the angle is referred to magnetic north). According to equation (9.1.9), δ has opposite signs for fore- and backsights and thus is not eliminated when forming the difference between back- and foresight. Therefore, instruments whose compensator is sensitive to magnetic influences, automatically produce an inclined horizon. This may cause effects which cannot be neglected for precision levelling (*Rumpf* and *Meurisch* 1981).

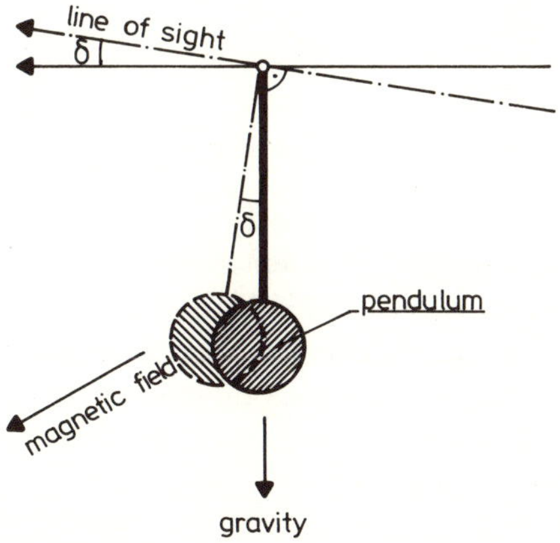

Fig. 9.1.43 Influence of the magnetic field on the line of sight

There is no measuring arrangement that can effectively eliminate the influence of this error. Therefore, it is advisable to check the instruments on a test line prior to commencing high precision levelling. If disturbing influences caused by the magnetic field are detected, the instrument should be returned to the manufacturer for service. Newer instruments are designed such that disturbing influences of the earth's magnetic field are avoided (*Leitz* 1984).

9.1.4.3 Automatic and Spirit Levels

A survey of instrument manufacturers who produce both types of levels still shows a surprisingly large fraction of spirit levels of the total sales. Construction levels with spirit bubbles still are very important in developing countries. They are simple to operate, require little service and few repairs, while the main advantage of automatic levels, namely saving of labour and time, has virtually no bearing in these countries. Engineers' and precision levels with spirit bubbles are also still selling well, because even in the U.S.A. the breakthrough of the automatic level is far less than in central Europe.

9.1.5 Levelling Rods

9.1.5.1 Simple Levelling Rods

They are 3 to 5 m long and 6 to 8 cm wide, usually made of dried pine wood and often strengthened by cross-ribs. Oil paint and varnish protect them against humidity. For easier transportation they are made as either folding or extension rods.

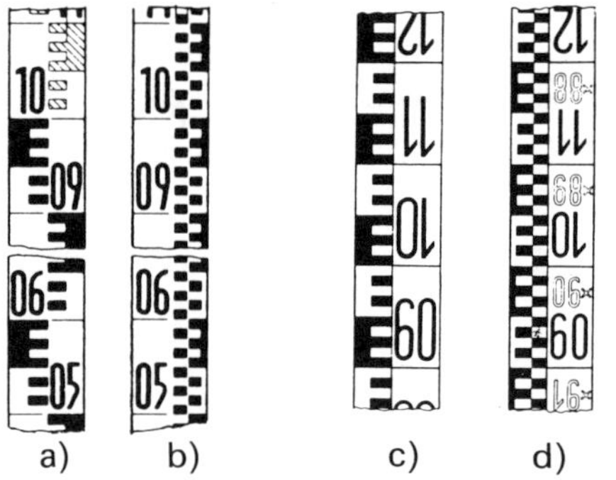

Fig. 9.1.44 Simple levelling rods (selection)

The front of a simple rod (Fig. 9.1.44 a)) is divided into centimetre sections and numbered in decimetres such that the numerals are positioned in the decimetre section in which the centimetre values are read. This graduation is often referred to as "*E*" graduation. Numerals and graduation are often black for odd metre and red for even metre values. The zero point of the graduation should coincide with the bottom surface of the rod. Rod b) has the so-called "checker board graduation" which helps in compensating estimation errors which occur because of halation, causing the white fields to appear larger than the black ones. Rods designed for use with reversed image telescopes – i.e., most spirit levels – have upside down numerals (Fig. 9.1.44 c) and d)) which appear right side up when viewed through those telescopes. Rods designed for upright image telescopes – i.e., most automatic levels – have upright numerals.

Rod d) carries a *decadic* graduation, i.e., in addition to the regular numbers, there are the so-called "decadic remainders". For instance, $-1.417 = -10 + 8.583 = \times\ 8.583$, where the "×" indicates a negative unit. As illustrated in section 9.2.2.3, "decadic" numbers simplify the computations, because subtraction is replaced by addition.

Several instrument manufacturers have developed rods which are especially suited to use with their levels.

9.1.5.2 Rods for Precision Levelling

Rods for precision levelling carry 1 cm graduation lines or are divided into $^1/_2$ cm sections (Fig. 9.1.45 a) b) c)).

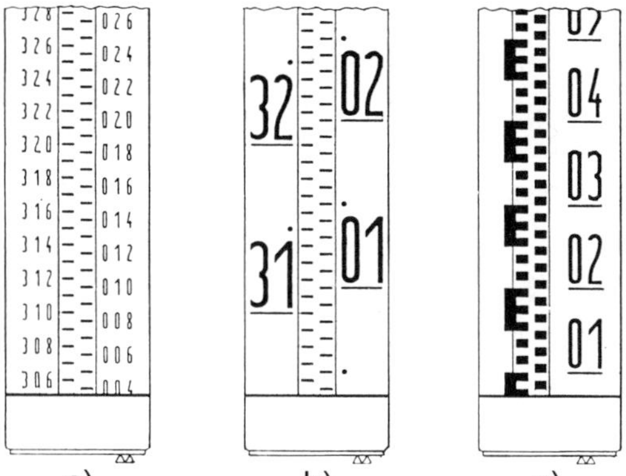

Fig. 9.1.45 Precision rods (selection)

The lines in a) and b) are especially suited for precision levelling with a parallel plate micrometer, where the cross hair is set precisely onto the border between adjacent centimetre sections. As a precaution against the effects, an invar tape carries the graduation lines. It is placed in a groove in the centre of the rod and kept under tension with a spring.

Rods a) and b) carry two graduations, one of which starts with zero, while the other one is shifted by an uneven amount, the so-called "rod constant". Such rods are referred to as reversible rods. Rods with two graduations practically eliminate reading and estimation errors, provided both scales are read. Furthermore, two parallel level loops can be carried out (section 9.2.2.3), which results in a noticeable gain in accuracy.

Rod c) carries a half centimetre graduation. Thus, elevation differences obtained with these rods have to be divided by two during calculations.

9.1.5.3 Rod Accessories

Each rod has to be equipped with a bull's-eye level (Fig. 8.1.1) which has to be checked regularly and after longer travels. Furthermore, two range poles are needed for precision levelling to properly brace the rod.

Cast iron base plates, usually equipped with three short steel spikes, are solidly placed on the ground to serve as turning points. Wooden pegs, iron pipes, plugs or other point monumentations can also be used as rod stations.

9.1.5.4 Rod Calibration

(1) *Scale Comparison for Levelling Rods*

For checking the scale of levelling rods, a 1 m calibration ruler with graduation extending by 2 mm on both sides of the index lines is most suitable. The separation of two decimetre lines covering 1 metre is measured for about 10 different sections to an accuracy of 0.1 mm. If the rod has a line graduation, both edges of a line are to be read. The mean of these individual determinations is used as the "mean rod metre" for the levelling calculations.

When carrying out precision bench mark levelling using wooden rods, it is essential that they be checked daily as the rod length may vary by several tenths of a millimetre per metre primarily due to humidity.

Invar rods should be checked about once a month and after shock suffered, for instance, during transport.

For simple level loops, deviations of the rod length of up to 0.2 mm per metre can be tolerated, unless large elevation differences are to be handled.

(2) *Calibration of Precision Rods with Rod Comparators*

The method described in (1) above provides the average scale of a levelling rod, however, positional errors of individual graduation lines are not detected. Such deviations of the lines from their correct position can amount to as much as 50 μm. These errors are only partially compensated within the mean rod metre.

If elevation differences are to be determined with accuracies of 1 : 10000 to 1 : 100000 then the deviations of individual graduation lines must be known so that the proper corrections can be applied during the elevation. Such high accuracy requirements occur, for instance, when observing

- crustal movements (0.1 to 0.5 mm per year)
- foundation settlements of large industrial complexes (<1 mm)
- first-order vertical networks ($\leqq 0.3$ mm/km).

Individual errors of all graduation lines can be determined virtually automatically with a laser interferometer which replaces the secondary normal. The structure of a rod comparator monitored by a computer is illustrated in Fig. 9.1.46. A computer monitors the measuring sequence and also records and processes the measured data.

The data are stored such that they can easily be transferred into another computer used for automatic evaluation of levelling (section 9.1.6). A motor drives the levelling

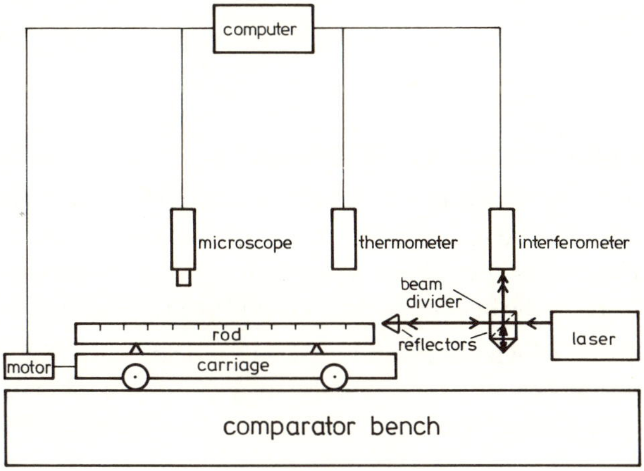

Fig. 9.1.46 Computer monitored rod comparator

rod past a measuring microscope. The measuring microscope scans each graduation line and determines its centre, then generates an impulse which causes the recording of the position as determined by the interferometer. The measuring microscope in other comparators converts the light-dark fields of the rod into a corresponding voltage signal. This analogue intensity profile is subsequently digitized so that the positions of the graduation lines can be mathematically determined in the computer. Furthermore, there are comparators which require a rough setting onto the graduation lines followed by a visual fine setting (*Tille* 1984).

The units of the secondary length normal and of the rod under calibration are both dependent on meteorological conditions, which therefore must be measured and considered.

Grabe and *Lebowsky* (1979) estimate a standard deviation of about 30 μm for the graduation line correction. This estimate is based on errors caused by monitoring, line detection, and evaluation as well as by the support of the rod.

More recently, graduation lines are etched into invar tapes by laser energy which is monitored by interferometry. This produces standard deviations of less than 5 μm for graduation lines (*Schlemmer* 1984).

9.1.6 Data Recording with Mobile Terminal; Automatic Data Flow

A data terminal – often referred to as electronic fieldbook – consists of a microprocessor, program storage and one or several data storage devices. A keyboard is used to input data as well as auxiliary information (e.g., point number, meteorological data) for identification and classification purposes.

A small program provides

- user commands
- mathematical checks
- unadjusted elevation differences along the levelling line.

The capacity of the data storage is chosen such that one week's measuring data from a levelling party can be stored. The information stored is such that it can be directly transferred to a mainframe computer. Thus, after manual input an automatic data flow is achieved which leads to the final adjustment resulting in a record of elevations.

Data terminals are available which are either based on fixed software (e.g., the Geomemo – Niv and Micronic 445 systems) or else freely programmable (e.g., GRE 3). The latter can also be used for other tasks, such as traversing, polar surveys, etc.

The Wild GRE3 data terminal is shown in Fig. 9.1.47.

Fig. 9.1.47 GRE 3 (*Wild*) data terminal

9.1.7 Refraction Models for Levelling

The line of sight is slightly deflected from the horizon because the refractive index n near the ground changes with elevation changes (Fig. 9.1.48). In order to consider this phenomenon, numerous models have been developed.

One group of models utilizes as input parameters the terrain slope α and the temperature change Δt as a function of the elevation h:

$$d_{b,} = -\cot^2 \alpha k \cdot \int_{h_i}^{h_b, h_f} \Delta t \cdot \mathrm{d}h \qquad (9.1.10)$$

with $k \approx -1 \cdot 10^{-6}$, and

$d_{b,f}$ = refraction for back- and foresight
h_i = instrument height
h_b and h_f = rod reading for back- and foresight.

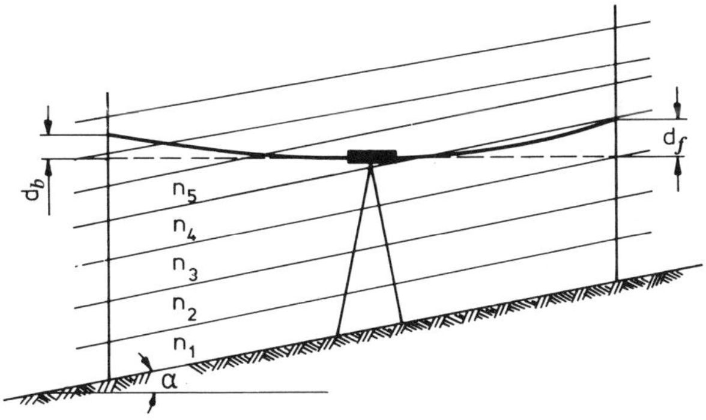

Fig. 9.1.48 Deflection of the line of sight due to refraction

Kukkamäki (1938) formulated the first model which describes the temperature t near the ground as a function of the elevation:

$$t = a + b \cdot h^c \tag{9.1.11}$$

where a, b, and c are constants, derived from temperature measurements. This leads to the following expression for the temperature difference:

$$\Delta t = b(h^c_{b,f} - h^c_i)\,. \tag{9.1.12}$$

Other models are based on different meteorological parameters and terrain information. *Angus-Leppan* (1979), for instance, uses cloudiness, soil humidity, zenith angle and azimuth of the sun, terrain slope, and azimuth of the levelling line.

A comparative study of a sizeable number of different models was published by *Heer* (1984). In view of the planned redefinition of the North American vertical datum, refraction models are being thoroughly investigated in the U.S.A. (*Holdahl* 1981, 1983; *Whalen* and *Strange* 1983).

9.2 Levelling Methods

9.2.1 Reference Surface and Bench Marks

9.2.1.1 Reference Surface and Vertical Networks

In spite of the definition given in section 1.2, the sea level elevation of a point is obtained in practice by summation of observed elevation differences. All these elevations are based on a reference surface, commonly the *mean sea level* (MSL) which is usually defined in relation to a point placed in a geologically solid area and monumented in bed rock. The MSL elevation of this *"origin" of the elevation system* is usually determined by long periods of observations of tidal gauges.

In large countries, bounded by different oceans, e.g., the U.S.A. or Canada, it is difficult to maintain a common mean sea level. Due to the extremely long lines involved to connect tidal stations, a general adjustment of the whole vertical control network together with ocean tie-in is required. While the redefinition of the North American vertical datum is now in progress, present elevations in the U.S.A. are still based on the fundamental vertical network of 1929. The latter consists mainly of levelling lines measured along railroad tracks. Recent investigations have shown significant errors due to refraction (also section 9.1.2.3) which were left uncorrected during the measurements for the 1929 adjustment. Thus the planned redefinition may lead to significant elevation changes in certain areas of North America (*Holdahl* 1983).

In Germany, MSL is defined as being 37,000 m below a bench mark at the old Berlin Observatory and monumented by the "normal bench mark of 1912", located about 40 km east of Berlin. The elevations of thousands of other bench marks are referred to this one and called "normal niveau (NN)" elevation. Precision levelling between the Amsterdam tidal gauge and Berlin ensured that this surface coincides within 1 dm with the average height of the North Sea. The vertical network has the following characteristics (Fig. 9.2.1)

1) The primary, or first-order net, consists of 30 to 80 km long levelling loops measured with highest precision, and evaluated by rigorous least-squares adjustment. The elevations represent the separation from the reference surface along the plumb line, and are referred to as NN heights. The International Metre is the measuring unit.

2) This primary network shall remain unchanged if possible, but shall keep pace with technical advances. This includes consideration of gravity acceleration at the earth's surface (also section 9.2.2.4), improved monumentation, geological considerations, re-observations and new adjustments, especially by including surface gravity and continental connections to the systems of the neighbouring countries.

3) Densification consists of:

a) First-order lines, measured with the same accuracy but fit into the primary network after its adjustment.

b) The second-order network, consisting of loops of up to 20 km diameter.

c) The third-order network consisting of loops of up to 10 km diameter.

d) The fourth-order level lines which supplement the higher order nets. These lines should not exceed 4 km in length.

4) The orthometric correction, dependent on the local gravity acceleration, is applied to the NN heights.

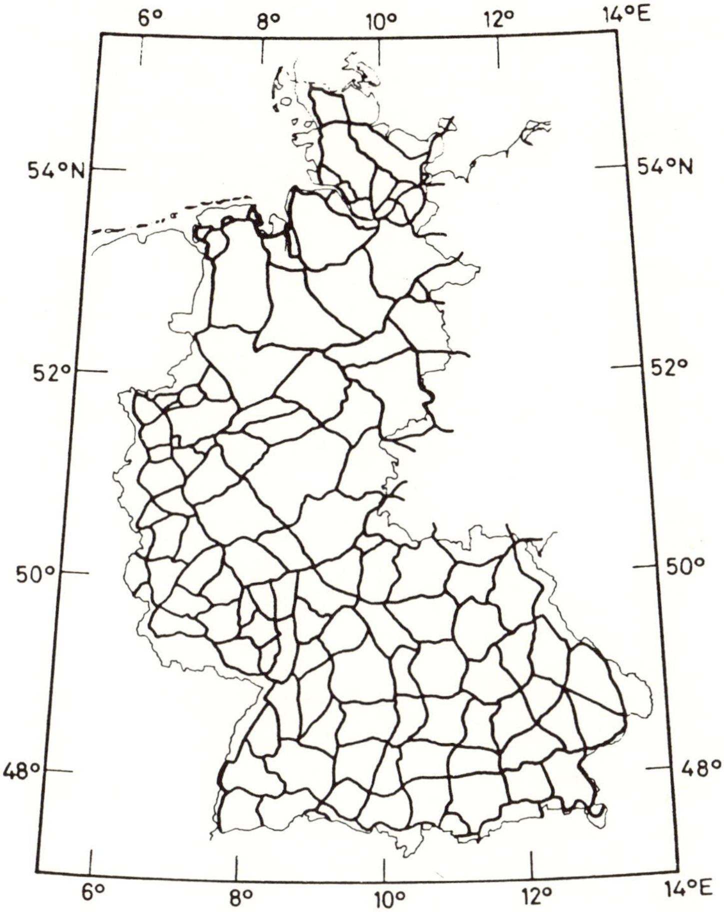

Fig. 9.2.1 Vertical network of Germany

9.2.1.2 Monumentation of Bench Marks

Bench marks are normally monumented above ground and are freely accessible. They usually consist of a plug with a spherical or conical head which clearly defines the highest point. Levelling rods can be placed directly onto this point. These plugs may be placed in walls, bed rock or specially placed pillars.

As illustrated in Figs. 9.2.2 and 9.2.3, the following monuments are normally used:

Wall plug – used in walls, bedrock, etc.

Pillar plug – used in granite or concrete pillars.

If possible, bench marks are placed on vertical surfaces. Occasionally underground monuments are used for scientific purposes.

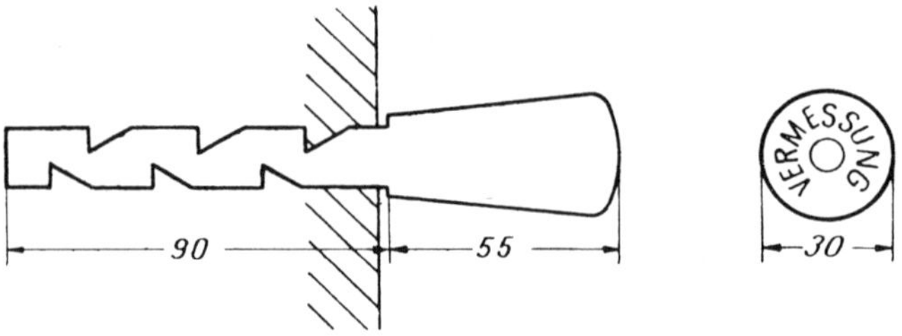

Fig. 9.2.2 Wall plug

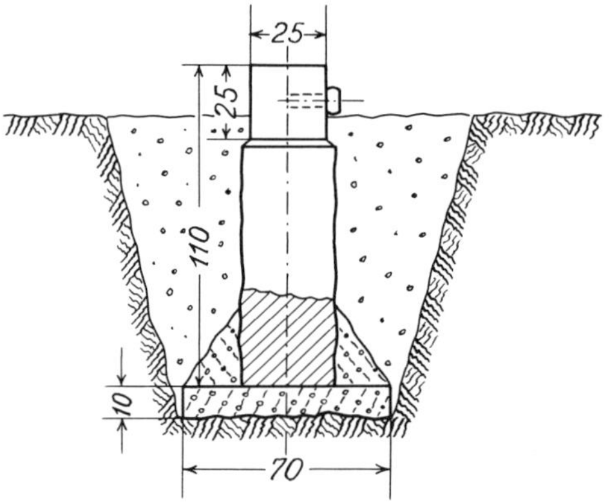

Fig. 9.2.3 Pillar plug

9.2.2 Bench Mark Levelling

9.2.2.1 General Rules

Bench mark levelling is used to densify an existing primary vertical network. The selection of instrument, rods, and procedure depends on the required accuracy.

To eliminate the influences of the earth curvature, refraction and residual instrumental errors, equal back- and foresights are to be used. To accelerate the work, two levelling rods should be utilized.

After solidly setting up the tripod, the instrument is secured on the tripod head. The tripod plate should be roughly level, so that the vertical axis is nearly plumb when the foot screws are turned equally. (For instruments with levelling tripods see section 2.4.3.) The levelling is improved by using foot screws and bull's-eye bubble. Spirit levels then need be properly levelled, using foot screws or tilting screw as appropriate, while automatic levels do not require further levelling. The instrument remains on the tripod when moving to the next station. Spirit levels should be carried in such a manner that their vertical axis remains approximately vertical. Automatic levels, however, should be inclined to such an extent that the movable part of the compensator is arrested on one side rather than banging back and forth.

When placing the level into its container, all screws should be loosened, and slightly tightened after the instrument sits snugly. For longer transports, padded crates should be used.

The rod bubbles should be regularly checked and adjusted, especially after transport. It is important during transport to protect the graduated side of the rod.

9.2.2.2 Simple Levelling

In order to determine the elevation of a point B above the (curved) reference surface using bench mark A, the elevation difference $\Delta h_{A,B}$ is to be measured by simple levelling. For this, the distance AB (Fig. 9.2.4) is subdivided by "turning points" into 70 to 100 m long sections, and the individual elevation differences h_1, $h_2 \ldots h_n$ are observed and added together.

First the instrument station S_1 is selected halfway between A and TP_1 (the first turning point). Then the level is set up at S_1 and levelled. The rod man erects the rod at A, watching that the bull's-eye bubble is centred. The observer then reads the backsight (B_1) after centring the spirit bubble or making sure that the compensator swings freely. The rod man preceeds now to TP_1 where he places a ground plate and then erects the rod. The observer now reads the foresight (F_1), after having checked and recentred the spirit bubble if required. The difference between back- and foresights provides the elevation difference, namely $h_1 = B_1 - F_1$, which completes the work at station S_1.

Now the rod at TP_1 is carefully rotated, and the instrument is moved to S_2. Again, the backsight B_2 and – after moving the rod to TP_2 – the forsight F_2 are read. This provides $h_2 = B_2 - F_2$. This is continued. Therefore

$$\Delta h = h_1 + h_2 + \ldots = \Sigma h = \Sigma \text{backsights} - \Sigma \text{foresights}.$$

If the elevation difference is positive, than the terrain rises, while it falls if Δh is

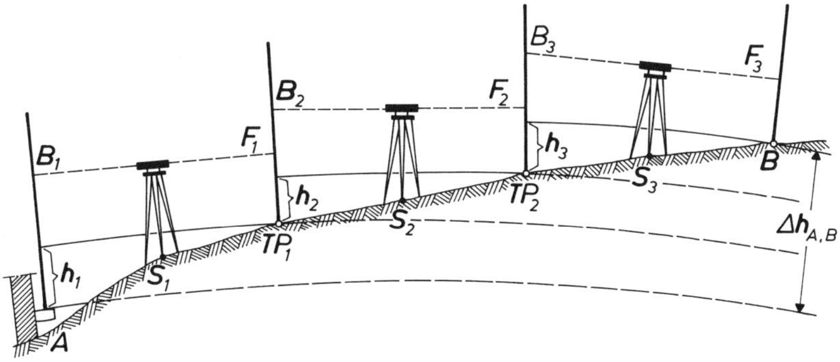

Fig. 9.2.4 Differential levelling

negative. In order to check the result, one can either return to the origin, thus levelling a closed loop, or else continue to level to a known bench mark.

Simple levelling with a construction level easily yields a standard deviation of ± 5 to 10 mm per kilometre, however, as mentioned in section 1.4.9, an error of 2 to 3 times this amount can occur occasionally. If the mentioned accuracy is achieved, the closure error is distributed to the observation stations proportional to the sighting distances. If these values are exceeded, a blunder has to be expected, which can only be eliminated by resurveying the whole loop.

In preparation for such a case, it is useful to occasionally utilize a well defined point (curb, hydrant, manhole, etc.) as a turning point.

The upper limit of the range is 50 m when reading millimetres at the rod. If only centimetres have to be read, up to 100 m sighting distances can be used. For steeper slopes, one is often forced to shorter sights because of the terrain. Possibly unequal sighting distances may be corrected at the following stations provided that the sum of all backsight distances equals to the sum of all foresight distances. This, however, requires concentration and should be avoided if feasible. The following Table 9.2.1 also serves as an example for note keeping.

Example

Table 9.2.1 Differential levelling

Pt.	d	B	F	$h = B - F$ +	 −	MSL elevation
1	2	3	4	5	6	7
A	40	1.524				52.475
	40		1.217	0.307		0.307
TP_1	35	0.723				52.782
	35		1.894		$\overset{+1}{1.171}$	−1.170
TP_2	50	1.234				51.612
	50		0.654	0.580		0.580
B	45	2.421				52.192
	45		1.945	$\overset{+1}{0.476}$		0.477
TP_3	30	0.635				52.669
	30		0.513	0.122		0.122
E						52.791
		6.537	6.223	$\overset{+1}{1.485}$	$\overset{+1}{1.171}$	0.316
		0.314		$\overset{+2}{0.314}$		

The conventional type of recording, using HI (height of instrument), is just as efficient for levelling lines, but leaves any computations using ground readings completely unchecked.

9.2.2.3 Engineer's Levelling

a) *Levelling with double scale rods*

If the required accuracy is higher than commonly achievable by simple levelling with a construction level, then an engineer's level is used in conjunction with a self checking method, such as levelling with double scale rods (section 9.1.5.2). If both scales are read, there is the check that their difference has to be constant to within ± 1 mm. If, in addition, the spirit level is newly centred or the compensator activated prior to the second reading, then the levelling is checked as well. Based on section 1.4, the standard deviations of reading and setting are reduced by a factor of $\sqrt{2}$.

In order to counteract possible systematic errors (section 9.1.2.4), it is advisable to use two rods and to observe in the sequence B_1, F_1, F_2, B_2, i.e.

- backsight to scale I on rod A
- foresight to scale I on rod B
- foresight to scale II on rod B
- backsight to scale II on rod A.

As an independent check, the elevation differences are formed for scales I and II separately and then compared. Differences should be less than ± 1 mm. By carefully operating, a standard deviation of ± 2 to ± 3 mm per km can easily be achieved with this method.

Recording and calculations become simpler if decadic rods (Fig. 9.1.44d)) are used. As the following sample shows, one reads the backsight naturally, and the foresight decadic, records both, and obtains h by simple addition. The following rules have to be considered:

1) The cross indicates a negative unit at the place where it appears:

   $2.103 - 1.417 = 2.103 + \times 7.583 = 0.686$.

2) If h is negative, the terrain falls:

   $2.103 - 2.366 = 2.103 + \times 7.634 = \times 9.737$.

3) A cross replaces a 9 to its right:

   $\times 9.737 = \times 0.737 = -0.263$.

Example:

The elevation of point B between two bench marks A and E is to be determined by levelling using double scale rods with decadic rods.

Pt.	d	scale I B_I; F_I	$h_I = B_I - F_I$	scale II B_{II}; F_{II}	$h_{II} = B_{II} - F_{II}$	$h = \frac{1}{2}(h_I + h_{II})$	MSL elevation
1	2	3	4	5	6	7	3
A	40	2.465		6.500			127.568
	40	× .873	2.338	× 5.837	2.337	2.338	2.338
TP_1	35	1.827		5.863			129.906
	35	× 8.914	0.741	× 4.879	0.742	$\overset{+1}{0.742}$	0.743
B	50	1.066		5.101			130.649
	50	× .412	0.478	× 5.378	0.479	0.478	0.478
TP_2	45	0.505		4.540			131.127
	45	× 8.762	× .267	× 4.727	× .267	$\overset{+1}{\times .267}$	× .268
TP_3	30	1.358		5.393			130.395
	30	× .907	1.265	× 5.873	1.266	1.265	1.265
E							131.660
		4.089	4.089	4.091	4.091	$\overset{+2}{4.090}$	4.092

b) *Levelling with double turning points*

If rods with double scales are not available, a self-checking method using double turning points can be used. In this approach, each rod man operates with two ground plates, placing them close to each other, and setting the rod first on one then the other. In this manner two parallel level loops are obtained, which – although recorded together – are separately computed. The readings are checked if the difference of the foresights at station S_i equal to the difference of the backsights at station S_{i+1} (Fig.

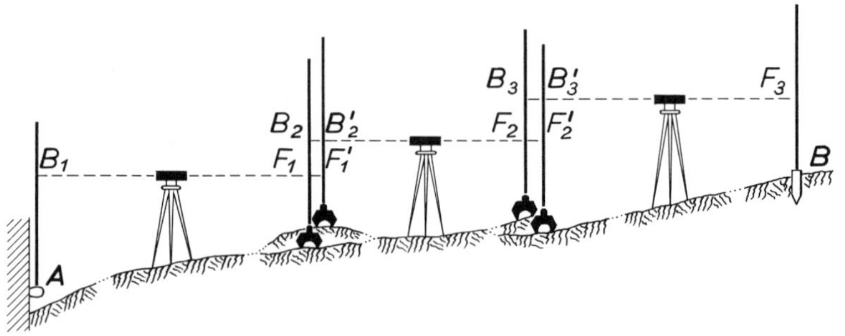

Fig. 9.2.5 Levelling with double turning points

9.2.5). Thus the check is only available at the next station. An instantaneous check is available if a ground plate with two nipples of known elevation difference is used instead of two ground plates (Fig. 9.2.6). As for levelling with double scale rods, the spirit bubble is newly centred for each reading.

Fig. 9.2.6 Ground plate with two nipples

The accuracy is practically the same as for levelling with double scale rods, but the method is a bit more cumbersome.

c) *Three-wire levelling*

When working with a level equipped with stadia hairs (i.e., three horizontal "wires"), three wire levelling can be performed. Readings are taken on all three cross hair lines, resulting in virtually three executions of the same job. Inherent checks help in preventing blunders, and the averaging of the results increases the accuracy.

9.2.2.4 Precision Levelling

To reach the highest accuracy, levels of very high and highest accuracy are to be used. Maximum sighting distances are 40 m for instruments with 40 × magnification, or 30 m for 30 × magnification. Back- and foresight distances are to be equal within one metre. Furthermore, the line of sight shall nowhere be closer to the ground than 50 cm. To improve both accuracy and precision, precision rods (invar rods) with two graduations shifted against each other are used, as they provide a check of the readings at each station (section 9.1.5). As a check, the elevation differences on both graduations are compared as well as the difference between readings to the same point which should differ by the same constant amount. All readings are to be repeated if the difference exceeds 0.3 to 0.4 mm. Since each levelling line is to be observed back and forth, a summation of the elevation differences for each path provides a further check (example at the end of this section).

A data terminal (electronic fieldbook) is usually used for recording. With the recorded data, an automatic data flow leads to final elevations by means of a least-squares adjustment (section 9.1.6).

Each levelling line is subjected to *systematic and random errors*. Systematic errors are usually smaller than the random ones at each station. However, even if they amount to just a fraction of the random errors, they are subjected to a rather dangerous error propagation for longer lines (section 9.2.6.2). Thus the systematic component may cause a resulting total error which is larger by several magnitudes than its random component. Therefore, one has to steadily try to counteract systematic errors in precision levelling.

The most dangerous *systematic errors* are:

1) One sided estimation errors when reading the rod. These are counteracted by measuring the small intervals on the rod rather than estimating them. In years gone by, this was achieved by reading with an offcentred bubble, i.e., setting of the horizontal cross hair line onto the graduation line with the tilting screw, and reading of the bubble path which together with the known sensitivity provides the angular deviation from the horizon. Nowadays, a parallel plate micrometer is used (section 9.1.2.5).

2) Graduation and zero errors of the rods. This is counteracted by regular calibration of the rods (section 9.1.5.4).

3) Inclined rods. This is counteracted by frequent checking of the rod levels (i.e., comparison against a plumb line). The rods are to be braced during observations.

4) Temperature influences on measuring equipment. This is counteracted by providing time for the instrument to reach the field temperature. As a rule of thumb, one minute waiting period is used for each °C of temperature difference.

Furthermore, the level is to be protected from direct sunlight, using an umbrella at the station, and a hood during transport. It is also important to touch the instrument as little as possible.

For highest accuracy requirements, the temperature of the invar tapes of the precision rods has to be measured in order to be able to apply scale corrections if needed (*Zippelt* 1983). It should be kept in mind that direct exposure to sunlight may cause temperature changes of the tapes of up to 5°C.

5) Sinking of the instrument during measurements. This is counteracted by setting up on solid ground and by using double scale rods in the B_1, F_1, F_2, B_2 sequence (section 9.2.2.3). This also eliminates other time dependent influences.

6) Influence of the magnetic field of the earth. This is counteracted by checking the behaviour of the instrument on a longer test line (section 9.1.4.2).

7) Residual compensator errors in automatic levels. This is counteracted by levelling the instrument with the telescope pointing backward at the first station, and forward at the second one, and continuing this alternate pointing all along (section 9.1.4.2).

8) Refraction influence. This is counteracted by keeping the line of sight at least 50 cm above the ground. Levelling in mountainous terrain has to be performed at suitable times when the temperature gradients are small. Corrections, based on refraction models (section 9.1.7), are to be applied as well.

The most critical *random errors* are:

1) Uncertain cross hair setting due to air movements. This is counteracted by working on overcast days, and only in the morning and evening on hot days.

2) Wandering of level bubble. This is counteracted by making sure that the bubble has come to a complete rest prior to taking the readings. If needed, the vertical axis has to be checked.

3) Sinking and other changes of the rods. This is counteracted by carefully placing the ground plates and by keeping the ground plates and bottom of the rods clean.

4) Different sighting distances. This is counteracted by measuring the distances, e.g., with the stadia hairs, in addition to pacing. Discrepancies can be corrected at the next stations.

5) Faulty bottom surfaces of the rods. This is counteracted by setting the rods always in the same position on the ground plates. Possible zero differences between rods are to be determined, or else one has to finish with the same rod as used for the first reading.

When the above points are properly considered, the mean elevations from back and forth measurements should have a standard deviation of ± 0.3 to 0.5 mm per km (section 9.2.5.2).

The final sea level elevations are obtained after a least-squares adjustment. The influence of local differences in gravity acceleration are considered by introducing gravity reductions into the adjustment (*Torge* 1980; *Vaníček* and *Krakiwsky* 1986).

Example: See Table 9.2.2.

The elevation difference between bench marks 1 and 10 is to be determined by precision levelling. The measurements are carried out with double-scale rods. The instrument has a least count of $^1/_{100}$ mm. The elevation differences are continuously added together. The cumulative station differences in $^1/_{100}$ mm provide a further check of the observations.

Often an electronic fieldbook is used for precision levelling. Evaluation and checks are then performed at each station automatically by the microprocessor of the electronic fieldbook.

9.2.3 Profiles and Cross Sections

9.2.3.1 Profiles

A profile is a vertical section through the earth's surface along a specific line, which usually is the projected centre line for roads and railroads and a parallel to it for canals.

The work commences with stationing of the centre line. Depending on the purpose and terrain, stakes are placed every 25, 50, or 100 m as well as at principal points of a curve, abrupt changes in slope and at intersections with roads, water courses, etc. The centre line stake is usually at terrain level, while a second, higher stake nearby carries

Table 9.2.2. Precision levelling with double scale rods

Stn.	Dist.	Dir.	Foreward Date: 11.07.79 Time: 11:40				Rod 3343 A, B		Stn.	Dist.	Dir.	Back Date: 11.07.79 Time: 12:10			
			left		right		Σ diff. in 1/100 mm					left		right	
	m		cm		cm					m		cm		cm	
	10	b	406	110	+ 104	578				15	b	433	869	+ 132	337
		f	− 453	189	− 151	662					f	− 488	818	− 187	289
1			− 47	079	− 47	084	5	3	1			− 54	949	− 54	952
		b	494	822	+ 193	292					b	357	048	+ 55	505
	18		447	743	+ 146	208				20		302	099	+ 0	553
		f	− 402	859	− 101	350					f	− 559	588	− 258	062
2			44	884	+ 44	858	26	20	2			− 257	489	− 257	509
		b	502	801	+ 201	279					b	351	418	+ 49	880
	18		547	685	+ 246	137				20		93	929	− 207	629
		f	− 394	091	− 92	568					f	− 559	371	− 257	852
3			153	594	+ 153	569	25	39	3			− 465	442	− 465	481
		b	535	814	+ 234	301					b	374	219	+ 72	692
	17		689	408	+ 387	870				17		− 91	223	− 392	789
		f	− 367	589	− 66	071					f	− 564	199	− 262	669
4			321	819	+ 321	799	20	36	4			− 655	422	− 655	458
		b	571	722	+ 270	210					b	357	478	+ 55	951
	20		893	541	− 592	009				20		− 297	944	− 599	507
		f	− 353	330	− 51	809					f	− 567	590	− 266	067
5			540	211	+ 540	200	11	40	5			− 865	534	− 865	574

		b	563	744	+ 262	232					b	354	424	+ 52	899
	20		1103	955	+ 802	432				20		− 511	110	− 812	675
		f	− 353	601	− 52	085					f	− 572	310	− 270	790
6			750	354	+ 750	347	7	45	6			−1083	420	−1083	465
		b	562	291	+ 260	768					b	368	765	+ 67	232
	17		1312	645	+1011	115				17		− 714	655	−1016	233
		f	− 372	345	− 70	834					f	− 537	923	− 236	398
7			940	300	+ 940	281	19	53	7			−1252	578	−1252	631
		b	562	280	+ 260	740					b	395	630	+ 94	100
	20		1502	580	+1201	021				18		− 856	948	−1158	531
		f	− 353	985	− 52	449					f	− 504	030	− 202	500
8			1148	595	+1148	572	23	53	8			−1360	978	−1361	031
		b	560	971	+ 259	458					b	400	011	+ 98	486
	20		1709	566	+1408	030				18		− 960	967	−1262	545
		f	− 358	651	− 57	128					f	− 491	960	− 190	432
9			1350	915	+1350	902	13	50	9			−1452	927	−1452	977
		b	490	070	+ 188	553					b	453	184	+ 151	653
	15		1840	985	+1539	455				10		− 999	743	−1301	324
		f	− 435	081	− 133	550					f	− 406	113	− 104	580
10			1405	904	+1405	905	−1	48	10			−1405	856	−1405	904

$\Delta h_{\text{fore}} = 14.05904$ m Distance 0.35 km $\Delta h_{\text{back}} = -14.05880$ m the station number. Stationing is usually carried out in hectometres, which means that points at 50 m, 175 m, 273 m ... are numbered as 0 + 50, 1 + 75, 2 + 73 .... All distances are counted along the centre line, starting from the origin of the specific project.

The elevations are measured with a construction level. The sea level elevation of starting and terminal points are tied into the vertical net by levelling to a bench mark. The full station stakes are utilized as turning points, while any intermediate points are observed as ground readings. As a check, the profile line is closed at a bench mark, or else returned to close the loop. Closure errors are distributed proportional to the sighting distances, as for simple levelling.

The example in Table 9.2.3 may serve as an illustration of the recording.

9.2.3.2 Cross Sections

Cross sections are straight vertical sections through the earth's surface normal to the centre line. They are staked with simple tools as indicated in section 8.2. At deflection points the angle is bisected, while in curves it is normal to the tangent. The lateral extension of cross sections is between 20 and 50 m, depending on the purpose and type of terrain. Cross sections are needed for the geometric design work and also provide the data for earthwork computation. Therefore, they have to be measured wherever there are significant changes in the terrain and where the direction of the centre line changes. The lateral distances are measured in decimetres from the centre line and recorded in a hand sketch (not to scale, Fig. 9.2.7). The elevations are determined with a construction level which is usually set up in such a manner that the whole cross section can be measured from one station. The height of the instrument is determined by reading the rod at the centre line point. The rod readings are entered into the cross section sketch. If, as in Fig. 9.2.7, the whole profile cannot be measured from one station, then the instrument has to be set up again, and another reading at the centre line point is required. It is possible to obtain measuring and computational checks by an independent second survey with a slightly different height of instrument. However, the direct comparison between terrain and rod readings make large blunders unlikely, which is why a second survey is not usually carried out. Occasionally, cross sections are measured with a horizontal rod (Fig. 9.1.3), if the slope is extremely steep or in dense forest. Cross sections in rivers are tied to the shore. A measuring string or line with metre sections is used for distances. For narrow rivers, this line is stretched horizontally and serves as the height reference. The water surface is used as the reference for wider rivers, and the depth is measured with poles. The water surface itself is determined by tying the top of posts near the shore into a bench mark and then measuring down along this pole.

Remarks to the field notes: Starting at the origin A, the instrument height for each station is obtained by adding the backsight reading to the MSL elevation of the back

Table 9.2.3
Profile record

Pt.	d	back-sight	ground reading	fore-sight	H.I.	MSL elev.	corr. (mm)
1	2	3	4	5	6	7	8
A	50	1.415				49.675	
	50			1.290	51.090		
TP_1	45	0.420				49.800	3
	45			1.655	50.220		
0^{+0}	50	1.390				48.565	6
0^{+25}			3.175		49.955	46.780	6
0^{+50}			2.955			47.000	6
0^{+75}			0.595			49.360	6
	50			0.700			
1^{+0}	50	2.040				49.255	9
1^{+25}			1.545		51.295	49.750	9
1^{+50}			3.235			48.060	9
1^{+75}			0.820			50.475	9
	50			0.415			
2^{+0}	30	1.005				50.880	12
	30			2.090	51.885		
TP_2	40	1.870				49.795	15
	40			1.235	51.665		
E						50.430	18
		8.140		7.385		50.448	
			0.755			0.773	

point. By subtracting foresights or ground readings from this instrument height, the elevation of turning points or intermediate points is obtained. Thus for perfect levelling, the given MSL height of the end point E should be obtained. A closure error within the required accuracy is afterwards distributed to each station. The elevations of intermediate points are not controlled by this page check, they can, however, be checked by using the rise-fall approach as mentioned in section 9.2.2.2. Only a second independent survey will check the actual measurements.

9.2.3.3 Plotting of Profiles and Cross Sections

Based on field notes (Table 9.2.3) and field sketch (Fig. 9.2.7), profiles and cross sections are plotted on graph paper. The profile shown in Fig. 9.2.8 is based on a reference line which is parallel to the horizon and a multiple of 10 metres above the reference surface. The centre line is usually plotted in the same scale as the planimetric plans with a 10 times or higher scale increase for elevations. Thus the vertical characteristics are displayed prominently. The x-coordinates are the station numbers along the centre line. Common scales for planimetry are 1 : 1000 to 1 : 5000, with 1 : 100 to 1 : 250 for elevations.

The designed structure is also entered (usually in a different colour), so that elevation differences between the local terrain and the planned structure can be obtained from the profile.

For cross sections (Fig. 9.2.9), a uniform scale for planimetry and elevations is

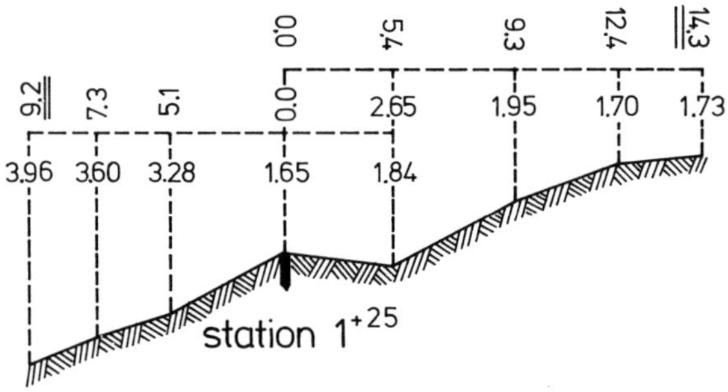

Fig. 9.2.7 Cross section sketch

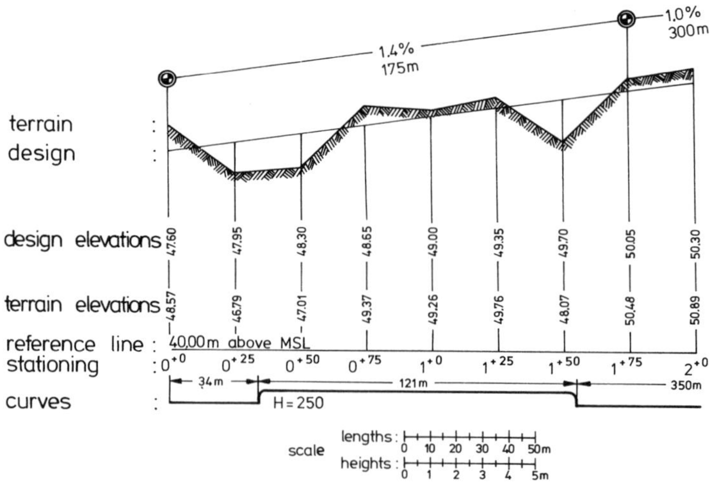

Fig. 9.2.8 Profile plot

chosen, usually the same as the height scale in the profile. The reason for this is the direct determination of cross sectional areas as well as having true angular relationships which help in the design.

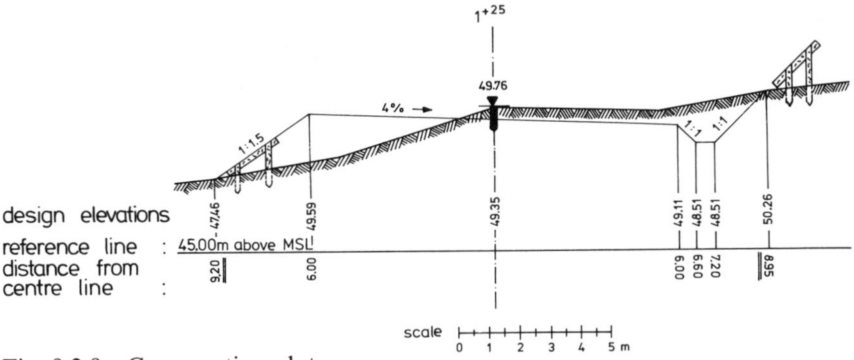

Fig. 9.2.9 Cross section plot

The centre line stake is transferred from the profile, then the instrument height is lightly drawn, and all readings are plotted as negative offsets. For space reasons, it is common to raise the reference line more than for the profiles. By superimposing the design cross section (use of templets) the cross sectional area representing cut and/or fill is defined and can be determined.

The earthwork is usually computed as the product of the average cross sectional area between two subsequent cross sections, times their separation (section 14.2).

9.2.4 Area Levelling

(see also section 14.1.3.2)

For certain types of planning and construction (e.g., sports fields, drainage systems, etc.), profiles and cross sections provide insufficient coverage. Instead, contour lines are required in the plan. They are plotted based on points which are locally coordinated in both positions and height. In flat terrain levelling is used, while stadia methods (4.2) are preferred in more varied terrain.

9.2.4.1 Positioning

Positioning depends on the purpose of the plans and the terrain characteristics. Usually, available plans or maps are sufficient.

a) If a site plan with numerous points is available (boundary markers, etc.), some of these points are selected for levelling, and necessary additional ones for heights are referenced from the existing points.

b) If only the boundaries of a large lot are known, then profiles are measured across the lot whose end points are tied into the boundaries (Fig. 9.2.10).

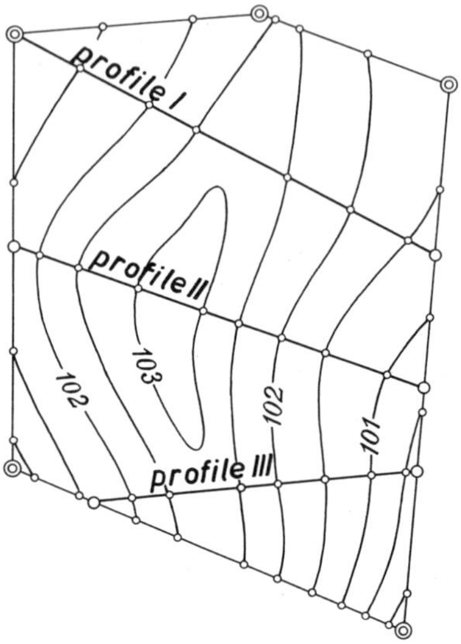

Fig. 9.2.10 Tie-in of profiles

c) If there is no plan available, the area is covered with a rectangular grid, usually with equal separation (Fig. 9.2.11) of 10 or 50 metres. This is especially useful if earthwork volumes have to be computed. Significant terrain points can be referenced within this grid frame. The points are usually staked and numbered.

d) In a somewhat more varied terrain, the important points are determined by stadia levelling (section 9.1.2.5), which utilizes the same kind of point selection as regular stadia surveys.

9.2.4.2 Elevation Measurements

Elevation measurements consist of level lines tied into bench marks, using characteristic terrain or profile points as either turning points or ground readings. Simple levelling with a construction level is sufficient for this purpose. In method d) (section 9.2.4.1) stadia distances and angles are measured as well.

Due to the lower accuracy requirements, sighting distances of up to 300 m can be used.

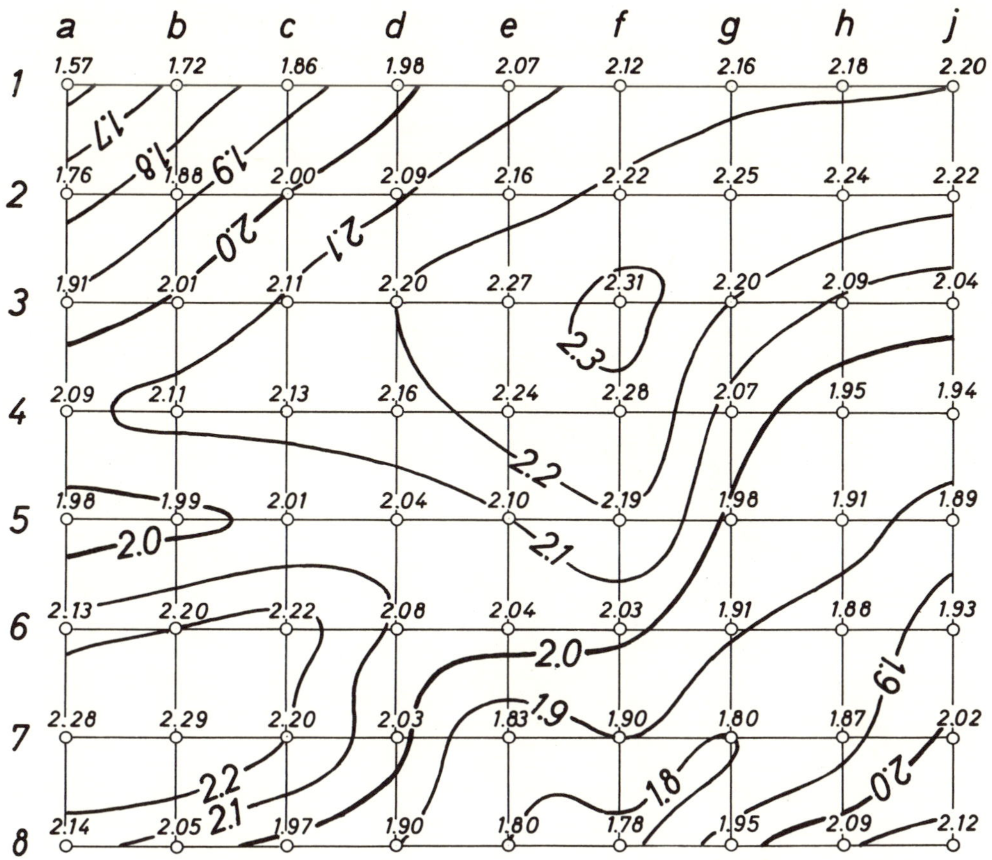

Fig. 9.2.11 Levelling grid

9.2.4.3 Plotting of Contour Plans

First the positions of all points are plotted onto the planimetric plan. The elevations are noted beside each point in preparation for the generation of contour lines. Contour lines are horizontal curves which connect points of equal elevation.

Depending on the required accuracy, they are plotted at 0.1, 0.5, or 1 m intervals in flat areas, and 2.5 or 5 m in moderately undulating terrain in such a manner that points of these elevations are interpolated from the measured terrain points and then connected by a smooth curve (Figs. 9.2.10 and 9.2.11).Interpolation is performed along the line of steepest slope by either numerical or graphical means. In Fig. 9.2.12 two terrain points (1) and (2) with $h_1 = 63.3$ and $h_2 = 68.5$ are located at a distance of 28.3 mm from each other on the plan. It is desired to obtain the intersections of line (1), (2) with the contours 64.0, 65.0, ... 68.0, thereby assuming that the straight line connection between (1) and (2) represents the terrain (linear interpolation).

If in Fig. 9.2.12, the distance from (1) to the contour 64.0 is denoted as S_4, to the contour 65.0 as S_5, etc., then the following values are obtained by simple slide rule or calculator operation:

$$S_4 = \frac{28.3}{5.2} 0.7; \quad S_5 = \frac{28.3}{5.2} 1.7; \quad S_6 = \frac{28.3}{5.2} 2.7; \text{ etc.}$$

or

$$S_4 = 3.8; \quad S_5 = 9.2; \quad S_6 = 14.7; \quad S_7 = 20.1; \quad S_8 = 25.6 \text{ mm}.$$

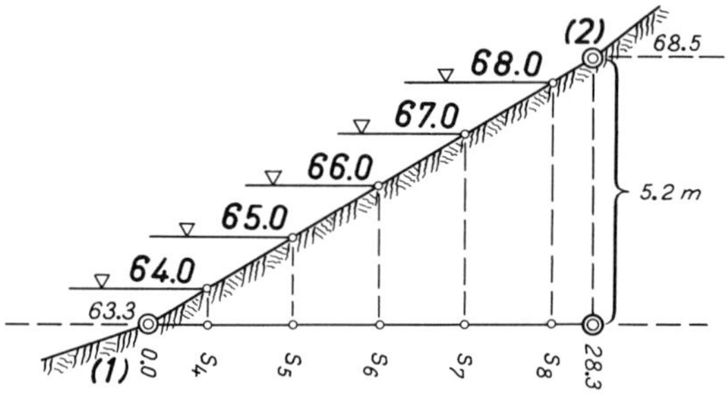

Fig. 9.2.12 Interpolation of contours

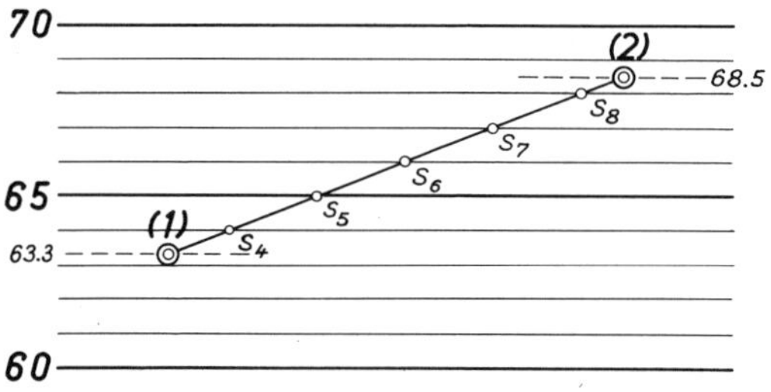

Fig. 9.2.13 Graphical interpolation

For the graphical interpolation (Fig. 9.2.13) a sheet of transparent millimetre graph paper is placed on top of the plan in such a manner that point (1) obtains a value corresponding to 3.3, and point (2) to 8.5. Then a ruler is placed along points (1) and (2), and the points of intersection between ruler edge and respective graph lines is pricked onto the plan and subsequently numbered.

Occasionally doubts arise as to the best course of a contour line. It is therefore recommended to prepare a field sketch during the survey, indicating the approximate course of contours as well as the direction of maximum slope to aid the interpolation.

Contours with even elevations are drawn as thicker lines to provide a better overview and to serve as index contours.

9.2.5 Levelling Methods for Special Cases

9.2.5.1 River Crossing

When levelling across a river it is impossible to set up the instrument in the middle between the two rods. In this case, rods or target on the other shore are simultaneously observed from both sides. By taking the mean of the observed values, the influence of the earth's curvature is completely eliminated, while the influences of refraction and residual instrument errors are reduced if a strictly symmetric observation program is carried out (*Drodofsky* 1960, 1964; *Kakkuri* 1966, 1969). If possible, the terrain should be similar on both sides of the river, as should the position of the sun, the water level, especially in tidal rivers, air and water temperature, and wind direction. Instrumental shortcomings are counteracted by exchange of instruments between shore stations. This method, however, requires extended observation time and a large number of calculations. Zeiss Oberkochen developed river crossing equipment for river crossings of medium length (1 to 2 km), which eliminates instrument adjustment errors, reduces reflection influences, as well as considerably shortens the observational and computational efforts. This equipment consists of four Ni 2 levels, two each mounted onto two base plates which can be levelled with the aid of a bull's-eye bubble. Each level is equipped with a rotating wedge attachment. Two sets of targets or illuminated reflectors on tribracks for the far shore and two levelling rods for closure at bench marks at the near shore complete the equipment (Fig. 9.2.14).

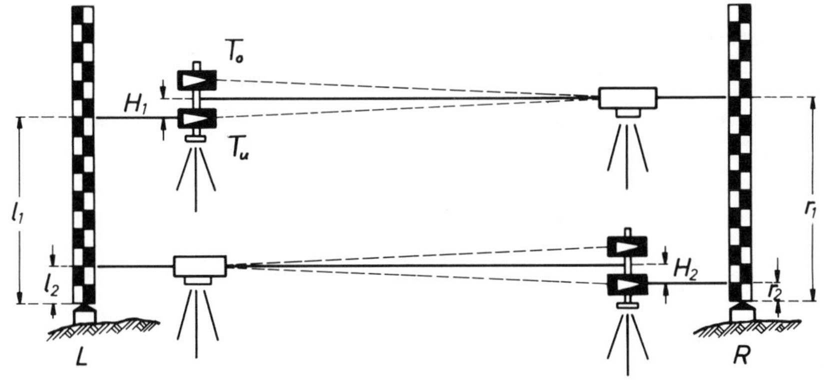

Fig. 9.2.14 Instrument arrangement for river crossing

The method is based on the following idea:

(1) If two levels focussed to infinity are brought into collimation (section 9.1.2.4), then their collimation axes are parallel. The zenith distances of the two collimation axes complement each other to 180°, in other words, one instrument is inclined

upward from the true horizon by the same amount as the other one is tilted downward.

(2) If two Ni 2 levels have been brought into this position, the tilt of the collimation axes remains because of the compensators, even if the telescope is rotated. Thus the instruments can be directed towards a rod and read. Thus the mean of the values obtained with the two instruments represents the true horizon.

(3) Since the anallactic point of the Ni 2 telescope falls into the horizontal axis of the Ni 2, the height of the horizon of these instruments is equal to the arithmetic mean of their horizontal axis heights.

Considering first the right part of Fig. 9.2.14, the items (1) to (3) above lead to the following observation procedure:

1st step: After attaching the rotating wedges to the two Ni 2 levels fastened to one base plate, collimation is achieved solely by operating the rotating wedges. The tilts of the collimation axis are read to 2″ at the scales of the rotating wedge attachments and denoted as n_1 or n_2 for the first or second instrument, respectively. Subsequently the height of the horizontal axis for each level is determined by reading the rod at a nearby bench mark.

2nd step: At the left shore, the targets T_o (upper target) and T_u (lower target) are placed ideally vertically above each other, and their elevation difference b is measured.

Then level 1 is used at the right shore to sight to T_o and T_u using the rotating wedge and reading the inclinations o_1 and u_1. Similarly, level 2 provides o_2 and u_2. The same measuring sequence is carried out possibly simultaneously by the other crew at the other shore. Finally, on both shores the rod at the nearby bench marks are read again. This completes one set of river crossing observations, which, as a rule, is followed by several other sets.

The *evaluation* is first carried out separately for each station. Fig. 9.2.15 illustrates the sights taken with both levels 1 and 2 to the station at the opposite shore. o_1 and u_1 represent the arithmetic means of all inclinations measured with level 1 to T_o and T_u, while o_2 and u_2 represent the corresponding values for level 2. n_1 and n_2 are the means of all collimation readings taken with levels 1 and 2.

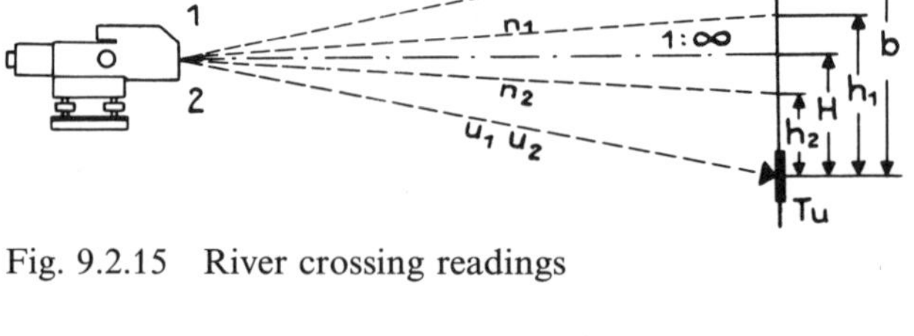

Fig. 9.2.15 River crossing readings

The calculations are simplified by the introduction of the elevations h_1 and h_2, which represent the elevations above T_u of the mean collimation rays n_1 and n_2. According to Fig. 9.2.15, the following relations are obtained:

$$h_1 = b\frac{n_1 - u_1}{o_1 - u_1}; \quad h_2 = b\frac{n_2 - u_2}{o_2 - u_2}. \tag{9.2.1}$$

As noted in concept (2), the arithmetic mean of n_1 and n_2 represents the true horizon of the two Ni2 levels located at the right shore. This horizon intersects the target plane at

$$H = \frac{h_1 + h_2}{2}. \tag{9.2.2}$$

This holds for elevations measured from right to left (index 1) as well as reverse (index 2). Thus h_1 and h_2 are known without requiring the distance s between the station.

h_1 and h_2 however, are the result of one set of river crossing observations. The number of repetitions depends on the weather and on the required accuracy.

The results have to be corrected for the earth's curvature and refraction, according to *Drodofsky's* formula

$$dH = \frac{1 - k}{6.37} s\,ds, \tag{9.2.3}$$

k is the refraction constant, dH is the elevation correction in millimetres, while s represents the sighting distance in kilometres, and ds the difference between the sighting distances in metres. Using the target separation b, s can be computed as

$$s = \frac{10000\,b}{o_1 - u_1} = \frac{10000\,b}{o_2 - u_2}. \tag{9.2.4}$$

In average atmospheric conditions a standard deviation of ± 1 mm/km is given by *Drodofsky* for 6 to 8 sets. Practical measurements by *Kakkuri* produced ± 12 mm for crossing a 931 m wide river.

The same method (using rotational wedge attachments) can also be used for precision levelling. However, it is not suited for crossing wide rivers or ocean inlets.

Heavy spirit levels with 5″ bubbles and specially designed targets should be used for this. Hydrostatic levelling is more accurate but significantly more expensive. Finally, precision theodolites with good vertical circles and index compensation can be utilized as well.

9.2.5.2 Motorized Precision Levelling

Precision levelling over long distances is extremely time consuming. Attempts to replace it by trigonometric heighting (section 10.6) enjoyed only limited success in mountainous terrain, because of the influence of refraction which, in spite of over 100 years of scientific investigations, cannot and probably will not be known to complete satisfaction. For over a decade, Prof. Peschel at the University in Dresden, GDR, has been trying to overcome both difficulties (slow progress in precision levelling, and insufficient accuracy with trigonometric heighting) by motorized precision levelling.

According to the Dresden method, people and peripheral material are transported in regular cars. Two specifically equipped small vehicles are used for instrument and rod transport, the level vehicle, and the rod vehicle. Both instrument and rods are set up and operated from within the vehicle.

a) The rod vehicle (Fig. 9.2.16) carries on its reinforced left door the rod base of 10 kg mass, which can be hammered onto the road surface by the driver. The rod support is mounted on the roof from where the rod is placed onto the rod base. The driver can perform this without leaving his seat and also set the rod into a vertical position by activating some steering devices.

b) The instrument vehicle (Fig. 9.2.17) was developed from a jeep-type vehicle. One of the three extra long tripod legs is lowered onto the road through an opening in the floor of the vehicle, while the other two are outside the vehicle. On top of the tripod is a spherical head which facilitates a fast rough levelling of the instrument. Metre

Fig. 9.2.16 Rod vehicle

Fig. 9.2.17 Instrument vehicle

counter, two-way radio, recording device, traffic sign, and optionally an instrument for determining vertical temperature differences complete the inventory of this vehicle.

All the equipment has been developed in steps under steady testing in both research and practical projects. Standard deviations of less than ± 0.2 mm/km have been continuously achieved during the last decade.

Angus-Leppan (1983) reports on further developments of motorized levelling in Australia. One important feature of his system is the combination of rod and instrument on one vehicle, thus requiring only two vehicles for levelling. Of the possible methods for measuring the refraction, a new concept of refraction by reflection is favoured as it is simple and accurate. The resulting procedure is a motorised, automated EDM-height traverse which can progress in steps of several hundred metres, whether the topography is flat or steep.

9.2.6 Levelling Accuracy

Elevation differences obtained by levelling are influenced by systematic and random errors. By using procedures described in section 9.2.2.4, systematic errors are reduced to such an extent that primarily random errors are left. Section 9.2.6.1 deals with the effect of random errors, while the influence of residual systematic errors on the final result is covered in section 9.2.6.2.

9.2.6.1 Propagation of Random Errors and the Standard Deviation for 1 km Levelling

Since the total elevation difference for levelling is equal to the sum of individual elevation differences, the standard deviation increases with the square root of set-ups. If L is the total length of the levelling line and z the average sighting distance, then $L/2z$ set-ups are required. Thus with σ_ε being the standard deviation of a single elevation difference, the standard deviation of the total elevation difference becomes

$$\sigma_L = \sigma_\varepsilon \sqrt{L/2z} = \sigma_\varepsilon \sqrt{n} \tag{9.2.5}$$

where n is the number of set-ups. As a rule, the standard deviation of a 1 km line is used as the reference, rather than the standard deviation of a particular elevation difference. Thus, the standard deviation of an n-kilometre long line is

$$\sigma_L = \sigma_{1\,\mathrm{km}} \sqrt{n}\,. \tag{9.2.6}$$

The standard deviation per kilometre can be computed from differences between back and forward measurements, or from closure errors on bench marks or within loops. In the following formulae R denotes the distance between two neighbouring bench marks, and L denotes the "line", the sum of distances between junction points (where several lines meet). A loop passes over several bench marks and junction points and returns to its origin.

The following specific formulae have been developed for the *determination of the standard deviation per kilometre.*

a) *From differences d_i between forward and return survey:* with n distances $R_1, R_2 \ldots R_n$ observed to and from, the standard deviation per km for a one kilometre measured distance R_i is

$$s_{\mathrm{Niv/km}} = \sqrt{\frac{1}{2n} \Sigma \frac{d^2}{R[\mathrm{km}]}} \tag{9.2.7}$$

and for the mean value from forward and return measurements:

$$\sigma_{\mathrm{Niv/km}} = \pm \frac{s_{\mathrm{Niv/km}}}{\sqrt{2}}\,. \tag{9.2.8}$$

For lines, R is to be replaced by L.

b) *Standard deviation per kilometre for closed loops using residuals r_i:*

A closed loop of length ΣL is measured yielding a closure residual r. For one residual, the standard deviation per kilometre becomes

$$\sigma_{\mathrm{Niv/km}} = \pm \frac{r}{\sqrt{\Sigma L[\mathrm{km}]}}\,. \tag{9.2.9}$$

One residual, however, is statistically not very useful, thus a standard deviation per kilometre, determined from n residuals is more suitable:

$$\sigma_{\mathrm{Niv/km}} = \pm \sqrt{\frac{1}{n} \Sigma \frac{r^2}{\Sigma L[\mathrm{km}]}}\,. \tag{9.2.10}$$

If in both cases the measurements were taken in one direction only, than the standard deviation per kilometre refers to one measurement. If the mean of two measurements is used in the above formula, then the standard deviation per kilometre is valid for the mean of two measurements as well.

c) *Standard deviation computed from forward and return measurements and from loop closures.*

If several loops, as illustrated in Fig. 9.2.18, are measured forward and return, then the standard deviation per kilometre double levelling is computed separately according to equation (9.2.8) as well as according to equation (9.2.10). If for the majority of loop closures the standard deviation per kilometre is larger than for forward and return measurements, one has to assume that systematic errors have not been properly eliminated. Then the measuring procedure is to be checked according to the criteria given in section 9.2.2.4.

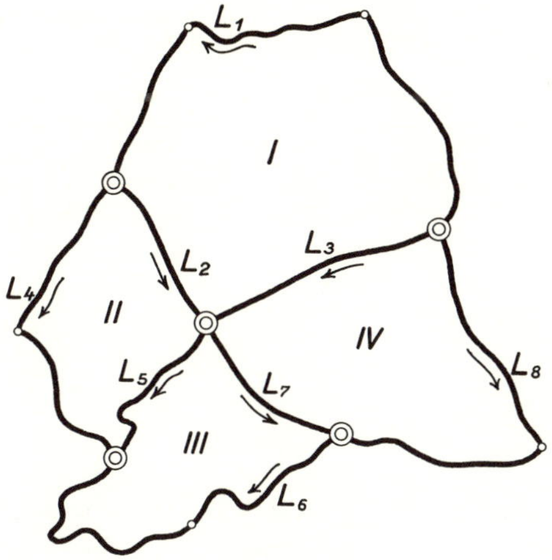

Fig. 9.2.18 Levelling network

Examples:

To a): Forward and return levelling for the lines L_1 and L_2 of the first-order network shown in Fig. 9.2.18 resulted in the following differences:

line	L in mm	d in mm	dd/L
L_1	108.2	−2.4	0.05
L_2	30.4	1.0	0.03
L_3	41.1	−8.1	1.60
L_4	61.3	13.5	2.97
L_5	30.5	−4.8	0.75
L_6	73.5	8.7	1.03
L_7	35.0	−4.5	0.58
L_8	83.1	−0.3	0.00
			7.01

This leads to the following standard deviations per kilometre according to equations (9.2.7) and (9.2.8):

$$s_{\text{Niv/km}} = \sqrt{\frac{1}{2.8} \cdot 7.01} = 0.66 \text{ mm} ,$$

$$s'_{\text{Niv/km}} = \frac{0.66}{\sqrt{2}} = 0.47 \text{ mm} .$$

To b): The same network provided the following closure errors:

loop	$[L]$ in km	d in mm	$dd/[L]$
I	179.7	−0.31	0.00
II	122.3	−9.00	0.66
III	139.2	5.25	0.20
IV	159.6	6.73	0.28
			1.14

This according to equation (9.2.10) leads to:

$$s'_{\text{Niv/km}} = \pm\sqrt{\frac{1}{4}1.14} = \pm 0.53\ \text{mm}\,.$$

To c): Since the results according to equations (9.2.8) and (9.2.10) are not significantly different, one can assume that systematic errors have been properly eliminated.

9.2.6.2 Propagation of Random and Systematic Errors

As a rule, the remaining systematic errors are much smaller than the random ones, however, their propagation is worse. The standard deviation of a levelled elevation difference $\Delta\Delta H$ along a line L is, according to *Pelzer* (1984):

$$\sigma_L = \sigma_\varepsilon\sqrt{n}\sqrt{1+nq^2} = \sigma_\varepsilon\sqrt{n}\sqrt{1+\frac{q^2}{2z}L} = \sigma_\varepsilon\cdot k\cdot\sqrt{n} \tag{9.2.11}$$

with

$$q = \frac{\sigma_\Delta^2}{\sigma_\varepsilon^2}\,,$$

which is the ratio of the variances of the systematic and random errors of one set-up. The scale factor k is a function of the length L of the levelling line, as obvious in

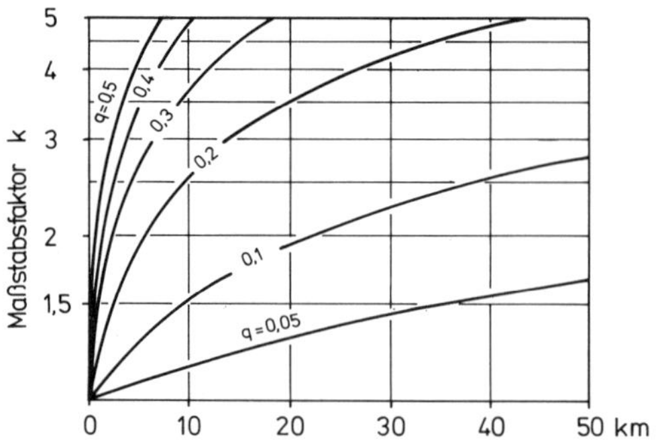

Fig. 9.2.19 Scale factor k as function of the length of the levelling line

equation (9.2.11). Values for this scale factor can be extracted from the diagram (Fig. 9.2.19); e.g. $k \approx 2$ for $q = 0.1$ and $L = 20$ km. In this case the standard deviations as obtained from random errors are already doubled because of the systematic errors, even though the latter are smaller by a factor 10 for a given set-up. Assuming that σ_ε amounts to approximately 0.1 mm, the following requirement appears to be justified (*Pelzer* 1984):

$$\sigma_\Delta \leqq \frac{\sigma_\varepsilon}{10} = 0.01 \text{ mm} .$$

9.2.6.3 Tolerance Limits for Bench Mark Levelling

Based on the guide lines mentioned in section 9.2.1.1, tolerance limits were established which amount to about 3 times the expected standard deviation. The main values are:

a) Difference d_1 from forward and return levelling between two bench marks:

first-order: $d_1 = \pm 2\sqrt{R}\,[\text{km}]$ mm

second-order: $d_1 = \pm 3\sqrt{R}\,[\text{km}]$ mm

third-order: $d_1 = \pm 5\sqrt{R}\,[\text{km}]$ mm .

b) Difference d_2 between new determination and published elevation difference:

first-order: $d_2 = \pm(2 + 2\sqrt{R})$ mm

second-order: $d_2 = \pm(2 + 3\sqrt{R})$ mm

third-order: $d_2 = \pm(2 + 5\sqrt{R})$ mm

fourth-order: $d_2 = \pm(2 + 6\sqrt{R})$ mm

where R equals the single distance in kilometres. The accuracy for engineering levelling depends solely on the purpose of the job. Thus general tolerance limits are useless, and the surveying engineer has to select his own tolerances to meet the given specifications.

9.2.6.4 Accuracy of Area Levelling

The error propagation given in the previous sections are valid for the transfer of elevation from one station to the next. The area levelling depends on terrain form and point density. If the influence of error propagation has been properly taken into account for the instrument stations, then one needs to protect oneself only against blunders when carrying out a common area levelling project.

10. Trigonometric Heighting

10.1 Basic Equation for Trigonometric Heights

Geometric levelling is the simplest and most accurate method to determine elevation differences. Occasionally, however, it cannot be applied, e.g., for the determination of the elevation of a tower, or else is not economical, e.g., when crossing mountain ranges. For such cases, trigonometric heighting is applied. It is also generally recommended when elevation differences of lower accuracy are required, because it is more economical.

To determine the elevation difference between points A and B by trigonometry, the horizontal distance S^H or the slope distance S^R between the points is needed, and the vertical angle from one station to the other has to be measured. As indicated in Fig. 10.1.1 either the vertical angle α or its complement to 90°, the zenith distance $z = 90° - \alpha$ is measured. The basic equations for trigonometric heights are

$$H_2 - H_1 = \Delta H = S^H \cot z + i - t \tag{10.1.1a}$$

$$H_2 - H_1 = \Delta H = S^R \cos z + i - t \tag{10.1.1b}$$

where z is the zenith distance measured at A, i is the height of the horizontal axis of the theodolite above the ground point (= instrument height), and t is the target height of the target B above the ground point. This equation is only valid for short distances. Earth curvature and bending of the sighting ray due to refraction have to be considered for longer distances.

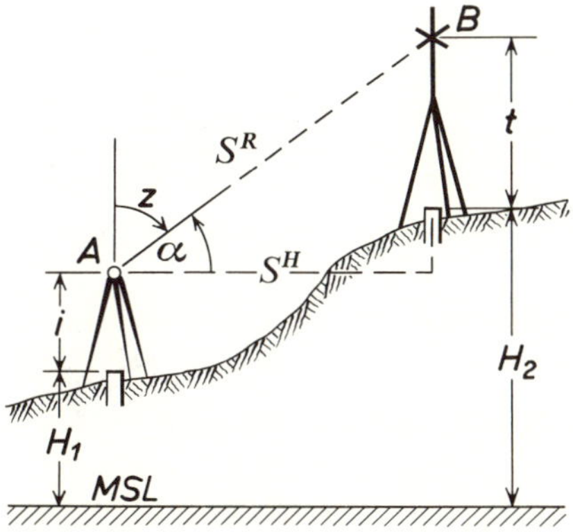

Fig. 10.1.1 Trigonometric heighting

Trigonometric heighting is gaining in importance, since the slope distance s^R can be measured very accurately and fast with EDM instruments.

10.2 Theodolite Parts Designed for Vertical Angle Measurements

Theodolites need two parts for vertical angle measurements, namely the vertical circle, and a reading device for it.

10.2.1 The Vertical Circle

The diameter of the vertical circle is usually smaller than for the horizontal circle. It is attached and centred onto the horizontal axis such that it follows all tilting motions of the telescope. It can be fixed at a certain place with the aid of a clamping screw and slightly moved with a fine motion screw. Although the least count is the same as for the horizontal circle, the accuracy is slightly lower because of the smaller diameter of the circle except for some construction theodolites.

Older instruments are equipped with circles graduated counter clockwise in 360° or else twice in 180° to provide vertical angles. Newer instruments are graduated clockwise from 0 to 400 gon (0° to 360°) such that zenith distances are read (Figs. 10.2.1 and 10.2.2).

The reading device, commonly denoted as the height index, has two functions. It serves for reading of vertical angles or zenith distances at the vertical circle, but also

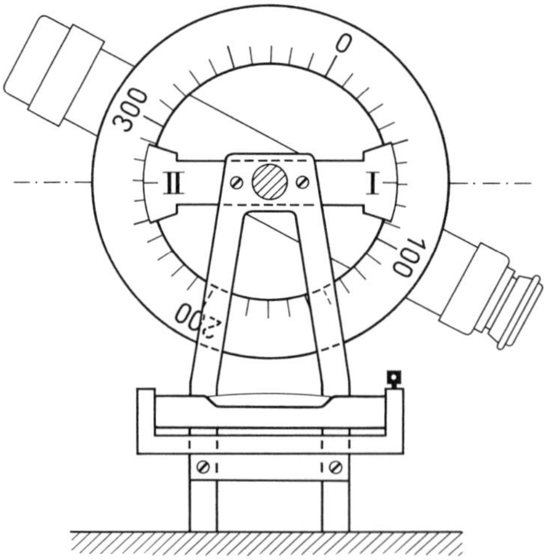

Fig. 10.2.1 Level bubble at standard

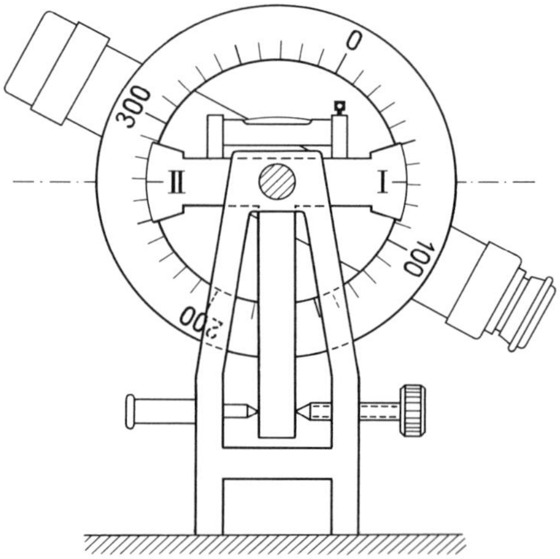

Fig. 10.2.2 Level bubble at height index

as a reference to the direction of gravity. This reference is established either manually with the aid of a spirit level or else automatically through a pendulum system (compensator).

10.2.2 The Height Index

This is monitored by a spirit level. It has either one or two reading locations plus a tubular level which has to be adjusted such that a vertical angle of 0 gon or a zenith distance of 100 gon is read for a horizontal line of sight. The height index has to be brought into the proper position with the adjusted level prior to reading the vertical angle with vernier or microscope as for the horizontal circle. Newer microscope theodolites display both vertical and horizontal angles in a reading eyepiece next to the telescope (sections 3.3 and 3.4).

Instruments with a spirit level for setting the height index are subdivided into instruments with a level at the telescope standards and with a level at the index (Figs. 10.2.1 and 10.2.2).

Construction theodolites and older instruments usually have a level at the standards or an alidade level. This level is rigidly fastened onto the standard or the alidade (except for small adjustment possibilities) and can only be centred with the aid of footscrews. The small theodolite 80 by Zeiss, Jena is an example of such a design.

Engineer's and precision theodolites normally have the spirit level directly connected to the height index. With such instruments the height index can be rotated within limits around the horizontal axis of the theodolite without changing the position of the vertical axis of the instrument. This rotation of the height index is monitored by the spirit level such that the correct vertical angle is read when the level bubble is centred, provided the instrument is properly adjusted.

Instruments with height index bubbles are not only more accurate but also easier to operate than instruments with alidade levels. For optical theodolites the height index is replaced by a prism which transmits the reading into the reading eyepiece.

Theodolite telescopes are sometimes equipped with a reversible level bubble (Fig. 10.2.3). Vertical angles can be measured with the aid of this bubble by first setting the telescope into the horizontal plane and reading the vertical circle. The amount read is then subtracted from the reading obtained when sighting to the target.

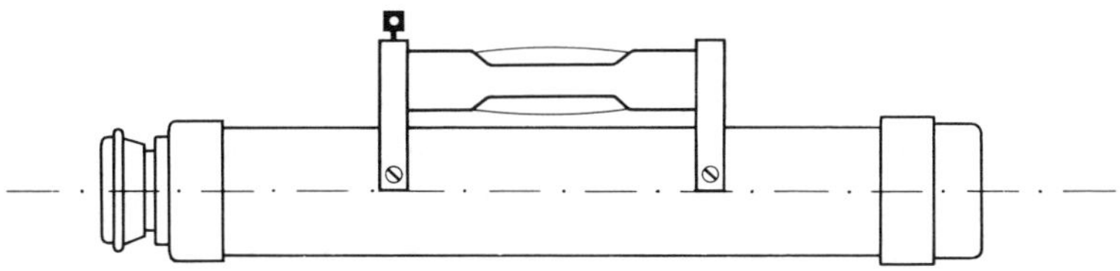

Fig. 10.2.3 Telescope bubble

10.2.3 The Automatic Height Index

This has an optical element (*the compensator*) like the automatic level, which is under gravity influence and thus monitors the reading device. As in Fig. 10.1.1, the vertical angles or zenith distances are therefore referred to gravity, while the vertical axis and the microscope parts attached to it cannot be brought accurately enough into the plumb line with the aid of the platen level alone. Therefore, the remaining inclination of the vertical axis – or its component in the direction of the line of sight – has to be compensated before correct vertical angles are obtained.

Either a freely swinging pendulum or liquid compensators are used for this purpose. One distinguishes here between mechanical (physical) pendula, which directly lead to the direction of gravity (section 10.2.3.1) and pendula which show the residual inclination between the vertical axis and the gravity direction magnified or reduced instead of at 1 : 1 and thus not gravitate into the plumb line (section 10.2.3.2). Different solutions also exist for the liquid compensators (section 10.2.3.3). In the following, a few examples are given for the most important types with the compensator illustrations being largely schematic for better understanding. If two illustrations of the same compensator are shown, the left one shows the compensator with the plumb vertical axis. If only one illustration is given, a slightly inclined case is illustrated.

10.2.3.1 Free Swinging Pendula

These compensators are found in System Askania TU theodolites (Fig. 10.2.4). In it the microscope optics are attached to solid supports, while a prism (Fig. 10.2.4, lower right) is pendulum suspended and monitors the imaging of the graduated circle onto the focal plane of the microscope.

The FT 1 A and FTRA Fennel instruments (Fig. 10.2.5), which since 1978 are built by the optical company Theis in Wolzhausen, have a similar arrangement. Again a pendulum prism is suspended within the optical train of the microscope, while the other elements and objective lenses are solidly attached.

The K1-A engineer's theodolite by Kern also has a pendulum without angular magnification. As shown in the left part of Fig. 10.2. 6, which illustrates the situation for a plumb vertical axis, the reading location is imaged onto the height index by means of the reading optics with a double objective lens. The upper of the two objective lenses is attached to the lower part of a pendulum which has the same length as the radius of the vertical circle and which is suspended as shown on the figure.

In the right part of Fig. 10.2.6, the vertical angle α is to be measured with the vertical axis inclined by β to the plumb line. Since the pendulum elements come to rest below the point of suspension, and since the rays are parallel between the two objective lens elements, the section of the circle is imaged at the correct location of the reading index.

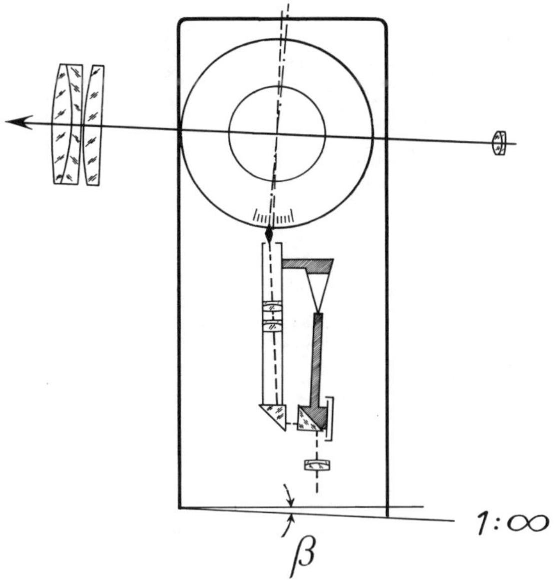

Fig. 10.2.4 *Askania* TU system

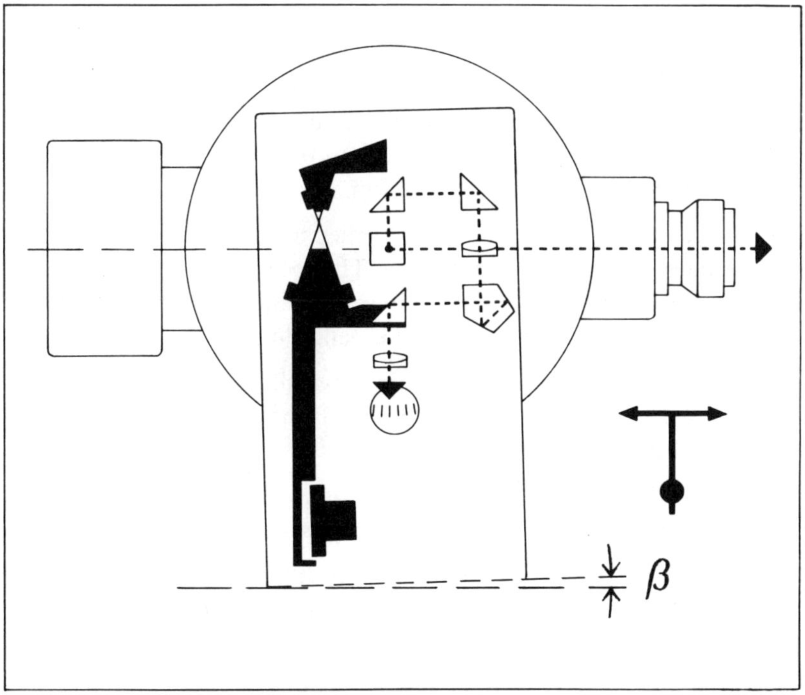

Fig. 10.2.5 *Fennel* FTIA and FTRA system

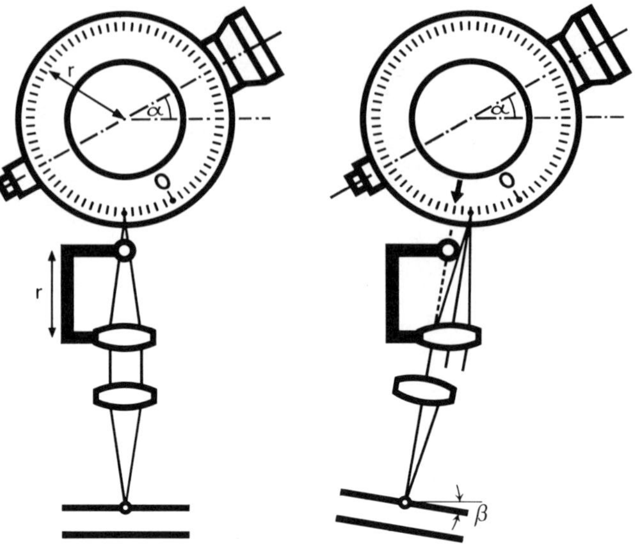

Fig. 10.2.6 *Kern* K1-A

10.2.3.2 Pendula with Angular Magnification

Pendula with angular magnification $n > 1$ are used as compensators of the Zeiss-Oberkochen theodolites. As illustrated in the schematic Figs. 10.2.7 and 10.2.8, the V-shaped joint quadrilateral forms the basis for the Th4 and RTa4 with the X-shaped one for the Th2. They work like mechanical lever transmissions, where their moving parts settle to a larger inclination than the residual inclination between vertical axis of theodolite and plumb direction.

Since, in both cases, the centre of gravity of the movable parts is above the rotational centre, these parts resemble a metronome. The location of the centre of gravity (S), which determines the factor n of the angular magnification, can be changed in height with the aid of a weight shifted along the pendulum rod.

When the rays pass through the parallel plates within the comparator (shown as white surfaces in Figs. 10.2.7 and 10.2.8) the image of the circle graduation is shifted such that the residual inclination of the vertical axis is compensated.

Compensators with angular magnification (n up to 30 times) have a longer period; the frequency is lower, and the pendulum is thus less sensitive to disturbing oscillations.

When computing the theoretical pendulum length from the period as for a physical pendulum, the Th2 compensator results in a theoretical pendulum length of approximately 2 metres.

For the 020 theodolite by Zeiss, Jena the whole microscope with the exception of the reticule is fastened onto a spring suspended pendulum with angular magnification

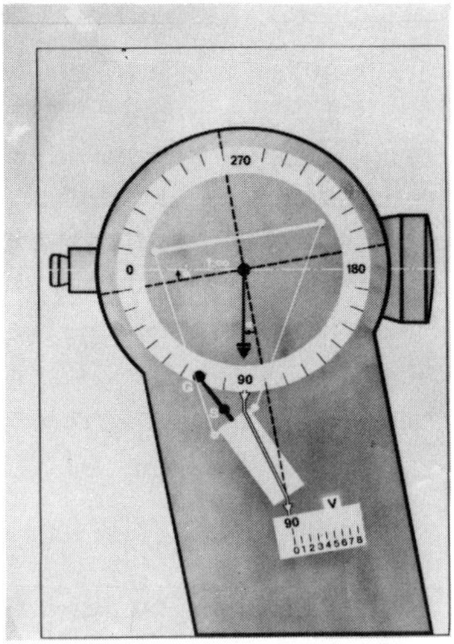

Fig. 10.2.7 *Zeiss*, Oberkochen Th4 and RTa4

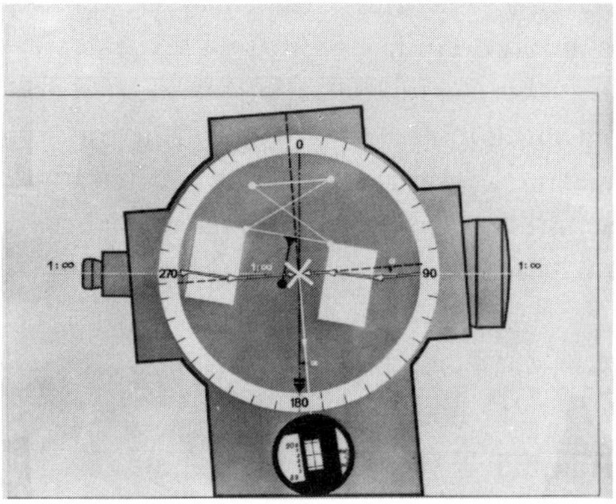

Fig. 10.2.8 *Zeiss*, Oberkochen Th2

$n < 1$. As illustrated in the left part of Fig. 10.2.9, the objective lens O and the two prisms P_1 and P_2 are mounted onto a plate P which is attached to one of the telescope standards by a jointed spring suspended at Z. The reticule with height index is attached to the horizontal axis K and imaged through the two prisms and the objective lens O onto the graduation of the vertical circle. If the instrument is correctly set up, the centre line of the plate coincides with the plumb line LL.

The right part of Fig. 10.2.9 shows the situation with the vertical axis inclined against the plumb line by the angle α; the suspension point Z of the pendulum is moved from LL to the line $B'B'$ and the plate rotates around M, just below Z.

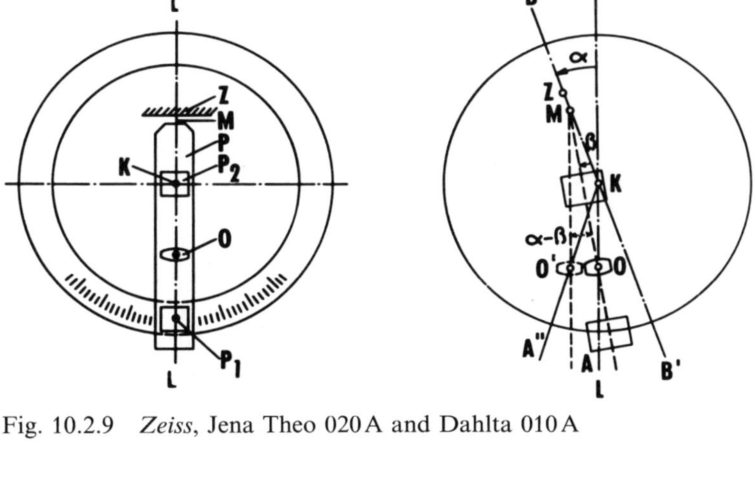

Fig. 10.2.9 *Zeiss*, Jena Theo 020A and Dahlta 010A

If the plate would be influenced only by gravity, the objective lens would swing to position O' and the reticule of the height index K would be projected to A''. This reticule in K, however, has to be imaged at point A lying on LL to give the correct reading. Therefore, the objective lens has to lie in LL or, differently expressed, the pendulum cannot swing by the total amount α in Z, but by the slightly smaller angle β. For this the flexibility of the jointed spring has to be reduced, which is achieved by selecting material with a specific module of elasticity. Furthermore, the moment of inertia of the cross section of the spring, the weight of the swinging parts, and the separation between the centre of gravity and the point of suspension have to be properly chosen.

10.2.3.3 Liquid Compensator

A liquid compensator is used in the T1A theodolite by Wild, Heerbrugg. On its way to the reading index, the image of the vertical circle passes through a container filled with transparent silicon oil. If the vertical axis of the instrument is perfectly plumb, then the bottom of the container and the surface of the silicone oil act as a plane

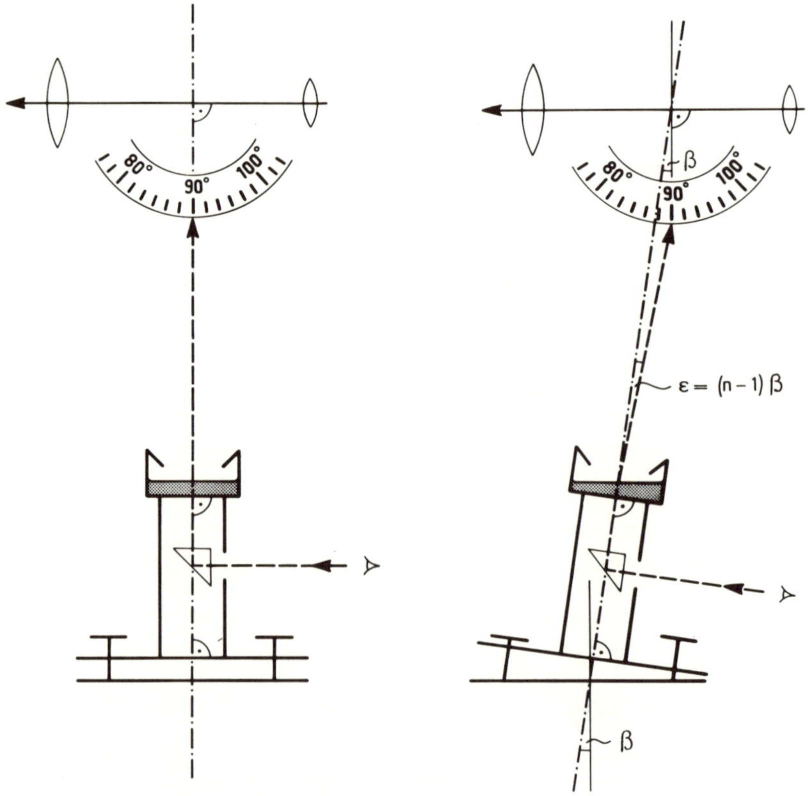

Fig. 10.2.10 *Wild* T1A liquid compensator

parallel plate where the rays penetrate without refraction. If, however, the vertical axis is inclined by the angle β (Fig. 10.2.10, right part), then the liquid forms a wedge with wedge angle β. Based on a refractive index n for silicone oil, rays which pass through are deflected by an angle $(n-1)\beta$, which compensates the influence on the vertical circle reading caused by the inclination of the vertical axis.

The liquid compensator which was developed by Kern for the DKM 2-A precision theodolite is very interesting. Its main characteristic is the utilization of total reflection which occurs at the boundary between an optically denser material and its less dense neighbouring material. As illustrated in the left part of Fig. 10.2.11, the liquid is in a closed glass container mounted on the standard. Below its bottom surface there is a symmetric trapezium prism. A ray entering through a pentagon prism and a lens hits the trapezium prism perpendicular to its side. With liquid and glass having practically identical coefficients of refraction, the ray is deflected upward into the liquid and totally reflected at its surface which is always horizontal.

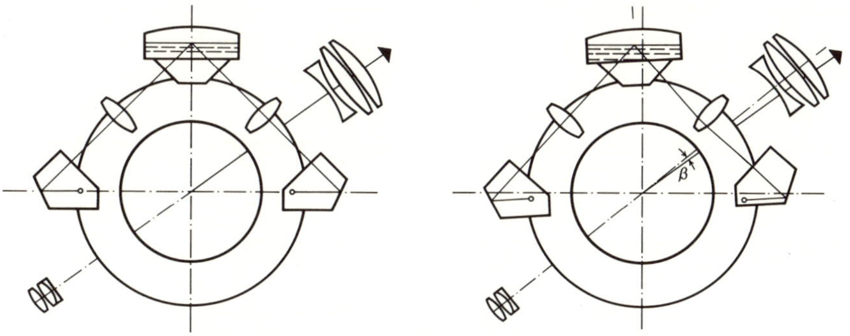

Fig. 10.2.11 *Kern* DKM 2–A liquid compensator

When the vertical axis is inclined by a small angle β, as shown in the right part of Fig. 10.2.11, all optical elements fastened to the standard necessarily follow the motion. However, at the point of total reflection the angle between incident and emerging rays is now $90^\circ + 2\beta$ instead of 90°. Thus, the ray coming from reading mark I, which consists of a line only, is imaged in the area of reading mark II with a shift of 2β. Therefore, the inclination β is compensated if the mean of both readings is taken.

The operational range of automatic height index compensators is about ± 0.005 gon ($\pm 15''$), which is significantly smaller than the range of height index bubbles. Furthermore, automatic height index compensators have a nonlinearity in the per mille magnitude. Therefore the theodolite has to be very carefully levelled for precise vertical angle measurements.

10.3 Measuring of Vertical Angles

10.3.1 Measuring Arrangement

Unless specifically stated differently, it is assumed from now on that the theodolite has a height index bubble or a compensator, and a vertical circle subdivided into 400 gon (360°) such that zenith distances are obtained. The telescope position with the vertical circle at the left side as viewed from the observer is called "direct", or position I, the other one is referred to as "reverse" or position II. If two reading indices are present, they are called index A and B, with index A being nearest to the observer in the direct telescope position. When vertical angles are required, they have to be calculated using zenith distances.

Vertical angles, like horizontal angles, are observed in direct and reverse to eliminate instrument errors. While the circle remains stationary and the indices move for horizontal angle measurements, the situation is reversed for vertical angles. Furthermore, two directions have to be measured for the determination of a horizontal angle, while the vertical angle is obtained by measuring one direction, namely by setting and reading in direction of the target. The other side, namely the reference direction is the horizontal direction for vertical angles or the zenith direction for zenith distances. Both are determined at the instrument with the aid of spirit levels or compensators.

10.3.2 Calculation of Zenith Distances and Index Correction

In order to determine zenith distances, it is necessary that the height index is oriented with respect to the telescope such that for sighting into the zenith direction a reading of $Z = 0$ is obtained.

Unfortunately, the position of the collimation axis with respect to the vertical circle is somewhat changed when adjusting the cross hairs. Furthermore, a height index, which is monitored by levelling a height index bubble or a platen bubble, suffers a small shift with respect to the vertical axis when the bubble is adjusted. Primarily, these two influences cause a reading of $z = \zeta$ instead of $z = 0$ when sighting to the zenith. The angle ζ which affects all readings of the vertical circle in the same manner is called the *index error*. It is basically a zero error which theoretically can assume any value, and which also occurs with instruments with compensators even though its causes are somewhat different.

In practice, the direction Z to the zenith is not known. However, it is possible to determine the zenith distance z together with the index error ζ by measuring in direct and reverse. When a zenith distance is measured in one telescope position, it is read along increasing numbers of the graduation (Fig. 10.3.1, reading A_{I} in left part), while it is read against the positive sense of the graduation in the other telescope position (reading A_{II} in right part of Fig. 10.3.1).

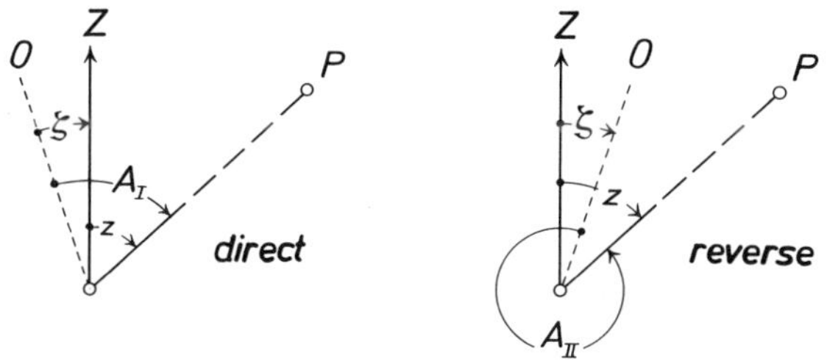

Fig. 10.3.1 Zenith distance measurement in direct and reverse telescope positions

This leads to the following relationships:

$$\text{left part} \quad z + \zeta = \quad + A_{\text{I}} \tag{10.3.1}$$

$$\text{right part} \quad z - \zeta = 360^\circ - A_{\text{II}} \,. \tag{10.3.2}$$

Forming equations (10.3.1) + (10.3.2) and (10.3.1) – (10.3.2) we obtain the equations for the zenith distance and the index error as

$$2z = (360^\circ + A_{\text{I}}) - A_{\text{II}} \tag{10.3.3}$$

$$2\zeta = (A_{\text{I}} + A_{\text{II}}) - 360^\circ = -2v_z \,, \tag{10.3.4}$$

where $v_z = -\zeta$ is the *index correction* which is preferred by many practitioners. Both equations can be evaluated in two ways:

1st way: Compute z according to equation (10.3.3) and, as a check, $\zeta = -v_z$ with equation (10.3.4). The parentheses in equations (10.3.3) and (10.3.4) point to the simplest mathematical approach.

2nd way: Form the sum $(A_{\text{I}} + A_{\text{II}})$ and bring it to 360° by correcting A_{I} and A_{II} by half of the difference to 360° each. These corrected readings provide the desired zenith distance z and its complement to 360° as well as the index correction v_z which is half the difference (see numerical example in section 10.3.5)

Therefore, a zenith distance is determined as follows: After levelling the theodolite with the platen level, the vertical circle is brought to the left (direct telescope position) and the centre horizontal cross hair line is set onto the target. After levelling the index bubble or letting the compensator come to rest, reading A_{I} is taken. The same steps are repeated for the reverse telescope position, and reading A_{II} is taken. Then z and v_z are computed using either equations (10.3.3) and (10.3.4) or else the adjustment to 360°. Although v_z is not really required, it is determined, because v_z remains nearly equal for all vertical angle measurements taken at the same instrument station, which provides a measuring check.

Generally, the following has to be considered: If the vertical circle is only read at one index, as is the case for most modern stadia theodolites, a possible graduation eccentricity of the vertical circle is not eliminated when measuring in direct and reverse. For highest accuracy it is therefore advisable to work with reciprocal sights.

10.3.3 Adjusting the Index Error

For ease of calculations, it is desirable to have a small index error. For secondary measurements, which are performed in one telescope position only, it should be completely corrected. The following procedure is utilized.

For instruments with an index bubble, z is first determined by measuring to a well defined target. Afterwards the target is again sighted – for simplicity's sake in the direct mode – and the index is moved with the fine motion screw until the theoretically correct reading appears. The index bubble is now levelled in this position with the aid of its adjustment screws. For a theodolite with coincidence micrometer, the desired reading is set at the micrometer drum and the graduation lines of the vertical circle are coincided with the aid of the fine motion screw. Then the bubble is adjusted.

Even for theodolites with automatic height index, the desired reading is first determined. Then the cross hair is set onto the target while the desired reading is set with the adjustment screws, keeping the cross hair on the target (Askasia and Wild instruments); or else the desired reading is set and the cross hair is brought onto the target with the aid of adjustment screws (Zeiss, Jena instruments).

For theodolites with a platen level, again the desired reading is determined according to section 10.3.2. To eliminate the index error, first the platen bubble is carefully levelled. Then the desired reading is set at the vertical angle and the cross hair is set onto the target with the aid of the reticule adjustment screws, or else the cross hair is set onto the target, and the desired reading is obtained by shifting the index.

10.3.4 Accuracy of Vertical Angle Measurements

The standard deviation of an observed zenith distance can be computed using multiple determinations of ζ. Based on the law of error propagation (section 1.4.4) we have according to equations (10.3.3) and (10.3.4):

$$\sigma_\zeta^2 = \tfrac{1}{4}(\sigma^2 + \sigma^2) = \sigma_z^2$$

where σ is the standard deviation of an observation in one telescope position.

If one observes s zenith distances in n full sets each, at one station, then $\bar{\zeta}$ is computed from the $n \cdot s$ observations each providing $v_z = -\zeta$. With $v_\zeta = \bar{\zeta} - \zeta$, the standard deviation of an index error ζ and a zenith distance z measured in direct and reverse becomes according to section 1.4.5:

$$\sigma_\zeta = \pm\sqrt{\frac{\Sigma v_\zeta \cdot v_\zeta}{s \cdot n - 1}} = \sigma_z\,. \tag{10.3.5}$$

This measuring arrangement provides $n \cdot s$ determinations of ζ, and n determinations of each zenith distance. Therefore the standard deviations of the mean values are

$$\sigma_{\bar{\zeta}} = \pm \frac{\sigma_\zeta}{\sqrt{n \cdot s}}; \qquad \sigma_{\bar{z}} = \pm \frac{\sigma_z}{\sqrt{n}}. \tag{10.3.6}$$

The accuracy of vertical angle measurements depends less on the quality of the vertical circle and of the reading device but rather on how carefully the height index bubble is levelled, or the compensator is brought to its rest position. As a rough average, a standard deviation of ± 2 mgon (6.5″) can be expected for a vertical angle measured direct and reverse with an engineer's theodolite. For theodolites with only one reading index, an additional systematic error of about ± 3 mgon (10″) can be caused by eccentricity of the vertical circle. For higher accuracy requirements, it is therefore necessary to either determine this systematic error and to eliminate it, or else to use a precision theodolite with coincidence microscope (section 3.3.3) and thus also significantly reduce the random error.

Example:

Observations of the vertical circle and computations according to the second way mentioned in section 10.3.2:

Instrument: Electronic theodolit of medium accuracy

Target	Set	Telescope Position	Vert. Circle Reading	ζ	z $400 - z$	mean z	v_ζ	v_ζ^2
TP *B*	1	I	87°38′38″	−6.5″	87°38′31″	87°38′32″	−2.5″	6.25″
		II	272°21′35″	−6.5″	272°21′29″			
			360°00′13″		360°00′00″			
	2	I	87°38′41″	−8.0″	87°38′33″		−1.0″	1.00″
		II	272°21′35″	−8.0″	272°21′27″			
			360°00′16″		360°00′00″			
TP *C*	1	I	88°05′57″	−11.5″	88°05′46″	88°05′45″	+2.5″	6.25″
		II	271°54′26″	−11.5″	271°54′14″			
			360°00′23″		360°00′00″			
	2	I	88°05′54″	−10.0″	88°05′44′		+1.0″	1.00″
		II	271°54′26″	−10.0″	271°54′16″			
			360°00′20″		360°00′00″			
								14.5″

Mean: $\bar{\zeta} = -9.0''$
or $v_z = +9.0''$

$$\sigma_\zeta = \sqrt{\frac{14.5}{2 \cdot 2 - 1}} = 2.2''$$

$$\sigma_{\bar{\zeta}} = \frac{2.2}{\sqrt{2 \cdot 2}} = 1.1''; \qquad \sigma_{\bar{z}} = \frac{2.2}{\sqrt{2}} = 1.6''$$

10.4 Trigonometric Heights Over Short Ranges

Short range in this context means distances, where the influence of the earth's curvature and refraction can be neglected.

10.4.1 Determination of the Height of a Tower Using an Auxiliary Horizontal Triangle

Given is the mean sea level elevation H_a of a point A; wanted is the elevation H_t of a nearby tower (Figs. 10.4.1 and 10.4.2). The problem consists of two steps:

1st step: Since the projection of the top of the tower usually falls inside the building, it is necessary to indirectly determine the horizontal projections of the distance from the instrument to the top of the tower. As illustrated in Fig. 10.4.1, a base $AB = b$ is laid out near the tower. The horizontal angles α and β are observed at A and B. This provides the needed distance as

$$s = b\frac{\sin\beta}{\sin(\alpha+\beta)}\,. \tag{10.4.1}$$

As a check for s, a second base c can be used.

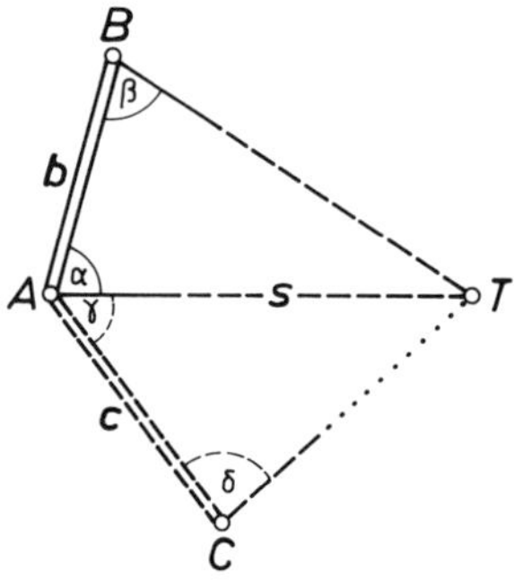

Fig. 10.4.1 Auxiliary horizontal triangle

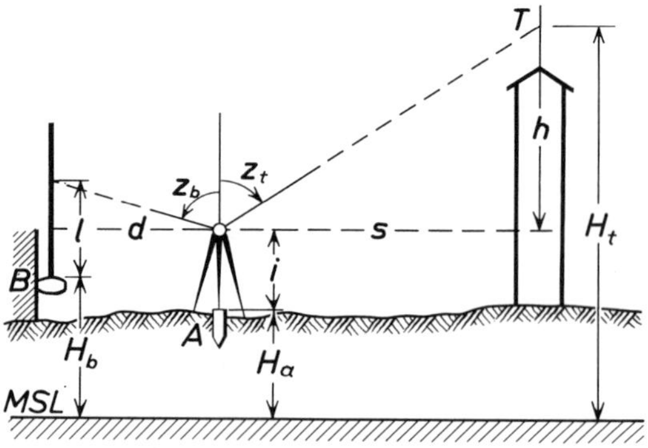

Fig. 10.4.2 Determination of the height of a tower

2nd step: After measuring the instrument height i (between ground point and horizontal axis), the elevation of the tower becomes

$$H_t = H_a + i + s\cot z_t\,. \tag{10.4.2}$$

If H_a is unknown, but the elevation H_b of a bench mark B at a distance d is available, then the instrument horizon $(H_a + i)$ is obtained with the aid of a levelling rod. The rod is either read with centred telescope bubble, or else the value l is set at the rod and z_b is measured (Fig. 10.4.2). In this case we get:

$$H_t = H_b + l - d\cot z_b + s\cot z_t\,. \tag{10.4.3}$$

As a check, it is possible to transfer the height from another point. The shape of the auxiliary horizontal triangle is to be selected according to the guidelines given in section 6.1.2. The length of the side towards the tower should be such that z_t does not become less than 70 gon ($\sim 60°$).

10.4.2 Accuracy of Trigonometric Heights over Short Range

Prefix: SI – units for accuracy estimations. Of the SI units described in section 1.3.2, the mathematically most important one is the radian, which is defined as the ratio of arc b to radius r of a circle around the vertex of an angle. Its unit 1 rad equals $b/r = 1$. If an angle is given without dimensions just as a number, especially as a multiple or fraction of π, then the angle is in radians.

To convert to gons, the following holds:

$$1\ \text{rad} = \frac{200}{\pi}\ \text{gon}; \qquad 1\ \text{gon} = \frac{\pi}{200}\ \text{rad} = 15.7\ \text{m rad}$$

or to degrees:

$$1\ \text{rad} = \frac{180°}{\pi}; \qquad 1° = \frac{\pi}{180}\ \text{rad}\,.$$

When differentiating for accuracy evaluations, it is advisable to use radians for the differentials $\mathrm{d}\alpha$ and the standard deviations σ_α. Afterwards the results are converted to gon or degrees of arc using

$$\mathrm{d}\alpha(\text{rad}) = \mathrm{d}\alpha(\text{gon})\frac{\pi}{200}; \quad \mathrm{d}\alpha(\text{gon}) = \frac{200}{\pi}\mathrm{d}\alpha(\text{rad})$$

$$\mathrm{d}\alpha(\text{rad}) = \mathrm{d}\alpha(°)\cdot\frac{\pi}{180}; \quad \mathrm{d}\alpha(°) = \frac{180}{\pi}\cdot\mathrm{d}\alpha(\text{rad})$$

where $\mathrm{d}\alpha(\text{rad})$ means "$\mathrm{d}\alpha$ in radians" or $\mathrm{d}\alpha(\text{gon})$: "$\mathrm{d}\alpha$ in gon", etc.

By giving the radian a prime position, the formerly often used conversion factor ϱ becomes unnecessary. With modern pocket calculators it is simpler to input $200 : \pi$ or $180 : \pi$ than all the digits of ϱ.

The accuracy of trigonometric heighting depends on whether the influences of the earth's curvature and refraction are in a favourable ratio to the accuracies of vertical angle and distance. As mentioned in section 10.5.1, the influence of the earth's curvature and refraction amount to about 0.7 mm, 2.8 mm, and 18 mm for 100 m, 200 m and 500 m, respectively. To estimate the influences of measuring inaccuracies, equation (10.1.1 a) is differentiated. With $\Delta H = H_2 - H_1$ and d_z in radians we get

$$\mathrm{d}\Delta H = \cot z \,\mathrm{d}s - \frac{s}{\sin^2 z} \cdot \mathrm{d}z + \mathrm{d}i - \mathrm{d}t \tag{10.4.4}$$

which leads to the following standard deviations

$$\sigma_{\Delta H}^2 = (\cot z \,\sigma_s)^2 + \left(\frac{s}{\sin^2 z}\sigma_z\right)^2 + \sigma_i^2 + \sigma_t^2 \,. \tag{10.4.5}$$

If one assumes that the zenith distance was measured in two independent sets, then according to section 10.3.4 a measuring accuracy of $\sigma_z = 2\ \mathrm{mgon}/\sqrt{2} \approx \pm 1.4\ \mathrm{mgon}$ can be expected. Then σ_s is to be chosen such that the first term has the same magnitude as the second one. If one selects $\sigma_s = \pm 0.0015\sqrt{s}$ then the following amounts in millimetres are obtained:

z \ s [°] [gon]		$\cot z\,\sigma_s$ 100 m	200 m	500 m	$\frac{s}{\sin^2 z}\sigma_z$ 100 m	200 m	500 m
81	90	2.4	3.4	5.4	2.4	4.8	12.0
63	70	7.6	10.8	17.1	2.9	5.9	14.7
45	50	15.0	21.2	33.6	4.7	9.4	23.5

According to this table, observations with $\sigma_z = 1.5$ mgon and $\sigma_s = 0.0015\sqrt{s}$ result in errors of the same magnitude. Only for very steep sites it is necessary to measure the distances more accurately. If the zenith distances are measured more or less accurately, then the accuracy of the distance measurements has to change at the same ratio. Furthermore, i and t have to be determined with the appropriate care.

Introducing the numerical values for σ_z and σ_s into equation (10.4.5) and using ± 2 mm for σ_i and σ_t, then the following standard deviations in millimetres can be expected for the observed elevation difference:

z \ s [°] [gon]		100 m	200 m	500 m
81	90	4.4	6.5	13.4
63	70	8.5	12.7	22.7
45	50	16.0	23.4	41.1

$\sigma_z = 4.9'' = 23\ \mu\mathrm{rad}$
$\sigma_s = 1.5\sqrt{s_m}\ \mathrm{mm}$
$\sigma_i = 2\ \mathrm{mm}$
$\sigma_t = 2\ \mathrm{mm}$

For distances up to 200 m, these amounts are roughly four times the effects of the earth's curvature and refraction, while both are of the same magnitude for 500 m.

Therefore, short range is considered up to about 250 m. This distance, however, is rarely exceeded for determinations of tower or building heights and for trigonometric levelling.

Therefore, the values given in the table above serve directly for estimating the feasible accuracies for trigonometric heighting as discussed in section 10.4.1.

10.5 Trigonometric Heighting Over Long Ranges

10.5.1 Earth Curvature and Refraction

As illustrated in Fig. 10.5.1, it is necessary to use circular arcs for the distance AD and the line of sight AB for long ranges. To simplify the treatment of this, instrument height i and target height t are omitted in Fig. 10.5.1.

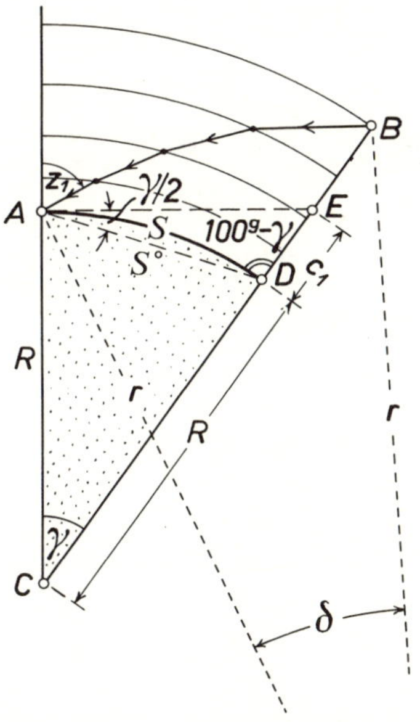

Fig. 10.5.1 Earth curvature and refraction influence

Due to the *earth's curvature*, a tangential plane to the earth in A intersects a normal to the earth's surface in B at a distance c_1, above the earth. It should be noted in Fig. 10.5.1 that the angle DAE equals half of the centre angle γ which is $S/2R$. Then with $S = S^0$ and $\sin\gamma/2 \approx S/2R$ we can apply the sine rule and obtain

$$c_1 = \frac{S\sin\gamma/2}{\sin(\pi/2-\gamma)} = \frac{S\sin\gamma/2}{\cos\gamma} \approx \frac{S\gamma/2 + \ldots}{1+\ldots} \approx \frac{S^2}{2R} + \ldots \,. \qquad (10.5.1)$$

With this formula the following values are obtained:

for	100 m	200 m	500 m	1000 m	5 km	10 km
$c_1 =$	0.8 mm	3.2 mm	2.0 cm	7.9 cm	1.96 m	7.9 m.

Refraction is primarily caused by the fact that the density of the air decreases with increasing elevation. If one considers the air as being a series of layers with decreasing density (Fig. 10.5.1), then a ray originating at A is continuously refracted towards the denser medium. The curve generated in this manner can be considered in a first approximation as a circular arc with the radius r. Experience has provided an average of $r \approx 8\,R$. It is, however, customary in practice to use the *refraction coefficient* $k = R/r$ instead of r.

For easier explanation of the circumstances, the upper part of Fig. 10.5.1 has been redrawn in Fig. 10.5.2. It should be noted that γ and the centre angle δ for the arc AB are both extremely small angles. It is further apparent that the zenith distances deviate little from 100 gon, especially in flat and lightly hilly terrain. Finally, it is sufficiently accurate to set $R \approx R + H$.

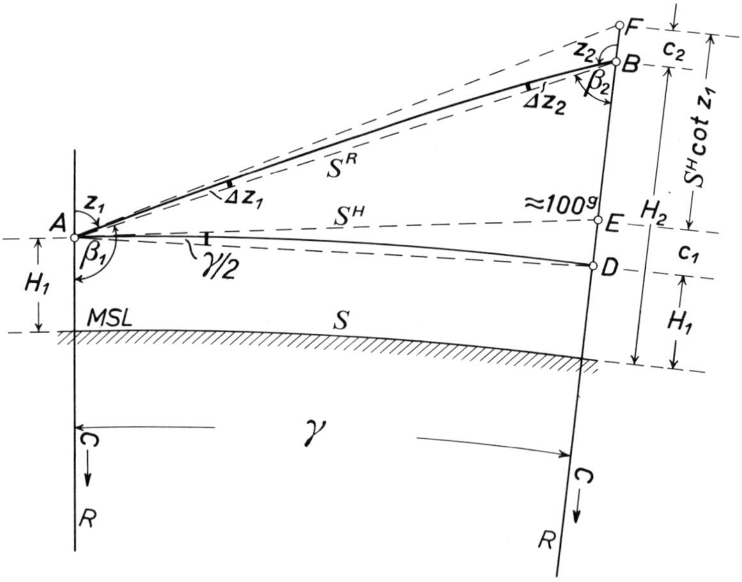

Fig. 10.5.2 Refraction in trigonometric heights

Thus without loss in accuracy the following is considered equal: $AD \approx AE \approx AB \approx AF \approx S$.

Thus, similar to equation (10.5.1) we get with $\Delta z_1 = {}^1/_2\,\delta$:

$$c_2 \approx \frac{S^2}{2r} \approx \frac{kS^2}{2R}\,. \tag{10.5.2}$$

The values for c_2 are only about $^1/_8$th of the respective values for c_1 and have the opposite sign as evident in Fig. 10.5.2. Thus the effect of the earth's curvature is reduced by roughly $^1/_8$th because of refraction.

For $r \approx 8R$, the refraction coefficient k is approximately 0.13. k, however, is dependent on the air density, which in turn is a function of air pressure, temperature, and humidity. k is further influenced near the ground by different terrain forms, vegetation and by other rather complex influences. k is therefore subjected to regional and time variations which even under favourable conditions may amount to ± 0.04. The uncertainty is even worse for low sightings above water surfaces, forests and industrial plants. In alpine terrain, k remains rather stable for sights well above the ground. Even during one day the variations of k are significant, with the smallest changes occurring in the afternoon. With higher clearance above terrain or treetop level, the variations become smaller.

10.5.2 Elevation Differences Using One-Sided Zenith Angles

From Fig. 10.5.2 it is apparent that

$$\Delta H = H_2 - H_1 = S^H \cot z_1 + c_1 - c_2 \,. \tag{10.5.3}$$

By inserting the values from equations (10.5.1) and (10.5.2) and adding instrument height i and target height t as in equation (10.1.1a) we get

$$\Delta H = H_2 - H_1 = S^H \cot z + (1-k)\frac{(S^H)^2}{2R} + i - t \,. \tag{10.5.4}$$

With $k = 0.13$ and $R \approx 6370$ km, and using S^H in kilometres in the correction term, e.g. equation (10.5.4) becomes more suited for pocket calculators as

$$\Delta H = H_2 - H_1 = S^H \cot z + 0.068\,(S^H_{\text{km}})^2 + i - t \,. \tag{10.5.5}$$

For highest accuracy, the whole computation is referred to the mean height $H_m = {}^1/_2(H_1 + H_2)$ and one considers in the second term of equation (10.5.5) that S^R should have been used instead of S^H in equation (10.5.3). This leads to the complete formula for trigonometric heighting for long ranges:

$$H_2 - H_1 = S^H\left(1 + \frac{H_m}{R}\right)\cot z - (1-k)\frac{(S^H)^2}{2R\sin^2 S^H} + i - t \,. \tag{10.5.6}$$

In this case, the $H_2 - H_1$ is the distance of point B from the sphere through A (with radius $r + H_1$) along its normal. For reasons mentioned in section 10.5.1, it is advisable to observe one-sided zenith distances in the afternoon.

Using the slope distance S^R, equation (10.5.4) changes to

$$\Delta H = H_1 - H_1 = S^R \cos z + 1 - k)\frac{(S^H)^2}{2R} + i - t \tag{10.5.4a}$$

with

$$S^H \approx S^R \sin z\,.$$

10.5.3 Elevation Differences Using Reciprocal Zenith Angles

In triangle ABC (Fig. 10.5.2), the angle BAC may temporarily be referred to as β_1, and the angle CBA as β_2. Then the plane tangent rule provides

$$\frac{(R+H_2)-(R+H_1)}{(R+H_2)+(R+H_1)} = \frac{\tan{}^1/_2(\beta_1-\beta_2)}{\tan{}^1/_2(\beta_2+\beta_2)}\,. \qquad (10.5.7)$$

Because of

$$\beta_1 = 200\text{ gon} - (z_1 + \Delta z_1);\quad \beta_2 = 200\text{ gon} - (z_2 + \Delta z_2)\,,$$
$$\tan{}^1/_2(\beta_1+\beta_2) = \cot{}^1/_2\gamma$$

we get

$$H_2 - H_1 = 2R\left(1 + \frac{H_2+H_1}{2R}\right)\frac{\tan{}^1/_2(z_2+\Delta z_2 - z_1 - \Delta z_1)}{\cot{}^1/_2\gamma}\,.$$

Again we set $H_m = {}^1/_2(H_2 + H_1)$ and obtain with $\tan\gamma \approx S^H/R$ for the small angle γ:

$$2R:\cot\gamma/2 = 2R\cdot\tan\gamma/2 \approx S^H\,. \qquad (10.5.8)$$

If, furthermore, the light path is considered to be symmetrical, then $\Delta z_1 = \Delta z_2$ and the final formula for trigonometric heighting with reciprocal sights becomes

$$\Delta H = H_2 - H_1 = S^H\left(1 + \frac{H_m}{R}\right)\tan\frac{z_2 - z_1}{2}\,. \qquad (10.5.9)$$

Using slope distances S^R the corresponding equation is found by calculating the half difference of the two reciprocal height differences ΔH_{21} (equation (10.5.4a)) and assuming $k_1 = k_2$:

$$\Delta H = (\Delta H_{12} - \Delta H_{21})/2 = \frac{S^R}{2}(\cos z_1 - \cos z_2) \qquad (10.5.9a)$$

The assumption of a symmetric light path is only justified when the terrain profiles at A and B are relatively uniform, when the atmospheric conditions at both sides correspond, and when the reciprocal zenith angles are observed at the same time.

Only under these conditions are the influences of the earth's curvature and refraction cancelled for reciprocal sights according to equation (10.5.9).

Generally the most favourable observation conditions exist when the instruments at A and B are set up several metres above the terrain or the tree top level, and when the measurements are taken on overcast days or at hours prior to isothermal conditions in the evening.

10.6 Refraction Coefficient from Reciprocal Sights

Under the conditions mentioned at the end of section 10.5.3, it is possible to compute k from the observed zenith angles. In triangle ABC of Fig. 10.5.2 we get, with $\pi = 200$ gon:

$$z_1 + \Delta z_1 + z_2 + \Delta z_2 = \pi + \gamma\,. \qquad (10.6.1)$$

Considering the light path AB as circular arc and using equation (10.5.8) we get:

$$\Delta z_1 = \Delta z_2 \approx \frac{S}{2r} \approx \frac{S \cdot k}{2R} \approx {}^1/_2 \gamma \cdot k\,.$$

After inserting into equation (10.6.1):

$$z_1 + z_2 + \gamma \cdot k = 180^\circ + \gamma\,.$$

Or, since z_1 and z_2 are measured in degrees

$$\gamma(k-1) = (180^\circ - z_1 - z_2)$$

and finally with

$$\gamma = \frac{S}{R}\frac{180^\circ}{\pi}$$

$$k = 1 - \frac{z_1 + z_2 - 180^\circ}{180^\circ/\pi} \cdot \frac{R}{S}\,. \qquad (10.6.2)$$

This formula provides reasonable results only under the favorable conditions mentioned in section 10.5.3.

Numerical Examples for Sections 10.5.2 to 10.6

– One-sided zenith distance, according to equation (10.5.5):

Observed: $z_1 = 88^\circ 48' 27''$; $i = 1.42$ m; $t = 6.10$ m; $S = 1578.1$ m

$$\Delta H = 1578.1 \cdot 0.020\,816 + 0.17 + 1.42 - 6.10 = 28.34 \text{ m}$$

– Reciprocal zenith angles, according to equations (10.5.9) and (10.6.2):

$z_1 = 88^\circ 58' 39''$; $z_2 = 91^\circ 02' 09''$; $S = 1578.1$ m

$$\Delta H = 1578.1 \cdot 0.017\,965 = 28.35 \text{ m}$$

$$k = 1 - \frac{0.0148}{180^\circ/\pi}\,\frac{6370}{1.58} = 0.063$$

10.7 Reduction of Zenith Angles to Station Origin

For trigonometric heighting over long ranges, often towers are used as station and target. It is often necessary to use different theodolite set-ups for observations to different directions from one tower. Also, if the tower is the target, various targets at different heights may be sighted from different stations. In such cases, it is advisable to reduce all observations to a common station origin, usually the horizontal axis of the theodolite set up with the most observations. This reduction is equivalent to centring eccentric horizontal angles. The formulae are best explained with the aid of Figs. 10.7.1 and 10.7.2: B_1 and B_2 represent two auxiliary observation stations, while z_1 and z_2 represent two auxiliary targets. Meaning and signs for i and t are evident in the figures. The zenith angle observed at B_1 or B_2 is z', while the desired one is denoted as z.

In the upper triangle in Fig. 10.7.1, $\sin\delta_s = \sin z' \cdot i/s'$. Since $s' = s\sin z \approx s\sin z'$ and i does not exceed a few metres

$$\delta_S = \frac{i}{S}\frac{180^\circ}{\pi}\sin^2 z \quad \text{and} \quad z = z' - \delta_S . \tag{10.7.1}$$

Similarly we get for the target reduction using Fig. 10.7.2:

$$\delta_z = \frac{t}{S}\frac{180^\circ}{\pi}\sin^2 z \quad \text{and} \quad z = z' + \delta_z . \tag{10.7.2}$$

For flat terrain $\sin^2 z \approx 1$, and thus

$$\delta_S = \frac{i}{S}\frac{180^\circ}{\pi} \quad \text{and} \quad \delta_z = \frac{t}{S}\frac{180^\circ}{\pi} . \tag{10.7.3}$$

Similarly for the δ values in the lower triangles in Figs. 10.7.1 and 10.7.2, the opposite sign is used in the same equations.

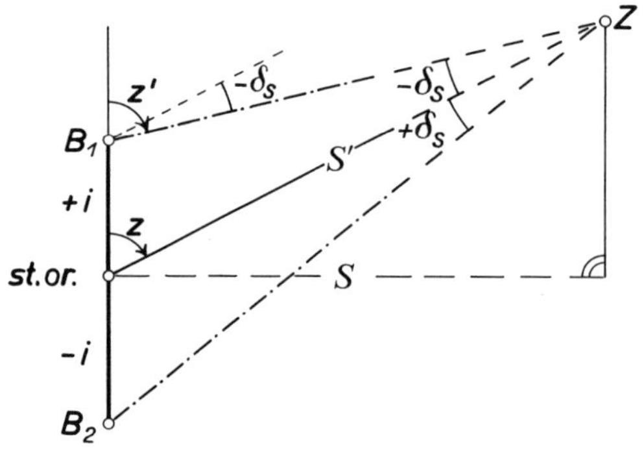

Fig. 10.7.1 Reduction at station

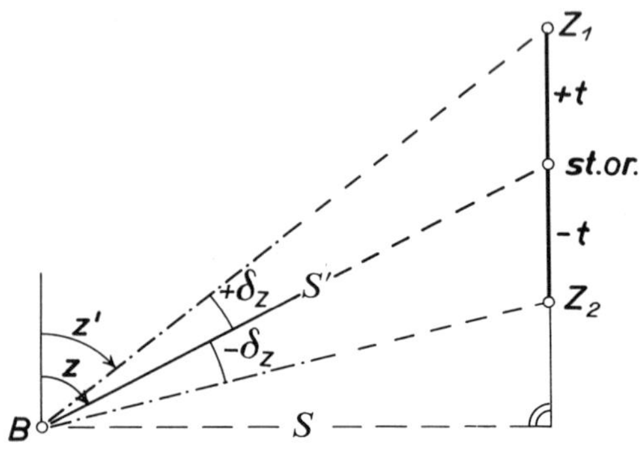

Fig. 10.7.2 Reduction at target

10.8 Accuracy of Trigonometric Heights Over Long Ranges

To find the influence of the measuring elements on the result, equation (10.5.4a) is differentiated:

$$\mathrm{d}\Delta H = \cos z \, \mathrm{d}S^R - S^H \mathrm{d}z - \frac{(S^H)^2}{2R} \mathrm{d}k + \mathrm{d}i - \mathrm{d}t \qquad (10.8.1)$$

with $S^H \approx S^R \sin z$.

The influence of $\mathrm{d}z$ increases with S^H, that of $\mathrm{d}k$ with $(S^H)^2$. $\mathrm{d}S^R$ will only affect the results of very steep sightings. Applying the law of error propagation to equation (10.8.1) leads to the following standard deviations:

$$\sigma^2_{\Delta H} = \cos^2 z \, \sigma^2_{S^R} + (S^H)^2 \sigma^2_z + \left(\frac{(S^H)^2}{2R}\right)^2 \sigma^2_k + \sigma^2_i + \sigma^2_t \, . \qquad (10.8.2)$$

If EDM instruments with standard deviations as small as $\sigma_{R^R}[\mathrm{mm}] = \sqrt{3^2 + (2S^R[\mathrm{km}])^2}$ are used, the first term cancels out. With $\sigma_{\Delta H}$, σ_i and σ_t in millimetres σ_z in microrads and S^H in kilometres, the following is obtained:

$$\sigma^2_{\Delta H} = (S^H)^2 \sigma^2_z + (80(S^H)^2)^2 \sigma^2_k + \sigma^2_i + \sigma^2_t \, . \qquad (10.8.3)$$

To get a numerical handle on this, let us assume that each zenith angle was observed four times each in direct and reverse with an electronic theodolite, which according to section 10.3.4 provides a standard deviation of the mean value of $\sigma_{\bar{z}} = \pm 0.3$ mgon $= \pm 5$ µrad. According to section 10.5.1, we assume $\sigma_k = \pm 0.04$ and use $\sigma_i = \pm 5$ mm and $\sigma_t = \pm 5$ mm. Based on, e.g., equation (10.8.3), the following results in centimetres are obtained:

S^H [km]	Error contribution in [cm] from: zenith angle	refraction	i and t	Total error in [cm]
2	± 1.0	± 1.3	±0.7	± 1.9
3	1.5	2.9	0.7	3.3
4	2.0	5.1	0.7	5.5
5	2.5	8.0	0.7	8.6

Even under favourable conditions, the influence of refraction increases so rapidly beyond 5 km, that the result is hardly suitable for medium accuracy requirements.

10.9 Trigonometric Height Traversing

The low speed and systematic errors of geometric levelling with its horizontal lines of sight are reason enough for trying to replace geometric levelling by a trigonometric method with measured slope distances and vertical angles. There are two basic types of trigonometric height traversing:

- Leap-frog traversing with balanced backward and forward sight distances (Fig. 10.9.1a).
- Traversing with reciprocal measurement of vertical angles (Fig. 10.9.1b).

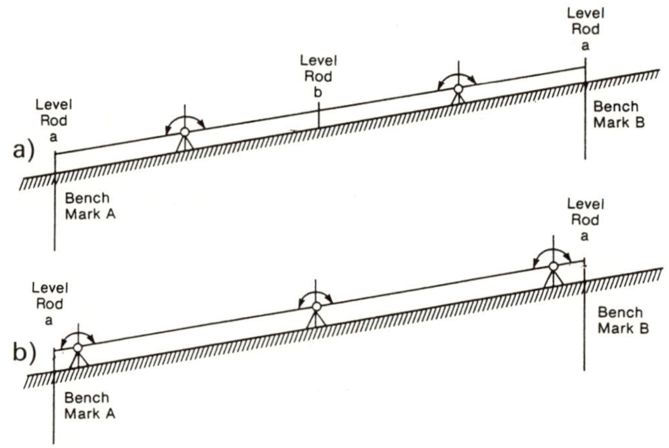

Fig. 10.9.1 Trigonometric height traversing.
a) Leap-frog mode of field operation. b) Reciprocal mode of field operation.

The targets in trigonometric height traversing can always be placed at the same height above the ground. Thus, the sight distances are not limited by the inclination of the terrain and the systematic errors from refraction are expected to become random because the back- and foresight lines pass through similar layers of air when traversing along long slopes. By extending the sight distances to a few hundred metres, the number of set-ups per kilometre is minimized. Consequently, the accumulation of errors due to the sinking of the instrumentation is reduced.

The height difference Δh_{AB} between two bench marks A and B is

$$\Delta h_{AB} = \sum_{A}^{B} \Delta H_{ij} + h_A - h_B \tag{10.9.1}$$

where ΔH_{ij} are the individual mean height differences, e.g., according to equation (10.5.9a), and h_A and h_B are the mean height differences between the trunnion axis of the theodolite and the bench marks at A and B, respectively.

Field tests show that standard deviations of $\leqslant 2\ \text{mm}/\sqrt{\text{km}}$ are achievable, even along inclined terrain, at a speed of 10 km/day when using sights no greater than 200 m for the leap-frog and 250 m for the reciprocal method *Chrzanowski* et al. (1985).

Each of the methods has advantages and disadvantages. If there is a large density of bench marks, the reciprocal method is slowed by the necessity of performing separate connecting surveys to the bench marks. The leap-frog method, however, requires a time consuming reconnaisance for balancing the forward and backward sight distances. The reciprocal method is less affected by the cumulative effects of refraction.

Additional test results are reported by *Rüger* and *Brunner* (1981; 1982), *Whalen* (1985), *Becker* (1985).

11. Barometric Heighting

Barometric heighting is only rarely used in the well surveyed countries of central Europe. However, it is extremely important for developing countries, where it often provides the only terrestrial means to obtain elevations for maps of all scales within a reasonable time period. Barometric heighting has become indispensible for planning and execution of large agricultural and resource developments, as well as for extended engineering structures. These have provided new impetus for its further development.

11.1 Physical Fundamentals

The *air pressure*, which on average is equal to the pressure of a 760 mm high mercury column, decreases with increasing sea level elevation. The reduction in pressure, however, is not constant per unit elevation difference, but less at higher elevations. Thus at sea level a reduction in pressure of 1 mm of mercury represents an increase in elevation of about 10 metres, while it represents about 14 metres on 2000 m high mountains. Furthermore, the air pressure depends on the air temperature and to a lesser extent on humidity and geographic latitude, because it is influenced by the meteorological conditions of the atmosphere. Therefore, elevation differences, rather than absolute elevation, are determined by air pressure measurements.

Air pressure is measured by comparison with an opposing pressure. Mercury barometers utilize the pressure of a mercury column for this, while aneroid barometers use the tension of a spring connected to a vacuum box. Gas chamber barometers utilize the known tension of the enclosed gas, and bypsometers the tension force of vapour when boiling water. Originally the height in millimetres of a mercury column served as a measure for air pressure. The pressure of 1 mm of mercury, denoted as 1 Torr results in

$$\begin{aligned} 1\ \text{Torr} &= 13\,595.1 \cdot 9.806\,65 \\ &= 133\,322.4\ \text{kg}\,\text{m}^{-1}\,\text{s}^{-2} = 1.333\,224\ \text{mbar} \end{aligned} \qquad (11.1.1)$$

where the density of mercury is taken as 13.5951 g/cm^3 and the gravity acceleration as 980.665 cm/s^2.

The SI unit for pressure is the Pascal (Pa), with the millibar (mbar) as the derived unit:

$$\begin{aligned} 1\ \text{mbar} &= 100\ \text{Pa} = 100\ \text{N}/\text{m}^2 \\ &= 100\ \text{kg} \cdot \text{m}\,\text{s}^{-2}/\text{m}^2 = 0.750\ \text{Torr}\,. \end{aligned} \qquad (11.1.2)$$

Using these conversions, the use of instruments with Torr scales does not generate much difficulty when converting to mbar. Thus these changes have virtually no influence on the measuring techniques. The method still consists of reading air pressure differences from which metric elevation differences are computed. To save on these computations, some barometers carry an additional scale graduated in elevation differences in metres. Since, as already mentioned, the pressure reduction is not linear, such an elevation scale gets narrower with increasing height. There are a few barometers on the market carrying a linear elevation scale (see sections 11.4.4 and 11.6).

11.2 Mercury Barometers

There are two distinctive types, mercury pipe barometers, and container barometers (Figs. 11.2.1 and 11.2.2).

The *mercury pipe barometer* consists basically of a glass pipe whose upper end is closed, while the lower end is U-shaped and open (Fig. 11.2.1). A vacuum is generated, and mercury is fed into the pipe. This mercury is pressed upward in the closed-off branch until its pressure balances the air pressure. As a measure for the air pressure the length l of this column is measured in millimetres, thus giving the air pressure in mm Hg or in Torr. To measure l, a fixed or movable ruler is fastened to the barometer. With a fixed ruler, l is obtained as the difference of the readings at both mercury surfaces. If the scale can be moved, the zero line is set onto the lower mercury mirror, and l is read directly on the upper one. If the U-shaped part of the pipe is

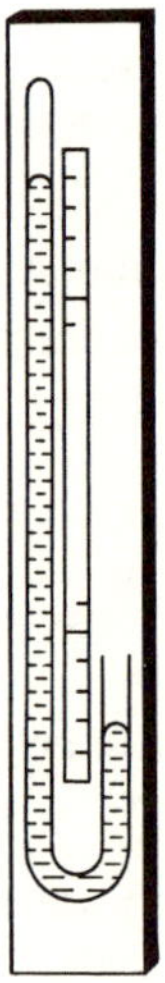

Fig. 11.2.1 Mercury pipe barometer

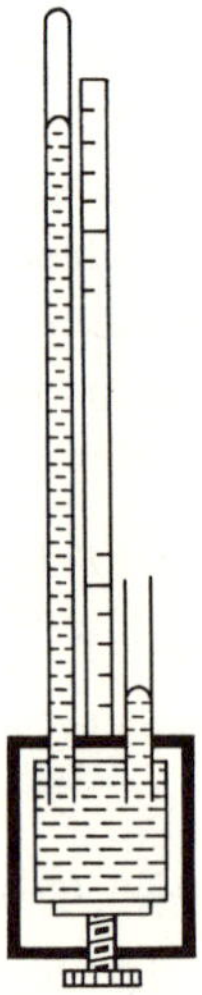

Fig. 11.2.2 Container lift barometer

replaced by a container with a movable bottom, then both mercury surfaces can be raised or lowered by the same amount. l can then be determined at different positions of the fixed scale, thus not only providing a check for the readings but also a refined value for l as the mean of all observations. This type is called a *container lift barometer* (Fig. 11.2.2).

Most manufacturers distinguish their barometers by their use rather than by their design type. *Normal barometers* are used for calibration and for observation on primary stations. They have to be completely controlled. Only the container lift barometer has this possibility. *Station barometers* are used during barometric heighting at a centrally located station for monitoring changes of the air pressure caused by meteorological conditions. Aneroid barometers are suitable for this purpose (section 11.4). They are to be read to ± 0.1 Torr $\approx \pm 0.13$ mbar.

Prior to using a mercury barometer, one has to check that there is a vacuum in the closed branch. The instrument is carefully inclined until the mercury column hits the upper wall closure. If a sharp light sound is obtained, then there is no air in the pipe. If, however, a low sound is heard, air has seeped in and needs to be removed. The mercury column has to be vertical for the measurements. Therefore, barometers often carry a simple plumbing device.

11.3 Corrections of the Mercury Barometer Readings

11.3.1 Temperature Correction

Both mercury and scale are temperature dependent. Therefore, the readings are reduced to $t = 0\,°C$. The reduced value B_0 at $t = 0\,°C$ of a reading B is, after neglecting higher order terms:

$$B_o = B(1 - qt + mt)$$

where q and m are the temperature expansion coefficients of mercury and the scale, respectively. Rearranged, this gives

$$v(t) = B_0 - B = -B(q - m)t\,. \tag{11.3.1}$$

In this equation, $q = 0.000\,182$, and for a brass scale $m = 0.000\,019$, or $q - m = 0.000\,163 = 1:6140$. With these values, equation (11.3.1) becomes

$$v(t) = -\frac{B}{614} \cdot \frac{t}{10}\,. \tag{11.3.2}$$

11.3.2 Capilar Depression

The mercury mirror in narrow pipes is somewhat lower due to adhesive forces. Furthermore, its surface is curved (meniscus) with the edges being lower by k. The amount $v(k)$ by which the vertex is too low is the capilar depression, which is directly dependent on the height k and the diameter d of the pipe (Fig. 11.3.1).

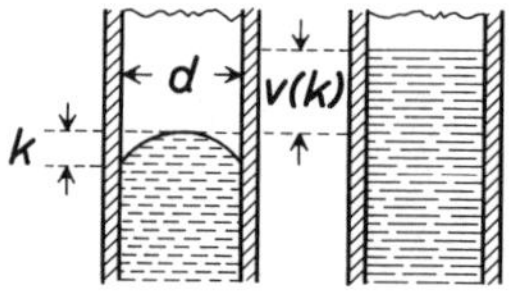

Fig. 11.3.1 Capilar depression

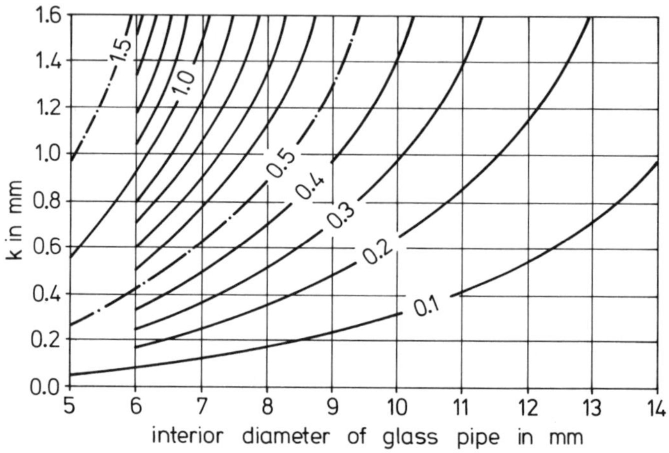

Fig. 11.3.2 Determination of capilar depression

Since the capilar depression does not occur in pipes with a diameter of more than 20 mm, it can be empirically determined by comparison with a mercury barometer with 20 mm diameter and tabulated or represented in diagram form (Fig. 11.3.2). After measuring the internal diameter of the pipe and the height with a millimetre ruler, both values are used in their diagonal to find $v(k)$.

11.3.3 Gravity Reduction

The weight force of a body equals the product of its mass m and the gravity acceleration g. The value for g in turn depends on the geographic latitude φ and the sea level elevation H. Therefore, the observations are reduced to sea level at 45° latitude using

$$g = g_0^{45}\left(1 - \beta\cos 2\varphi - \frac{2H}{r}\right), \tag{11.3.3}$$

where β is a gravity dependent coefficient. Based on the equilibrium condition

$$B \cdot g = B_0^{45} \cdot g_0^{45}$$

we obtain

$$B_0^{45} = B\left(1 - \beta\cos 2\varphi - \frac{2H}{r}\right), \tag{11.3.4}$$

and

$$B_0^{45} - B = -B \cdot \beta \cdot \cos 2\varphi - \frac{2HB}{r} = v(\varphi) + v(H) \tag{11.3.5}$$

or

$$v(\varphi) = -0.002\,64 \cos 2\varphi \cdot B\,, \tag{11.3.6}$$

$$v(H) = -0.000\,000\,3\; H \cdot B\,. \tag{11.3.7}$$

11.3.4 Station Correction

If the reading of a barometer after correction for $v(t)$ and $v(k)$ is compared to a normal barometer at the same location, a constant difference appears, which is the station correction of the barometer in question. It is determined by comparing different pressure readings and taking the mean.

If the direct reading of a mercury barometer is denoted by Q, the barometer reading reduced to $t = 0\,°\mathrm{C}$ and $\varphi = 45°$ at mean sea level becomes

$$B_0 = Q + v(t) + v(k) + v(\varphi) + v(H) + v(S)\,. \tag{11.3.8}$$

Numerical Example: A container barometer with a pipe diameter of 10 mm and a brass scale provided a reading at the vertex of the meniscus of 710.0 Torr. The temperature was read as $t = +14.1\,°\mathrm{C}$ and k was 1.3 mm. The observation station has a geographic latitude of $\varphi = 48°40'$ and a mean sea level elevation $H = 660$ m. The station correction was determined as $v(s)$ -0.20 Torr. This leads to:

$$\begin{aligned}
&v(t) &&= -(710:614) \cdot 1.41 = -1.61 \text{ Torr},\\
&v(k) \text{ (from Fig. 11.3.2)} &&= +0.38 \text{ Torr},\\
&v(\varphi) &&= -0.002\,64 \cdot \cos 97°20' \cdot 710 = +0.24 \text{ Torr},\\
&v(H) &&= -0.000\,000\,3 \cdot 660 \cdot 710 = -0.14 \text{ Torr},\\
&B_0 &&= 710.0 - 1.61 + 0.38 + 0.24 - 0.14 - 0.20 = 708.7 \text{ Torr}\\
& &&= 944.9 \text{ mbar.}
\end{aligned}$$

11.4 Aneroid Barometers with Membrane Box

11.4.1 The Membrane Box

This is a vacuum box with a rippled membrane on top, as first suggested by L. Vidi, of France. A strong spring whose tension force is equivalent to the air pressure prevents the box from being crushed. Within the measuring range, the membrane moves proportionally to the changes of the air pressure by about 0.01 mm per 1.3 mGal. Since the air pressure is required to about $+{}^1/_{10}$th of this amount, the movements have to be measured to an accuracy of 0.001 mm. In order to be able to read this amount on a scale it is either mechanically or optically magnified and/or microscopically measured.

11.4.2 The Naudet Barometer

The *Naudet* barometer (Fig. 11.4.1) transfers the movements of the membrane with the aid of a lever system *BCDEF* and a link-chain *G* onto the index *Z* which moves across the scale *S*. A thermometer is located inside the housing to eliminate temperature influences caused by the metal parts of the barometer (section 11.5.2). To reduce the temperature sensitivity inside the box, compensation barometers operate with thin air (65 to 130 mbar) in the box, or else with a lever consisting of two different metals, which is connected to the spring. The latter causes a curvature of the area which counteracts the direction of the air pressure for rising temperature.

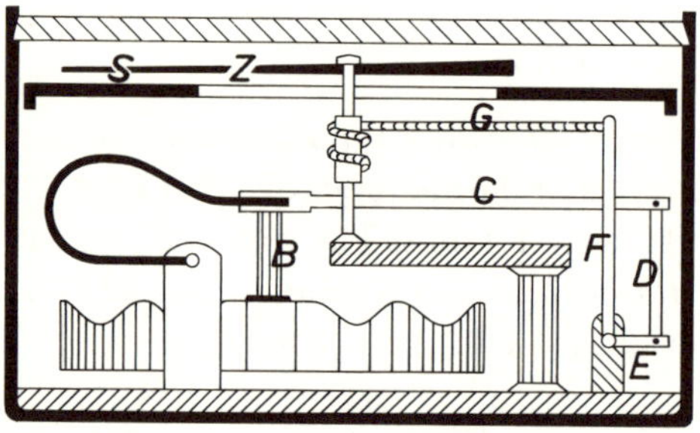

Fig. 11.4.1 The *Naudet* barometer

11.4.3 The Paulin Barometer

The *Paulin* barometer (Fig. 11.4.2) developed in Sweden, also has a solely mechanical transmission. The movements of the membrane are transferred onto a lever *a* which is held by a leaf spring and a spiral spring b_1. The linear motions of the levers are transformed into rotations by a tension joint with spring bands. These rotations are not directly read on a scale but act because of a measuring screw drum onto a second spiral spring b_2 attached to *a* until the normal tension marked by an index (zero reading) is obtained. The movement of the measuring screws can be estimated on a graduated gauge to about 0.025 mbar. The company also offers gauges graduated in metres.

The Paulin barometer is very sensitive to inclinations. For fine measurements it is therefore placed on a tripod and levelled with a bull's-eye level.

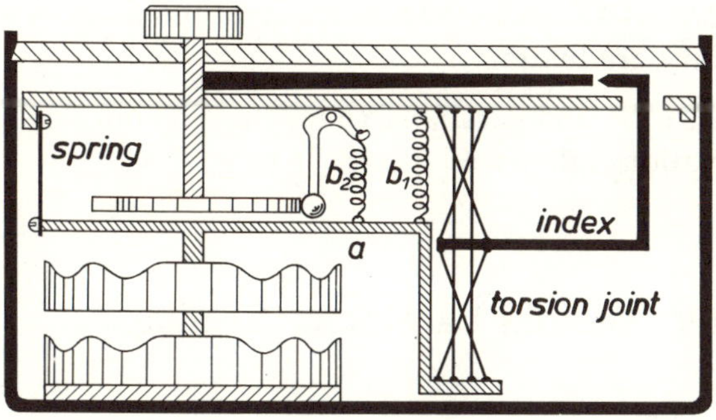

Fig. 11.4.2 The *Paulin* barometer

11.4.4 Altimeter

The 3B4 altimeter by Thommen, Waldenburg (Switzerland), is also equipped with membrane boxes. In spite of this, it is in a class by itself. Instead of air pressure variations it displays direct (linear) elevation differences and is thus referred to as an altimeter. It is easily recognized by having a linear graduation in metres instead of in millibars, where a line interval of 0.8 mm represents an elevation difference of 2 metres with a reading accuracy of 0.5 m. Thus the altimeter is not influenced by the change from Torr to mbar. Three boxes of copper, which behave linearly (Fig. 11.4.3), react very quickly to any pressure changes and are the principal measuring elements. A very sensitive transfer mechanism on stone bearings provides a measuring range of

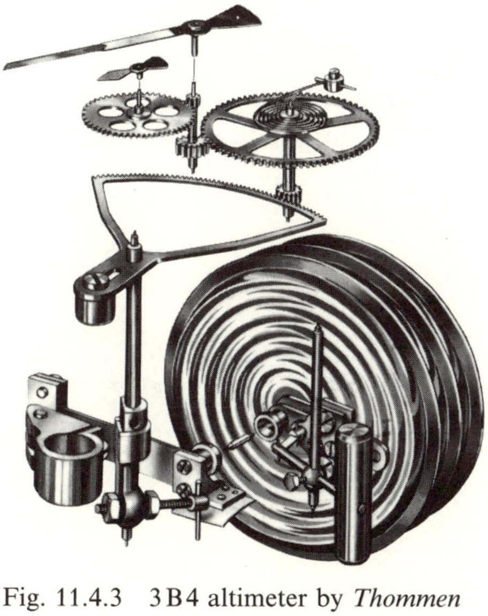

Fig. 11.4.3 3B4 altimeter by *Thommen*

~400 m to selectively +3000, +4000 and +5000 m for eight full index revolutions, which are read at an auxilliary graduation.

A good temperature compensation and parallax free readings permit reliable measurements even for large elevation differences.

11.5 Corrections to the Aneroid Barometer Readings

11.5.1 The Reduction Formula

The aneroid readings have to be reduced because of temperature correction $v(t)$, graduation $v(g)$, and zero error. $v(t)$ can be taken as linear, since only a relatively small range has to be considered. Thus only the temperature coefficient a need be determined. $v(g)$ can also be represented by a linear coefficient b, indicating whether the graduation is too wide or too narrow. The zero correction c, which is usually determined for the graduation line 1000, is a constant. Thus the barometer reading reduced to $t = 0\,°\mathrm{C}$ becomes

$$B_0 = F + at + b(1000 - F) + c$$

where F is the original reading of the aneroid barometer.

In order to determine a, b, and c, the readings F are compared to reduced readings Q_0 (to $t = 0\,°\mathrm{C}$) of a calibrated mercury barometer taken at different pressures and at different temperatures. It is best to first determine a alone, and then b and c together.

11.5.2 The Temperature Coefficient

The aneroid barometer and a calibrated mercury barometer are subjected to vastly different temperatures at a constant sea level elevation. This can be done in a cooling

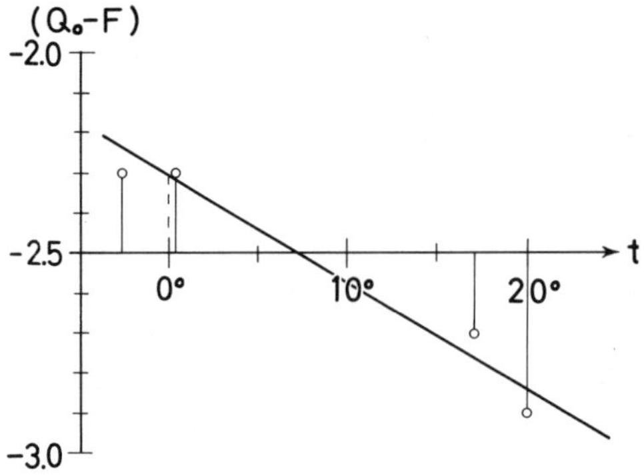

Fig. 11.5.1 Determination of temperature coefficient

chamber, or else in the winter, first outdoors and then in a room which is being slowly heated.

The mercury readings Q are reduced to $t = 0\,°C$, resulting in Q_0 in mbar. In a rectangular coordinate graph, the differences $Q_0 - F$ are plotted against the temperature t. A best fitting straight line is drawn (Fig. 11.5.1), where a is equal to the slope of this line. For a good aneroid barometer a should be less than 0.15.

Example: Temperature coefficient for a simple aneroid barometer:

Q_0	F	$Q_0 - F$	t
mbar	mbar	mbar	°C
1001.4	1003.7	−2.3	− 2.7
1002.5	1004.8	−2.3	0.3
1001.0	1003.7	−2.7	17.1
1002.0	1004.9	−2.9	20.0

At $t_1 = 0\,°C$ and $t_2 = 20\,°C$ readings of -2.32 and -2.84 are scaled from the graph. Therefore $a = (-2.84 + 2.32) : 20 = -0.026$.

11.5.3 Graduation Coefficient and Zero Correction

These are determined by comparing several reduced readings F_0 of an aneroid barometer to the corresponding reduced readings Q_0 of a mercury barometer for different pressures. The air pressure variation is obtained by either observing over a long period of time, or on a mountain hike, or in a slow pressure chamber (barometer test apparatus). Again, a graphical evaluation is common, with $(1000 - F_0)$ plotted against $(Q_0 - F_0)$ and a best fitting straight line (Fig. 11.5.2).

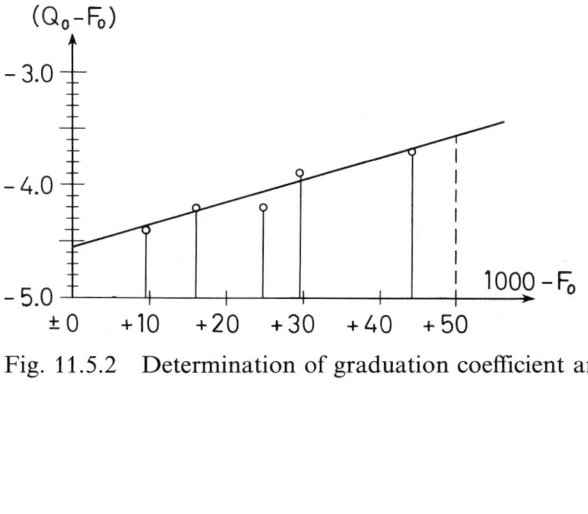

Fig. 11.5.2 Determination of graduation coefficient and zero correction

Example: The same instruments used in the example in section 11.5.2 were used ($a = -0.026$).

Q_0	F	t	$F_0 = F + at$	$Q_0 - F_0$	$1000 - F_0$
mbar	mbar	°C	mbar	mbar	mbar
986.3	991.2	21.1	990.7	−4.4	9.3
979.9	984.6	21.1	984.1	−4.2	15.9
971.3	976.0	21.1	975.5	−4.2	24.5
966.6	971.0	21.1	970.5	−3.9	29.5
952.3	956.5	21.1	956.0	−3.7	44.0

The slope of the straight line provides the graduation coefficient as

$$b = [-3.55 - (-4.55)] : 50 = +0.020 .$$

The zero correction (i.e., for a barometer reading of 1000 mbar) is $c = -4.55$ mbar.

Thus the air pressure B_0 reduced to 0 °C is obtained with this aneroid barometer in mbar according to

$$B_0\,[\text{mbar}] = F - 0.026\,t + 0.020\,(1000 - F_0) - 4.45$$

where $F_0 = F - 0.26\,t$. This equation holds for temperatures between 0 °C and 20 °C as well as for air pressures between 950 and 1000 mbar. The manufacturers provide the following equation for the Askania microbarometer:

$$B = p_i + AC + a(K - 293)$$

where p_i represents the absolute pressure in the measuring range i, A the scale reading, C the scale value for 20 °C = 293 K, a the temperature coefficient, and K the temperature measured in Kelvin. For each measuring range p_i, C, and a are determined during calibration.

11.5.4 Elastic After Effects

Ideally, the deformation of a measuring body held by an elastic spring depends on the load only. Actually, elastic and permanent after effects can be observed, especially with simple aneroid barometers, which appear as additional expansion or binding (hysterese). For tubular springs which are produced of a homogeneous metal piece, these after effects are small. All springs show signs of aging which become smaller after long use. Therefore the constants a, b, and c are valid for a limited time only and should be determined yearly.

11.6 Determination of Elevation Differences Using Barometer Measurements

11.6.1 Jordan's Barometer Formula

Considering an air cylinder normal to the sea level surface with unit cross section, there is an air pressure p at height H, and an air pressure $p - \mathrm{d}p$ at height $H + \mathrm{d}H$ (see Fig. 11.6.1). The weight of the differential piece of the air column equals the gravity acceleration times its mass, which in turn equals density times volume. Thus the equilibrium condition can be written as

$$\mathrm{d}p = -\varrho g \mathrm{d}H . \tag{11.6.1}$$

With ϱ_0 and p_0 representing density and pressure of a gas at 0 °C, and ϱ and p the corresponding values at t °C, then we get, according to the laws by Boyle and Gay-Lussac:

$$\frac{\varrho}{\varrho_0} = \frac{p}{p_0} \cdot \frac{1}{1 + \alpha t} \tag{11.6.2}$$

where α is the temperature expansion coefficient of air.

Substitution into equation (11.6.1) and rearrangement gives:

$$\frac{\mathrm{d}p}{p} = -\frac{\varrho_0}{p_0} \cdot \frac{g}{1 + \alpha t} \mathrm{d}H .$$

Integration of both sides leads to

$$\ln p = -\frac{\varrho_0}{p_0} \cdot \frac{g}{1 + \alpha t} H + C , \tag{11.6.3}$$

where g and t were considered as being constant, and C is the integration constant. To elimate the latter, the pressure is measured at two stations, namely

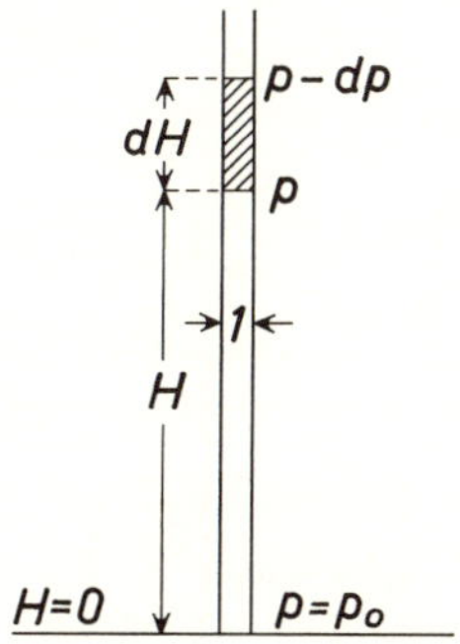

Fig. 11.6.1 Pressure-elevation relationship

at a lower station $H = H_1$; reading $p = B_1$
and at the upper station $H = H_2$; reading $p = B_2$.

The difference between upper and lower stations becomes

$$\ln B_1 - \ln B_2 = \frac{\varrho_0}{p_0} \cdot \frac{g}{1+\alpha t}(H_2 - H_1)$$

with $\mathrm{Mod}(\ln B_1 - \ln B_2) = \lg B_1/B_2$ we get

$$H_2 - H_1 = \Delta H = \frac{p_0}{\mathrm{Mod}\,\varrho_0} \cdot \frac{1+\alpha t}{g} \lg \frac{B_1}{B_2} \tag{11.6.4}$$

which is the basic equation of barometric heighting. Since B_1 and B_2 appear as a fraction, this equation is independent of the measuring units of air pressure.

Since temperature and gravity acceleration vary with increasing elevation and increasing or decreasing geographic latitude, *Jordan* et al. (1956) introduces average values as simple means of considering these:

$$t_m = {}^1/_2\,(t_1 + t_2); \quad H_m = {}^1/_2\,(H_1 + H_2)\,.$$

Thus the temperature is considered to be constant (isotherm). The changes in g and p_0 are covered in similar manner by

$$\frac{1}{g} = \frac{1}{g_0^{45}}(1 + \beta \cos 2\varphi_m)\left(1 + \frac{2H_m}{r}\right)$$

and

$$p_0 = g_0^{45} \cdot 0.760 \cdot 13.5951$$

(13.5951 is *Jordan's density value for mercury*).

The air density and its carbon dioxide content is covered by replacing

$$\frac{1}{\varrho_0} \text{ by } \frac{1}{\varrho_0}\left(1 + 0.377\frac{e_m}{p_m}\right)\frac{1}{1.0002\,g_0^{45}}$$

where $\dfrac{e_m}{p_m} = \dfrac{\text{average partial pressure of water vapour in air}}{\text{average air pressure}}$.

The known coefficients are combined to one constant:

$$K = \frac{p_0}{\mathrm{Mod.}\,\varrho_0 \cdot g_0^{45}} = \frac{g_0^{45} \cdot 0.760 \cdot 13.5951}{0.434\,295 \cdot 0.001\,293 \cdot 1.0002\,g_0^{45}} \approx 18\,400\,.$$

This lead to *Jordan's complete barometer formula*

$$\Delta H = 18\,400 \lg \frac{B_1}{B_2}(1 + \alpha t_m)\left(1 + 0.377\frac{e_m}{p_m}\right)(1 + \beta\cos 2\varphi_m)\left(1 + \frac{2H_m}{r}\right) \tag{11.6.5}$$

where $\alpha = 0.003\,665$, $\beta = 0.002\,64$; $r = 6\,370\,000$ m.

To simplify the application of equation (11.6.5), *Jordan* et al. (1956) prepared barometric height tables for central Europe. These tables, however, are hardly suited for other areas since the meteorological conditions are nearly always very different.

However, it is generally useful to have some knowledge of barometric height slopes. Based on *Jordan's* central European formula, the following differential relationship between elevation differences and pressure differences exists:

$$\mathrm{d}H = -18.464(1 + 0.0037\, t_m)\frac{\mathrm{Mod}}{B}\,\mathrm{d}B\,.$$

By replacing differentials by finite small quantities, this leads to

$$\Delta H = -\frac{8019}{B}(1 + 0.0037\, t_m)\,\Delta B\,. \tag{11.6.6}$$

The coefficient for ΔB is the barometric height step, valid for the (mean) barometer reading B, i.e., the elevation difference in metres which corresponds to our unit of pressure difference at a certain pressure. Some of the key values are:

t_m \ B	1020	1000	980	960	940	920	900 mbar
−10°C	7.57	7.72	7.88	8.04	8.22	8.39	8.58
0	7.86	8.02	8.18	8.35	8.53	8.72	8.91
10	8.15	8.32	8.49	8.66	8.85	9.04	9.24
20	8.44	8.61	8.79	8.97	9.16	9.36	9.57
30	8.73	8.91	8.09	9.28	9.48	9.68	9.90

11.6.2 Measuring Elevation Differences with Barometers

Jordan's barometric height formula (11.6.5) is only valid under the condition that the air pressure at the upper and lower ends of the air column is measured simultaneously and that the temperature decreases uniformly from bottom to top. None of these conditions is strictly fulfilled. Therefore the measurements have to be arranged such that the meteorological influences are circumvented by closing at fixed points, or else that they can be mathematically reduced or eliminated by using observations at locally fixed station barometers.

Mercury container barometers or precise aneroid barometers are used as station barometers, while field points are usually measured with simple aneroid barometers. It is advisable to remain at each point for a few minutes so that the instrument can react to the actual air pressure. The following quantities are read: air pressure, internal temperature of the barometer, air temperature, and time of reading. If possible, the path is selected such that the air pressure increases or falls steadily. Turning points, i.e., from increasing pressure to decreasing one, should be avoided. If this is impossible, such points should be occupied with the instrument for at least 15 minutes before the readings are taken.

11.6.3 Observation Methods

These methods, discussed in the following sections, are primarily designed to eliminate the inertia of the aneroid barometers, as well as possible pressure changes due to regional meteorological changes. Many methods have been devised. One distinguishes primarily between interpolation methods, where the heights of new points are obtained by interpolation between two known points, and methods where elevation differences to one known point are determined with the aid of the barometric height formula. Section 11.6.3.1 presents an example of an interpolation method, while sections 11.6.3.2 and 11.6.3.3 deal with precision methods using measurements relative to one known point.

11.6.3.1 Barometric Heighting by Interpolation

To determine the elevation of points whose elevation lies between two known points, A and E, one travels on a steadily climbing or falling path with one or several aneroid barometers. The barometer readings are taken at A, the new points 1, 2, 3 ... and at E, which should not be farther than one hour away from A. Unless significant changes in air temperature are expected, it is not necessary to measure it.

The elevations are computed by interpolation between A and E according to the barometric readings. For this the empirical height step

$$\delta H = \frac{H_E - H_A}{B_A - B_E} \tag{11.6.7}$$

is used. With F being the readings at the aneroid barometer, the elevation differences ΔH_i are computed as

$$\Delta H_{i,k} = \delta H \cdot \Delta F_{i,k} \, . \tag{11.6.8}$$

Example: Measurement with an aneroid barometer

Point	F	ΔF	$\Delta H = \delta H \cdot \Delta F$	H
	mbar	mbar	m	m
A	959.5			548.3
1	962.3	2.8	24.9	573.2
2	965.5	3.2	28.4	601.6
3	967.2	1.7	17.8	616.7
E	969.2	2.0	17.8	634.5
		9.7	86.2	

The empirical height step is in this case

$$\delta H = \frac{634.5 - 548.3}{969.2 - 959.5} = \frac{86.2}{9.7} = 8.887 \,.$$

If it is desirable to eliminate possible systematic temperature or barometer changes, then each elevation difference is measured three times (step procedure) by arranging the observations in the following order: $A-1-A-1$, $1-2-1-2$, etc.

11.6.3.2 Barometric Heighting with Field and Station Barometers

To obtain the terrain form, the height of a number of points, whose positions are scaled from a planimetric map, is to be determined. This requires one station and at least one field barometer. The station barometer is placed at a central location with known elevation, which is not subjected to temperature or air pressure anomalies. Time, pressure, and instrument temperature are read every 15 minutes. Since only pressure changes are required, temperature correction only need be applied. These pressure changes observed at the station barometer are usually plotted such that they can be easily obtained for any point in time.

The field observer moves with the field barometer from point to point, reading the barometer, as well as instrument and air temperature, and the time. At the beginning and end of the measurements, and, if possible, every 2 hours, the field barometer is compared to the station barometer, to see whether the station correction remains constant.

Example: Barometric heighting with field and station barometers (see Table 11.6.1)

In this example, the readings of station and field barometers have already been reduced according to sections 11.3.3 and 11.3.4. The difference between station and field barometer observations, which averages – 1.6 mbar, has been applied as a constant reduction to the station barometer readings. The time dependent pressure changes observed at the station barometer are added to the field barometer readings with opposite sign.

11.6.3.3 Step and Leap Methods

Let us assume that the heights of a stretched line of points between two known terminals have to be determined as accurately as possible using two barometers of equal quality. In this case, the elevation difference of two neighbouring points is obtained by simultaneously observing at both stations. The movement of instruments can be done in various ways:

a) In the *step method*, the second barometer is always brought to where the first one is and then both are read at the same time. This leads to the following measuring arrangement:

Table 11.6.1

Station barometer			Field barometer							Height computation					
Time	Barom. reading	Pressure change	Pt.	Time	Barom. reading	Air-temp.	Corr.	Reduced field bar read.		$\frac{B_i + B_k}{2}$	t_m	Height step	$B_i - B_k$	ΔH	MSL elev.
	St	Δ_{St}			F	t	$-\Delta_{St}$	B							
h min	mbar	mbar		h min	mbar	°C	mbar	mbar		mbar	°C	m	mbar	m	m
9 15	999.6	0.0	A	9 15	999.6	19.0	0.0	999.6							103.4
9 30	999.6	0.0													
										999	19.0	8.59	1.9	16.2	119.2
9 45	999.4	−0.2	1	9 41	997.5	19.0	0.2	997.7							
										996	18.9	8.61	3.4	29.3	148.9
10 00	999.3	−0.3	2	9 53	994.0	18.8	0.3	994.3							
										993	18.5	8.63	2.8	24.2	173.1
10 15	999.2	−0.4	3	10 09	991.1	18.2	0.4	991.5							
⋮	⋮	⋮	⋮	⋮	⋮										⋮
11 00	998.8	−0.8	A	11 03	998.7										

Points	A	1	2	3	E
	I/II				
	II	I			
		II	I		
			II	I	
				II	I
Time ↓					II/I

b) In the *leap-frog method*, developed by Bell Telephone Laboratories in the U.S.A., the second barometer is placed only onto every second point and compared with the first one:

Points	A	1	2	3	E
	I/II				
	II	I			
		I	II		
			I/II		
			II	I	
				I	II
Time ↓					I/II

This method has been quite successful.

Example: Observation with Askania microbarometers using the leap-frog method.

	Observations with barom. I and II (one way)				Computation of elevation differences			
Time	Pts.	Barom. I	Barom. II	Air temp.	I–II II–I	ΔB	Height step	ΔH
h min		mbar	mbar	°C	mbar	mbar	m	m
12 31	A	1002.63	1002.63		0.00	·	·	·
12 37	$1-A$	1001.77	1002.63	8.5	−0.86	−0.86	8.26	7.1
12 43	$1-2$	1001.70	1000.61	8.0	−1.09	−1.06	8.26	8.8
12 47	2	1000.61	1000.58	·	0.03	·	·	·
12 52	$3-2$	999.15	1000.55	8.0	−1.40	−1.43	8.26	11.8
12 57	$3-E$	999.21	996.55	7.5	−2.66	−2.68	8.26	22.1
13 02	E	996.57	996.59	·	−0.02	·	·	·

The same points have been observed on the return route. The average difference in the results was 0.2 m.

11.7 Determination of Elevation Differences Using Altimeter Measurements

11.7.1 Formula for Altimeters with Linearly Graduated Height Scale

According to section 11.1, altimeters are instruments of the Thommen type with a linearly graduated metric height scale, which directly provide normed heights without the need of pressure determination. Normed heights are barometric heights based on the international height scale, which would be read in metres in an ideal normal atmosphere (NA), provided the latter were available. The norms for this normal, or standard atmosphere, were established in 1924 by the *Commission Internationale de Navigation Aerienne (CINA)* as:

Yearly mean of air pressure at sea level		$p_0 = 760$ Torr
		$= 1013.325$ mbar
Air temperature at sea level		$t_0 = 15\,°C$
	or	$T_0 = 288\,K$
Temperature drop per km up to 11 km elevation		$a = 6.5\,°C/km$
Density of dry air with a carbon dioxide content of 0.03%		$\varrho_0 = 0.001\,266\ g/cm^3$
Gravity acceleration		$g_0 = 980.62\ cm/s^2$

Contrary to Jordan's formula (section 11.6.1), there is no isothermal requirement, and the norm relation between air pressure p and norm height h_{NA} is given by

$$p - p_0\left(\frac{T_0 - ah_{NA}}{T_0}\right)^n \quad \text{with} \quad n = \frac{g_0 T_0 \varrho_0}{ap_0} = 5.256\,.$$

Solving this equation for h_{NA} gives

$$h_{NA} = \frac{T_0}{a}\left(1 - \left(\frac{p}{p_0}\right)^{1/n}\right). \qquad (11.7.1)$$

Using the CINA norm values and p in [mbar], we get for h_{NA} in [m]:

$$h_{NA} = 44\,307.69 - 11\,874.31 \cdot p^{0.190259}\,. \qquad (11.7.2)$$

This norm height $(h_{NA})_i$ (equation (11.7.2)) measured at point P_i at the pressure p_i is directly displayed with a Thommen's altimeter (section 11.4.4).

The norm height difference h_{NA} between two points P_1 and P_2 is obtained as

$$\Delta h_{NA} = \Delta h = h_2 - h_1, \qquad (11.7.3)$$

if the mean temperature at the mean height h_m is $t_m = t_0 - ah_m$ and if the humidity is 0% and the mean gravity g_m^h is equal to the normal gravity g_0. Since these conditions are usually not fulfilled, the corrections for temperature as well as for vapour pressure, gravity acceleration and sea level elevation as in Jordan's formula (equation

(11.6.5)) are applied to the norm height difference h_{NA} before changing to sea level elevations H_i. This leads to the complete altimeter formula

$$\Delta H = \Delta h_{NA}\left(1+\frac{\Delta t_m}{T_E}\right)\left(1+\delta\frac{e_m}{p_m}\right)(1+\beta\cos 2\varphi_m)\left(1+\frac{2h_m}{r}\right). \tag{11.7.4}$$

where $\Delta t_m = t_m - (t_0 - ah_m)$ and h_m equals the mean elevation. Furthermore, as in equation (11.6.5), $T_E = 273\ K$, $1/T_E = 1/273 = 0.003\,665$, $\delta = 0.337$, $\beta = 0.002\,64$, φ_m = mean geographic latitude, and $r = 6\,370\,000$ metres.

11.7.2 Moeller's Simplified Formula

To simplify its application, equation (11.7.4) is rewritten as (*Moeller* 1962)

$$\Delta H = \Delta h_{NA} + v_{\Delta t} + v_e + v_\varphi + v_k\,, \tag{11.7.5}$$

with the value for a from section 11.7.1 we get

$$v_{\Delta t} = \Delta h_{NA}\frac{\Delta t_m}{T_E} = 3.665\ 10^{-3}(t_m - (t_0 - ah_m))\,\Delta h_{NA} = \text{temperature correction.}$$

Then we use

$$\Delta h'_{NA} = \Delta h_{NA} + v_{\Delta t}$$

and find

$$v_e = \Delta h'_{NA}\delta\frac{e_m}{p_m} = \Delta h'_{NA}0.377\frac{e_m}{p_m}, \qquad = \text{humidity correction}$$

$$\left.\begin{aligned} v_\varphi &= \Delta h'_{NA}\beta\cos 2\varphi_m = \Delta h'_{NA}2.64\cdot 10^{-3}\cos 2\varphi_m \\ v_h &= \Delta h'_{NA}\frac{2h_m}{r} = \Delta h'_{NA}0.314\cdot 10^{-6}\,h_m\,. \end{aligned}\right\} = \text{gravity correction}$$

Because of equation (11.7.3), h_{NA}, t and e have to be measured at both the lower and upper stations.

Unlike for Jordan's formula, these corrections are not tabulated. However, the computations of the corrections with Moeller's formula is not as cumbersome as it appears, and can easily be performed on a pocket calculator.

For the temperature correction, the values are usually smaller than with Jordan's formula, because for Moeller's formula they are proportional to the temperature difference rather than the temperature itself. The determination of e_m for v_e causes some difficulties. For this the dry temperature t and the wet temperature t' are measured with an aspiration-psychrometer. Then a can be computed using Sprung's formula.

$$e = E' - c\frac{t-t'}{1006.66}p \tag{11.7.6}$$

where c is to be taken as 0.5 above water, and 0.43 above ice. The saturation vapour pressure is tabulated and usually comes with the psychrometer. For the determination of single elevations, the correction terms can be determined by slide rule, while for larger areas, tables or diagrams are prepared for fast and easy operation. The use of computers of course is most efficient as the programming is fairly simple for this problem.

11.7.3 Elevation Differences with Altimeters

Barometer or altimeter heights are very common in developing countries, where they are essential for engineering and agricultural projects, as well as for the first stage of a control survey, e.g., a vertical network based on altimeter measurements.

In order to improve the observation accuracy, the following rules should be considered for all altimeter measurements:

1) Measure all elevation differences forwards and backwards; read the altimeter to 0.5 m.

2) Observe simultaneously with 2 or 3 altimeters; find the mean value of the results and round to the nearest decimetre.

3) Select the locations of the new points such that their elevation differences are proportional to the time intervals of the observations.

4) Utilize a station altimeter if the measuring period is extended or if the pressure changes significantly.

5) For longer distances and large elevation differences it is desirable to have a few intermediate points with higher accuracy.

11.7.4 Special Observation Methods for Altimeters

The methods discussed in sections 11.6.3.1, 11.6.3.2 and 11.6.3.3 are also well suited to measurements with altimeters. Again, one distinguishes between interpolation between two points, and connection to one point with subsequent elevation difference determination using the altimeter equation (11.7.5). The computation, however, differs somewhat from equation (11.6.4). Therefore, an example each for interpolation section 11.7.4.1 and connection section 11.7.4.2 is provided.

11.7.4.1 Elevation Determination by Interpolation

The following interpolation formula for altimeters was given by Moeller:

$$H_i - H_A = (H_E - H_A)\frac{h_i - h_A}{h_E - h_A} = \Delta H \frac{\Delta h_i}{\Sigma \Delta h_i} \tag{11.7.7}$$

where

h_i = measured elevation difference between neighbouring points

Σh_i = measured total elevation difference between fixed points

H = given elevation difference between fixed points.

With the corrections v_i, this leads to $\Sigma v_i = H - \Sigma \Delta h_i$ and

$$v_i = \frac{\Sigma v_i}{\Sigma \Delta h_i} \Delta h_i . \qquad (11.7.8)$$

Example:

Point	h_i [m]	Δh_i [m]	v_i [mm]	ΔH_i [m]	H_i
A	250.0				264.0
		199.0	5.8	204.8	
1	449.0				468.8
		297.5	8.6	306.1	
2	746.5				774.9
		281.5	8.1	289.6	
3	1028.0				1064.5
		135.5	3.9	139.4	
E	1163.5				1203.9
Σ or diff.	913.5	913.5	26.4	939.9	939.9

The last row serves as a computational check with sums and differences. This leads to the computation of the v_i column:

$$\Sigma v_i = 939.9 - 913.5 = +26.4 \quad \text{and} \quad v_i = \frac{26.4}{913.5} \cdot \Delta h_i .$$

11.7.4.2 Altimetry with Connection to One Known Point

The following numerical example contains a segment of an altimeter survey in the Black Forest ($\varphi_m = 48°$), with altimeter heights being computed using equation (11.7.5).

Example:

Date	Time	Point	Δh_i [m]	(h_i) h_m [m]	t_m [°C]	t'_m [°C]	$t_m - t'_m$ [°C]
1	2	3	4	5	6	7	8
28.7.	10^{15}	A		(264)			
			201.7	365	19.4	16.2	3.2
	10^{33}	1		(466)			
			101.6	517	18.7	15.4	3.3
	10^{55}	2		(568)			
			197.5	667	16.4	13.8	2.6
	11^{10}	3		(766)			
	Δ or Σ		500.8	502	54.5	45.4	9.1

$v_{\Delta t}$ [m]	$\Delta h'_i$ (4 + 9) [m]	v_e [m]	v_φ [mm]	v_h [m]	ΔH (10 + 11 + 12 + 13) [m]	H
9	10	11	12	13	14	15
						264.0
5.0	206.7	1.3	−0.1	0.0	207.9	
						471.9
2.6	104.2	0.6	0.0	0.0	104.8	
						576.7
4.1	201.6	1.2	−0.1	0.0	202.7	
						779.4
11.7	512.5	3.1	−0.2	0.0	515.4	515.4

In this example the following notation was used:

Δh_i = elevation difference computed as the mean of readings on three altimeters, which had previously been corrected for instrumental errors (graduation, temperature) using calibration curves.

$$\left.\begin{aligned}(h_i) &= (h_{i-1}) + \Delta h_i \\ h_m &= (h_{i-1}) + {}^1/_2\, \Delta h_i\end{aligned}\right\} \text{rounded to full metres}$$

$$\left.\begin{aligned}t_m &= \text{mean dry temperature} \\ t'_m &= \text{mean wet temperature}\end{aligned}\right\} \text{using aspiration psychrometer}$$

The computation of $v_{\Delta t}$, v_e, v_φ and v_h is based on equations (11.7.5) and (11.7.6).

Note:

If terrain surveys (according to section 11.6.3.1) and point determinations (according to section 11.6.3.2) are to be performed with altimeters, the following should be considered:

For terrain surveys with field and station altimeters (one base method), Δh_{NA} is formed from the corrected altimeter readings of the field group using points in the order of measurements by applying station corrections, caused by air pressure changes as determined from the station altimeter readings.

For the step and leap methods, however, the norm height difference Δh_{NA} is to be formed using the altimeter readings taken simultaneously at the lower and upper points. Therefore, the corrected readings of both groups are to be first reduced to the known elevation of the starting point, and then the norm height differences for the whole line are to be found by forming the mean of both parts.

11.8 Accuracy of Barometric Heights

To get a theoretically correct error formula, the law of error propagation would have to be applied to the complete Jordan formula (11.6.5) or the altimeter formula (11.7.4). Humidity and gravity corrections, however, have little influence on the total error. For this error evaluation, a modified Jordan formula, suited for central Europe, is therefore used, namely

$$H_2 - H_1 = \Delta H = K(1 + \alpha t_m)(\lg B_1 - \lg B_2)$$

where $K = 18464$.

The quantities B_1, B_2 and t_m are measured and the total differential with respect to them is found to obtain their influence on the desired elevation difference H.

$$\mathrm{d}\Delta H = \frac{\partial \Delta H}{\partial B_1}\mathrm{d}B_1 + \frac{\partial \Delta H}{\partial B_2}\mathrm{d}B_2 + \frac{\partial \Delta H}{\partial t_m}\mathrm{d}t_m\,, \tag{11.8.1}$$

with
$$\frac{\partial \Delta H}{\partial B_1} = \frac{\mathrm{Mod}\,K}{B_1}(1 + \alpha t_m)\frac{8019}{B_1}(1 - 0.0037\,\mathrm{t}_m)\,, \tag{11.8.2}$$

$$\frac{\partial \Delta H}{\partial B_2} = \frac{\mathrm{Mod}\,K}{B_2}(1 + \alpha t_m) = \frac{8019}{B_2}(1 + 0.0037\,t_m)\,, \tag{11.8.3}$$

$$\frac{\partial \Delta H}{\partial t_m} = K(\lg B_1 - \lg B_2)\alpha = 0.0037\,\Delta H\,. \tag{11.8.4}$$

After substituting the values from equations (11.8.2) to (11.8.4) in (11.8.1) and transferring to standard deviations we get

$$\sigma_{\Delta H}^2 = \left(\frac{\partial \Delta H}{\partial B_1}\right)^2 \sigma_{B_1}^2 + \left(\frac{\partial \Delta H}{\partial B_2}\right)^2 \sigma_{B_2}^2 + \left(\frac{\partial \Delta H}{\partial t_m}\right)^2 \sigma_{t_m}^2 . \tag{11.8.5}$$

Because of $\sigma_{B_1} = \sigma_{B_2} = \sigma_B$ we get with the numerical values of equations (11.8.2) to (11.8.4):

$$\sigma_{\Delta H}^2 = \left(\frac{8019}{B_1}\right)^2 \sigma_B^2 + \left(\frac{8019}{B_2}\right)^2 \sigma_B^2 + (0.0037\,\Delta H)^2 \sigma_{t_m}^2 \tag{11.8.6}$$

or, with δ_B denoting the influence of the barometer error, and δ_t of the temperature error:

$$\sigma_{\Delta H}^2 = \delta_B^2 + \delta_B^2 + \delta_t^2 . \tag{11.8.7}$$

In the field, the air temperature can at best be obtained to $\pm 1\,°\mathrm{C}$. In elevations up to 1000 m, i.e. with an air pressure of 900 to 1000 mbar, a temperature error of $\pm 1\,°\mathrm{C}$ amounts to a height error of about $\pm 0.0037\,\Delta H$ or roughly $\pm 0.4\%$ of ΔH (equation (11.8.4)). Thus with $\sigma_t = \pm 1\,°\mathrm{C}$ and δ_t representing its influence on the determination of ΔH, the following relation exists between ΔH and δ_t:

– for $\Delta H =$	10	20	50	100	200	500	10000 m	
$\delta_t =$	0.04	0.07	0.18	0.37	0.74	1.85	3.70 m .	(11.8.8)

To determine the effect δ_B of a standard deviation σ_B of the barometer, one obtains from the table of height steps in section 11.6.1 that

$$\sigma_B = \pm 0.1 \text{ mbar causes } \delta_B = \pm 0.8 \text{ m} . \tag{11.8.9}$$

Introducing the values for δ_t from equation (11.8.8) and for δ_B from equations (11.8.9) into (11.8.7) and again using $\sigma_t = \pm 1\,°\mathrm{C}$, the standard deviation $\sigma_{\Delta H}$ of an elevation difference determination ΔH assumes the following values for $\sigma_B = \pm 0.1$, ± 0.05, and ± 0.02 mbar:

σ_B in mbar \ ΔH	10	20	50	100	200	500	1000 m
±0.10	1.2	1.2	1.2	1.2	1.4	2.2	3.9 m
±0.05	0.6	0.6	0.7	0.7	1.0	2.0	3.8 m
±0.02	0.3	0.3	0.4	0.5	0.8	1.9	3.7 m

(11.8.10)

For norm heights with altimeters the following error formula is obtained using equation (11.7.4):

$$\sigma_{\Delta H}^2 = \sigma_h^2 + \sigma_h^2 + (0.0037\,\Delta h)^2 \sigma_{\Delta t_m}^2 . \tag{11.8.11}$$

The values for σ_h correspond to the first two terms in equation (11.8.6). The last term of the Jordan error formula (equation (11.8.6)), however, has σ_{t_m} as the standard deviation of the mean of the air temperature readings t_1 and t_2 at lower and upper stations, while in equation (11.8.11), σ_{t_m} – similar to Δt_m in equation (11.7.4) – represents the standard deviation of the temperature difference between measured mean temperature and norm temperature, as described in section 11.7.1. As a rule, this difference is not very significant for practical work, so that the third terms of equations (11.8.6) and (11.8.11) are nearly equal. Therefore, equation (11.8.7) and the appropriate values can also be used for altimeter work.

Thus with $\sigma_h = 0.5$ m, $\sigma_{\Delta t_m} = \pm 1\,°C$ and $\sigma_{e_m} = \pm 1$ mbar for the effective vapour pressure, which are practically obtained values, the accuracies listed in equation (11.8.10) under ± 0.05 mbars are obtained for altimetry. Comparing this equation with equation (11.8.7), based on temperature errors only, the following becomes evident: The uncertainties for smaller elevation differences are primarily caused by uncertainties reading of the barometers, while the uncertainty of temperature determinations becomes the main error source for elevation differences of 500 to 1000 m.

It is thus desirable to cover larger elevation differences by measuring in sections of up to 200 m elevation difference. Generally an error of 1 to 2 m can be expected for measurements with a simple aneroid barometer ($\sigma_B = \pm 0.1$ mbar) for elevation differences up to 200 m. This value agrees with numerous practical results.

Measurements with precision aneroid barometers of the Paulin or Thommen type (± 0.05 mbar or ± 0.5 m) or the Askania microbarometer (± 0.02 mbar) provide results as shown in the second and third rows of equation (11.8.10). The use of precision barometers, therefore, is only justified for small elevation differences. Even for small elevation differences, the air pressure changes due to elevation are often exceeded by local pressure variations caused by heat radiation, soil moisture, wind, etc. For precision work with aneroid barometers, it is therefore necessary to chose the time of observations with calm air, stable meteorological conditions, and overcast sky. If, in addition, the barometers are protected from shock and vibrations, a standard deviation of ± 0.3 to 0.4 m can be expected.

12. Three-Dimensional Positioning

12.1 Topographic Surveying with Conventional Means

12.1.1 Fundamentals

As a basis for the construction of railroads, highways, bridges, dams, power plants, pipelines, and similar engineering projects, plans in 1 : 500, 1 : 1000 to 1 : 10000 are required showing both planimetry and heights.

12.1.1.1 Elevation Points and Contour Lines

In this context, elevation points are terrain points whose position in a local or global coordinate system and height above a reference surface (e.g.,MSL) are determined by direct measurements. If such a point is also a prominent terrain point, such as a mountain top, depression, road intersection, etc., it is often shown as a point on the map together with an elevation number, a so-called geopotential code.

Contour lines are curves which connect points of equal height above the reference surface. They are fictitious lines separated by equal amounts along the plumb line and normal to the direction of maximum slope at any point. Their orthogonal projection onto the horizontal map plane provides a contour map. The elevation difference between two contour lines measured along the plumb line is called a contour interval. It remains constant for a contour line system (i.e., within one map).

The steeper the terrain, the closer the contour lines on the map. Therefore, when selecting the contour interval, terrain slope and map scale are of primary concern, with accuracy of the measurements and required detail as secondary consideration. If special terrain details fall between contour lines, as is often the case in flat terrain, auxiliary contours with smaller interval are added at the respective areas.

12.1.1.2 Requirements for Contour Plans

As mentioned, the purpose of contour lines is the representation of the terrain. For this, several aspects need be considered. The engineer, who uses the map as a basis for design and construction of a building, is primarily interested in the geometry of the contour plan. He needs to extract individual elevations (usually by interpolating) in order to be able to compute inclinations and slope angles of the terrain surface as well as its curvatures. Finally, he often bases his mass computations on the contour plan.

He therefore requires large scale plans up to 1 : 1000 with contour intervals of 1 metre or less. Furthermore, all man-made terrain features, such as dams, curbs, and ditches, which usually can only be incompletely represented by contours, need be presented by slope hatches supplemented by numbers.

The contour intervals for other scales are selected according to the following rule of thumb: Contour interval in metres equals the scale number divided by 1000, i.e., 5 m for a 1 : 5000 map.

On the other hand, a geographer or geologist is primarily interested in the morphological content of a map. He expects to recognize and interpret terrain forms from the contour lines. Thus the lower limit for contour plans is 1 : 5000, because the terrain forms become obstructed by details at larger scales.

In addition to engineers and morphologists, there is a large group of map users who are interested in obtaining a visual image of the terrain from the map (e.g., hikers, mountain climbers, soldiers, etc.). These users require a comprehensive overview, which is obtained by the density of the contours if the proper scale is selected.

12.1.1.3 Collection of Terrain Data

The most commonly used method for terrain data collection is photogrammetry, which provides contours as well as individual terrain elevations. However, because of the high costs of aerial photography, this method is only economical for larger projects. For small projects or for areas with dense vegetation cover which cannot be penetrated by photography (e.g., dense forests), a terrestrial approach is needed. In

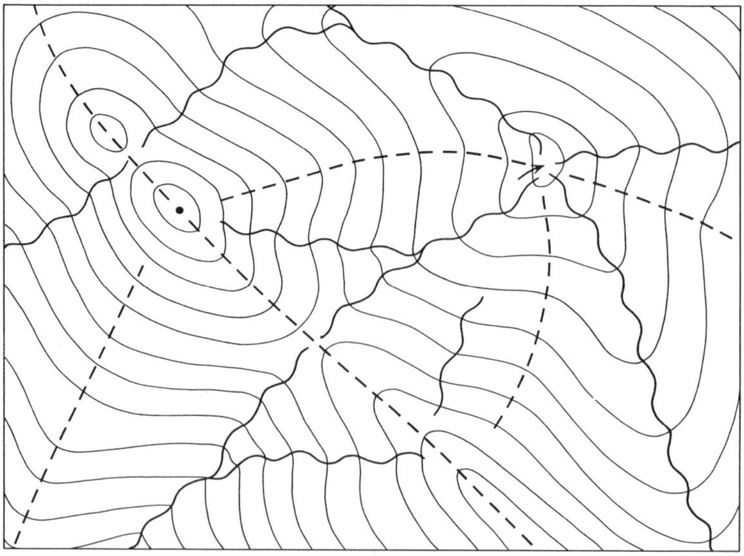

Fig. 12.1.1 Characteristic terrain lines

this case, a number of characteristic terrain points are determined in position and height, while contours are derived by means of interpolation. Unlike photogrammetric maps, neighbouring contours are not independent in this case.

The selection of terrain points is therefore very important. For this, first an overview of the area is used to obtain its characteristics forms (Fig. 12.1.1). The characteristic terrain lines are occupied, such as break lines and water courses as well as the tops of hills and bottoms of depressions and saddle points. Furthermore, lines of changing slope, edges, and cliffs as well as artificial slope changes are measured. All points should be terrain points.

The separation of neighbouring points depends largely on the terrain form, but also on contour interval and map scale, which in turn has a certain relationship to the required accuracy. For scales of 1 : 5000 to 1 : 2500, the point density varies between 300 and 700 points per km^2. This means an average (for 500 points) point separation of about 45 m in the field.

For 1 : 1000, up to 2500 points are used which represent about a 20 m point separation. The experienced topographical surveyor as a rule needs fewer points than the beginner who has a tendency to measure as much as possible.

The technical methods for terrestrial topographical surveys are stadia, compass tacheometry and plane tabling. The first two are referred to as numerical tacheometric methods, because position and height of selected terrain points is determined numerically, while contour lines are constructed in the office after completion of the field work. With a plane table, position and height of the points are determined in the field on a large scale plan, and the contours are plotted while viewing the terrain.

12.1.2 Special Instrumentation

In the previous chapters, a rather clear distinction was made between horizontal or planimetric, and vertical or height measurements. Basic surveying instruments are designed for use in either one of these types of measurements. By combining different procedures and instruments, three-dimensional results are obtained.

The basic ingredient for topographical surveying is the need to simultaneously obtain data on changes in horizontal and vertical directions, combined with the distance. Thus an instrument which provides horizontal and vertical angles plus distance is most suited for this type of work, just as is equipment which permits the measurement of positional changes in both planimetry and height.

Tacheometers were primarily designed for topographical surveying; their basic design and operation principles have been covered in sections 4.2 for optical (stadia) theodolites, and in 12.2 for electronic tacheometers. Therefore, only a few special instrument types, such as self-reducing tacheometers, compass instruments, and plane table with alidade, are covered here.

12.1.2.1 Self-reducing Tacheometers

The reading of vertical angles, followed by the reduction of slope distances to the horizontal and the computation of elevation differences is rather cumbersome and time consuming. It is thus not surprising that self-reducing tacheometers have been developed on the idea of the stadia theodolite long before the advent of electronic tacheometers. Although the latter have been gaining more and more acceptance, there are still many self-reducing tacheometers in use, so that some brief remarks on their basic design appear justified.

The earliest types of self-reducing tacheometers operate with *mechanic reduction*, where the stadia reduction formulae are solved through mechanical constructions. Slope screw, contact and tangent tacheometers belong to this category. Details on these instruments may be found in *Deumlich* (1982).

The more commonly used type of self-reducing tacheometer is the *diagram tacheometer*, based on the ideas of Professor Hammer (Stuttgart) and the instrument manufacturer Fennel (Kassel), who built the *Hammer-Fennel* tacheometer.

The principal element was a diagram, mounted on the standard at the horizontal axis. It was imaged through a prism system into the eyepiece, thereby following the tilt of the telescope. Thus the horizontal distance and positive or negative elevation difference could be directly read on the distance or height curve.

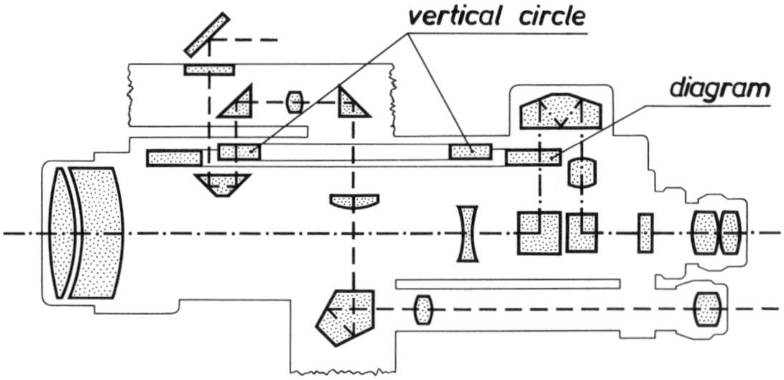

Fig. 12.1.2 Optical train of the *Zeiss*, Jena Dahlta

The Norwegian engineer Dahl proposed to etch the diagrams onto the vertical circle. This eventually led to the development of the Dahlta 010A which is still manufactured by Zeiss, Jena (Figs. 12.1.2 and 12.1.3). Other diagram tacheometers were developed by Wild (RDS), Zeiss, Oberkochen (R Ta 4), and Kern (Kl-Ra and DK-RV). The illustrations 12.1.3 to 12.1.5 should suffice to provide an idea on the reading of such instruments. Again, the reader is referred to special literature, such as *Deumlich* (1982) for details.

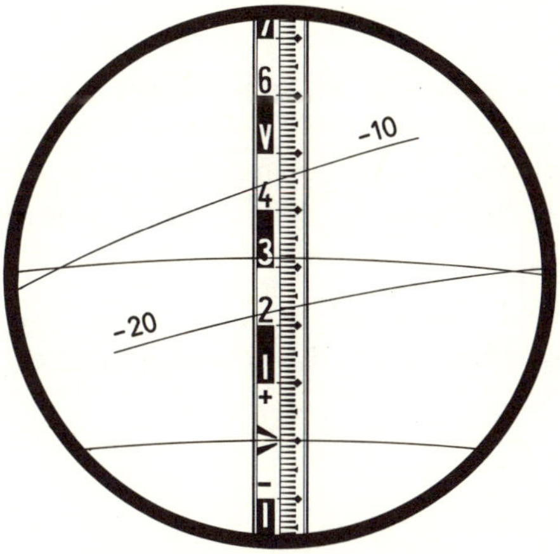

Fig. 12.1.3 Field of view of the *Zeiss*, Jena Dahlta
$\Delta H = 0.454 \cdot (-10) = -4.54$ m
or $\Delta H = 0.227 \cdot (-20) = -4.54$ m

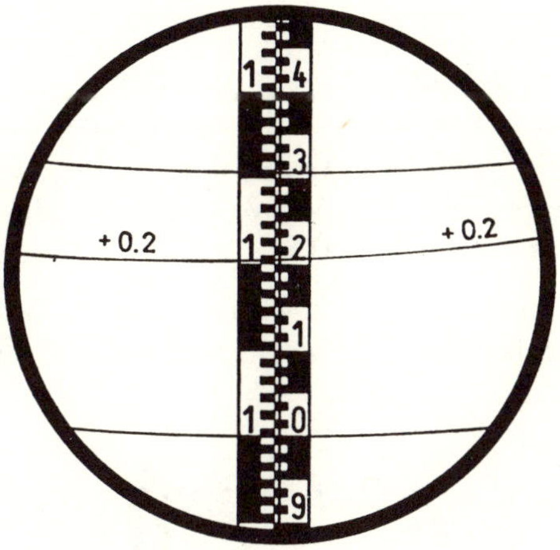

Fig. 12.1.4 Field of view of the *Wild* RDS
$D = 30.6$ m
$\Delta H = 20.4 \cdot (+0.2) = 4.08\ m$

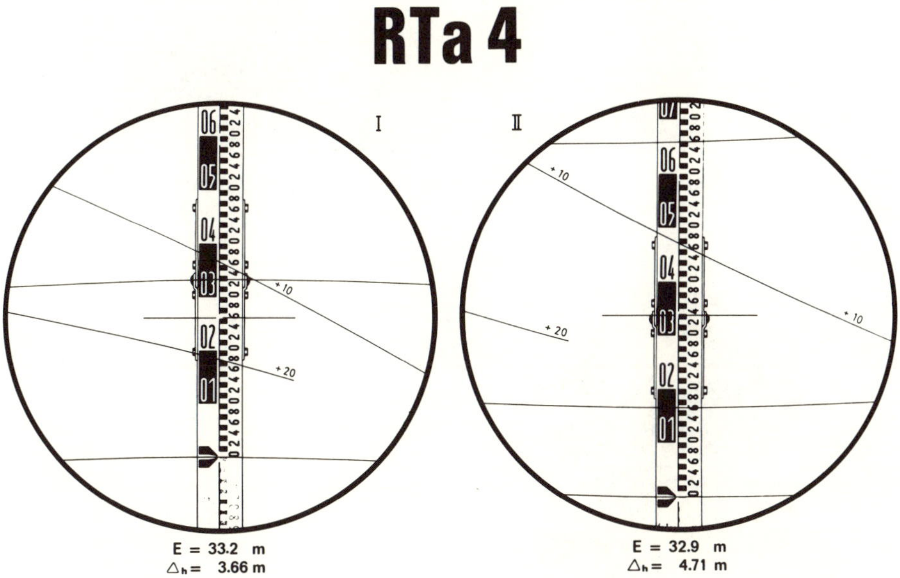

Fig. 12.1.5 Field of view of the *Zeiss*, Oberkochen RTa 4 (direct (I) and reverse (II) telescope positions)

The accuracy of self-reducing tacheometers is equivalent to stadia theodolites, because possible losses during the reduction process are compensated by improved optics and finer lines for the diagrams.

12.1.2.2 Compass Instruments

The simplest instrument of this type is the *stock compass* where a sighting device (diopter) is connected to the zero mark of the graduated circle. This circle is usually graduated anticlockwise, so that the magnetic azimuth can be directly read on the compass needle. A bull's-eye level completes the instrument, which is used for horizontal measurements (Fig. 12.1.6).

If the diopter is replaced with a telescope with stadia hairs and vertical circle, a *stadia compass* is obtained. Such instruments are being manufactured with various degrees

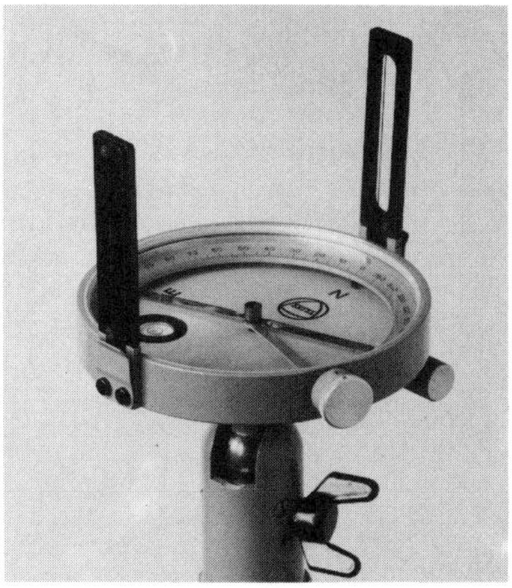

Fig. 12.1.6 Diopter compass

Fig. 12.1.7 Stadia compass

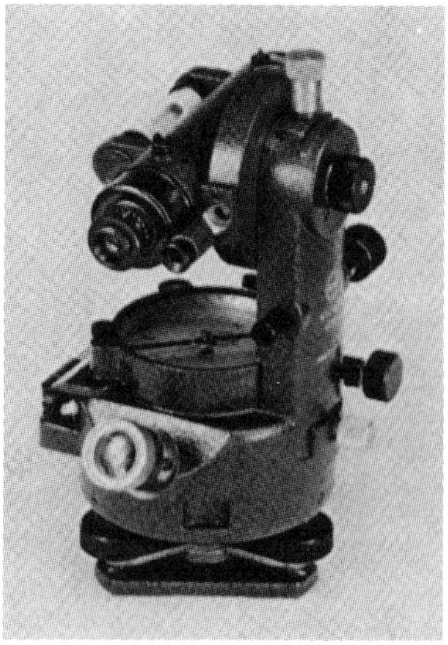

Fig. 12.1.8 Compass tacheometer

of sophistication. Furthermore, *compass attachments* are available for regular optical theodolites.

All these instruments provide direct magnetic azimuths, which combined with the stadia distance are used for topographic surveys for which their limited accuracy (usually 5′ to 10′ based on the accuracy of the needle readings) is sufficient. Figs. 12.1.7 and 12.1.8 illustrate some of these instruments.

12.1.2.3 Plane Table and Alidade

The *plane table* is the oldest topographic instrument still in use today. It consists of a table connected with a jointed head to a tripod plus the alidade which can be freely moved on the table (Fig. 12.1.9). The *alidade* consists of a base plate with graduated ruler carrying the telescope with vertical circle. Older alidades are equipped with simple stadia hairs, while most newer ones are self-reducing tacheometers (Fig. 12.1.10).

Fig. 12.1.9 Plane table and alidade

Auxiliary equipment consists of cross-level, orientation compass, parallel ruler, as well as bull's-eye level and plumb bob attachment for levelling or centring of the plane table.

Prior to measuring, the plane table and alidade have to be checked and adjusted if necessary for collimation error, horizontal axis error, and index correction. The procedures are similar to the ones used for theodolites and are described in section 3.6.

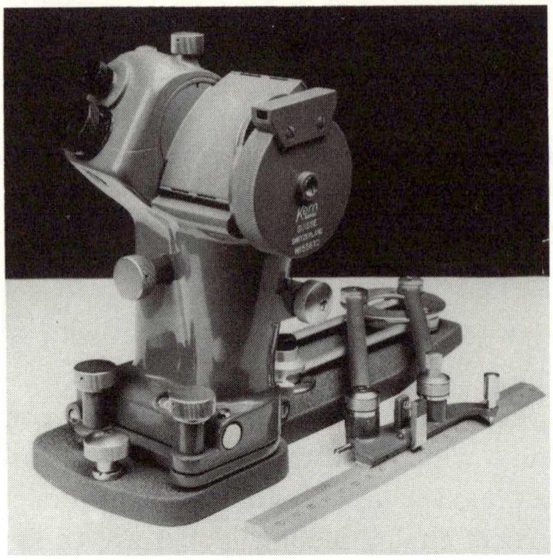

Fig. 12.1.10 Self-reducing alidade by *Kern*

12.1.3 Topographic Measurements with the Stadia Theodolite

12.1.3.1 Fundamentals

Often it is necessary to supply contour lines for existing planimetric plans in a desired scale. Then it is only necessary to measure terrain forms. For this, characteristic terrain points are selected and plotted onto the plan. Their height may be determined by geometric levelling in flat areas, but are usually determined by stadia means.

If a planimetric map is not available or is incomplete, then position and height of the terrain points is determined by stadia. A framework of control points is required in both position and height. If the area is extensive, triangulation points with known elevation usually provide this framework. These are densified by traverses or stadia traverses, which also have to be determined vertically. If possible, the traverse points are selected such that they can serve as stations for the stadia survey, and therefore should provide clear sights to the surrounding area. If needed, additional observation stations have to be set out, usually by polar means. Since the range of optical tacheometers is limited to about 120 m for reasonable accuracy, observation stations should not be farther apart than 150 to 200 m. For stadia surveys along a narrow corridor (roads, etc.), usually a traverse or primary tacheometer traverse is used as the basic frame.

12.1.3.2 Measuring and Computing Stadia Traverses

Like regular traverses, stadia traverses are connected to higher order control points if possible. Primary stadia traverses are usually measured prior to the actual stadia survey with the highest accuracy achievable by tacheometric means. The elements of secondary stadia traverses, however, are measured in conjunction with the actual topographic survey and with the same accuracy. For primary stadia, however, the traverse angles β_i are measured in direct and reverse (example in section 3.7.2). As a check, the horizontal circle is slightly rotated for the observations in the reverse telescope position. The legs S_i are measured optically (section 4.2) or with an EDM instrument, both forwards and backwards. Legs which are longer than 100 m are subdivided when measuring by optical means.

The elevation differences are observed at the same time when measuring the distance. Fig. 12.1.11 illustrates the first case, where H_a and H_b represent the MSL elevations of station and target, respectively. Furthermore, i represents the height of the horizontal axis above the ground point, and h the elevation difference between the horizontal axis and the rod reading m of the centre cross hair or the reference line (g) of the diagram tacheometer. Therefore, the elevation difference of the ground points is

$$H_b - H_a = h + (i - m) \quad \text{or} \quad H_b - H_a = h + (i - g)\,. \tag{12.1.1}$$

If one sets the measure i at the rod, the expression in parenthesis become zero. The same holds if the zero mark is set to i at rods with telescopic leg. If this is not possible because of obstructions in sight, then a value m or g is set on the rod which differs from i by a round amount.

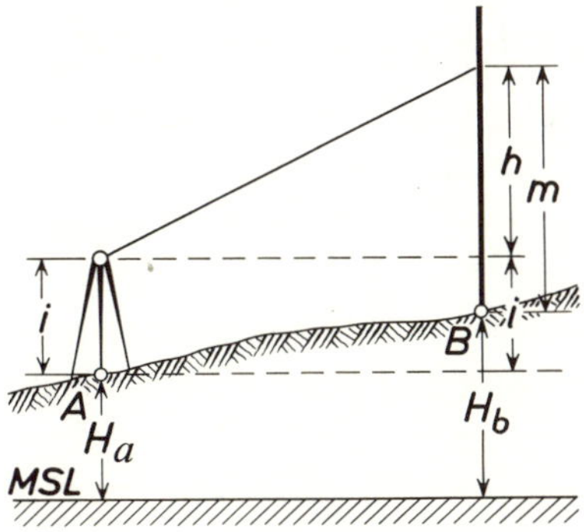

Fig. 12.1.11 Stadia measurement

If the distance has to be subdivided at a point W_1, it is advisable to sight to the same line at the rod from both A and B, which leads to

$$H_b - H_a = (h_a + i_a) - (h_b + i_b)\,. \tag{12.1.2}$$

As a check, elevation difference and distance is determined with a second intermediate point W_2 (Fig. 12.1.12).

If the elevation of one or both of the terminal points of a traverse is unknown, it is to be obtained as shown in Fig. 12.1.11 from a nearby bench mark. In this case, H_a represents the unknown elevation of the traverse point, while H_b is the known bench

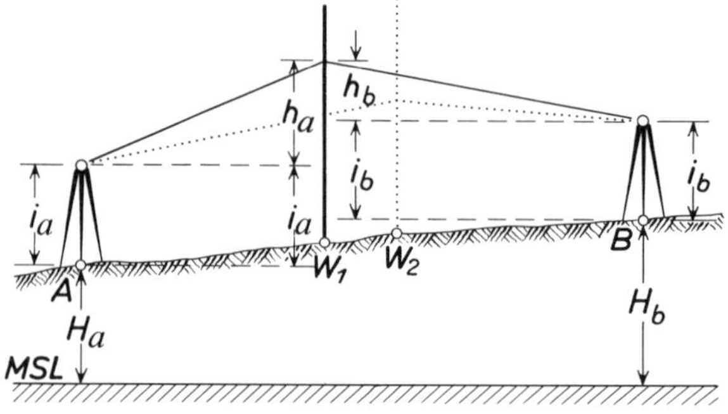

Fig. 12.1.12 Intermediate point

mark elevation. This leads to

$$H_a = H_b - h - (i - m) \quad \text{or} \quad H_a = H_b - h - (i - g) \,. \tag{12.1.3}$$

The computation of primary stadia traverses starts with a field book evaluation. First the mean values of all traverse angles are computed. Then horizontal distances and elevation differences are found, with the mean values recorded in the notes.

The x, y coordinates of the traverse stations are computed as in section 6.8.2. If the measurements were taken carefully, the angular closure should be similar as for lower order traverses, while the longitudinal error should be within $\pm 1.5 \text{ m} \sqrt{S\,[\text{km}]}$.

The sea level elevations of the stations are computed in a separate column, either in the field book or with the traverse computation. For this, the average elevation differences are added over the whole traverse and then compared against the known elevation difference between the terminal stations. If the closure does not exceed $\pm 0.2 \text{ m} \sqrt{S\,[\text{km}]}$, it is distributed to the elevation differences proportional to the length of each leg. By adding all these elevation differences, the known elevation of the end point has to be obtained as a check.

Explanation for the example on page 449

Order of entries at ⊙ 2:

a) Horizontal angles: Observation of the direction to 1 and 3 in both telescope positions; the difference of the mean direction is used to compute the traverse angle, which is then entered and framed.

b) Distance and vertical angle measurement (backwards):

   1) The height i of the horizontal axis is measured and noted in column 1, then the stadia rod is held at 1, and the centre cross hair is set roughly at the value i. Then the lower stadia hair is set to the nearest full decimetre, and a reading is taken on the upper one. Then the upper stadia hair is set to the nearest full decimetre, and a reading is taken at the lower one. Both rod sections are focused and noted. The difference should not exceed 2 mm.

Example: Primary stadia traverse with stadia theodolite

		Horizontal circle		Rod reading		Slope distance	Vertical circle	Horizontal distance	Elevation-difference
Stn.	Targ.	Traverse angle		o_1	o_2	$100 \cdot l_m$	z_I		$i-m$
		direct	reverse	u_1	u_2	ΔE	z_{II}	E	h
		β mean		$l_1 = o_1 - u_1$	$l_2 = o_2 - u_2$	S	α	E mean	ΔH mean
(i)	(m)	gon		m	m	m	gon	m	m
	A	2.180	202.112						
⊙ 1	2	276.482	76.405	1.769	1.8	56.80	105.88		−0.02
(1.48)	(1.50)	274.302	274.293	1.2	1.233	0.06	294.13	56.38	−5.22
		274.2975		0.569	0.567	56.86	−5.88	56.30	−5.22
				1.668	1.7	56.65	94.11		0.00
	(1.40)			1.1	1.135	0.06	305.90	56.22	5.22
	1	0.690	5.307	0.568	0.565	56.71	+5.90		
⊙ 2	3	118.442	123.058	1.756	1.8	75.70	104.13		0.00
(1.40)	(1.40)	117.752	117.751	1.0	1.042	0.10	295.89	75.49	−4.89
		117.7515		0.756	0.758	75.80	−4.12	75.58	−4.92
				1.859	1.9	75.90	95.83		−0.04
	(1.50)			1.1	1.141	0.10	304.20	75.67	4.98
	2	1.003	203.867	0.759	0.759	76.00	4.18		
⊙ 3	4	226.348	29.221	1.997	2.0	119.30	102.48		0.06
(1.46)	(1.40)	225.345	225.354	0.8	0.811	0.25	297.56	119.37	−4.61
		225.3495		1.197	1.189	119.55	−2.46		

Example: Primary Stadia Traverse with RDS

Stn.	Targ.	Traverse angle		e-Curve	$e-g$	ΔE	$i-g$
		direct	reverse	h-Curve	c_h	E	$(h-g)\cdot c_h$
		β mean		g-Curve	$h-g$	E mean	ΔH mean
(i)		gon		m	cm	m	m
⊙ 1	23	0°38.2′	180°38.2′				
	2	176°03.0′	356°03.2′	2.042	69.2	0.08	0.02
(1.37)		175°24.8′	175°25.0′	1.802	−0.1	69.28	−4.52
		175°24.9′		1.35	45.2	69.23	−4.50
				2.091	69.1	0.08	0.00
				1.849	0.1	69.18	4.49
⊙ 2	1	1°12.1′	181°34.9′	1.4	44.9		
	3	179°59.5′	0°22.7′	1.767	66.7	0.07	0.30
(1.40)		178°47.4′	178°47.8′	1.559	−0.2	66.77	−9.18
		178°47.6′		1.1	45.9	66.82	−8.88
				1.868	66.8	0.07	0.23
				1.633	0.2	66.87	8.66
⊙ 3	2	0°02.7′	180°02.5′	1.2	43.3		
	4	245°17.4′	65°17.7′				
(1.43)		245°14.7′	245°15.2′				

Note: The field of view of the RDS is illustrated in Fig. 12.1.4

$g =$ basic curve

$h =$ height curve with height factor $c_h = 0.1; 0.2 \ldots$

$c =$ distance curve

$E = (e-g)\cdot 100 + \Delta E$

$\Delta H = (h-g)\cdot c_h + (i-g)$

2) For the vertical angle measurement to 1 in the direct telescope positions, the centre cross hair is set exactly to the value i or else to a value m (noted in column 2) which differs from i by a round amount. The index bubble has to be centred prior to reading the vertical circle. The zenith distance is computed according to section 10.3.2, and the vertical angle $\alpha = 100$ gon $- Z$ $(90° - Z°)$ is entered.

c) Distance and vertical angle measurements (forwards) are carried out as in b) 1) and b) 2).

d) Compute E and h as in sections 4.2.2.1 or 10.1. and take the mean values using the reversed sight. This value is entered and framed.

The recording of observations with a self-reducing tacheometer is evident from the example on p. 450.

12.1.3.3 Survey of Terrain Points

At the observation stations the directions, distances and vertical angles, or elevation differences to the selected terrain points are measured together with the elements of the secondary stadia traverses. It is desirable to get by with as little data as possible. For a scale of 1 : 1000, horizontal and vertical circles are read in only one position to, at the most, 0.01 gon (30″). Since the average instrument height is 1.4 m, the rods are read at 1.40 m with an accuracy of 1 mm or 0.5 cm. For scales of 1 : 5000 and less, 0.1 gon (5′) or 1 cm are sufficient.

Instead of traverse angles, it is also possible to directly observe the azimuths. For this, the theoretical azimuth of the reference sight is set as computed from coordinates, using the repetition device. Then the sights to the next point of the secondary stadia traverse and to the terrain points are geodetic azimuths. Similarly, at the next station the obtained azimuth ±180° is set for the backsight. This provides a check at the end of the traverse where the closing azimuth has to match its theoretical value within the expected accuracy. It is advisable to check the readings at the stadia hairs by comparing them with the centre cross hair reading.

As for the primary stadia traverse, the elevation differences h_i between station and target are computed and centred in the field book.

Example: Topographic survey with stadia theodolite

	Rod reading		Rod section	Horiz. circle	Vert. circle	Dist.	Elev. diff.	Elev.
	o	m	$l=o-u$	Direction	z		$(i-m)$	
	u	$^1/_2(o+u)$	$100\,l+\Delta E$	r	α	E	ΔH	H
	m	m	m			m	m	m
			Station 27 $i=1.42$					114.64
26				62° 19.5′				
1	1.837	1.4	0.872		87° 53.1′		0.0	
	0.965	1.401	87.60	44° 19.5′	2° 06.9′	87.5	3.2	117.8
2	1.613	1.4	0.425		84° 53.3′		0.0	
	1.188	1.400	42.72	112° 52.1′	5° 06.7′	42.4	3.8	118.4
3	1.753	1.4	0.704		91° 56.6′		0.0	
	1.049	1.401	70.73	176° 33.2′	−1° 56.6′	70.6	−2.4	112.2
4	2.581	2.3	0.563		83° 52.3′		−0.9	
	2.018	2.300	56.57	193° 43.5′	6° 07.7′	55.9	5.9	119.6

Example: Topographic Survey with RDS

Target	Direction	$E =$ $100 \cdot l_c$	$100 \cdot l_h$	c_h	$\Delta H =$ $c_h \cdot 100\, l_h$	H
		m	m	Const.	m	m
	Station 27 $i = 1.45 = g$					183.04
26	0° 00.0′					
1	22° 11.1′	37.4	20.2	0.1	2.02	185.06
2	77° 51.0′	49.5	14.9	−0.1	−1.49	181.55
3	125° 43.3′	17.2	8.0	0.2	1.60	184.64
4	158° 01.3′	11.7	14.5	−0.5	−7.25	175.79

Again, this is based on equation (12.1.1), however $(i - m)$ or $(i - g) < 0.1$ m is neglected.

In the field, a sketch (Fig. 12.1.13) is made which shows the approximate location of the measured points.

Terrain break lines as well as the direction of maximum slopes are shown. Furthermore the terrain form is indicated by short curve sections at selected points, and

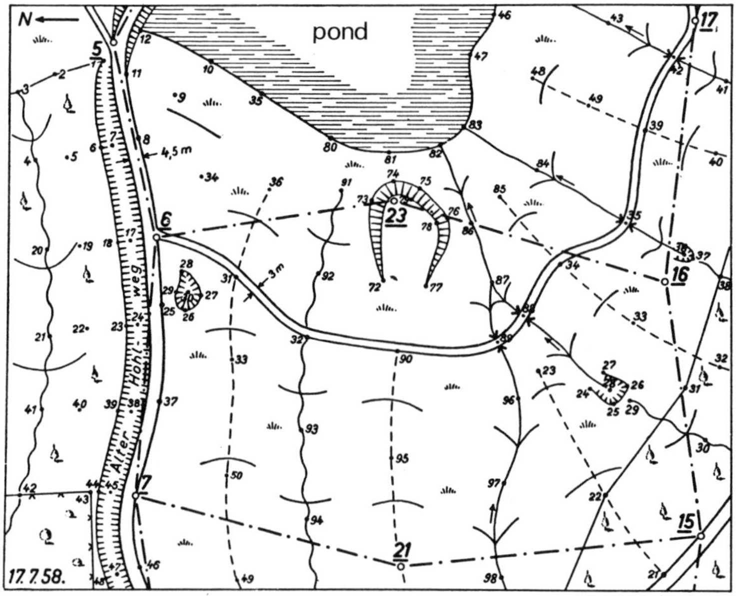

Fig. 12.1.13 Field sketch

dams, ditches, retaining walls, etc., are also included. A good field sketch is just as important as good observations for plotting the final contour plan.

12.1.4 Topographic Survey with Recording Electronic Tacheometers

With these instruments (see, e.g., Fig. 12.1.14) the work can be accelerated significantly, because horizontal and vertical angles as well as slope distances are recorded automatically (section 12.2). Generally, the instruments have small computers which immediately provide horizontal distances and elevation differences. If the field work is properly planned, the traverses are measured in conjunction with the topographic survey. If the terrain is reasonably free of obstacles, a widely spaced traverse net is sufficient as a framework because of the long range of the EDM instruments. Legs and angles are measured with the regular reflectors supplied with the instrument. For the terrain survey, light rods of 1.5 to 8 m in length, equipped with reflectors, are used and set vertical with the aid of a bull's-eye level. The selection of the terrain points is the main job of the party chief. By utilizing a special classified point numbering system, he also can emphasize the relative importance of certain points. For larger areas it is advisable to utilize walkie-talkies for communication.

Fig. 12.1.14 Recording Elta 40 electronic tacheometer (*Zeiss*)

As described in section 8.5.2, an automatic data flow for plotting and for data banks is obtained with this type of instrumentation. This is clearly the most efficient approach.

12.1.5 Plotting of the Terrain Points

First the observation stations are plotted. After all stations are plotted, directions and distances to the terrain points are plotted, and the MSL elevation in decimetres is marked in pencil for each point.

Several small instruments are available for plotting of directions and distances combining angular protractor and scaled ruler. They are usually made of plexiglass as full or semi-circles or else as geometry triangles. The polar coordinator by Ott, Kempten (Fig. 12.1.15) is especially suited for plotting stadia points, as well as for reading polar coordinates. The instrument consists of a pointing ruler coupled with a single counter. The ruler can freely rotate around the pole, which is a pin. The measuring roll moves proportional to the rotational angle. The angles are read at a counter disk, counter drum and vernier, and the distances at the vernier of the ruler. The instrument is available for scales 1 : 250, 1 : 500, 1 : 1000, 1 : 2000.

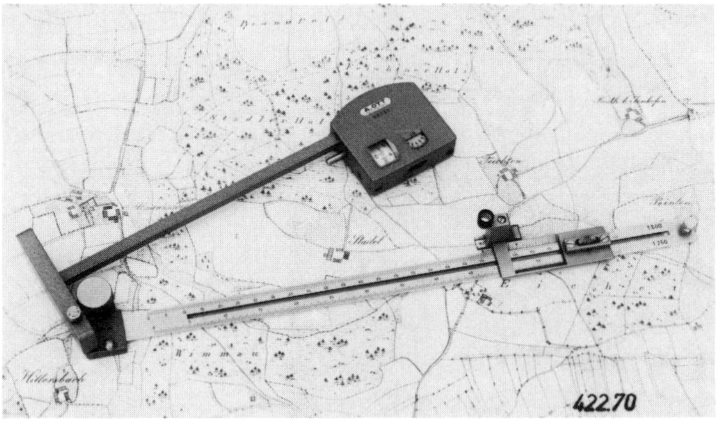

Fig. 12.1.15 Polar coordinator by *Ott*

After all points have been plotted, the plan is first drawn in pencil then the contour lines are designed with careful consideration of the field sketch.

With the development of automatic plotting, the above mentioned methods are nowadays used only for isolated surveys in small areas.

Larger projects, especially the graphical representation of the surveying results, virtually exclusively utilize automatic plotting methods, as described in detail in section 8.5.2.

12.1.6 Stadia Survey with Stadia Compass

Like a stadia theodolite, the stadia compass is used for numerical work. The observation stations are determined by compass traverses in this case. The rotations between the magnetic and geodetic systems need be discussed, as these traverses have to be fitted into a geodetic coordinate system.

12.1.6.1 Fundamentals

Since the magnetic north pole does not correspond to the geographic north pole, a magnet needle does not exactly point towards north, but rather in the direction of the magnetic meridian ("magnetic north").

Between the directions to magnetic north, geographic north (or "true north"), and grid north, which is parallel to the principal meridian of the geodetic reference system (section 5.3), the following relations exist (Fig. 12.1.16).

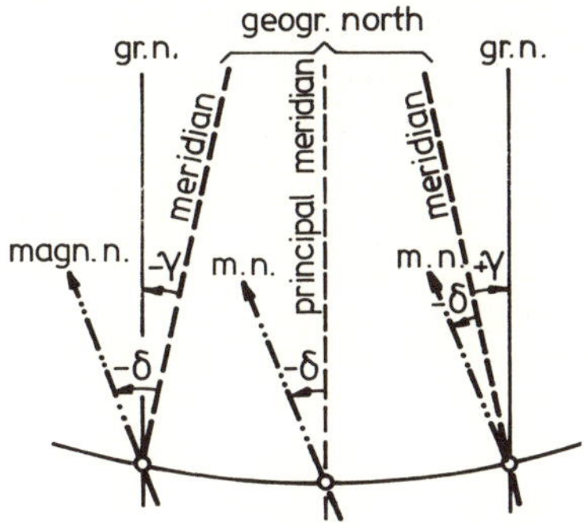

Fig. 12.1.16 Geographic, magnetic, and grid north

a) The angle δ between the directions to magnetic north and to geographic north is called the *declination*. δ is counted clockwise starting from geographic north. If the northern end of the magnetic needle lies east of the geographic north direction, a positive declination is present, while it is negative when pointing west.

b) The angle γ between the directions to geographic north and grid north is called the *meridian convergence*. It is counted clockwise from the local meridian, which means that γ is positive when the point is located east of the principal meridian, and negative when it is west. An approximation for γ is $\gamma = l \sin \varphi$, where l represents the longitude difference between local and principal meridian, and φ represents the geographic latitude of the observation point.

c) The angle between the directions to magnetic north and grid north is referred to as the *needle deflection*. It is the difference between the declination and meridian convergence.

Curves which connect locations of equal magnetic declination are called *isolines*. Their course does not lead straight to the magnetic pole, but is disturbed by local and regional influences. It is also subject to continuing secular changes and daily periodic variations. In Germany, the secular changes reduce the declination by approximately 0.1° per year. The daily period amounts to about 4′ in winter, and 8′ to 10′ on hot summer days, where the minimum occurs in the morning and the maximum near noon.

Most maps give an indication of the needle deflections and its yearly variations on the edge of the sheet.

12.1.6.2 Determination of the Deflection of Sight

The values taken from isolines or similar diagrams are usually not accurate enough for surveying work. Therefore, when using a stadia compass, the magnetic azimuth α_m of a coordinated point P_1 to another coordinated point P_2 is measured in the morning, at noon, and in the evening. After computing the geodetic azimuth, $\alpha_g - \alpha_m$ is formed. The resulting value contains declination and meridian convergence as well as possible instrumentatal errors and is thus called the deflection of sight.

If this deflection of sight is added to the observed compass asimuths α_m, thereby considering variations during the course of a day, the resulting geodetic azimuths are sufficient for the computation of compass traverses (Fig. 12.1.17).

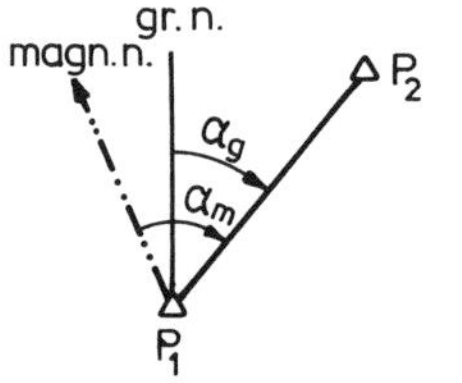

Fig. 12.1.17 Magnetic and geodetic azimuths

12.1.6.3 Measurement and Computation of Compass Traverses

A compass traverse is a short stadia traverse, where compass directions are measured instead of traverse angles, i.e., the magnetic azimuths of the traverse legs. In practice, this is achieved by setting up at every second station only and measuring compass direction, distance and vertical angle to the preceeding as well as to the next station. If the deflection of sight has been determined at both terminal stations, the scheme as illustrated in Fig. 12.1.18 is used, namely the instrument is set up at stations with odd numbers, and the range pole or rod at stations with even numbers.

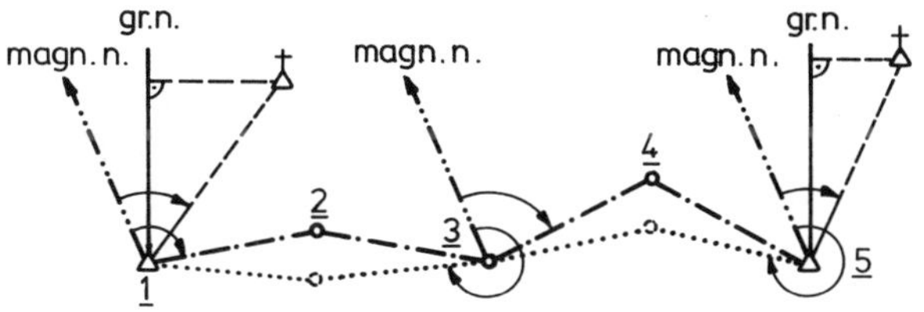

Fig. 12.1.18 Compass traverse

Distances and vertical angles are measured only once. If a check is desired, double stations, as indicated in Fig. 12.1.18, are used.

As with stadia traverses, the coordinates of the primary compass traverses are computed, while secondary traverses are plotted onto transparent paper and then adjusted graphically. When computing the traverses, the deflection of sight has to be applied to all compass directions. Then the traverse is computed as usual. The elevations are computed as for stadia traverses. Compass traverses are preferred in difficult terrain, especially in dense forests.

12.1.6.4 Accuracy of Compass Traverses

The somewhat low accuracy of the compass directions is largely compensated by the favourable error propagation. When setting up at every second station, there are no centring errors. Furthermore, directional errors cause a rotation of the respective leg only, rather than of the whole remaining traverse, thus they are summed according to the square root law (section 1.4.4). Furthermore, since a directional error causes a positional error which increases with the length, it is also favourable due to the short legs (50 m). Finally, if the observer takes care of his equipment, and if he avoids getting too close to rail lines, iron fences, and hydro lines, then the closing error will hardly be larger than for stadia traverses.

12.1.7 Plane Table Surveying

12.1.7.1 Preparations

Before the draughting base (i.e., plastic transparent foil, or white draughting paper with aluminium sheet centre) is attached to the plane table, the mapping grid and all locally available control points are plotted. If there are not enough of these points that can be used as stations for the survey, additional points are determined by stadia traverse. If a plan view of the area is already available, a scaled copy of it is placed onto the plane table.

12.1.7.2 Centring and Orienting a Plane Table

The so prepared plane table has to be centred, levelled and oriented at the observation station (Fig. 12.1.19). *A* and *B* are two locally monumented control points whose map positions *a* and *b* are plotted onto the draughting base fastened to the plane table. In order to centre the plane table above terrain point *A*, its map point *a* has to be vertically above. To achieve this, the centre point *c* of the plane table is coordinated with respect to the map line $a - b$. The offsets *x* and *y* are staked on the ground along and normal to the line *AB* (Fig. 12.1.19). This provides the terrain point *C* with the plane table being centred above using a plumb bob.

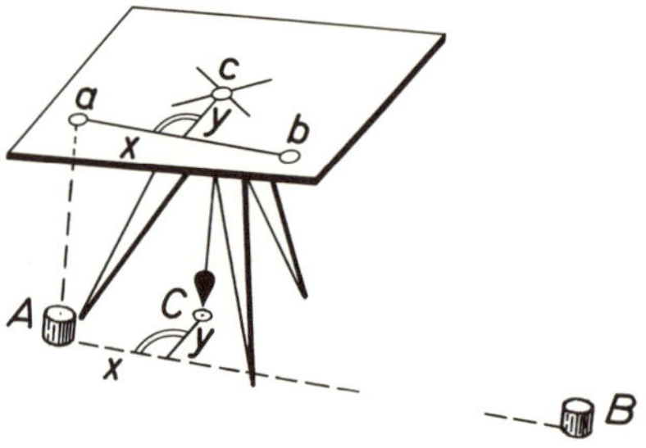

Fig. 12.1.19 Plane tabling

The plane table is then carefully levelled using a spirit level as indicator. Then the alidade edge is laid along the line $a - b$, and the plane table is rotated until point *B* appears in the centre of the cross hair. This centring is checked with the plumbing fork for *A*.

In order to improve the orientation, the alidade is placed at *a*, and additional control points plotted on the plan are sighted. The differences obtained are averaged and used for fine tuning of the orientation. Finally, magnetic north is determined and plotted using the orientation compass. This simplifies the orientation at subsequent observation stations.

If the MSL elevation of the terrain point is known, then the elevation of the horizontal axis of the alidade is determined by measuring the difference with a tape. Otherwise the elevation is determined as mentioned in section 12.1.3.2.

12.1.7.3 Determination of Observation Stations

If the given control and stadia points are not sufficient to cover the area, additional observation stations can be determined by graphical intersection or resection with the plane table.

For the intersection of a new point *P*, at least two control points *A* and *B* on the ground and their plotted positions *a* and *b* on the plane table have to be available.

First the plane table is centred at A and oriented by sighting towards B. Then the alidade, which is positioned at a, is directed towards P and the line $a - p$ is plotted. The same procedure is repeated at B. The intersection of the two lines provides the plotted position of point P (Fig. 12.1.20).

At least three known control points are required for resection. For this method, a sheet of transparent paper is fastened onto the plane table, and the directions to the three control points are plotted from a randomly selected point near the centre (Fig. 12.1.21). Then the transparent paper is loosened and shifted on the plane table until each of the three rays passes through the appropriate point on the base sheet. Then the origin of the rays is pricked through, thus providing the plotted location of the observation station.

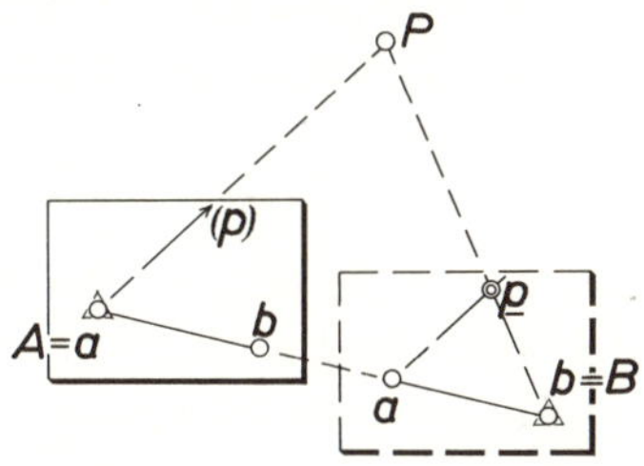

Fig. 12.1.20 Plane table intersection

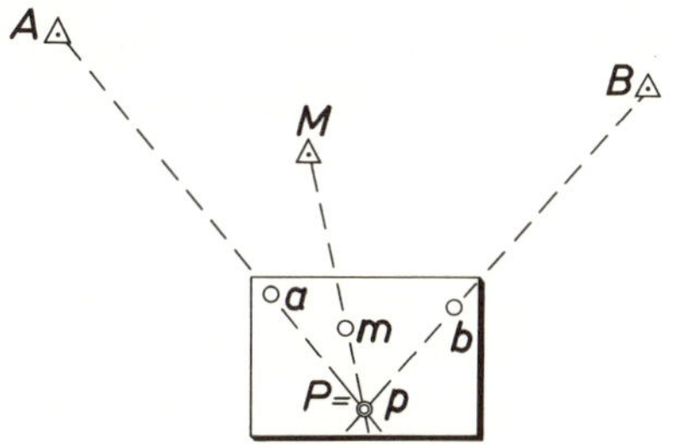

Fig. 12.1.21 Plane table resection

Since these graphical sections are rather time consuming, additional observation stations are determined by traverses and levelling, or stadia traverses or compass traverses, depending on map scale and required accuracy.

The elevation difference between control point and plane table station is obtained with the aid of the vertical angle plus the distance which is extracted graphically from the plane table plot. Since difficult terrain is not suited for plane table work, such areas are excluded and fitted in with compass traverses and stadia observations.

12.1.7.4 Determination of the Terrain Points

The points are selected as discussed in section 12.1.1. For a 1 : 5000 map, the plane table range is approximately 250 m in flat terrain, and 150 m in hilly terrain, while for 1 : 1000 the range is reduced to 100 to 150 m. Prior to commencing the measurements, the observer inspects the terrain and advises the rod man about the point locations. He then goes to the plane table and sends the rod man to the first point. After sighting to and reading the rod with the alidade, distance and elevation difference is determined as with a stadia theodolite. The horizontal distance is laid out along the ruler, the point is pricked onto the map sheet, and the MSL elevation is added in pencil. The rod man in the meantime, marks the point with a piece of paper, or a branch, etc., and moves to the next point where the whole sequence is repeated.

Once all points are measured, the contours are designed by interpolation as for stadia work. For the latter, the plane table is transferred onto a support pole. Then the observer moves from one terrain point to the next and complements the map free-hand drawing using paced distances and a hand held slope indicator.

In the office the lines are inked in, and appropriate names and vegetation symbols are added. Thus a complete map is generated practically in the field.

12.1.7.5 Advantages and Disadvantages of Plane Tabling

Plane tabling requires more field effort than numerical stadia work. However, an experienced plane table observer has a clear image of the contour lines by just observing the terrain, and thus uses the actual points as anchor points which provide the scale for his map. He thus needs far fewer points than for stadia measurements, and therefore recovers most of the time loss in this manner. However, he is bound to one fixed scale.

The arguments for and against plane tabling appear to be endless. As a rule, an observer who likes to measure and compute will prefer stadia, while an artistically inclined observer will lean toward the plane table. He can even get a plane table with alidade and small EDM attachment for faster operation.

12.1.8 Accuracy of Topographic Surveys

The end product of a topographic or stadia survey is a map. Since even for large scales the plotting accuracy is lower than the measuring accuracy, the scale defines the plotting accuracy which amounts to ± 0.2 mm for the position of pricked points, while it is about twice that amount in between. This means ± 0.2 or ± 0.4 m on the ground for 1 : 1000, and ± 1.0 or ± 2.0 m for 1 : 5000.

In addition to the planimetric accuracy, uncertainties in constructing contours affect the contour line accuracy. The standard deviation of a contour line is estimated using *Koppe's empirical formula*:

$$\sigma_h = \pm(a + b \tan\alpha) \tag{12.1.4}$$

where a and b are constants depending on the map scale, and α is the slope angle of the terrain.

Thus the standard deviation in height increases about proportionally to the slope of the terrain. For uniform slope, the height error can also be considered as the positional error of the contour line. For this, σ_h is multiplied by $\cot\alpha$, which leads to the standard deviation of the contour line position

$$\sigma_l = \pm(b + a \cot\alpha)\,. \tag{12.1.5}$$

The specifications for the German 1 : 5000 base map state that the standard deviation in height cannot exceed the values of equations (12.1.5) with $a = 0.4$ and $b = 5$. This leads to the following values for varying slopes:

Standard deviation according to Koppe's formula

α	m_h	m_l	α	m_k	m_l
	m	m	gon	m	m
0°	0.4	—	5°	0.8	10.0
1°	0.5	30	9°	1.2	7.5
2°	0.6	18	18°	2.0	6.2
3°	0.6	14	27°	2.9	5.8

12.2 Electronic Tacheometers

Electronic tacheometers are instruments which measure horizontal and vertical angles as well as distances by electronic means. Usually, the data are stored automatically in semiconductors. The electronic tacheometers are the first instruments which permit an automatic data flow from the generation of the field data to the production of maps or registers.

The development started with the code theodolites. Fennel built a tacheometer based on the 2 m subtense bars, while Kern applied the double image tacheometer principle. A photographic exposure replaced the circle reading. The circles carried code tracks for range readings. The separation of diametrical graduation lines was photographed as analogue quantity for the fine resolution to 0.0001 gon (0″3). The images were scanned by photo-electrical means with the results stored on punched tape.

The next generation of electronic tacheometers is based on the principle of electronic angle and distance measurement. The first instruments of this type were the AGA Geodimeter 700 and the Reg Elta 14 by Zeiss, Oberkochen.

12.2.1 Characteristics of Electronic Tacheometers

Electronic tacheometers can be distinguished by their design, type of data recording, scope of internal user software, range of the distance meter, and accuracy of distance and angle measurements.

There are two distinct types.

a) Instruments where the electronic theodolite forms the base and which is then expanded by the addition of an EDM and a recording unit. Depending on the manu-

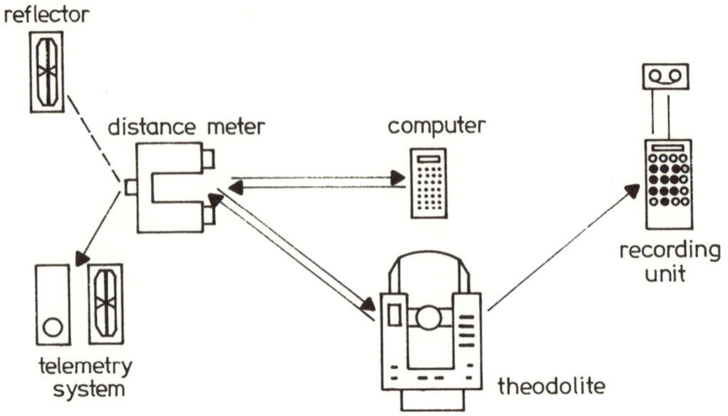

Fig. 12.2.1 Structure of a modular tacheometer (example: *Kern* DM 502/E 1).

Fig. 12.2.2 *Kern* D 15 tachometer

facturer, a computer and a telemetric system can be added (Figs. 12.2.1, 12.2.2 and 12.2.3)

b) Instruments where distance meter, angular measuring systems, recording systems and computer are integrated into one instrument. Depending on the manufacturer, a telemetry system can be included as well (Figs. 12.2.4 and 12.2.5).

Systems of type a) have the advantage that the individual modules can be utilized in

Fig. 12.2.3 *Wild* T 2000 tachometer

different combinations. Even for type b) instruments it is possible to exchange some parts, e.g., the storage for internal evaluation software, which can then be modified.

Electronic tacheometers are monitored and controlled by *microprocessors*. The automatic measuring sequence includes distance measurements, angle measurements, recording, computation of certain derived data (horizontal distance, coordinate differences between station and target, etc.) as well as static or continuous display of measured and computed quantities. The actual measuring task is thus reduced to: circle orientation (for incremental angular measurements), optical sighting, initiation of distance measurement, and initiation of angle measurement (code method).

The data are fed in blocks to the *recording unit*, with a block consisting of words. These may contain measuring values and information for further processing. Such information may be date, point numbers, type of point, number of station, etc., and can be input from a keyboard. Some of this information may be generated automatically by the recording unit. In Fig. 12.2.6 some typical block structures are illustrated.

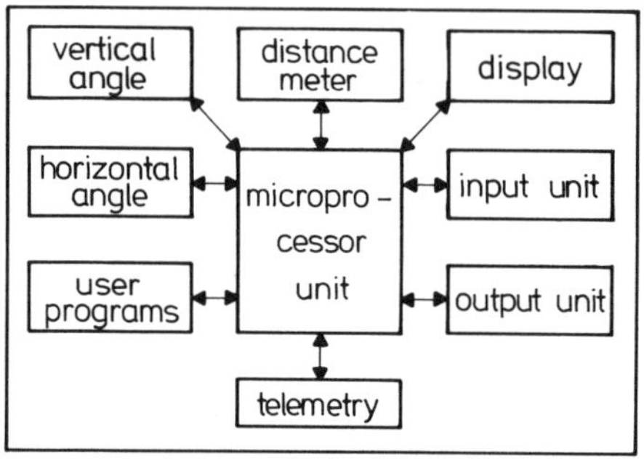

Fig. 12.2.4 Structure of an integrated electronic tacheometer

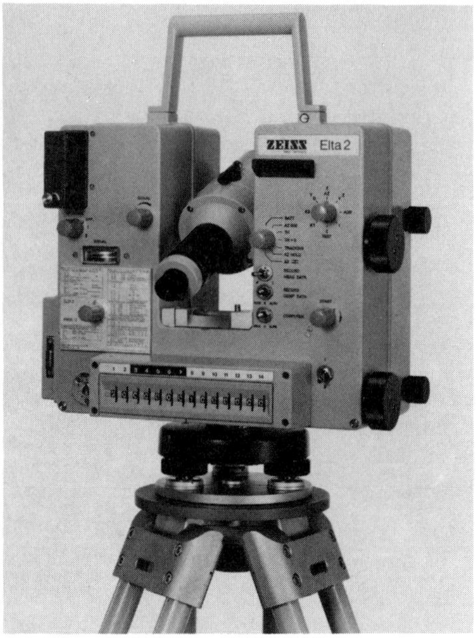

Fig. 12.2.5 *Zeiss* Elta 2

A block is always stored with a recording command. Different manufacturers use different definitions for the blocks. Therefore, there are difficulties in adaption if the software does not conform to the system.

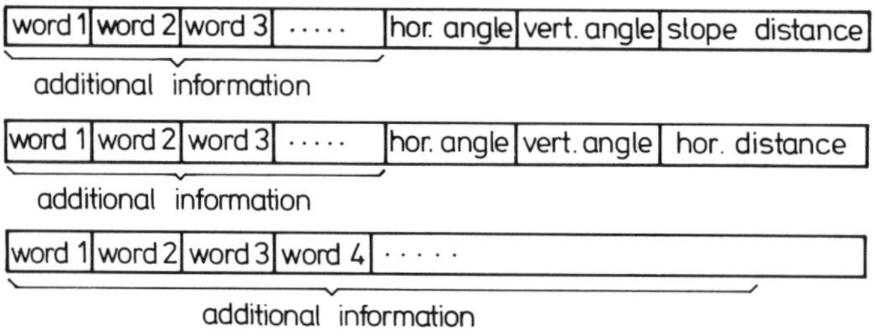

Fig. 12.2.6 Data block structures

Registration units, controlled by a microprocessor are capable of limited data handling and data management. This may include

- Reception of measured values from theodolite and distance meter
- Check of data transfer
- Matching of additional information to measured values
- Display of stored data
- Storage of a coordinate list for internal computation of staking values.

The recording units can be linked to a computer using an interface. However, the data transfer can also be accomplished by telephone through a modem or acoustic coupling (Fig. 12.2.7).

The *internal software* usually is concentrated on two aspects:

a) Consideration of instrumental constants (collimation correction, index correction, scale factor, zero correction).

b) Computation of parameters for staking or measuring of points (coordinates and elevation of instrument for free stationing, orientation of directions, polar coordinates for staking coordinate differences between station and target).

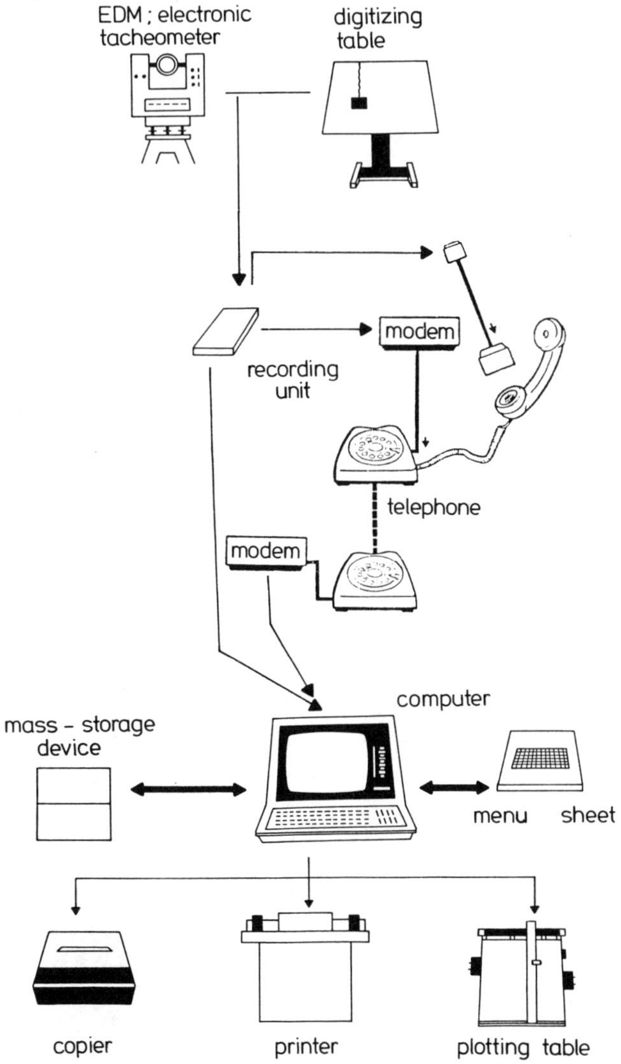

Fig. 12.2.7 Components of an interactive surveying and plotting system

Table 12.2.1 Electronic Tacheometers

Manu-facturer	Type	External structure*	Range No. of prisms	Std. dev. distance		Angle meas.	Std. dev. dir. + rev. H = horizontal V = vertical		Telescope magnif. opening	Tracking	Weight	Special features
			[m]	[mm]	[ppm]		[mgon]	[″]	[mm]	[s]	[kg]	
AGA	Geodimeter	b.	3600	±5	±3		H: ±0.6	2		0.4		Telemetry
Sweden	140		3				V: ±0.6	2				system
Hewlett Packard	HP 3820 A	b.	3000	±5	±5	Code-	H: ±0.6	2	30 ×	1.5…2.5	9.6	
U.S.A.						method	V: ±1.2	3	66			
Zeiss, Jena	RECOTA	b.	1500	±10		Code-	H: ±0.5	1.5			12	
GDR			3			method	V: ±0.5	1.5				
Kern	DM 502	a.	1700	±5	±5	Increment.-	H: ±0.6	2	32 ×		1.6	Telemetry
Switzerland	E 1		3			method	V: ±0.6	2	45		8.7	system
Keuffel &	Autoranger-	a.	4000	±5	±2	Increment.-	H: ±0.6	2	30 ×		13	
Esser, U.S.A.	Vectron		3			method						
Wild	Tachymat	b.	1000…1600	±5		Increment.-	H: ±0.6	2	25 ×	4	9.8	
Switzerland	TC 1		3			method	V: ±0.6	2	34			
	DI 4											
	DI 4 L											
	Di 20											
	T 2000					Increment.-	H: ±0.15	0.5	32 ×			
						method	V: ±0.15	0.5	42	0.3	8.9	

Table 12.2.1 Electronic Tacheometers (continued)

Manu-facturer	Type	External structure*	Range No. of prisms	Std. dev. distance		Angle meas.	Std. dev. dir. + rev. H=horizontal V=vertical		Telescope magnif. opening	Tracking	Weight	Special features
			[m]	[mm]	[ppm]		[mgon]	[″]	[mm]	[s]	[kg]	
Zeiss, Oberk. FRG	Elta 2	b.	2500 / 3 } II**	±10	+2} II	Code-method	H: ±0.2 V: ±0.2	6.6 0.6	30 × 60	1	13.5	
Zeiss, Oberk. FRG			2000 / 3 } I**	±5	±2} I							
Zeiss, Oberk. FRG	Elta 20	b.	2000 / 3 } II	±10	±2} II	Code-method	H: ±0.3 V: ±0.3	1 1	30 × 60	1	13.5	
			1600 / 3 } I	±5	±2} I							user program reduced in comparison to Elta 2
Zeiss, Oberk.	Elta 3	b.	1600 / 3 } II	±10	±2} II	Code-method	H: ±0.6 V: ±0.6	2 2	30 × 60	1	13.5	
			1200 / 3 } I	±5	±2} I							
Zeiss, Oberk.	Elta 4	b.	1600 / 3 } II	±10	±2} II	Increment.-method	H: ±1 V: ±1	3 3	25 × 40	1	6.5	
			1200 / 3 } I	±5	±2} I							

* a. modules, b. integrated system; ** II: measuring range II, I: measuring range I

The principle of the distance meters is described in section 4.4.1. The distances are selected in either feet or metres. The common ranges are 2 to 5 km, and the accuracies are ± 5 to 10 mm ± 1 to $6 \cdot 10^{-6}$ D (D = distance in km).

The *angles* are measured either with the code method or the incremental method (section 3.4.1). The result is displayed in degrees or in gon, whichever is selected. A directional resolution of 0.0001 to 0.0006 gon (0.3″ to 2″) is obtained with instruments using coded circles, while instruments with incremental systems plus interpolators reach 0.0002 to 0.001 gon (0.5″ to 3″). A detailed description of both angular and distance measuring methods is found, e.g., in *Kahmen* (1978). An internal telemetry system can be extremely helpful for tasks like coordinating and staking of points. These systems are described in section 4.4.7.1. Table 12.2.1 provides an overview of the technical data for a selection of electronic tacheometers.

12.2.2 Electronic Tacheometers and Interactive Surveying and Plotting Systems

A medium sized computer is the central unit of an interactive surveying and plotting system (Fig. 12.2.7). The computer must have a large working memory and high speed execution in order to solve larger adjustment problems.

A software package handles data management, computations, graphical work, as well as interactive dialogue between the computer and operator using a graphics terminal. The operator is then able to interpret the computations after certain program steps. The user-storage of the computer also contains data banks, e.g., a geodetic and a cartographic data bank.

The interactive mode is facilitated with a menu. This consists of various input fields to which particular functions, e.g., individual commands or sequence of commands, can be allocated. Automatic data flow is given where the data stored in the recording unit is lead to the computer directly or through the use of two telephones.

12.3 Inertial Positioning

The characteristics of gyros and their use for orientation purposes in surveying and navigation have been thoroughly discussed in section 3.8. There the emphasis was placed an orienting toward the north direction. This principle can be applied to other directions as well, which has led to the development of inertial platforms. The main components of an inertial platform are three precise gyroscopes that retain their spatial orientation of three mutually orthogonal axes.

Inertial platforms have been used in aerial navigation and are now also available for terrestrial surveying work. Such an inertial surveying system is capable of instantaneously providing position differences without requiring intervisibility of points.

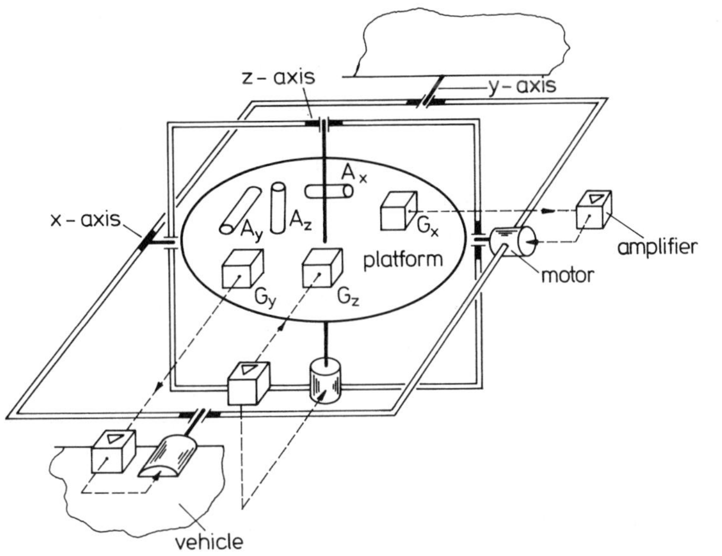

Fig. 12.3.1 Principle of an inertial platform with cardanic bearing, three gyroscopes (G_x, G_y, G_z) and three accelerometers (A_x, A_y, A_z)

The system basically consists of an inertial platform with three gyroscopes arranged such that a special orientation is obtained. Each axis contains a sensitive accelerometer which measures the acceleration component along the axis direction (Fig. 12.3.1).

After sampling these acceleration components at regular time intervals of a few microseconds, the difference of position travelled is obtained by double integration of acceleration with respect to time. All this is carried out in real time, by a computer which forms an integral part of the system.

The computer is also required for monitoring initial alignment, system calibration, and control functions, such as relating the inertial system which is fixed in space to an earth-related coordinate system which rotates in space. Corrections in the form of torque acting on the gyro axes are applied through servo motors driven by the computer. Furthermore, the influences of a number of time or distance related errors have to be reduced. To achieve this, it is necessary to stop the vehicle or land the helicopter every 3 to 5 minutes and determine the velocity, which theoretically should be zero. Therefore, this procedure is called *zero velocity update*. The measured velocity errors serve as input into a complex statistical filtering procedure used to compute the most likely estimates for corrections (*Joos* 1975).

12.3.1 Different Arrangements for the Orientation of an Inertial Platform

Three different arrangements are being used for the orientation of the platform axes:

- With the *space stablized platform*, the initial orientation is retained in space. The orientation of the carrier vehicle is continuously read with respect to the initial orientation of the axes. Servo motors compensate for disturbing forces and for systematic influences. The positional changes, derived from accelerations, refer to the inertial space and thus have to be transformed into geodetic coordinate differences. The Honeywell GEO-SPIN operates according to this principle.
- The *local level platform* is monitored by servo motors such that one axis coincides with the local vertical, the second axis points in the north-south direction, and the third in the east-west direction. Compensations for disturbances are required as well. The computational effort is small in comparison to a space stablized system, however, much more hardware is required. The Inertial Survey Systems (ISS) by Litton and by Ferranti operate according to this principle.
- The axes of a *strap down platform* retain their orientation with respect to the carrier vehicle. Movements of the vehicle cause moments which are detected at the gyro axes and used to compute the directions of the platform axes (strapped onto the vehicle) with respect to the inertial space. At the expense of higher demands on sensors, mathematic models, and computer capacity, the mechanical requirements are reduced. The accuracy of this system is not suitable for geodetic projects at present. However, recent research results are quite encouraging for the future of this approach.

12.3.2 Measuring Methods for Inertial Positioning

The use of inertial platforms can provide highly accurate positions if the following rules are obeyed (*Wickham* 1977; *Barr* and *Carriere* 1981):

- The measuring time has to be kept as short as possible, because system errors increase with the power of three of the time.
- The path between stations should be covered along a straight line with uniform speed, because orientation errors increase with major course changes.
- Zero velocity updates have to be performed at regular intervals of 3 to 5 minutes to provide systems control.
- The system has to be calibrated every five to six hours.
- After operating for one to one and a half hours, a known point should be visited for reference.

These rules can be followed if the measurements are performed along stretched traverses with equal legs and coordinated terminal points. As a rule, measurements are performed back and forth.

The measuring procedure starts with a system calibration which requires a half to two and a half hours, depending on the instrument system. Then the system is brought to the starting point of a traverse, whose coordinates are input into the system. Additional centring elements have to be provided if an eccentric set-up is used. Then the system is moved to the first unknown point. During the trip, momentary positions are being displayed at 16 to 20 cms cycles. This information is useful in locating the new point, provided its approximate coordinates are known. A system check is to be performed about every four minutes by interrupting the trip and going through a zero velocity update. This requires about half a minute. Data obtained during this procedure can be utilized for computing deflection components of the local vertical and gravity acceleration at the station. If the point is to be coordinated as well, a further half minute is required for it. Once the coordinated end point is reached, measure-

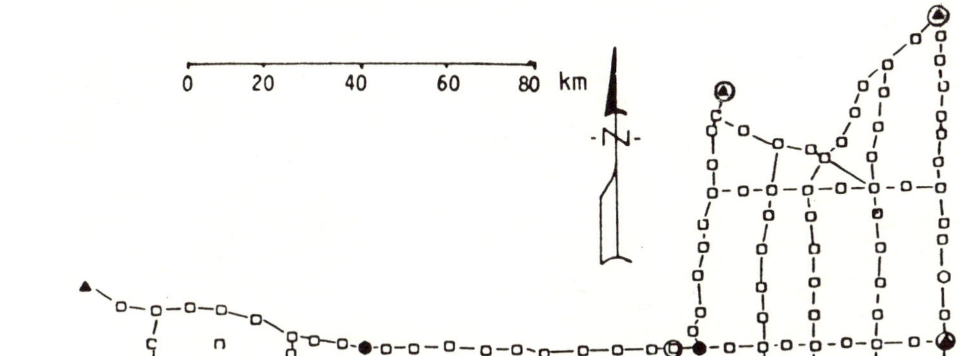

Fig. 12.3.2 Traverse network used for control determination in northern Alberta, Canada (*Schwarz* and *Gonthier* 1981);
▲ first-order control points; ● Doppler control points; ○ height control points; □ new points.

ments are taken there. Then the system compares the measured coordinates with the given ones and distributes the differences among the intermediate points according to an error distribution program. The error model of the instrument is being corrected as

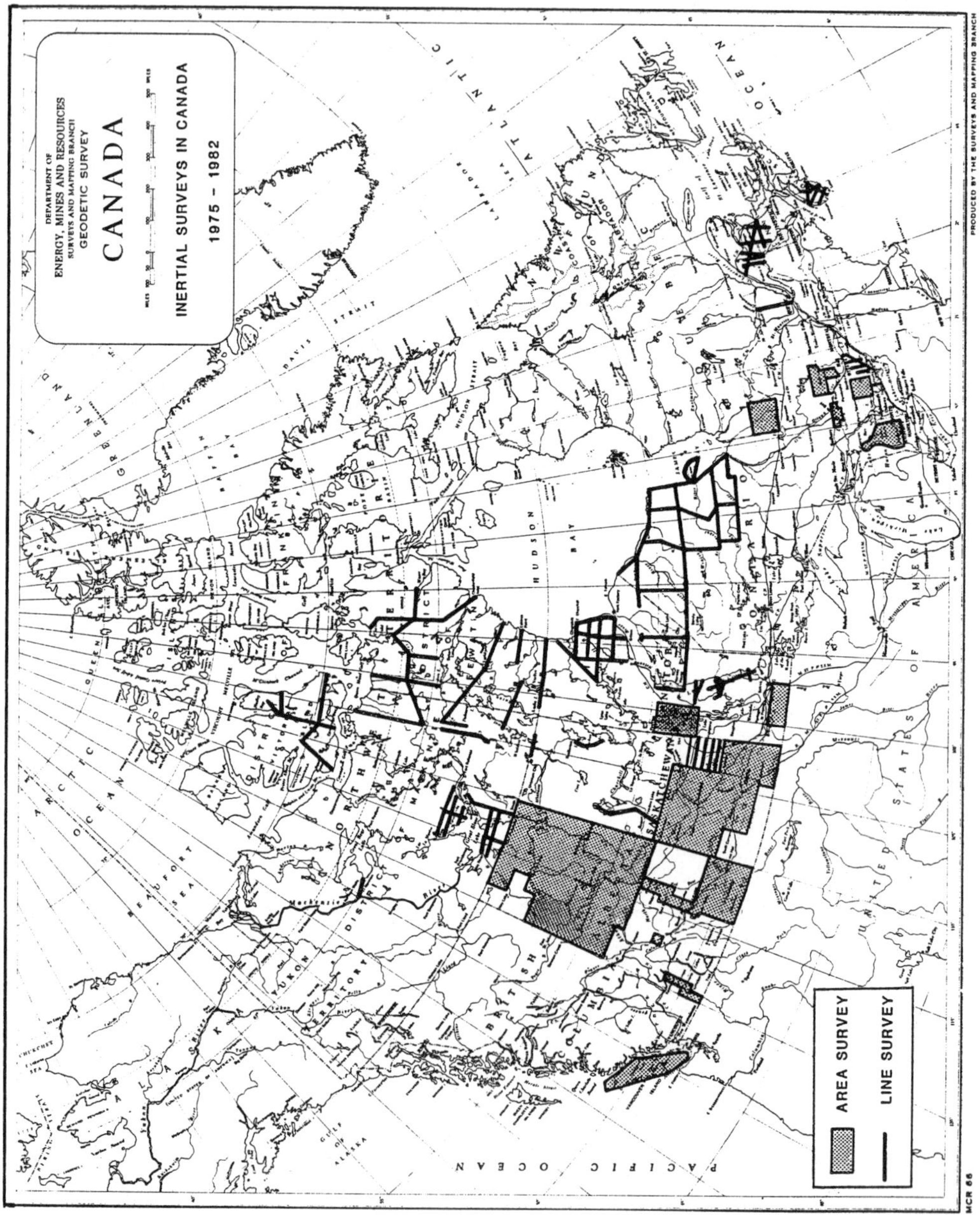

Fig. 12.3.3 Inertial surveys, carried out by the Canadian Geodetic Survey between 1975 and 1982

well. The return survey is conducted in the same manner. Results of forward and return surveys are averaged.

When using helicopters, a point separation of about 10 km can be accommodated within the four minute interval. Therefore points are open spaced accordingly. Then zero velocity update and coordinate determination are carried out at the same stations. Under these circumstances, a 200 km traverse can be measured in one day in both directions.

When using terrain vehicles, a zero velocity update has to be performed every three minutes if the same accuracy is to be reached as with helicopter surveys. This means that a traverse of 40 to 50 km can be measured in both directions on one day.

The average accuracy of a traverse point is $(0.1 + 7 \times 10^{-5} \cdot D)$ metres, where D denotes the separation to the nearest reference point. The accuracy refers to the semi-major axis of the 85% confidence ellipsoid. With a subsequent least-squares adjustment, the distance dependent term can be reduced by a factor 0.2 to 0.3 (*Schwarz* 1980).

For larger projects, it is advisable to determine the points with traverse nets. Fig. 12.3.2 illustrates a typical example (*Schwarz* and *Gonthier* 1981).

The net is arranged in three steps. First-order control points provide the initial framework. The first densification step utilizes satellite positioning with the Doppler method. All further points are tied in by inertial surveys. These points, as shown in Fig. 12.3.2, were coordinated to provide control for 1 : 50,000 topographic mapping in extended remote areas of Canada.

Fig. 12.3.3 provides a general overview of the activities of the Geodetic Survey of Canada from 1975 to 1982.

Comprehensive studies show in spite of high investment costs, that the use of inertial surveying systems is economically feasible if about 1000 new points are coordinated per year and if the system remains operational for seven years. Then the costs per point are significantly less than for conventional terrestrial techniques (*Babbage* 1977).

12.4 Positioning with Satellites

12.4.1 Orbits of Artificial Earth Satellites

At the beginning of the seventeenth century, the astronomer *Johannes Kepler* discovered the following natural laws, named after him and dealing with the orbiting of planets around the sun (see Fig. 12.4.1):

I The orbits of planets are ellipses with the sun located at one of the focal points.

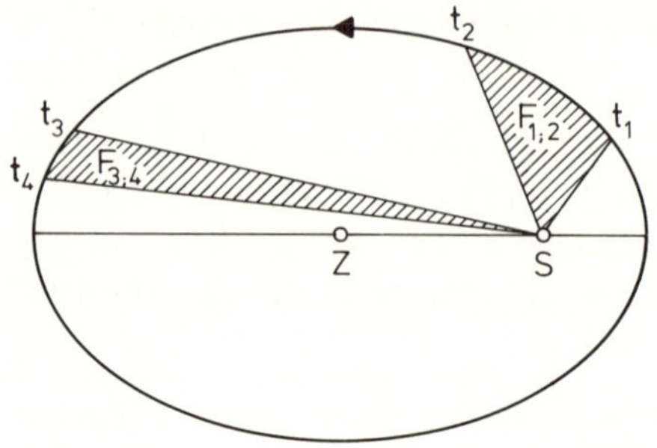

Fig. 12.4.1 Orbit of a planet around the sun S

II The radial vector (from the sun to the planet) scans equal areas in equal time segments; i.e., if $t_2 - t_1 = t_4 - t_3$, then $F_{1,2} = F_{3,4}$.

III The square of the times for completion of one orbit behave like the third power of the major semi-diameters of the orbits.

In a first approximation, the earth can be considered as a homogeneous sphere. Then *Kepler's laws* apply for undisturbed orbits of artificial satellites with the earth as the gravitational centre.

In celestial mechanics, the two-body problem leads to the following equation of motion:

$$\varrho = -GM/r^3 \varrho, \tag{12.4.1}$$

where M = mass of the earth
G = gravitational constant
ϱ = separation of the centres of mass for the earth and the satellite.

Equation (12.4.1) is a second-order vector differential equation with six integration constants. The motion of a celestial body with respect to its central body has, therefore, six degrees of freedom and thus can be defined by *six orbital parameters*.

These orbital parameters are represented in an equatorial coordinate system as shown in Fig. 12.4.2. The following symbols are used:

γ Equinox: on the line of intersection of the equatorial plane with the plane of the earth's orbit.
☊☋ Nodal line: line of interesection of orbital plane and equatorial plane.
☊ Ascending node: point of orbit where the satellite crosses the equatorial plane when moving north.
☋ Descending node: point of orbit where the satellite crosses the equatorial plane when moving south.
P Perigee: point of orbit nearest the earth.
A Apogee: point of orbit farthest from earth.

The centre of the earth (focal point of the orbital ellipse) is denoted by F, while Z represents the centre of the ellipse, and S the position of the satellite.

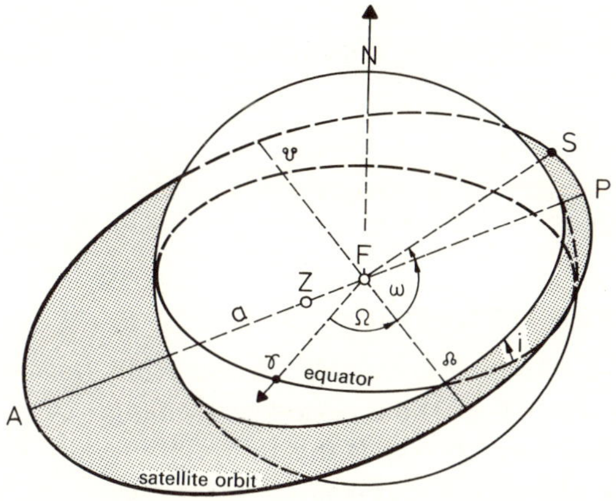

Fig. 12.4.2 Representation of a satellite orbit

In the previously described *equatorial coordinate system*, the position (or the orbit) of a satellite is defined by the following six orbital parameters:

a, b semi-major/semi-minor axes of the orbital ellipse
e numerical eccentricity: $e^2 = (a^2 - b^2)/a^2$
i inclination of orbital plane with respect to the equatorial plane
Ω right ascension[(1)] of the ascending node
ω argument of perigee
v true anomaly[(2)].

Deviations from sphericity, inhomogeneous masses, and external forces acting on the satellite cause *orbital disturbances*. "Kepler's ellipses" become spatial curves, which, while not closed, remain similar to an ellipse. Utilizing the location and velocity vectors, so-called instantaneous or osculating ellipses are defined for each orbit point. The actual satellite orbit is then the envelope of all osculating orbits. Its parameters are thus time dependent and, for practical applications, are obtained by series developments.

The ellipsoidal shape of the earth (flattening at the poles, equatorial bulge) causes the largest disturbance. These deviations in shape generate relatively large perturbations in Ω and ω, whereas the parameters a, e, and i remain relatively stable. The following approximations can be used:

---

(1) Right ascension α and declination δ are geocentric directional coordinates of celestial bodies. The equator is the basic circle with the equinox being the reference direction for α (0° to 360°). δ increases from the equator to the poles from 0° to 90° or to −90°.

(2) v cannot be represented as a direct function of time. Instead of the true anomaly, often the time T of transit of the satellite through perigee P is used, or else the mean anomaly $M = 360° (t - T)/U$. Here, t represents the time, and U the time for one revolution around the orbit.

$$\frac{\mathrm{d}\Omega}{\mathrm{d}t} = \frac{-9.96}{\left(\dfrac{a}{R}\right)^{7/2}(1-e^2)^2} \cos i \quad \text{degrees/day} \tag{12.4.2}$$

$$\frac{\mathrm{d}\omega}{\mathrm{d}t} = \frac{4.98}{\left(\dfrac{a}{R}\right)^{7/2}(1-e^2)^2} (\cos^2 i - 1) \quad \text{degrees/day} \tag{12.4.3}$$

with $R = 6378.14$ km as the radius at the equator.

The observation station usually obtains information on the orbits (*ephemeris*) for positioning. The ephemerides contain the orbital parameters, changes of those parameters, and deviations from them (corrections). If this information is directly transmitted by the satellite, it is referred to as the *broadcast ephemeris*.

With the aid of the orbital information, the satellite position can be computed within a rectangular coordinate sysτem. The X- and Y-axes lie in the equatorial plane, while the Z-axis coincides with the rotational axis, or passes through the pole (Fig. 12.4.3). The X-axis penetrates one selected meridian, which, for instance, can be the Greenwich meridian.

The rotational axis, equator, and meridian are defined from special measurements which are averaged. Thus, one refers to a mean rotational axis, a mean pole, a mean equator, and a mean meridian.

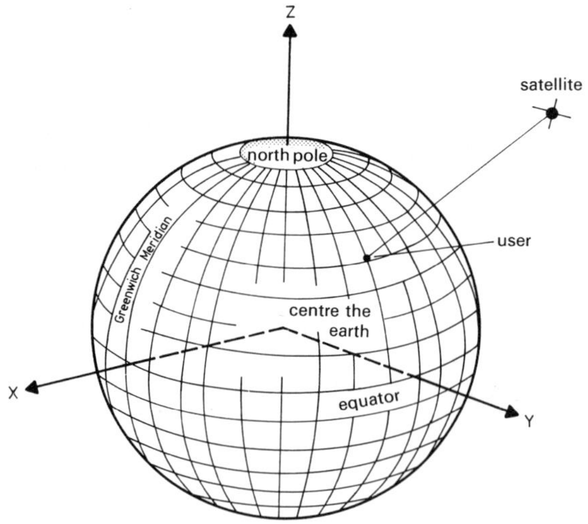

Fig. 12.4.3 Satellite reference systems

12.4.2 Satellite Systems for Positioning and Navigation

Satellite systems for positioning and worldwide navigation have been in the process of being established since about 1960. Usually distances (ranges) or range changes are measured between satellites and observation stations on the earth. These stations on the earth may be moving objects (vessels, vehicles) or static ones (geodetic points, object points).

With the ranges or range changes, three-dimensional coordinates of the ground stations can be computed. Satellite observations are also used for other tasks in geodesy, such as geoid determination, derivation of static and dynamic parameters of the earth.

12.4.2.1 The Transit Navigation Satellite System

The Transit navigation satellite system was initially established for worldwide navigation. It has been operational since 1964 for the U.S. Navy, and because of significant accuracy improvements it has been used for geodetic purposes since 1970.

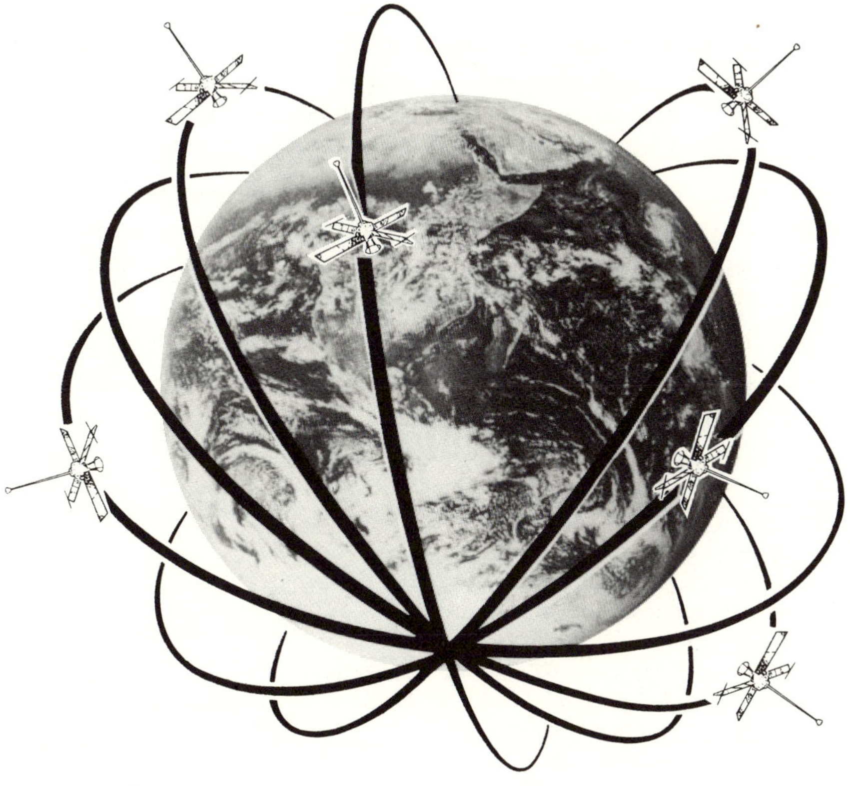

Fig. 12.4.4 Transit satellites on polar orbits

The satellites move along polar *orbits* at approximately 1000 km altitude (Fig. 12.4.4). These orbits are nearly circular with an inclination of $i = 90°$ against the equator. Presently, five satellites are available for geodetic measurements. Forty to 50 transits per day can be observed in mean latitudes, and 20 to 30 transits near the equator.

There are two types of satellites: the OSCAR satellites, and the NOVA satellites. The OSCAR satellites belong to an earlier series. The newer NOVA satellites are equipped with a compensation system which compensates for surface forces and thus permits a much more accurate forecast of the orbital positions.

Broadcast ephemeris are derived from observations of four tracking stations (Maine, Minnesota, California, and Hawaii). They are pre-computed, based on 36 hours of observations and fed to the satellites every 12 hours.

Precise ephemeris are computed from 48 hours of observations at 20 stations distributed all over the earth. They are subsequently provided to a limited user group as rectangular coordinates for minute intervals and as velocity components.

The ephemeris are computed in a special equatorial satellite reference system (Fig. 12.4.3). However, the broadcast ephemeris and the precise ephemeris are based on different reference systems.

The satellites continuously transmit two *carrier signals* of 150 and 400 MHz. These carriers serve as measuring signals as well as for the transfer of time and orbital information. A phase modulation is superimposed on the latter. A receiver, located at the ground station, accepts the signals and evaluates them.

The *position of the ground station* is computed from measured pseudo-range differences (see section 12.4.3.2). Basically, this type of positioning can be carried out with one signal. The second signal can then be used to determine the velocity of propagation within the ionosphere (see section 12.4.3.3).

The *error influences* occurring with one satellite transit are listed in Table 12.4.1.

Table 12.4.1 Error influences during one satellite transit

(1) Errors in the broadcast ephemeris	
– uncertainty of the earth gravity model	± 5 m
– incorrectly modelled surface forces	± 10–25 m[(1)]
(2) Refraction errors	
– ionospheric refraction of third order	± 1–3 m[(2)]
– neglected tropospheric effects	± 1–5 m
(3) Receiver errors	
– oscillator, "time jitter", variation of the phase centre of the antenna	± 1–6 m
	± 10–30 m

[(1)] for OSCAR satellites [(2)] for NOVA satellites

Depending on the selected evaluation model (see section 12.4.4), these error sources affect the result differently:

- For (absolute) *single point positioning*, all error sources listed in Table 12.4.1 affect the result. After approximately 50 satellite transits, an absolute coordinate accuracy of 3 to 5 m is obtained when using the broadcast ephemeris.
- For *relative positioning*, coordinate differences can be determined to an accuracy of 0.2 to 0.5 m by applying refined mathematical modelling and difference methods. This applies to both broadcast and precise ephemeris. As a rule, the results are obtained in the reference system of the ephemeris used and have to be transformed to a local geodetic reference system, i.e., for network densification work.
- A *satellite receiving station* consists of antenna, receiver, data recording unit, and process computer which handles the data and computes preliminary coordinates (see Fig. 12.4.5). A number of manufacturers offer Doppler receivers, such as Canadian Marconi, Montreal; JMR, California; and Magnavox, California. Newer instruments are equipped with a micro-processor system which facilitates fully automatic measurements even for longer observation periods.

Fig. 12.4.5 Doppler satellite receiver: *Magnavox* MX 1502

12.4.2.2 The NAVSTAR/GPS Satellite System

(a) Space segment

The NAVSTAR/GPS (NAVigation System with Time And Ranging/Global Positioning System) was developed primarily for military purposes. By the end of 1989, it will consist of 18 NAVSTAR satellites whose orbits are located within six orbital planes. The *orbits* are nearly circular and have an inclination of $i = 55°$ with respect to the equator. The satellite altitude is 20,000 km, requiring 12 hours for one complete orbital revolution (see Fig. 12.4.6). The satellite constellation is arranged in such a manner that four satellites can be observed simultaneously from any point on earth.

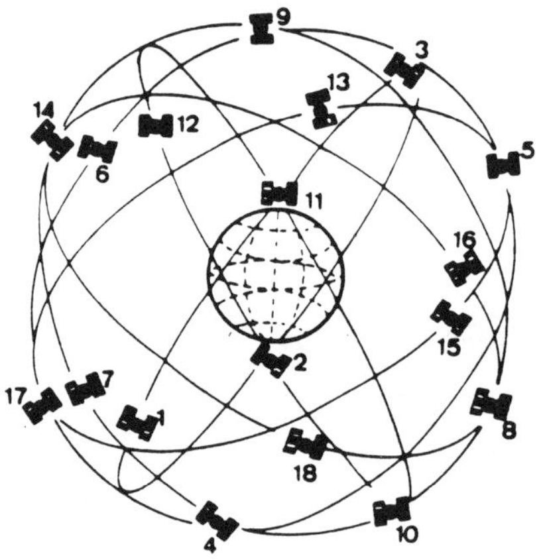

Fig. 12.4.6 Constellation for 18 GPS satellites

The GPS has its own *GPS time* which differs slightly from UTC (Universal Coordinated Time). This time difference is published regularly.

The satellites continuously transmit signals with a specific *signal structure*. There are two signals, *L1* and *L2*, with the following carrier frequencies:

$$L1 = 1575.42 \text{ MHz} = 154 \times 10.23 \text{ MHz}$$
$$L2 = 1227.6 \text{ MHz} = 120 \times 10.23 \text{ MHz}.$$

These are derived from the 10.23 MHz frequency with the aid of a precise atomic clock.

The *L1* signal is modulated with a *P-code* (Precise code) and a *C/A-code* ((binary) Coarse Acquisition code) which is a pseudo-range noise code, or *PRN code*. These codes can be described as rectangular signals whose amplitudes can randomly assume values of zero or one. The covariance function of this random sequence is a delta function. The *L2* signal is only modulated with a *P-code*. In addition, a data signal, which is a binary data stream of 50 bps (bits per second), is superimposed on both carrier signals.

The codes have two functions:

(1) Identification of the satellite. The code patterns are unique to each satellite and can be compared against a respective sample which is generated in the receiver.

(2) Measurement of propagation time. This is the time required for the signal to cover the distance from the satellite to the receiver.

Ther *P-code* is used for more accurate positioning. It has a repetition rate of 267 days and operates with 10.23 Mbps (Megabits per second), i. e., changes between 0 and 1 occur with a frequency of 10.23 MHz. However, each satellite transmits only a segment of the whole code, lasting 7 days. These segments are taken from different time segments of the total pattern and thus can be distinguished. They are re-started every Saturday at midnight. This is also the starting point for the counting of GPS time. The counter unit is an epoch (denoted with z) of 1.5 s duration. The system time of the week is transmitted every 6 s in the form of a transfer word HOW.

The less accurate *C/A-code* has a repetition rate of 1 ms. Again, each satellite has its own pattern. It is easier for the receiver to identify the C/A-code, because it has a shorter repetition rate. Once identification is completed, the transfer to the P-code can be achieved with the aid of the HOW word.

The superimposed *data signal* (navigation message) supplies to the user all the additional data required for positioning and navigation. This signal is formatted into five blocks of 6 s duration each (see Fig. 12.4.7). Each such block consists of ten, 30-bit words. The first word in each block is a telemetry word (TLM). This simplifies access to the data signal and contains further information for the four control stations. The second word in each block is the HOW word. It contains, among others, the number of the already counted epochs z and permits transfer from the C/A-code to the P-code.

The first block contains parameters for the clock correction and for the computations of propagation time delays of the signals in the atmosphere. Blocks 2 and 3 contain the parameters for the forecast of the ephemeris, while block 4 is reserved for special messages. Block 5 contains the almanac data. These data are composed of the follow-

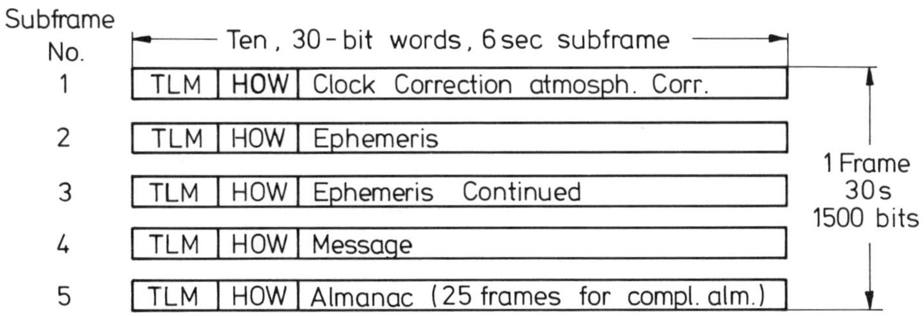

Fig. 12.4.7 Format of the data signal

ing information:

(1) Parameters on ephemeris, clock correction, and atmospheric corrections for all 18 satellites plus one back-up satellite (this corresponds to the content of blocks 1 to 3, but at a lower accuracy).

(2) Information on the status of the satellites. This information facilitates a fast selection of satellites which can be optimally used for positioning.

The almanac data can also be used to compute approximate distances to the satellites. Once these are known, it is easier for the receiver to identify the code signals transmitted by the satellite. Twenty-five blocks are needed to cover the total amount of almanac data. Further information on the signal structure can be found in, e. g., *van Dierendonck* et al. (1978) and *Spilker* (1978).

Positioning is based on the measurement of pseudo-range differences or pseudo-ranges (see section 12.4.3). The latter method is used for precise geodetic positioning in conjunction with phase comparison measurements.

Until 1985, the satellite ephemeris, which were made available for the evaluations, were referred to the World Geodetic System 1972 (WGS-72). Since 1986, WGS-84 has been used as the reference system.

Depending on the selected evaluation model, the individual error sources at the satellite, in the atmosphere, and at the receiver exert different influences on the results.

- The *real-time accuracy for positioning* of a mobile unit is in the vicinity of 10 m when using the P-code and 40 to 60 m for the C/A-code.
- After repeated measurements and averaging, *absolute positioning of a single fixed station* is possible to an accuracy of 3 to 5 m for coordination within the reference system of the ephemeris.
- For *relative positioning*, an accuracy of ± 0.5 to 1 cm $\pm 1 \times 10^{-6}$ for coordinate differences can be achieved by improved modelling and differential methods.

This requires that the pseudo-differences are obtained with the aid of phase comparison measurements with the carrier signals and also that the effects of the ionosphere are determined with two frequencies for larger point separation (> 5 km). Since the resulting values are based on the reference system of the ephemeris, usually a transformation to a local reference system follows (e. g., for network densification).

(b) Monitoring segment

This system is monitored by four *monitoring stations* located in Guam, Hawaii, Alaska, and Vandenberg, California. Each monitoring station receives the satellite signals, computes updated orbital data, and transmits these together with the clock corrections to the satellites where they are stored. These continuously updated data are re-transmitted by the satellite for the users.

(c) User segment

A satellite receiving station consists of an antenna and a receiver. Both units are linked by cable and are portable. A number of companies offer this instrumentation, e.g., Texas Instruments (TI-4100), U.S.A.; Macrometrics (Macrometer™), U.S.A.; WM Satellite Survey Company (WM 101), U.S.A. (see Fig. 12.4.8).

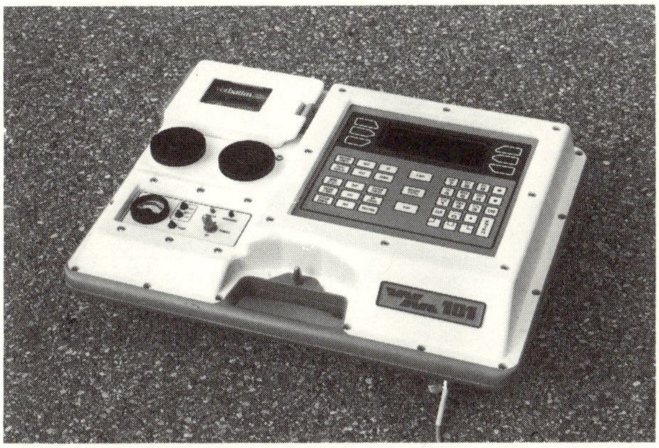

Fig. 12.4.8 GPS satellite receiver system WM 101 (WM Satellite Survey Company)

12.4.3 Observation Equations for Positioning

12.4.3.1 Positioning with Pseudo-range Measurements Between Satellites and Ground Stations

As shown in section 6.5, the x, y coordinates of a point can be uniquely determined in a plane if distances from two fixed points to the new point were measured. Similarly, simultaneous distance measurements between a new point N_i on the earth and three satellites S^j $(j = 1, 2, 3)$ whose positions $(X, Y, Z)^j$ are known, are sufficient for the determination of the (X, Y, Z) coordinates of this station in the satellite reference system (Fig. 12.4.9).

Each distance (range) defines a sphere with the respective satellite located in its centre. Thus three distances (ranges) define three spheres around three satellites. With the three spheres, the position of N_i can be determined by a spatial arc section: the station is located at the point of intersection of the three spheres. Two spheres intersect along a curved line, and the two intersecting lines of two pairs of spheres intersect in one point.

Open ranging systems are used where transmitter and receiver are located at different stations. In this case, both stations require very accurate clocks; ideally they should be synchronized with the system time of the satellite system. Synchronous with its clock, each satellite generates PRN-codes which are modulated onto the carrier signals and

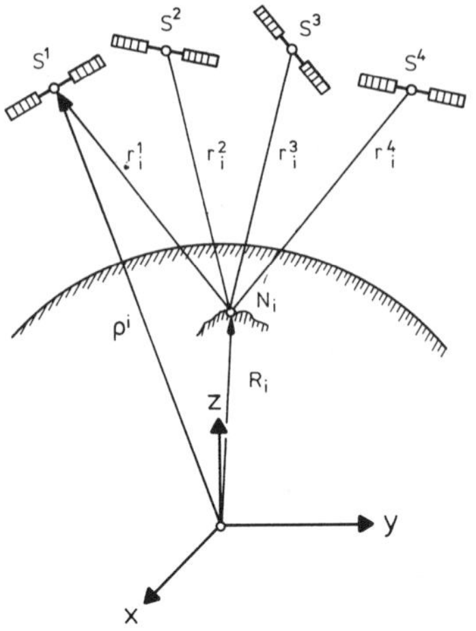

Fig. 12.4.9 Positioning with pseudo-distance measurements in the satellite reference system X, Y, Z

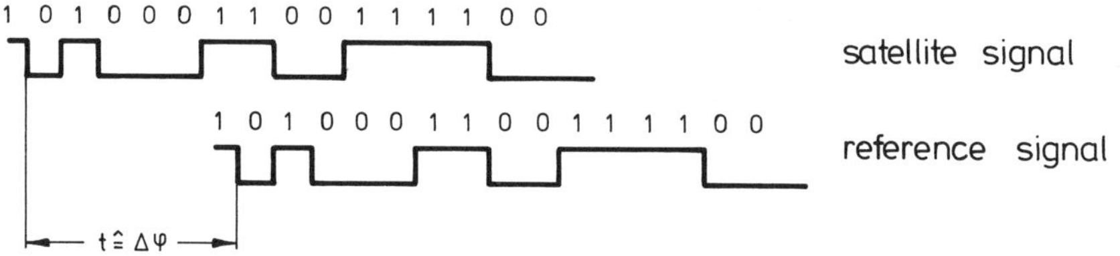

Fig. 12.4.10 Schematic representation of a segment of the PRN codes, generated in satellite and receiver

transmitted to the ground stations. The receiver accepts these signals and generates PRN-codes of the same pattern as reference signals, synchronous with its own clock (Fig. 12.4.10). The PRN-codes transmitted by the satellite show a time or phase shift (t or ϕ) because of the propagation time from satellite to ground station.

Now it is possible to measure distances by using codes with identical patterns, i.e., P-code or C/A-code. To do this, the reference signal and the received signal are shifted in time or phase until they become congruent. This can be checked within the receiver, e.g., with the cross-correlation function of both signals. It deviates from zero if both signals coincide in time, i.e., they are in phase. By multiplication of the propagation time t with the velocity c of the carrier signals, the distance between the jth satellite and the ith station becomes:

$$r_i^j = tc\,. \tag{12.4.4}$$

This leads to the following observation equation (see also Fig. 12.4.9):

$$r_i^j = |\varrho^j - R_i| = \{(X^j - X_i)^2 + (Y^j - Y_i)^2 + (Z^j - Z_i)^2\}^{\frac{1}{2}} \tag{12.4.5}$$

where $\varrho^j(X^j, Y^j, Z^j)$ represents the position vector of the satellite, and
$R_i(X_i, Y_i, Z_i)$ represents the position vector of the receiver,

both in the same spatial coordinate system. The three components X_i, Y_i, Z_i of the receiver station are unknown and can be uniquely determined if simultaneous range measurements were carried out to three satellites.

The method, as described, is based on the requirement that satellites and receiver are equipped with highly accurate clocks which do not deviate from the GPS time. Furthermore, it is assumed that the propagation velocity c can be determined accurately along the whole path. In reality, both clocks have deviations, and the propagation time can only be approximated. The satellite clock is a high quality atomic clock and thus has very small time errors. However, because of the costs involved, receivers are equipped with lower quality clocks, which means generally larger deviations from the system time. In total, these deviations manifest themselves as errors Δt in the propagation time. Thus a correction term $k_0 = c\Delta t$ is added to eqn. (12.4.5):

$$r_i^j = |\varrho^j - R_i| + c \cdot \Delta t = \{(X^j - X_i)^2 + (Y^j - Y_i)^2 + (Z^j - Z_i)^2\}^{\frac{1}{2}} + c \cdot \Delta t_i . \qquad (12.4.6)$$

The parameter Δt acts as an additional unknown aside from the station coordinates. This additional unknown can be determined by measuring a further distance to another satellite. Thus positioning can be uniquely accomplished under these simplified conditions by measuring simultaneously four ranges to four satellites (Fig. 12.4.9). Since, in this case, the actual ranges are no longer directly determined, the term pseudo-ranges is used.

Phase comparison measurements with the carrier signal can be directly taken, instead of using code measurements. As previously shown (see section 4.4.1.2), the phase difference between the signals generated in the transmitter and the receiver acts as a measure for the distance between transmitter and receiver, provided that both signals are synchronized. Since the wavelength of the carrier signals is smaller than that of the code signals, a more than one hundred times improved resolution for the distance measurements is obtained. In the receiver, the phase difference $\Delta\phi$ between the received and generated signals is actually measured. Similar to eqn. (4.4.7), by considering the clocks and propagation time errors (as in eqn. (12.4.6)), we get

$$\Delta\varphi = \frac{2\pi}{\lambda}\{|\varrho^j - \mathrm{R}_i| - a_i\lambda + c\Delta t_i\} . \qquad (12.4.7)$$

Distances are converted into phase differences by using the factor $2\pi/\lambda$; a_i denotes the integer number of the signal periods which occur in the range between the satellite and receiver, and the constant a is another additional unknown. It can be determined with the aid of additional observations. As a rule, a remains constant for each satellite within one observation period, even though several ranges to several satellite positions are being measured. Each receiver usually is equipped with a counter which records integer changes of a continuously.

The methods described with eqns. (12.4.6) and (12.4.7) are used for positioning with the NAVSTAR/GPS. By simultaneously observing four satellites at a time t, positioning is achieved in real time. For geodetic positioning, however, a larger number of observations, spanning a longer time period, are carried out. The coordinates are then computed with a least-squares adjustment, using the available overdeterminations.

12.4.3.2 Positioning with Pseudo-range Differences (Doppler Positioning)

Since the satellites move in relation to a station N_i on the earth (Fig. 12.4.11), their signals are received with a frequency shift because of the Doppler effect.

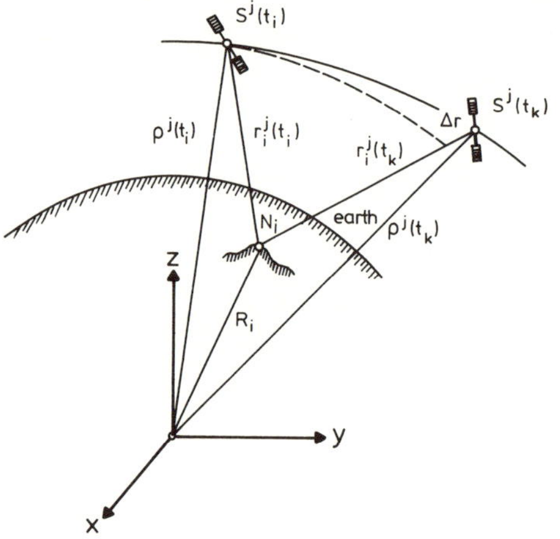

Fig. 12.4.11 Earth station and satellite orbit in the X, Y, Z satellite reference system

The receiver at station N_i does not determine the frequency shift itself but the integral over the frequency shift for a specific time interval, which is the so-called "Doppler count" N. This Doppler count is a measure for the range difference Δr between two subsequent satellite positions $S^j(t_i)$ and $S^j(t_k)$ at times t_i and t_k and the observation station N_i. By phase modulation, information on the orbital parameters (broadcast ephemeris) is superimposed onto the carrier signals. This permits computation of the time dependent satellite positions X, Y, Z within a terrestrial geocentric coordinate system. The receiver obtains this additional information.

The station coordinates in the terrestrial geocentric coordinate system are then computed with the aid of the following equation:

$$\begin{aligned}\Delta r &= |\varrho^j(t_k) - R_i| - |\varrho^j(t_i) - R_i| = \frac{c}{f_0}\{N_{ik} - (f_0 - f_s)(t_k - t_i)\} \\ &= \{(X^j(t_k) - X_i)^2 + (Y^j(t_k) - Y_i)^2 + (Z^j(t_k) - Z_i)^2\}^{\frac{1}{2}} \\ &\quad - \{(X^j(t_i) - X_i)^2 + (Y^j(t_i) - Y_i)^2 + (Z^j(t_i) - Z_i)^2\}^{\frac{1}{2}} \qquad (12.4.8)\end{aligned}$$

where $\varrho^j(t)(X^j(t), Y^j(t), Z^j(t))$ is the position vector of the satellite j at the time t;
$R_i(X_i, Y_i, Z_i)$ is the position vector of the receiver i;
f_0 represents a reference frequency, generated in the receiver, while
f_s is the frequency transmitted by the satellite, and
N_{ik} is the Doppler count.

Since the frequency f_s is only approximately known, the difference $(f_0 - f_s)$ represents an unknown in addition to the station coordinates. As a rule, a number of Doppler counts are observed, and all unknowns are determined by a least-squares adjustment. This method is applied to both the NAVSTAR/GPS and the Transit navigation satellite system.

12.4.3.3 Extensions to the Mathematical Models

(a) Influence of the ionosphere and troposphere.

On their way to the receiver, the satellite signals pass through the ionosphere and the troposphere. The troposphere is practically a neutral gas with extremely weak ionization and a refractive index that deviates very little from one. The ionosphere, however, is a much more ionized matter of extremely low pressure with refractive index values varying between zero and one. Figure 12.4.12 illustrates schematically the structure of the layers in the atmosphere and the behaviour of the signals. Because of different propagation velocities, the signal paths have different curvatures.

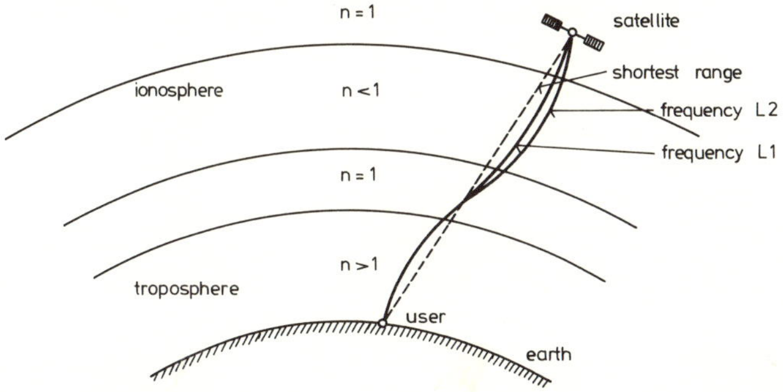

Fig. 12.4.12 Propagation of two wave signals *L1* and *L2* with different frequencies

The *influence of the ionosphere* on the propagation velocity depends on the degree of ionization along the signal path. Signals of different frequency are influenced differently.

For model computations, a vertical profile is utilized which considers the physical characteristics of the ionosphere and the vertical angle under which the satellite is being observed. Furthermore, a strong correlation with the time of day and season of the year can be observed. The average delay of the signal in the zenith direction can amount to 10 ns during the day and 50 ns at night. For small vertical angles, these values can increase by a factor of three. Because of the complexity of the situation, only marginal propagation time corrections can be obtained by developing time dependent correction models.

More accurate corrections can be obtained by measuring two signals (e.g., *L1* and *L2*) with different frequencies. In this case, first the difference $\Delta\tau$ of the propagation time delay of both signals is determined, which can then be used to compute the time delays τ_{L1} (or τ_{L2}) for the individual *L1* (or *L2*) signal. For the *L1* signal, we get:

$$\tau_{L1} = \Delta r \cdot k \tag{12.4.9}$$

with

$$k = 1 : \left[\left(\frac{f_{L1}}{f_{L2}} \right)^2 - 1 \right]$$

and

f_{L1} = carrier frequency of signal *L1*
f_{L2} = carrier frequency of signal *L2*
$\Delta\tau = \tau_{L2} - \tau_{L1}$.

The *influence of the troposphere* on the propagation time delay is a function of the profile for the refractive index, depending on altitude. Differences in frequency cause practically no influence. The distance corrections depending on the MSL elevation of the station and on the vertical angle for the satellite observation is given in Fig. 12.4.13. The following, therefore, is necessary for computing the distance correction:

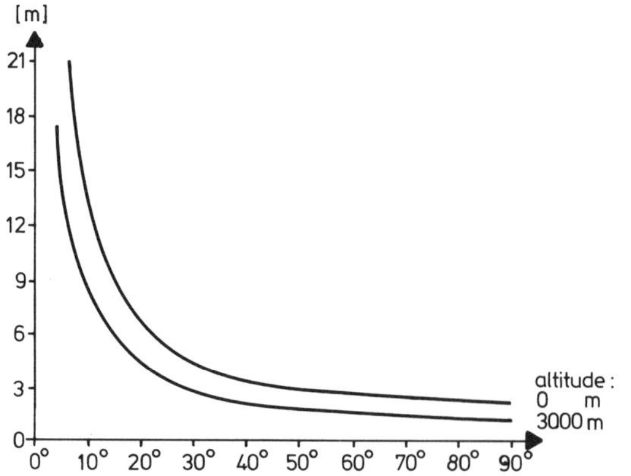

Fig. 12.4.13 Tropospheric range correction with respect to an atmosphere with refractive index $n = 1$

- Profile of the refractive index as a function of the altitude above the ground station.
- Meteorological parameters of the refractive index in the vicinity of the ground station.

- Vertical angle under which the satellite is observed (*Hopfield*, 1979; *Black*, 1978).

(b) Errors in the satellite clock and the receiver clock.

Even though the satellite clocks are highly accurate atomic clocks (accuracy 10^{-12} to 10^{-13}), there are some slight drifts. Figure 12.4.14 shows a typical example of how the time of the satellite clock can vary from the GPS time. The time of a satellite clock can be modelled by a second, or higher, order polynomial for high accuracy requirements. The GPS time, which varies linearly with time, is obtained by averaging the times observed with a group of atomic clocks. Deviations of the receiver clock can be modelled accordingly.

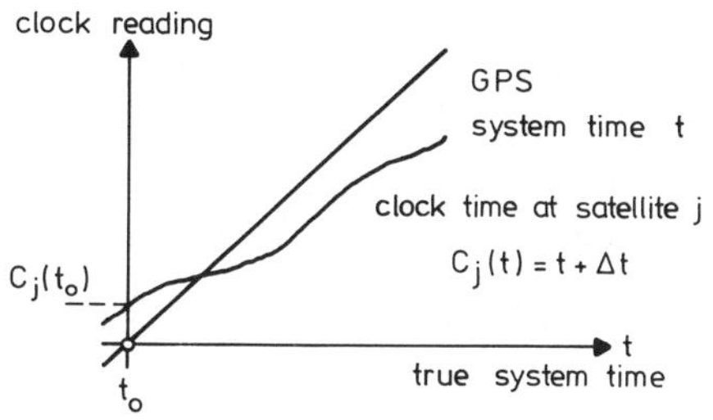

Fig. 12.4.14 Variation of the time of a satellite clock from the GPS system time

(c) Propagation along several paths.

In addition to the signals which reach the receiver along a direct path, signals reflected on the ground can be received under unfavourable conditions. These signals are superimposed on the antenna, which generates phase errors and thus range errors. Specially designed screening below the receiver antenna can eliminate this problem.

(d) Relativity effects.

Two relativity effects can cause small frequency deviations. One effect is caused by different gravity accelerations in the vicinities of the satellite and the station; another by different velocities of the user stations with respect to the satellite reference system. Both effects depend on the location of the user stations.

(e) Orbital errors.

Orbital errors are mainly caused by incorrectly modulated surface forces at the satellite. These are the atmospheric friction and the radiation pressure of the sun. These influences are time dependent and difficult to predict. Furthermore, uncertainties of the gravity model used and of the coordinates of the orbital prediction stations are superimposed. The effect of the orbital errors on the base measurements can be estimated with the following rule of thumb: Orbital errors influence base measurements in the ratio of base length to satellite altitude.

Example: Orbital errors of 20 m affect a 20 km baseline in the ratio of 1 : 1000 for a 20,000 km satellite altitude, i.e., they cause an error of ± 2 cm.

(f) Configuration.

The positioning accuracy also depends on the geometric arrangement of the satellites. When observing four satellites, for example, the optimal configuration would be one satellite at the zenith and the other three separated each by 120° and low on the horizon. An in-depth treatment of the optimization of satellite observations is given in *Wells* et al. (1985).

12.4.4 Evaluation Models for Absolute and Relative Positioning

It is basically possible to perform absolute positioning, i.e., absolute coordination of a single point, or else relative positioning, i.e., coordination of one point relative to another point. Depending on the approach, the error sources at the satellite, in the atmosphere, and at the receiver have different influences on the result.

For *kinematic positioning* (the point is in motion), only a single observation or a very limited number of measurements can be taken. This is especially true if the station moves rapidly, because the position is assumed to be stationary during the measuring process. In this case, the individual error sources will fully influence the result.

For *static positioning* (the point is fixed), it is possible to repeat measurements within an extended time period and, thus, improve the accuracy by averaging.

Further accuracy improvements are possible if the influences described in section 12.4.3.3 are modelled with *additional parameters* in the observation equations.

For *relative positioning*, much higher accuracies can be realized. At least two receivers are utilized for this approach. This permits the introduction of observation differences into the computations. These are free of all those error influences which, for single point measurements, appear with the same arithmetic sign and amount. The effect of forming differences can readily be noticed if eqn. (12.4.7) is transformed into the following form (see also *Wells* et al. (1985)):

$$\Delta\phi = \frac{2\pi}{\lambda}|\varrho^j - R_i| + a_r + \beta^s + \gamma_r^s. \qquad (12.4.10)$$

• α covers receiver based systematic influences, e.g., errors in the receiver clock and refraction errors in the vicinity of the receiver. The parameters of these influences are functions of the specific receiver and of the observation epoch, but not of the specific satellite.

• β covers satellite based systematic influences, e.g., errors in the satellite clock and refraction errors in the vicinity of the satellite. The parameters of these influences are functions of the specific satellite and of the observation epoch, but not of the specific receiver.

• γ covers systematic influences affecting the specific satellite/receiver pair; e.g., total number of full wavelengths and refraction along the path. The parameters of these influences are functions of the satellite and of the receiver, but not of the observation epoch.

It is possible to find different difference formulations, which are denoted by the following symbols:

∇ for differences between measurements from one receiver to the satellites;
Δ for differences between measurements from the receivers to one satellite;
δ for differences between measurements at different epochs.

Thus we can form three single differences, six double differences, and six triple differences. The operations for the single and double differences are summarized in Table 12.4.2 (*Wells* et al., 1985) and show which influences are eliminated in each case.

Table 12.4.2 Operations for single and double differences.

single differences		eliminated parameters
1 $\Delta\varphi \rightarrow \nabla(\Delta\varphi)$		α
2 $\Delta\varphi \rightarrow \Delta(\Delta\varphi)$		β
3 $\Delta\varphi \rightarrow \delta(\Delta\varphi)$		γ
double differences		
1 $\Delta\varphi \rightarrow \nabla(\Delta\varphi)$	$\rightarrow \Delta\nabla(\Delta\varphi)$	α, β
2 $\Delta\varphi \rightarrow \Delta(\Delta\varphi)$	$\rightarrow \nabla\Delta(\Delta\varphi)$	α, β
3 $\Delta\varphi \rightarrow \nabla(\Delta\varphi)$	$\rightarrow \delta\nabla(\Delta\varphi)$	α, γ
4 $\Delta\varphi \rightarrow \delta(\Delta\varphi)$	$\rightarrow \nabla\delta(\Delta\varphi)$	α, γ
5 $\Delta\varphi \rightarrow \Delta(\Delta\varphi)$	$\rightarrow \delta\Delta(\Delta\varphi)$	β, γ
6 $\Delta\varphi \rightarrow \delta(\Delta\varphi)$	$\rightarrow \Delta\delta(\Delta\varphi)$	β, γ

The formation of the double difference $\Delta\nabla(\Delta\phi)$ is illustrated with the aid of Fig. 12.4.15. Measurements have to be taken from two receivers to two satellites. As a first step, the single differences $\nabla(\Delta\phi)$ are formed for both stations, thereby eliminating α in each case. A second step eliminates β by forming the difference between the single differences. By forming the six possible triple differences, the parameters α, β and γ are eliminated.

Whether one uses specifically selected observation differences or applies additional parameters or a combination of both mathematical models, one has to investigate and decide from case to case (*Remondi*, 1984; *Wells* et al., 1985).

The results of these evaluations are initially within the reference system of the satellite ephemeris. Thus they usually have to be transformed into the local geodetic reference system. The relationship between these two systems is provided by the geodetic datum.

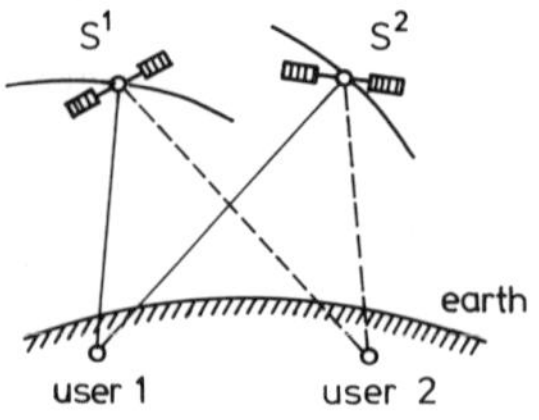

Fig. 12.4.15 Pseudo-range measurements for the double difference $\Delta\nabla(\Delta\phi)$

A geodetic datum consists of a set of parameters which define origin and orientation of another reference system x, y, z with respect to a fundamental absolute system X, Y, Z. In addition, a reference ellipsoid is usually introduced which serves as a reference surface for two-dimensional positioning. The set of parameters is then expanded by inclusion of the semi-major and semi-minor axes a and b of the ellipsoid. The coordinate systems of all practically used reference systems are approximately parallel. Therefore, the following 7 parameter transformation can be applied:

$$\begin{bmatrix} x \\ y \\ z \end{bmatrix} = \begin{bmatrix} X_0 \\ Y_0 \\ Z_0 \end{bmatrix} + (1+\mu) \begin{bmatrix} 1 & \varepsilon_3 & -\varepsilon_2 \\ -\varepsilon_3 & 1 & \varepsilon_1 \\ \varepsilon_2 & -\varepsilon_1 & 1 \end{bmatrix} \begin{bmatrix} X \\ Y \\ Z \end{bmatrix} \tag{12.4.11}$$

with X_0, Y_0, Z_0 being translation parameters,
ε_1, ε_2, ε_3 being rotational parameters, and
μ the scale factor.

In order to determine these parameters, at least three identical points have to be given.

13. Route Surveying

13.1 Basic Design Considerations

Design work for engineering projects, like highways, waterways, and railways, is generally divided into preliminary planning of a general nature and more detailed and specific preliminary design. Preliminary planning encompasses various socio-economic aspects, while preliminary design consists, among other aspects, of the determination of the most suitable route and staking from design plans in the field.

When selecting a route it is desirable to have as little slope as possible. Occasionally a line of constant slope is detected locally in the field with the aid of a hand level, usually, however, it is determined with the aid of a contour plan in the following manner (Fig. 13.1.1).

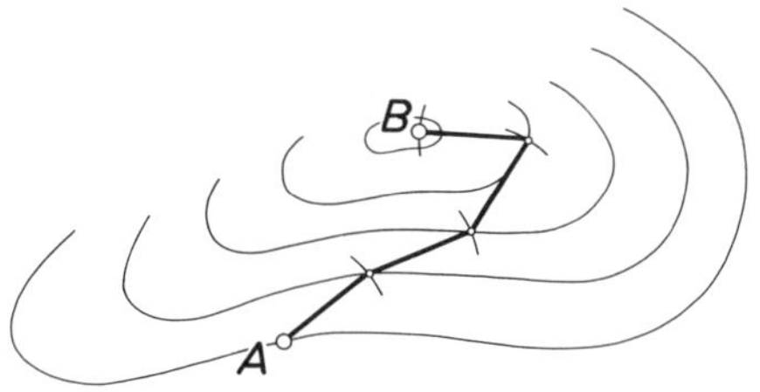

Fig. 13.1.1 Line of equal slope

Assuming a line of $p\ {}^0/_{00}$ slope is to be designed using a contour map with Δh as the contour interval. The elevation difference Δh corresponds to a horizontal distance $s_0 = 1000\ \Delta h$ for a slope of $1^0/_{00}$. Therefore for $p\ {}^0/_{00}$ the horizontal distance is $s_0 = 1000\ \Delta h/p$. (If the slope is given by the slope angle α instead of a per mille value then $s_0 = \Delta h \cos\alpha$.) This distance s_0 (at map scale) is taken on a compass and graphically carried from one contour line to the next, starting at the beginning of the route. This zero line forms the base of the design for position and elevation of the route. This design includes best fitting elements, such as straight lines, circles, and spirals. Special rulers are simple and efficient tools for the graphical design.

Once the design is completed, it has to be transferred onto the terrain. If the design is based on older plans or maps, a sufficient number of points have to be identified in the field as well as on the plan in order to stake the lines and curves.

Maps in 1 : 1000 to 1 : 5000 are most suited for this purpose. If the map coverage is insufficient, mapping is required, which is usually carried out by photogrammetry, but can also be accomplished with tacheometric or stadia measurements (section 12.1).

Often, it is necessary to determine a set of new fixed points (in planimetry and/or height). If this is the case, their location should be such that they can easily be used for staking. All have to be properly monumented and referenced because of possible damage during construction activities.

Different types of instruments can be used for staking:

- theodolite: for extending of straight lines as well as for lining in
- theodolite plus tape: staking of smaller projects with the polar method
- rectangle prism plus tape: staking of smaller projects with the orthogonal method
- electronic tacheometer: for staking of larger projects.

When staking, a basic difference to the usual surveying operations lies in the fact that measurements for points, lines, or elevations are often performed only once (i.e., not in direct and reverse) thus leaving instrumental errors uncompensated. Therefore it is advisable to check the staked values by a separate survey and correct them if needed (section 13.2.5).

Some electronic measuring devices, however, do have compensation of instrumental errors even for single measurements, based on calibration values which provide correction values by means of the internal microprocessor.

Different staking methods are described in the following sections. The final selection of the method depends on the route, available instruments, size of the project, scope of the work, and other external conditions.

13.2 Simple Staking with a Theodolite

13.2.1 Direct Staking of an Intermediate Point on a Straight Line

The theodolite is centred above one end point (Fig. 13.2.1). With a well plumbed vertical axis, the other end point is sighted, and the intermediate points are staked,

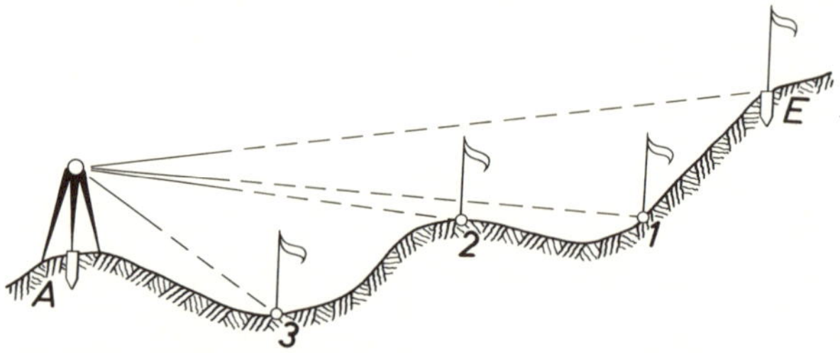

Fig. 13.2.1 Staking of intermediate points

working towards the instrument. If these intermediate points lie at different elevations, the telescope has to be tilted. It is then advisable to repeat the procedure in the reverse telescope position and use the mean of the two staked positions.

13.2.2 Extension of a Straight Line

The theodolite is set up at B (Fig. 13.2.2). After sighting to A, the telescope is plunged and P' is obtained. Then the telescope is rotated and brought back to A. After plunging, P'' is obtained. The point P in between P' and P'' is a point of the line AB.

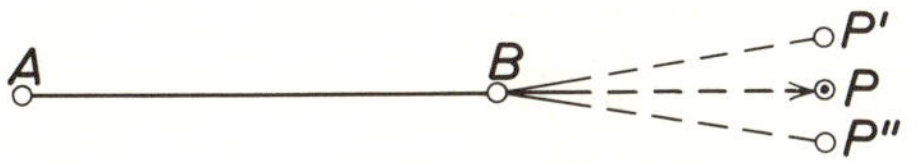

Fig. 13.2.2 Extension of straight line

13.2.3 Staking of an Intermediate Point if the End Points of the Line are not Accessible

The theodolite is set up at point P' which is approximately on the line AB and one measures the angle $AP'B$ (Fig. 13.2.3). $\varphi' = \sphericalangle AP'B - 180°$, and approximate values for a and b, as obtained from a map or by pacing the offset h or P' from the line AB, can be computed. This is best achieved by expressing the area of the triangle $AP'B$ in two ways, namely

$$2A = \overline{AB} \cdot h \approx (a+b) \cdot h$$

and $$2A = ab \sin\varphi' \approx a \cdot b \cdot \varphi' \cdot \frac{\pi}{180} = ab\varphi' \,(\text{rad})\,.$$

The right-hand sides of these equations provide

$$h = \frac{ab}{a+b} \cdot \varphi' \,(\text{rad})\,. \tag{13.2.1}$$

h is now measured toward the line AB. As a check, the angle at the newly staked point is measured. The procedure is repeated if it deviates from 200 gon.

If a and b are unknown, the angle φ'' is measured at a second approximate point P''. Then the distance PP' is obtained from the following ratio.

$$PP' : P'P'' = \varphi' : (\varphi'' - \varphi')\,. \tag{13.2.2}$$

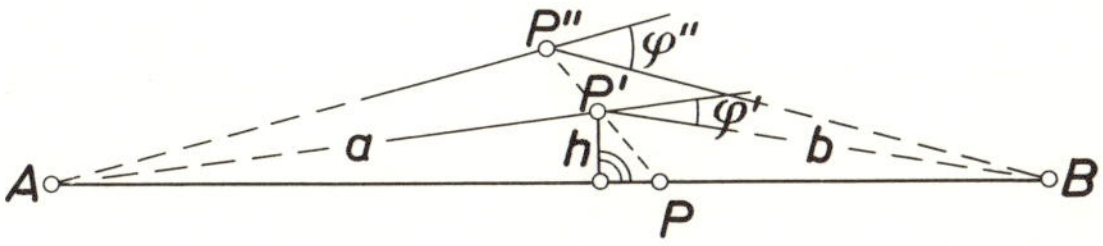

Fig. 13.2.3 Staking of intermediate point with inaccessible end points

13.2.4 Staking of a Straight Line from a Traverse

If terrain obstacles prevent the use of the methods described in sections 13.2.1 and 13.2.3, a traverse may be used which connects the points A and E of the traverse (Fig. 13.2.4). The coordinates are based on the first leg as the x-axis.

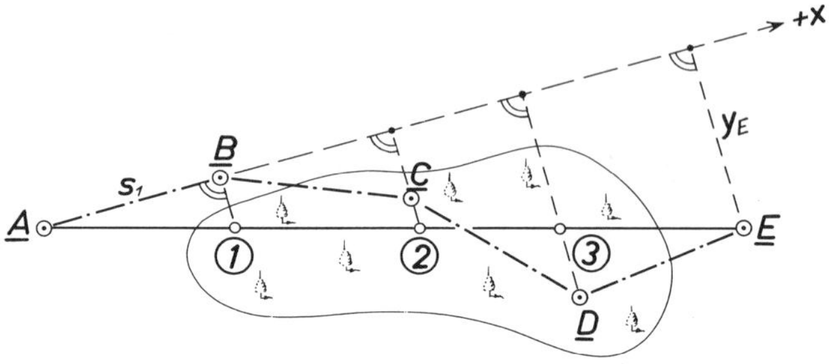

Fig. 13.2.4 Intermediate points by traverse

With $\alpha_{i,k}$ denoting geodetic azimuths, the staking values for 1, 2, and 3 from B, C, and D, respectively, are computed as

$$\sphericalangle CB① = \alpha_{B,1} - \alpha_{B,C}; \qquad B① = \frac{y_E}{x_E} x_B - y_B;$$

$$\sphericalangle DC② = \alpha_{C,2} - \alpha_{C,D}; \qquad C② = \frac{y_E}{x_E} x_C - y_C;$$

$$\sphericalangle ED③ = \alpha_{D,3} - \alpha_{D,E}; \qquad D③ = \frac{y_E}{x_E} x_D - y_D .$$

For the case illustrated in Fig. 13.2.4, $\alpha_{B,1} = \alpha_{C,2} = 180°$, while $\alpha_{D,3} = 0$ gon because $D3$ has the opposite sign. As a check, the angles can be measured at the staked points, or else the offsets from a second traverse can be measured.

13.2.5 Staking of a Given Angle

Two points A and B are given, and a point P is to be staked such that the angle ABP equals φ (Fig. 13.2.5).

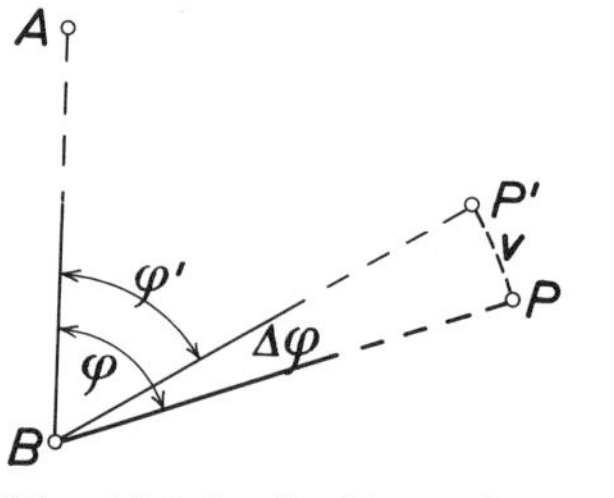

Fig. 13.2.5 Staking of an angle

The theodolite is set up at B and a first point P' is staked in the direct telescope position. Then the angle $ABP' = \varphi'$ is measured in direct and reverse with an accuracy corresponding to the specified requirements.

With $\Delta\varphi = \varphi - \varphi'$ and $BP = s$, the shift value from P' is obtained as

$$v = s \cdot \Delta\varphi \cdot \frac{\pi}{180^\circ} = s \cdot \Delta\varphi\,(\mathrm{rad})\,.$$

For highest precision, see section 14.1.2.

The same principle is utilized when staking with an electronic tacheometer. The conversion of route points into orthogonal or polar coordinates is treated in conjunction with spiral staking (section 13.7.3.10).

13.3 Staking of the Principal Points of a Circular Arc

Circles are staked with a series of points. For this, a few *principal points* are determined according to geometric formulae. Subsequently, *intermediate points (additional points)* are staked using simpler methods.

13.3.1 Staking of Symmetric Principal Points

Usually, the beginning and end points (A and B), and the midpoints S are considered as principal points (Fig. 13.3.1). For longer arcs, further subdivision leads to points B and D.

According to Fig. 13.3.1, the following relationships are used

$$\gamma = 180^\circ - \tau, \tag{13.3.1}$$

$$TA = TE = t = r\tan\gamma, \tag{13.3.2}$$

$$AM = MS = SN = NE = r\tan\gamma/2, \tag{13.3.3}$$

$$TS = \tan\gamma \cdot r\tan\gamma/2, \tag{13.3.4}$$

$$AF = {}^{1}/_{2}AE = r\sin\gamma. \tag{13.3.5}$$

$$FS = HS = p = r(1 - \cos\gamma), \tag{13.3.6}$$

$$MB = ND = r\tan\gamma/2\tan\gamma/4. \tag{13.3.7}$$

Computational checks:

$$TS = TA\tan\gamma/2, \tag{13.3.8}$$

$$(TS + r)\sin\gamma = t, \tag{13.3.9}$$

$$FS + r\cos\gamma = r. \tag{13.3.10}$$

The staking method depends on which parameters are given.

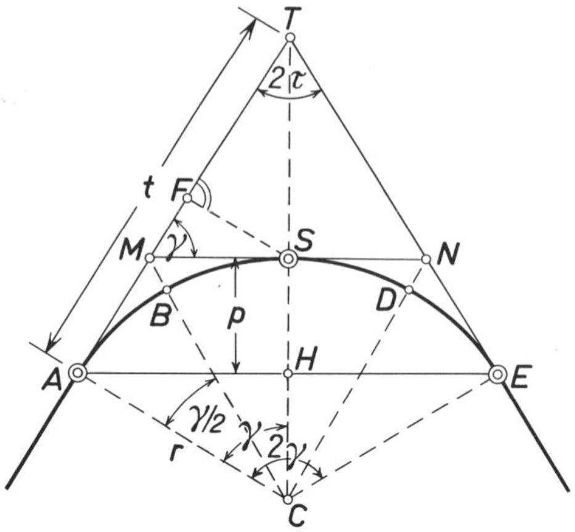

Fig. 13.3.1 Principal points on circular arc

a) If the directions of the two tangents are known, the angle 2τ is measured at T. Then γ and the distances $TA = TE$ are computed and staked along the tangents from T. There are several ways of staking S. The angle 2τ is bisected and the computed distance TS is staked along this line, or AF and FS are computed such that S is staked by orthogonal offset from the tangent, or else AM and NE are computed according to equation (13.3.3) with S lying halfway between M and N. One method can always be used to check another one. B and D are obtained in similar manner considering MN as being a common tangent for the two circular sections AS or SE.

b) If the length t of the two tangents TA and TE is given instead of r, then the latter is computed using equation (13.3.2). The rest follows the approach given in a).

c) If T is inaccessible (Fig. 13.3.2), an auxiliary line PQ is staked. After measuring b, φ and ψ, τ, TP and TQ are computed in triangle TPQ. If b, φ, and ψ cannot be measured, P and Q have to be connected by a traverse or a triangular system.

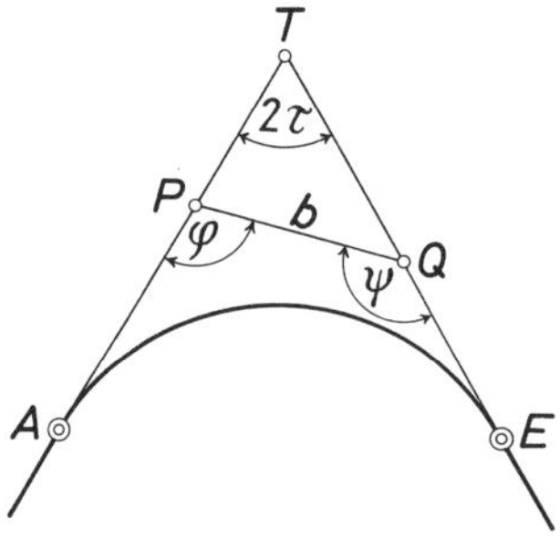

Fig. 13.3.2 Inaccessible point of intersection

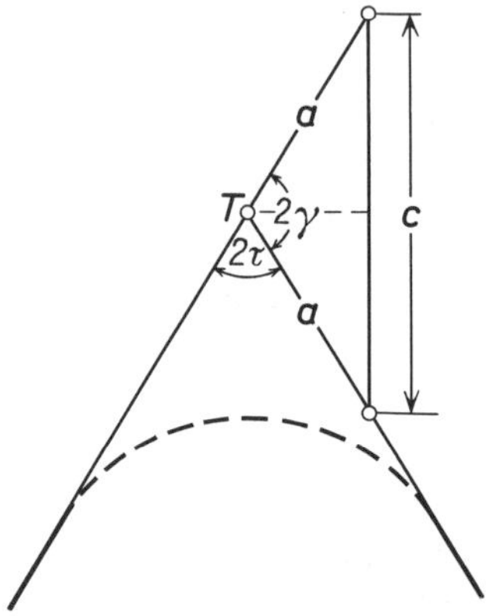

Fig. 13.3.3 Derivation of the angle of intersection

d) If a theodolite is not available for measuring 2τ (Fig. 13.3.3), then equal sections a are staked along the tangents from T. After measuring the distance c between these points, the angle γ is obtained as

$$\sin\gamma = c/2a\,.$$

e) If the arc is to pass through a given point K (Fig. 13.3.4), then 2τ, p and q are measured, and TA, TE, r, and γ are computed.

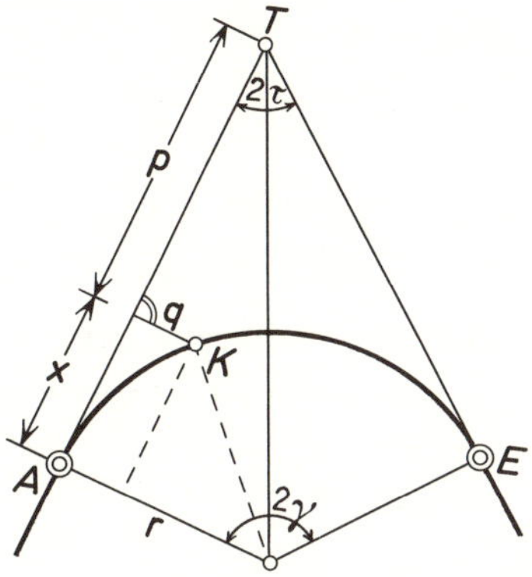

Fig. 13.3.4 Circular arc passing through fixed point

For this, the following regulations are used:

$$x^2 = r^2 - (r-q)^2 \quad \text{and} \quad (x+p) = r\cot\tau\,.$$

With $z = q\tan\tau$, we get

$$x = z \pm \sqrt{z^2 + 2pz - q^2}\,, \qquad (13.3.11)$$

$$TA = TE = x + p; \quad r = TA\tan\tau\,. \qquad (13.3.12)$$

Computational check:

$$x^2 + q^2 = 2rq\,. \qquad (13.3.13)$$

Using the positive sign in equation (13.3.11) provides the circle with the larger radius.

13.3.2 Staking of a Chord Traverse

This method is preferred when the arc is to be staked in a tunnel, forest, etc. (Fig. 13.3.5). First τ, A and E are staked as in section 13.3.1.

If d is the average width of the working corridor, then the longest possible chord s is computed from $s^2 \approx 4rd$. This provides the corresponding centre angle $\zeta = \dfrac{s}{r} \cdot \dfrac{180°}{\pi}$.

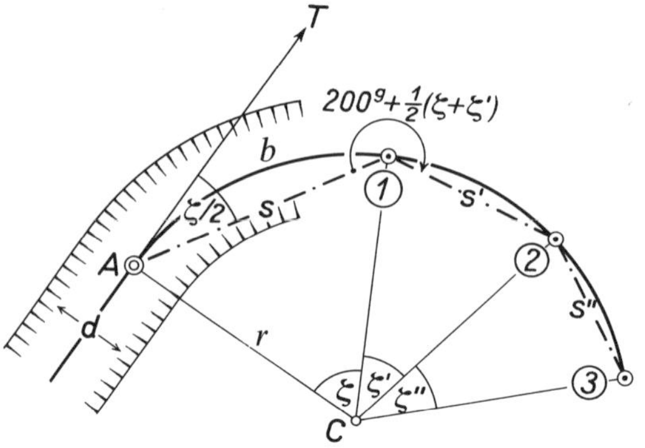

Fig. 13.3.5 Staking of chord traverse

Then several possible methods may be used:

a) If the points 1, 2 ... to be staked are separated by equal arc lengths b, which is desirable for stationing, then a round value for b is selected and the exact centre angle ζ and the chords are computed as

$$\zeta = \frac{b}{r} \cdot \frac{180^\circ}{\pi} \qquad \text{and} \qquad s = 2r \sin \zeta/2 .$$

Starting at A with the tangent as the x-axis, a traverse is staked with all logs equal to s, and the first azimuth TA1 being $\zeta/2$. All other traverse angles are $(180^\circ + \zeta)$.

b) If local conditions permit only steps of different lengths, then the centre angles ζ, ζ', ζ'' ... are computed from the chord lengths s, s', s'' ... . The traverse angles are then $180^\circ + {}^1/_2(\zeta + \zeta')$, $180^\circ + {}^1/_2(\zeta' + \zeta'')$, etc.

c) More accurate results than in a) and b) are obtained if a precision traverse is established near the actual arc and measured with forced centring. Using coordinates, the distances e between traverse point and centre of the circle are computed and compared with the radius r. The differences $(e_i - r)$ are then staked along the direction to the centre.

For tunnel staking, a preliminary layout is obtained using methods a) or b), which is then checked and improved, if necessary, using method c).

13.4 Staking of Intermediate Points

13.4.1 Rectangular Offsets from the Tangent

For this method, one of the principal points staked according to section 13.3 is used as the origin, and the tangent to the curve at this point as the x-axis. Usually, either equal x values or equal arc lengths are used.

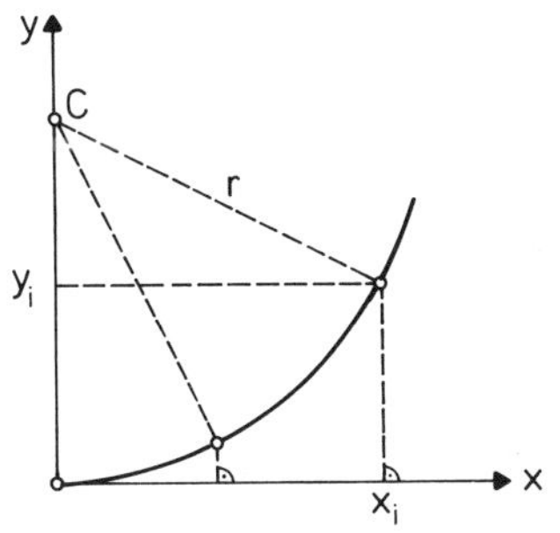

Fig. 13.4.1 Tangent offsets

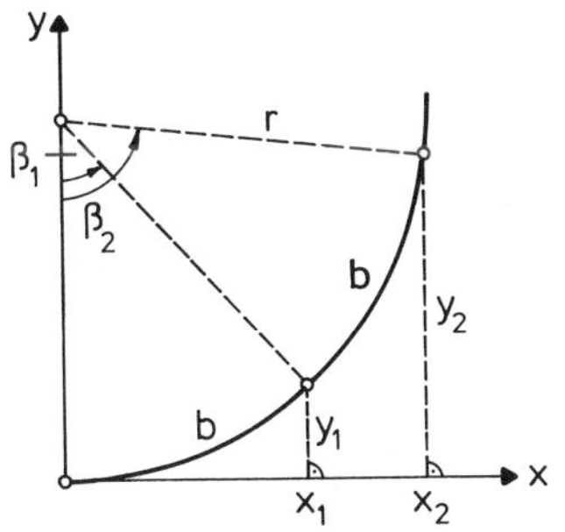

Fig. 13.4.2 Tangent offsets for equal arc lengths

Even x values (Fig. 13.4.1) are selected depending on the purpose of the curve (e.g., every 10 metres). The offsets are computed as

$$y_i = r - \sqrt{r^2 - x_i^2} = r - \sqrt{(r + x_i)(r - x_i)} \tag{13.4.1}$$

or
$$y_i = x_i \left\{ \frac{x_i}{2r} + \left(\frac{x_i}{2r}\right)^3 + \dots \right\} \tag{13.4.2}$$

or, derived from equation (13.4.1):

$$y_i = \frac{x_i^2}{2r} + \frac{y_i^2}{2r} + \dots . \tag{13.4.3}$$

For stationing, usually *equal arc lengths b* are used (Fig. 13.4.2). If the origin of the stationing coincides with the origin of the coordinate system, then the following equations provide the x, y staking values:

$$\begin{array}{lll} \beta_1 = \frac{b}{r} \cdot \frac{200 \text{ gon}}{\pi} = \frac{b}{r} \cdot \frac{180^\circ}{\pi} & x_1 = r \sin \beta_1 & y_1 = r(1 - \cos \beta_1) \\ \beta_2 = 2\beta_1 & x_2 = r \sin \beta_2 & y_2 = r(1 - \cos \beta_2) \\ \beta_3 = 3\beta_1 & x_3 = r \sin \beta_3 & y_3 = r(1 - \cos \beta_3) \\ \vdots & \vdots & \vdots \end{array} \tag{13.4.4}$$

For larger arcs, staking commences from both sides to avoid long ordinates.

Usually, however, the origin of the stationing does not coincide with the origin of the coordinate system (Fig. 13.4.3). This leads to *unequal arc lengths*.

The station of the origin is defined by the lengths of the previous route elements, while the station of the end point equals that of the starting point plus the arc lengths. Stations at given intervals of b metres (e.g. 20 m) are to be staked, using the following formulae:

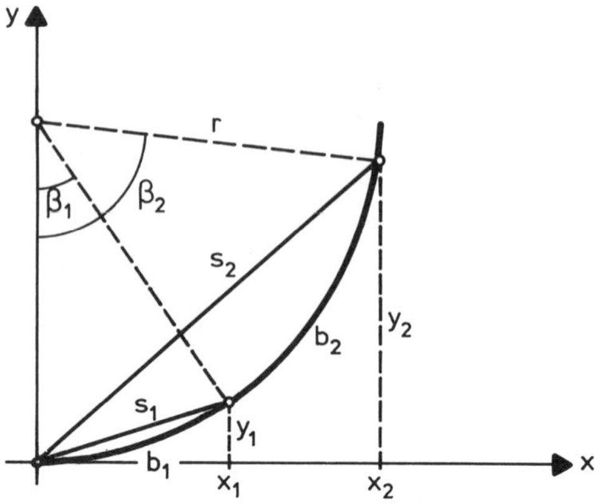

Fig. 13.4.3 Tangent offsets with unequal arc lengths

$$\begin{aligned}
\beta_1 &= \frac{b_1}{r} \cdot \frac{180^\circ}{\pi} & x_1 &= r \sin\beta_1 & y_1 &= r(1-\cos\beta_1) \\
\beta_2 &= \frac{b_1+b}{r} \cdot \frac{180^\circ}{\pi} & x_2 &= r \sin\beta_2 & y_2 &= r(1-\cos\beta_2) \\
\beta_3 &= \frac{b_1+2b}{r} \cdot \frac{180^\circ}{\pi} & x_3 &= r \sin\beta_3 & y_3 &= r(1-\cos\beta_3) \\
&\vdots & &\vdots & &\vdots \qquad (13.4.5) \\
\beta_{n-1} &= \frac{b_1+(n-2)b}{r} \cdot \frac{180^\circ}{\pi} & x_{n-1} &= r \sin\beta_{n-1} & y_{n-1} &= r(1-\cos\beta_{n-1}) \\
\beta_n &= \frac{b_1+(n-2)b+b_n}{r} \cdot \frac{180^\circ}{\pi} & x_n &= r \sin\beta_n & y_n &= r(1-\cos\beta_n).
\end{aligned}$$

13.4.2 Rectangular Offsets from the Chord

If the intermediate points are to be staked from the chord (Fig. 13.4.4), their coordinates are first referred to a tangent parallel to the chord as discussed in section 13.4.1.

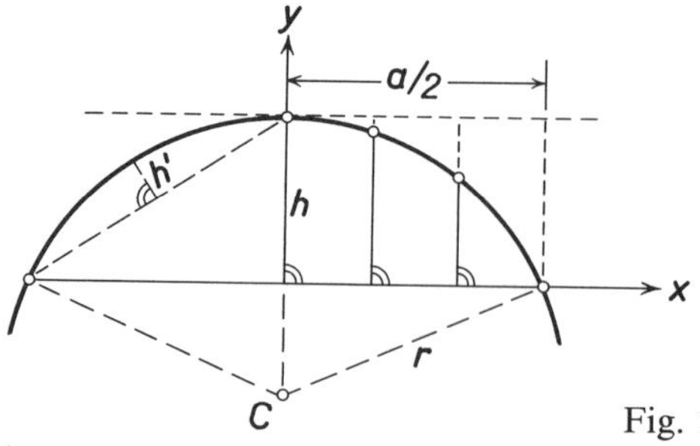

Fig. 13.4.4 Chord offsets

When switching to the chord, the x values along the chord remain the same as for the tangent, while the offset represents the difference to the value h, which is

$$h = r - \sqrt{r^2 - a^2/4}\,. \tag{13.4.6}$$

13.4.3 Polar Staking

Polar staking methods have become increasingly popular, especially with the availability of electronic tacheometers. A simple method can be derived using the starting point of the circle and the respective tangent. The peripheral angle ω is equal to the angle between the tangent and chord (Fig. 13.4.5).

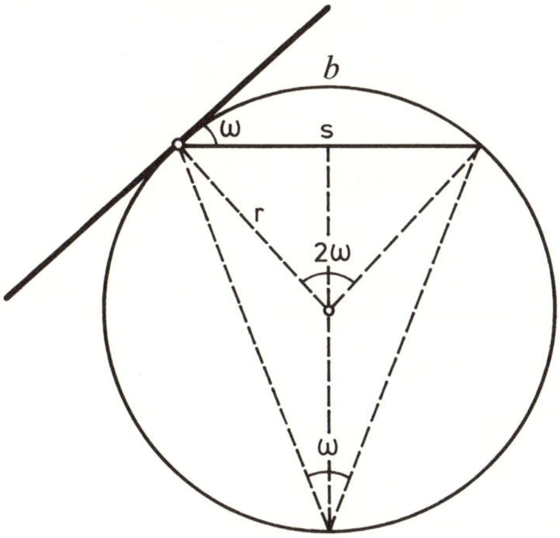

Fig. 13.4.5 Peripheral angle, centre angle and chord/tangent angle

Thus

$$\omega = \frac{\gamma}{n} = \frac{b}{2r} \cdot \frac{200\,\text{gon}}{\pi} = \frac{b}{2r} \cdot \frac{180°}{\pi} \tag{13.4.7}$$

$$s = 2r \sin\omega$$

where n denotes the number of equal arc lengths. For *equal arc lengths* the polar staking elements are determined with respect to the tangent (Fig. 13.4.6).

$$\left.\begin{array}{lll} P_1: & \omega = \dfrac{b}{2r} \cdot \dfrac{200\,\text{gon}}{\pi} = \dfrac{b}{2r} \cdot \dfrac{180°}{\pi} & s_1 = 2r\sin\omega \\ P_2: & 2\omega & s_2 = 2r\sin 2\omega \\ P_3: & 3\omega & s_3 = 2r\sin 3\omega \end{array}\right\} \tag{13.4.8}$$

If the arc is only one element of a longer route, its stationing does not usually start at the point of curvature. Thus *unequal arc lengths* are encountered. Again, the polar

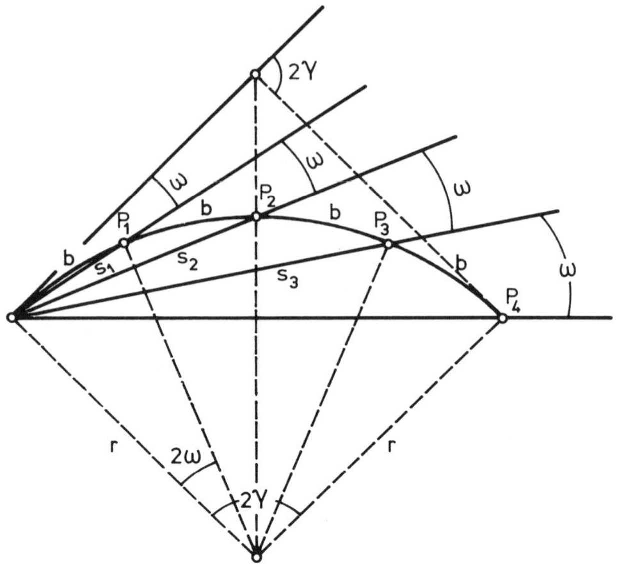

Fig. 13.4.6 Polar staking with chord/tangent angles plus chords

staking elements are determined with respect to the tangent (Fig. 13.4.3):

$$\left.\begin{aligned}
&P_1\colon\quad \omega_1 = \frac{b}{r}\cdot\frac{200\,\text{gon}}{\pi} = \frac{b_1}{r}\cdot\frac{180^\circ}{\pi} && s_1 = 2r\sin\omega_1 \\
&P_2\colon\quad \omega_2 = \frac{b_1+b_2}{r}\,\frac{200}{\pi} && s_2 = 2r\sin\omega_2 \quad s_2' = 2r\sin(\omega_2-\omega_1) \\
&P_3\colon\quad \omega_3 = \frac{b_1+b_2+b_3}{r}\cdot\frac{200}{\pi} && s_3 = 2r\sin\omega_3 \quad s_3' = 2r\sin(\omega_3-\omega_2) \\
&\vdots && \vdots
\end{aligned}\right\} \quad (13.4.9)$$

As a check, the chord lengths s_2', s_3' ... can be computed and measured in the field.

13.4.4 Staking with Deflection Angles

Chords of equal length require equal deflection angles (Fig. 13.4.7). Furthermore, the deflection angle equals the angle between the tangent and chord. These two facts provide the base of the following staking approach.

Depending on the purpose of staking, the arc length b between two neighbouring points is given. Then the deflection angle and the chord length can easily be computed.

The theodolite is set up at point A of the arc, and pointed toward point E. Then the telescope is turned by ω towards A, and the chord length s is measured from E with a

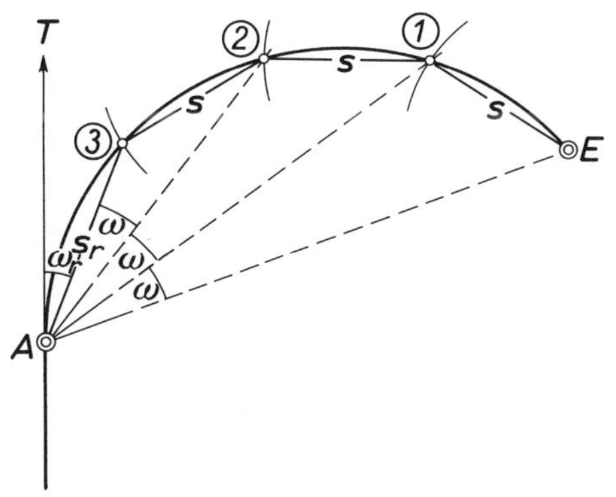

Fig. 13.4.7 Deflection angles

tape. By rotating the tape until the end point of s comes into the line of sight, point 1 is obtained. Now the telescope is again turned by ω, and s is set out from 1 to obtain 2, etc. Finally, a residual arc b_r with corresponding deflection angle ω_r and chord s_r are left. ω_r and s_r are measured and computed as a check using

$$\omega_r = \sphericalangle TAE - n\omega; \quad s_r = 2r\sin\omega_r\,. \tag{13.4.10}$$

To simplify the method, either round deflection angles or round chord lengths are used. The intermediate points are staked towards the instrument in order to avoid the accumulation of errors. This method provides good values with little measuring and computational effort. The working speed can be increased if the new points are roughly staked according to the second method (section 13.5). In this case, the approach is specifically suited for staking on top of dams.

13.5 Approximate Formulae for Practical Use

For rough staking and checking of already staked arcs, approximate formulae often simplify the work.

Approximate staking from the tangent (Fig. 13.4.1). The first term of equation (13.4.3) leads to

$$x \approx s; \quad y \approx \frac{s^2}{2r}\,. \tag{13.5.1}$$

Approximate staking from a secant (Fig. 13.5.1). This method requires that consecutive points have been previously staked.

According to equations (13.4.5) and (13.4.4) we get

$$y \approx s\sin 2\omega; \quad \sin\omega \approx s/2r; \quad \sin 2\omega \approx s/r$$

which leads to

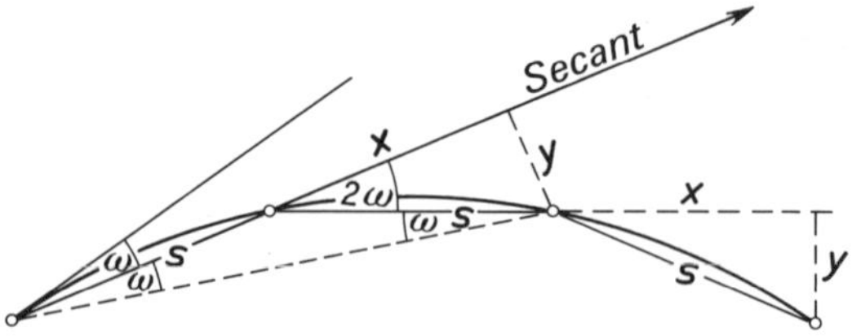

Fig. 13.5.1 Approximate staking from secant

$$x \approx s; \quad y \approx \frac{s^2}{r}. \tag{13.5.2}$$

Determination of the mid-ordinate (Fig. 13.5.2). Application of the binomial series to equation (13.4.6) results in

$$h = r - r\sqrt{1 - a^2/4r^2} \approx r - r(1 - a^2/8r^2 + \ldots)$$

or

$$h \approx \frac{a^2}{8r}. \tag{13.5.3}$$

Quarter method (Fig. 13.5.3). In order to derive the mid-ordinate h' for half the arc from the mid-ordinate h of an arc, a in equation (13.5.2) has to be replaced by $a/2$. Thus

$$h' \approx \frac{a^2}{32r} \approx {}^1/_4\, h. \tag{13.5.4}$$

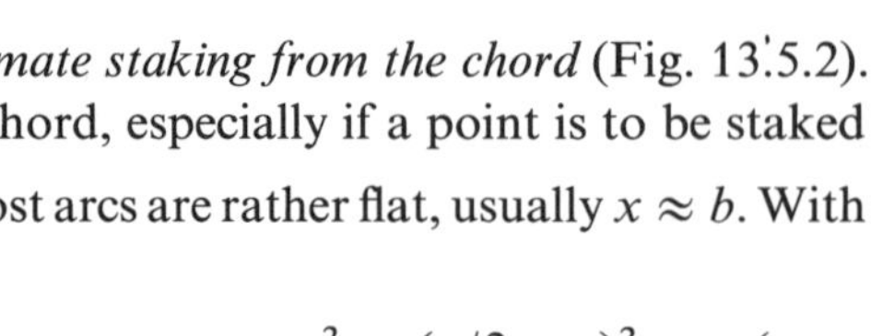

Fig. 13.5.2 Mid-ordinate and quarter method

Approximate staking from the chord (Fig. 13.5.2). Often it is most practical to stake from a chord, especially if a point is to be staked between two already staked ones.

Since most arcs are rather flat, usually $x \approx b$. With equations (13.5.1) and (13.5.3) this leads to

$$y = h - y' \approx \frac{a^2}{8r} - \frac{(a/2 - x)^2}{2r} \approx \frac{x(a - x)}{2r}. \tag{13.5.5}$$

13.6 Compound Curves

The position of a circular arc is defined by three parameters, such as the directions of the two tangents plus the radius. If additional conditions are imposed, such as the position of the points of a tangent, fixed points on bridges, etc., then the route may consist of circular arcs with different radii which share a common tangent. In most cases, it is necessary to determine the relative position of the route elements by indirect measurements. Furthermore, it is usually necessary to insert easement curves between straight lines and circular curves or between circular arcs with different radii (section 13.7). Finally, often it is necessary to deal with large changes in elevation. The computation of compound curves is thus rather difficult, and only the basic fundamentals (Fig. 13.6.1) are discussed.

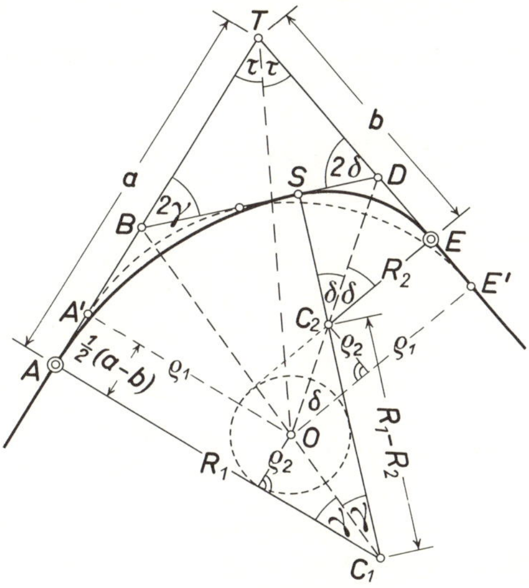

Fig. 13.6.1 Compound curve

For a compound curve, consisting of two circular arcs, the two tangents and the angle 2τ between them may be given together with three further pieces, usually taken from the two tangent sections a and b, the two radii of curvature R_1 and R_2, or the ratio between them. The following relations exist between these parameters.

First the two auxiliary circles, illustrated by dashed lines, are considered. These circles have a common centre at 0 which lies on the line bisecting the angle 2τ. The radii of these auxiliary circles are

$$\varrho_1 = \frac{a+b}{2}\tan\tau; \quad \varrho_2 = \frac{a-b}{2}. \tag{13.6.1}$$

According to Fig. 13.6.1, the centre angles 2γ and 2δ for the arc sections AS and SE are:

$$\tan\gamma = \frac{\varrho_2}{R_1 - \varrho_1}; \quad \tan\delta = \frac{\varrho_2}{\varrho_1 - R_2}. \tag{13.6.2}$$

Since 2γ and 2δ also occur at B and D

$$\tau + \gamma + \delta = 90° \tag{13.6.3}$$

and from Fig. 13.6.1

$$R_1 = \varrho_1 + \varrho_2 \cot\gamma; \quad R_2 = \varrho_1 - \varrho_2 \cot\delta. \tag{13.6.4}$$

The projection theories, which can be deducted from the figure, provide an independent computational check.

$$(R_1 - R_2)\sin 2\gamma + R_2 \sin 2\tau + b\cos 2\tau = a, \tag{13.6.5}$$

$$a\cos 2\tau + R_1 \sin 2\tau - (R_1 - R_2)\sin 2\delta = b. \tag{13.6.6}$$

If, for instance, a, b and R are given in addition to 2τ, then these equations are to be solved in the following sequence:

(13.6.1), left part of (13.6.2), (13.6.3), right part of (13.6.4),
(13.6.5) and (13.6.6).

Other starting values require a different sequence.

13.7 Easement Curves

13.7.1 Curvature and Length of Easement Curves

A route consisting of straight lines and circular arcs appears to be smooth. However, when changing from a straight line to a circle, or from one circle to another one with a different radius, there is a sudden change in curvature and in centrifugal force. To ease this transition, easement curves with varying radii of curvature are inserted in between.

The curvature relations are illustrated with a curvature diagram, where the length of the route is plotted along the abscissa axis, and the curvature k (the reciprocal value of the radius of curvature) as ordinate. Fig. 13.7.1 represents the curvature diagram of a route, consisting of two straight lines, a circular arc, and two easement curves.

The curvature lines of the two straight lines ($k = 0$) fall into the l-axis, while the curvature line of the circle (k = const.) is parallel to the l-axis. Thus straight lines with increasing or decreasing slopes, passing through the end points of the curvature lines of the straight lines and the circle provide a logical solution. The curvature of such easement curves is proportional to the length of the curve. The spiral, or clothoid, is the curve which has this characteristic. To obtain an equation for it, the origin of the l, k system may be shifted toward the (point of spiral) PS (see Fig. 13.7.1). With the

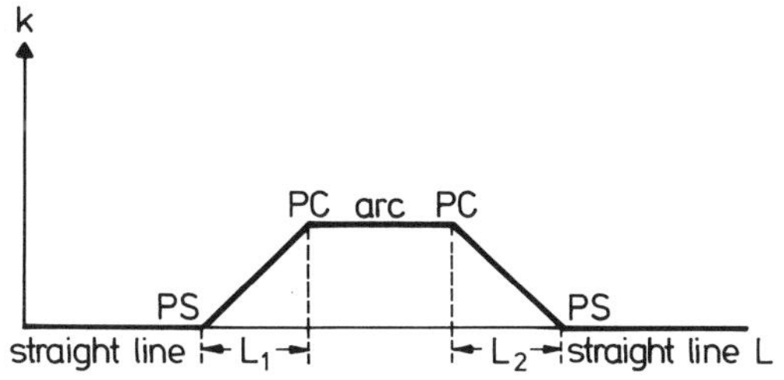

Fig. 13.7.1 Curvature diagram (PS: point of spiral; PC: point of curvature)

changing radius of curvature r and the slope of the straight line being denoted as $1/A^2$ we get

$$k = \frac{1}{R} = \frac{1}{A^2} L \quad \text{and} \quad A^2 = RL\,, \tag{13.7.1}$$

where L represents the total length of the easement curve, and R the radius of the circular arc.

In equation (13.7.1), A is squared, because it represents the product of two lengths. This shows that all clothoids are similar (i.e., one can be derived from another by similarity transformation). The clothoid with the parameter $A = 1$ is known as the unit clothoid (unit spiral). It is denoted by lower case letters, thus according to equation (13.7.1):

$$k = \frac{1}{r} \cdot l \quad \text{and} \quad r \cdot l = 1\,.$$

Based on local conditions, either A, and thus the curvature, or else the length L are selected when designing a route.

In order to obtain a guide for determining L, vertical considerations are used. In order to counteract the effect of centrifugal force, the outside edge of a railbed or a roadway is elevated. The amount of cross slope and of the super elevation as well as of the length L of the super-elevation ramp are a function of the design speed and of the radius of the main arc. It is desirable to use the same length for the easement curve, thus having transition in both horizontal and vertical directions begin and end at the same points. The actual values differ for railways and highways.

Operational consideration favours long straight lines for *railways*, thus having relatively short easements. The specifications of the German Federal Railways require that easement curves and super elevation ramps coincide in length. With V [km/h] being the maximum design speed, w [mm] the amount of super elevation, the following rules apply for R [m] and L [m]:

Super elevation $u = 8\ V^2/R$
where $20 \leqq u \geqq 150$ mm
cross slope of ramp: $1:10\ V$
length of easement curve: $10\ V \cdot u/1000$ for $V > 40$ km/h.

The following table is based on these specifications for $V = 160, 120, 90$ and 60 km/h. The last row for each design speed contains the smallest permissible radius of curvature.

V	R	L	u	V	R	L	u
km/h	m	m	mm	km/h	m	m	mm
160	2000	160	100	90	1300	45	50
160	1300	240	150	90	400	135	150
120	2000	75	60	60	2000	—	—
120	1300	110	90	60	1300	15	20
120	700	180	150	60	400	45	70
90	2000	30	30	60	180	90	150

In *highway* design, the spiral is used because the front wheels of a vehicle have to be turned at a constant rate when travelling at a constant speed. An abrupt change in curvature, e.g., from a straight line to a circle, cannot be handled with the vehicle, as the wheels should instantly be brought to a new position at the point where these elements meet. For divided highways, straight lines are avoided because they have a tiring effect on the driver. Therefore, long easement curves are preferred for super highways (e.g., turnpikes) which elevates the spiral from an easement curve to a primary design element. On the other hand, it is desirable to have longer straight lines on two-lane highways for better passing.

The minimum parameters for easement curves are based on three requirements:

1) The increase in lateral acceleration shall be

$$\mathrm{d}p/\mathrm{d}t < 0.5\ \mathrm{m/s^3}\,.$$

2) The additional cross slope Δg [%] because of super elevation cannot exceed the specifications.

3) The directional change caused by the spiral has to be noticeable.

These requirements are incorporated into the following formulae:

1) $A_{\min} = 0.17\sqrt{V^3}$

2) $A_{\min} = \sqrt{\dfrac{R}{g_{\max}} \cdot \dfrac{b}{2}(g_e + g_b)}$

3) $A_{\min} = R/3$

where b in metres represents the width of the road, and g_a in percent and g_e in percent are the respective cross slopes at the beginning and end of the spiral.

For different design speeds, the following values adhere to these requirements:

Design Speed V	30	40	50	60	80	100
minimum radius R [u]	35	70	120	180	350	600
minimum parameter A [u]	30	45	60	80	120	200
min. lengths of spiral L [u]	26	29	30	36	41	67

Although the spiral has simple geometric characteristics, as shown in equation (13.7.1), it is not as practical as one might think, because it cannot be staked using the mentioned equation. Therefore, substitute curves are often preferred, especially for short easement curves. Rural roads occasionally use a circular arc with twice the radius of the main arc as an easement curve. In railroad design, a cubic parabola is often substituted for a spiral.

13.7.2 The Cubic Parabola

This is most frequently used as a substitute for spirals. Fig. 13.7.2 illustrates the situation in an x, y system with the x-axis following the straight line route. In order to obtain the equation for the cubic parabola, the reciprocal radius of curvature $y''/(1 + y^2)^{3/2}$ is used for $1/R$. Since an easement curve is always flat, which means that y' is very small, it can be neglected and $1/R \approx y''$. The same reasoning leads to $x = l$.

This leads to a simplified equation after double integration:

$$y'' = \frac{d^2y}{dx^2} = \frac{1}{A^2}x; \quad y' = \frac{dy}{dx} = \frac{x^2}{2A^2}; \quad y = \frac{x^3}{6A^2}. \tag{13.7.2}$$

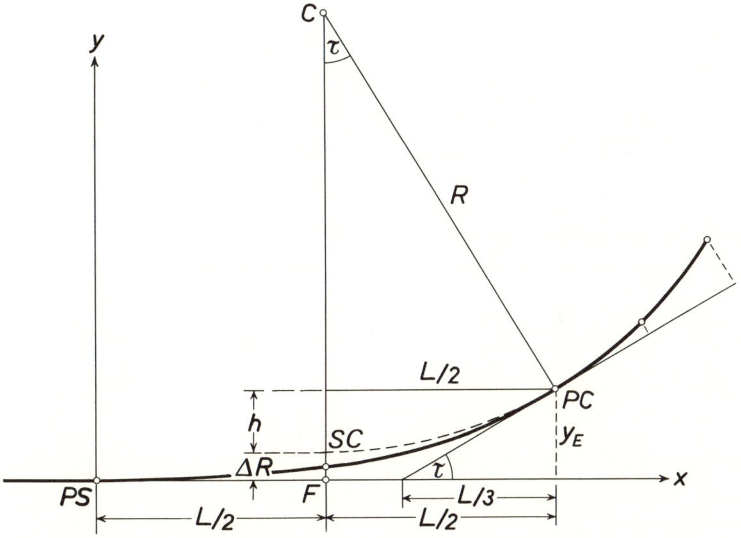

Fig. 13.7.2 Cubic parabola

The last equation describes the cubic parabola. Figs. 13.7.2 and 13.7.3 provide the following:

a) Without an easement curve, the circular arc would touch the tangent at F (Fig. 13.7.2). In order to gain space for the easement curve, the circular arc is shifted along the y-direction in F by the "throw" ΔR from the tangent towards the centre C. This "throw" ΔR is obtained with sufficient accuracy from the difference of the ordinate at $x = L$ and the grid ordinate h (equation (13.5.3)) for $L \leqq R/3$ as

$$\Delta R = y_E - h = \frac{L^3}{6LR} - \frac{L^2}{8R} = \frac{L^2}{24R} = \frac{L^3}{24A^2}. \tag{13.7.3}$$

b) In order to stake the easement curve near E in Fig. 13.7.3, the tangent distance TE is computed (equation (13.2.2)) for a circle with radius $(R + \Delta R)$. This provides point E in Fig. 13.7.3 which corresponds to F in Fig. 13.7.2.

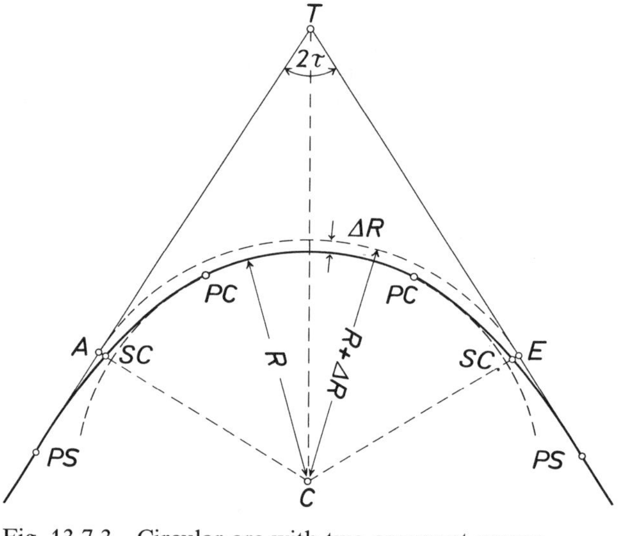

Fig. 13.7.3 Circular arc with two easement curves

c) The beginning of the easement curve is at $L/2$ from E along the extension of TE. Its centre point C is $L/2$ away from the PS on the ordinate of E, and $L/2$ away from SC which is the beginning of the circular arc. When neglecting the difference between L and X_E (see Fig. 13.7.2), the following formulae are obtained

$$\begin{aligned} &\text{At } PS\text{:} \quad x_A = 0; && y_A = 0, \\ &\text{At } SC\text{:} \quad x_M = \frac{1}{2}L; && y_M = \frac{L^3}{48A^2} = \frac{1}{2}\Delta R, \\ &\text{At } PC\text{:} \quad x_E = L; && y_E = \frac{L^3}{6A^2} = 4\Delta R. \end{aligned} \tag{13.7.4}$$

d) The aximuth τ of the tangent at SC (Fig. 13.7.2) is obtained from the centre equation (13.7.2), with $x = L$ as $\tan\tau = L^2/2A^2$ because $\mathrm{d}y/\mathrm{d}x = y' = \tan\tau$. The subtangent equals $y_E/\tan\tau$ or $L/3$ when using the value for y_E from equation (13.7.4). This relationship between arc and subtangent holds for any curve point.

e) The points of the circular arc can be staked either from the tangent at SC, or else from the main tangent. In the latter case the ordinates have to be increased by ΔR.

f) Since the approximation $1/y''$ for the curvature was used when deriving the cubic parabola, it does not have the desired curvature as presented in equation (13.7.1). Therefore, a slight jump in curvature occurs at SC, which, however, does not exceed 10% of R as long as $L \leqq 0.5\,R$. This can be avoided by using the "improved" cubic parabola.

$$y = \frac{x^2}{6RL}\left[1 + \left(\frac{L}{2R}\right)^2\right]^{3/2}. \qquad (13.7.5)$$

13.7.3 The Spiral (Clothoid)

13.7.3.1 The Unit Spiral (Unit Clothoid)

In order to stake a spiral from its tangent, rectangular coordinates are derived using equation (13.7.1) with the tangent being the abscissa axis, originating at PS.

As illustrated in Fig. 13.7.4 we have

$$\mathrm{d}\tau = \mathrm{d}l/r = l\,\mathrm{d}l/A^2$$

which, after integration, leads to

$$\tau = l^2/2A^2. \qquad (13.7.6)$$

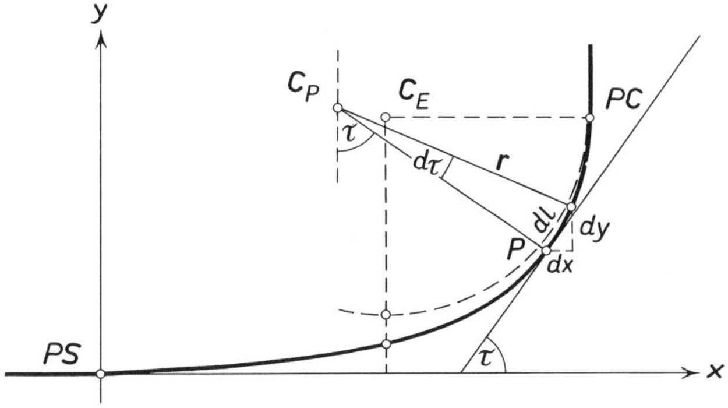

Fig. 13.7.4 Slope of the clothoid (spiral)

With $\mathrm{d}x = \cos\tau\,\mathrm{d}l$ and $\mathrm{d}y = \sin\tau\,\mathrm{d}l$, the *Fresnel integrals* are obtained from equation (13.7.6):

$$x = \int_0^l \cos \frac{l^2}{2A^2} \, dl; \quad y = \int_0^l \sin \frac{l^2}{2A^2} \, dl. \tag{13.7.7}$$

These integrals can be developed into power series, whose first terms represent the cubic parabola.

$$\begin{aligned} x &= \sum_{n=1}^{\infty} (-1)^{n+1} \frac{l^{4n-3}}{(4n-3)(2n-2)!\,2^{2n-2}} \\ y &= \sum_{n=1}^{\infty} (-1)^{n+1} \frac{l^{4n-1}}{(4n-1)(2n-1)!\,2^{2n-1}} \end{aligned} \tag{13.7.8}$$

These series converge slowly. However, for most tasks, a suitable accuracy is obtained when using the first few terms:

$$\begin{aligned} x &= l - \frac{l^5}{40} + \frac{l^9}{3456} - + \ldots \\ y &= \frac{l^3}{6} - \frac{l^7}{336} + \frac{l^{11}}{42240} - + \ldots . \end{aligned} \tag{13.7.9}$$

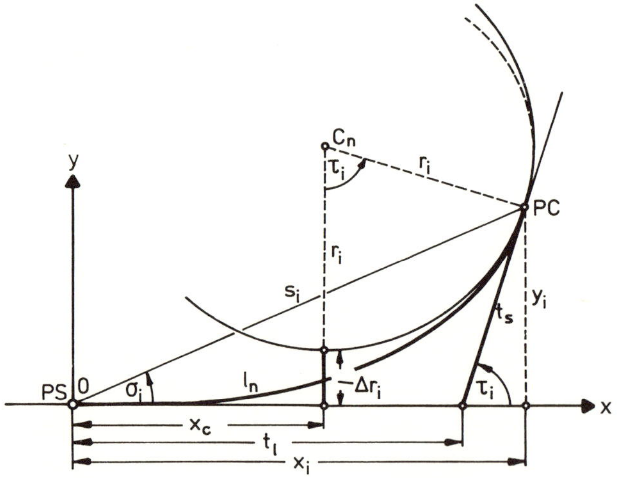

Fig. 13.7.5 Elements of the clothoid (spiral)

As evident from Fig. 13.7.5, the polar coordinates for the end point of the spiral PC become:

$$s = \sqrt{x^2 + y^2}; \quad \sigma = \arctan y/x. \tag{13.7.10}$$

The coordinates of the centre of curvature are

$$x_c = x - r \sin \tau; \quad y_c = y + r \cos \tau, \tag{13.7.11}$$

and the "throw"

$$\Delta r = y_c - r = y - r(1 - \cos\tau) . \tag{13.7.12}$$

Furthermore

$$t_s = \frac{y}{\sin\tau} \quad \text{and} \quad t_l = x - y\cot\tau . \tag{13.7.13}$$

13.7.3.2 Spirals (Clothoids) with Parameter A

The values for any spiral can be derived from the unit spiral by means of similarity transformation. All lengths of the unit spiral have to be multiplied by the parameter A:

$$\left.\begin{array}{llll} X = A \cdot x; & Y = A \cdot y; & X_c = A \cdot x_c; & Y_c = A \cdot y_c \\ L = A \cdot l; & R = A \cdot r; & S = A \cdot s; & \Delta R = A \cdot \Delta r \\ T_s = A \cdot t_s; & T_L = A \cdot t_l & & \end{array}\right\} \tag{13.7.14}$$

13.7.3.3 Geometry of a Spiral

As evident from Fig. 13.7.6, the radius of curvature is infinite at the origin ($L = 0$). Thus, the spiral is transferred into a straight line at the origin without any sudden change in curvature. This straight line also is a reversal tangent. With increasing arc

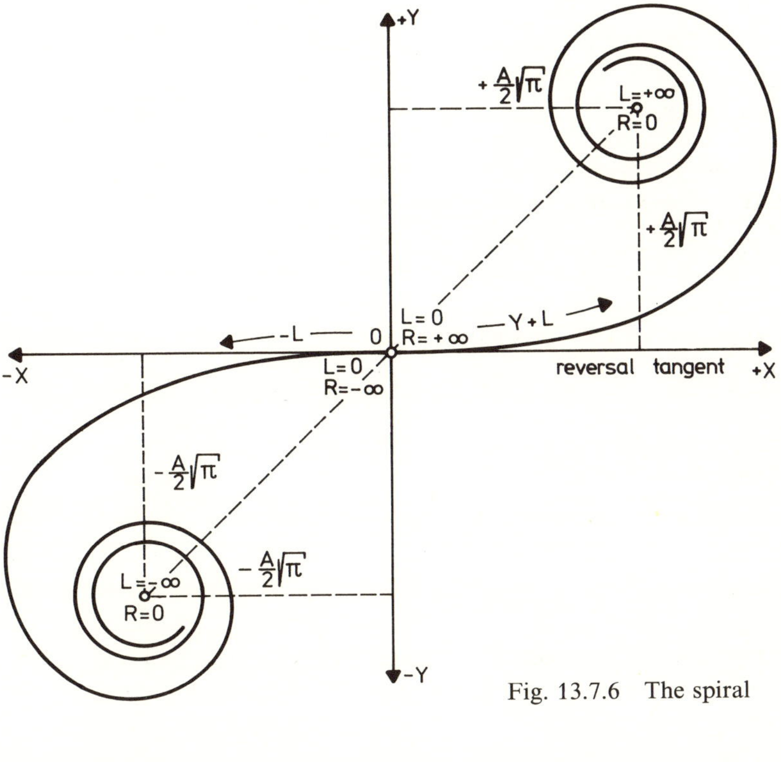

Fig. 13.7.6 The spiral

length L (positively or negatively) the radius of curvature is reduced from $R = \infty$ to $R = 0$. The transfer to a circular element occurs at a point where the radius of curvature equals the radius of the circle. Thus a spiral section can also be used as easement between two circular arcs. The relative position of the two circles fixes the spiral parameter A.

The following easements can utilize the spiral:

- spiral between a straight line and a circular arc
- spirals between two straight lines
- spiral as a reverse curve between two circles
- spiral as an egg line between two circles
- series of spirals with different parameters.

13.7.3.4 Spiral Tables

If an electronic computer is not available, spiral tables can be utilized (e.g., *Kasper* et al. 1968). Usually, the following tables are provided:

1) E-table (for unit spiral):

enter: l

obtain: τ[gon] or τ[°]; σ, r, Δr, x_c, x, y, $\frac{1}{r}$, $\frac{\Delta r}{r}$, t_s, t_l.

2) τ-table:

enter: τ

obtain: l, r, x_c, x, y, σ, t_s, t_l.

3) A-table:

enter: L (for $A = 15$ to $A = 3000$, i.e., 70 spirals)

obtain: τ, R, ΔR, X_c, X, Y, T_S, T_L.

4) L-table:

enter: L (for $A = 15$ to $A = 3000$)

obtain: X, Y.

5) R-table:

enter: A (for $R = 15$ to $R = 10000$)

obtain: L, τ, ΔR, X_c, X, Y, T_S, T_L.

13.7.3.5 Coordinate Transformations

When computing staking values, it is often necessary to carry out transformations between local and state coordinate systems. The azimuths increase clockwise for geodetic systems according to the graduation of theodolite circles. Coordinate systems for spirals, however, may be right or left handed depending on the specific curvature direction. In order to avoid confusion, the transformation formulae (also section 5.2) are given for both cases.

1) Two right-handed systems (left-hand curve, Fig. 13.7.7):

$$x = x_0 + aX + oY$$
$$y = y_0 + aY - oX \tag{13.7.15}$$

or

$$X = (x - x_0)a - (y - y_0)o$$
$$Y = (y - y_0)a + (x - x_0)o \tag{13.7.16}$$

with

$$o = q \cdot \sin\varphi \quad \text{and} \quad a = q \cdot \cos\varphi \,.$$

The scale factor q is usually equal to one.

2) Right-handed and left-handed systems (right-hand curve, Fig. 13.7.8):

$$x = x_0 + oX + aY$$
$$y = y_0 - oY + aX \tag{13.7.17}$$

$$X = (y - y_0)a + (x - x_0)o$$
$$Y = (x - x_0)a - (y - y_0)o. \tag{13.7.18}$$

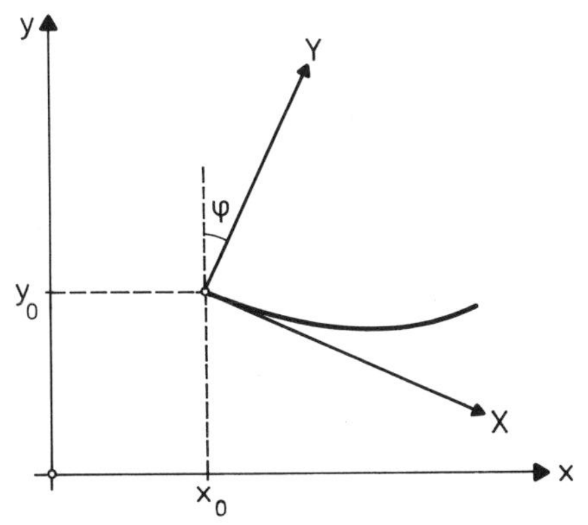

Fig. 13.7.7 Coordinate transformation for 2 right-handed systems

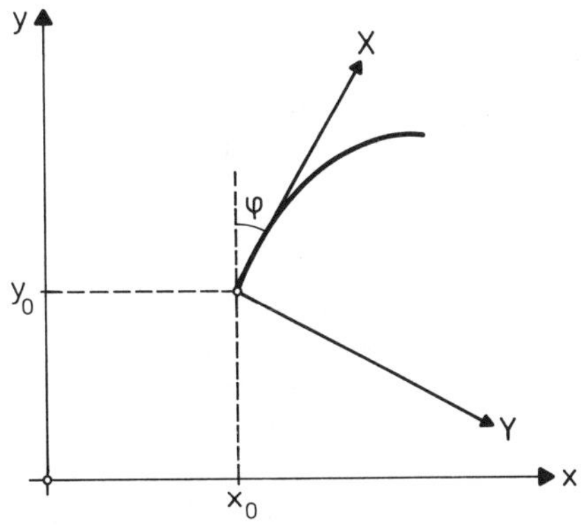

Fig. 13.7.8 Coordinate transformation for right- and left-handed systems

13.7.3.6 Spiral Between Straight Line and Circular Arc

Two values determine the spiral uniquely with respect to a given circle. Depending on the known elements, different problems have to be solved; e.g. the following value pairs may be given: R, A; R, L; $R, \Delta R$; R, τ; R, σ. The staking values are determined using the formulae given in sections 13.7.3.1 and 13.7.3.2 or else taken from a table.

Example:

Given: $A = 200$; $R = 160$ m.
Wanted: X, Y for the end point, plus T_S, T_L and τ.

1) Determination of the length l of the unit spiral: $l = A/R$.

2) With equation (13.7.9) x and y is obtained, which provides X, Y after multiplication with A.

3) Equation (13.7.6) provides τ.

4) Equations (13.7.13) give t_s and t_l, which again have to be multiplied by A to give T_S and T_L.

Thus: $l = A/R = 1.25$

$$x = 1.175832 \rightarrow X = 235.17 \text{ m}$$
$$y = 0.311602 \rightarrow y = 62.32 \text{ m}$$
$$t_s = 0.442511 \rightarrow T_S = 88.50 \text{ m}$$
$$t_l = 0.861634 \rightarrow T_L = 172.33 \text{ m}$$
$$\tau = 49.7359 \text{ gon} = 44.7623° .$$

13.7.3.7 Reverse Spiral: Egg Line

The reverse spiral connects two circles with opposite curvature.

For dynamic reasons, the two spiral branches should have the same parameter A. The location of the two circles then uniquely defines the parameter A.

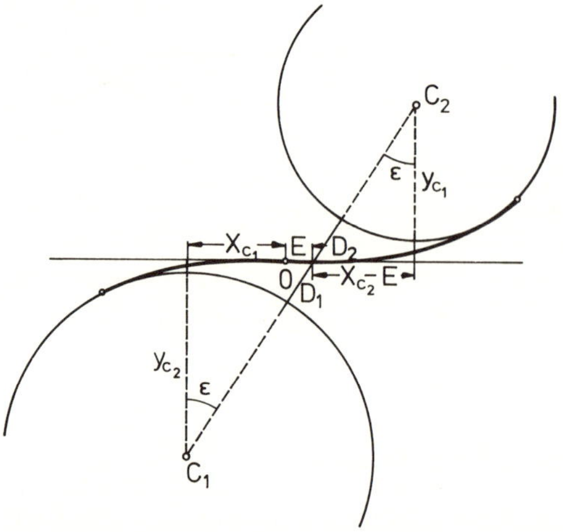

Fig. 13.7.9 Reverse spiral

A simple geometric solution is sufficient for many practical applications. R_1, R_2 and A are graphically determined on the map. Then the reverse spiral is plotted onto transparent paper. Afterwards, this line is fitted onto the map or plan. If a poor fit occurs, another reverse curve with corrected parameters is plotted.

The parameter A can also be determined with the aid of spiral tables (e.g., *Kasper* et al. 1968). Then the distance $D = D_1 + D_2$ (Fig. 13.7.9) has to be determined for the two circles. With it, a number of even parameters A can be selected, for which the staking values E and ε are tabulated. These values define the reversal point, which provides sufficient information for the computation of all other staking values.

The egg line connects two circles with the same sense of direction. The solution of this problem is similar to the previous one.

13.7.3.8 Series of Spirals

In many cases it is possible to connect circular arcs with a simple spiral. However, sometimes it is necessary to have several spirals in sequence. Often a 10% curvature jump is permissible. The coordination of spiral sections is best achieved by coordinate transformation (section 13.7.3.5) using orthogonal coordinates of the end points plus the tangent angles, all computed in local coordinate systems.

The parameters of spiral series can also be obtained mathematically from the individual segments, as shown in the following example.

The 4 points P_0, P_1, P_2, P_3 in Fig. 13.7.10 are to be connected by a spiral. These points are given by arc differences and the angles α and β, measured in the field or on a plan.

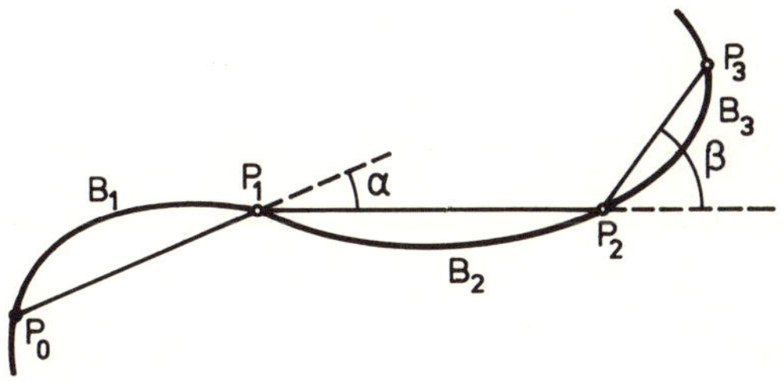

Fig. 13.7.10 Spiral passing through 4 given points

The spiral parameter is obtained from

$$A^2 = \frac{200\,\text{gon}}{6 \cdot \pi} \frac{(B_1 + B_2 + B_3)(B_1 + B_2)(B_2 + B_3)}{\beta(B_1 + B_2) \pm \alpha(B_2 + B_3)} . \tag{13.7.19}$$

Where B_i represents the arc length from the origin of the spiral to point P_i. The positive sign applies for a reverse spiral, the negative sign for an egg line (*Kasper* et al. 1968).

13.7.3.9 Computation of Additional Spiral Points

a) Spiral with increasing curvature

Given are the stationing L_0 of the origin of the spiral, the spiral parameter A, the coordinates x_0, y_0 of the origin in the given coordinate system, the azimuth φ of the tangent, and the stationing L_i of the spiral point P_i (Fig. 13.7.11).

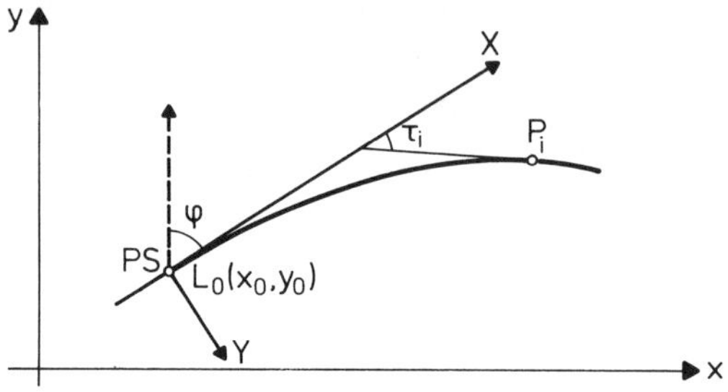

Fig. 13.7.11 Spiral points for increasing curvature (spiral bends to the right)

1) Computation of the arc length l_i and of τ_i for the given station on the unit spiral

$$l_i = \frac{L_i - L_0}{A}; \quad \tau_i = \frac{l_i^2}{2} \cdot \frac{200\,gon}{\pi} = \frac{l_i^2}{2} \cdot \frac{180^\circ}{\pi}.$$

2) Computation of the station coordinates with respect to the spiral tangent:

$$X_i = \left(l_i - \frac{l_i^5}{40} + \frac{l_i^9}{3456} - + \ldots\right) \cdot A$$

$$Y_i = \left(\frac{l_i^3}{6} - \frac{l_i^7}{336} + \frac{l_i^{11}}{42\,240} - + \ldots\right) \cdot A\,.$$

3) Transformation of the spiral coordinates into the given regional system.

If *the spiral bends to the right* (see equation (13.7.17)):

$$x_i = x_0 + X_i \sin\varphi + Y_i \cos\varphi$$
$$y_i = y_0 - Y_i \sin\varphi + X_i \cos\varphi\,,$$

and *if the spiral bends to the left* (equation (13.7.15)):

$$x_i = x_0 + X_i \cos\varphi + Y_i \sin\varphi$$
$$y_i = y_0 + Y_i \cos\varphi - X_i \sin\varphi\,.$$

b) Spiral with decreasing curvature

Given are the stationing L_E of the station PC, the radius R of the station PC, the spiral parameter A, the coordinates x_E, y_E of the station PC, the azimuth α_{t_E} of the tangent, and the stationing L_i of the spiral point P_i (Fig. 13.7.12).

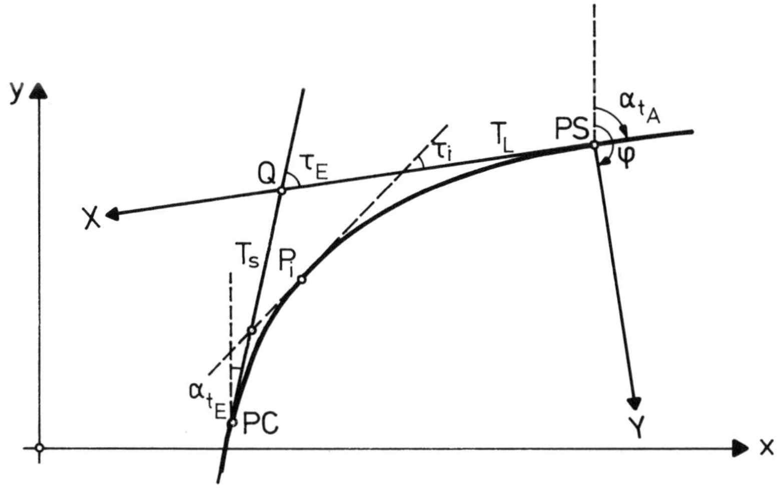

Fig. 13.7.12 Spiral points for decreasing curvature (spiral bends to the right)

1) Computation of the stationing L_A for the origin of the spiral

$$L_A = L_E + \frac{A^2}{R} = L_E + L_0 .$$

2) Computation of the tangent angle τ_E

$$\tau_E = \frac{l_0}{2}\,\frac{180}{\pi}; \quad l_0 = \frac{L_0}{A} .$$

3) Computation of the tangent lengths T_S and T_L

$$T_S = \frac{Y_E}{\sin\tau_E}; \quad T_L = X_E - Y_E \cot\tau_E$$

with

$$X_E = \left(l_0 - \frac{l_0^5}{40} + \frac{l_0^9}{3456} - + \ldots\right) \cdot A$$

$$Y_E = \left(\frac{l_0^3}{6} - \frac{l_0^7}{336} + \frac{l_0^{11}}{42240} - + \ldots\right) \cdot A .$$

4) Computation of the coordinates of PS by means of the traverse PS, Q, PC

$$x_0 = x_E + T_S \sin\alpha_{t_E} + \mathrm{T}_L \sin\alpha_{t_A}$$
$$y_0 = y_E + T_S \cos\alpha_{t_E} + T_L \cos\alpha_{t_A}$$

with

$$\mathrm{a}_{t_A} = \alpha_{t_E} + \tau_E \quad \text{(if the spiral bends to the right),}$$
$$\alpha_{t_A} = \alpha_{t_E} - \tau_E \quad \text{(if the spiral bends to the left).}$$

5) Computation of the arc length l_i

$$l_i = \frac{L_A - L_i}{A}.$$

7) Computation of τ_i

$$\tau_i = \frac{l_i^2}{2} \cdot \frac{200 \text{ gon}}{\pi} = \frac{l_i^2}{2} \cdot \frac{180^\circ}{\pi}.$$

8) Coordination of the station L_i with respect to the spiral tangent

$$X_i = \left(l_i - \frac{l_i^5}{40} + \frac{l_i^9}{3456} - + \ldots\right) A$$

$$Y_i = \left(\frac{l_i^3}{6} - \frac{l_i^7}{336} + \frac{l_i^{11}}{42\,240} - + \ldots\right) A.$$

9) Transformation of X_i/Y_i into the regional x, y-system.

If the spiral bends to the left (equation (13.7.15)):

$$x_i = x_0 + X_i \cos\varphi + Y_i \sin\varphi,$$
$$y_i = y_0 + Y_i \cos\varphi - X_i \sin\varphi,$$

with

$$\varphi = \alpha_{t_E} + \tau_E + 100 \text{ gon} = \alpha_{t_E} + \tau_E + 90^\circ.$$

If the spiral bends to the right (equation (13.7.17)):

$$x_i = x_0 + X_i \sin\varphi + \mathrm{Y}_i \cos\varphi,$$
$$y_i = y_0 - Y_i \sin\varphi + X_i \cos\varphi,$$

with

$$\varphi = \alpha_{t_E} - \tau_E + 200 \text{ gon} = \alpha_{t_E} - \tau_E + 180^\circ.$$

Examples:

Computation of additional spiral points.

1) Spiral with increasing curvature

a) Spiral bends to the right (Fig. 13.7.12)

Given:

$L_0 = 0.3 + 02.10$ km + m $= 302.10$ m
$A = 300$
$y_0 = 6468.36$ m
$x_0 = 5235.16$ m
$\varphi = 49.873$ gon $= 44.886^\circ$
$L_i = 0.5 + 27.32$ km + m .

Computed:

$l_i = 0.7507$
$\tau_i = 17.938$ gon $= 16.144^\circ$
$Y_i = 21.03$ m
$X_i = 223.43$ m
$y_i = 6608.98$ m
$x_i = 5409.30$ m

b) Spiral bends to the left

Given:

$L_0 = 0.5 + 39.85$ km + m
$A = 400$
$y_0 = 6999.85$ m
$x_0 = 1792.15$ m
$\varphi = 380.322$ gon $= 342.290°$
$L_i = 7 + 68.52$ km + m

Computed:

$l_i = 0.5717$
$\tau_i = 10.404$ gon $= 9.364°$
$Y_i = 12.43$ m
$X_i = 228.07$ m
$y_i = 7081.07$ m
$x_i = 2005.63$ m

2) Spiral with decreasing curvature

a) Spiral bends to the right

Given:

$L_E = 1.8 + 81.37$ km + m $= 1881.37$ m
$R = 800$ m
$A = 750$
$y_E = 5686.86$ m
$x_E = 8947.15$ m
$\alpha_{t_E} = 28.986$ gon $= 26.087°$
$L_i = 2.1 + 10.40$ km + m

Computed:

$L_A = 2.5 + 84.50$ km + m
$L_0 = 703.13$ m
$l_0 = 0.9375$
$\tau_E = 27.976$ gon $= 25.178°$
$Y_E = 101.58$ m
$X_E = 689.67$ m
$T_S = 238.77$ m
$T_L = 473.59$ m
$\alpha_{t_A} = 56.962$ gon $= 51.266°$
$y_0 = 6197.63$ m
$x_0 = 9421.57$ m
$l_i = 0.6321$
$Y_i = 31.48$ m
$X_i = 472.19$ m
$\varphi = 156.962$ gon $= 141.266°$
$y_i = 5877.62$ m
$x_i = 9072.93$ m

b) Spiral bends to the left

Given:

$L_E = 0.9 + 17.63$ km + m
$R = 300$ m
$A = 325$
$y_E = 5438.95$ m
$x_E = 0540.13$ m
$\alpha_{t_E} = 179.171$ gon $= 161.254°$
$L_i = 1.0 + 06.13$ km + m

Computed:

$L_A = 1.2 + 69.71$ km + m
$L_0 = 352.08$ m
$l_0 = 1.0833$
$\tau_E = 37.355$ gon $= 33.620°$
$Y_E = 67.19$ m
$X_E = 340.14$ m
$T_S = 121.35$ m
$T_L = 239.09$ m
$\alpha_{t_A} = 141.816$ gon $= 127.634°$
$y_0 = 5178.04$ m
$x_0 = 0768.47$ m
$l_i = 0.8110$
$Y_i = 28.67$ m
$X_i = 260.74$ m
$\varphi = 341.816$ gon $= 307.634°$
$y_i = 5359.96$ m
$x_i = 0579.49$ m

13.7.3.10 Staking Methods

Once the elements of a route are finalized, the coordinates of their starting and ending points are computed in the regional coordinate system as defined by fixed control points. Subsequently the staking data for intermediate points are computed. The staking can be carried out by orthogonal or polar methods. The staking data can be based on chords, secants, tangents, traverses, freely selected survey lines, or a randomly selected station.

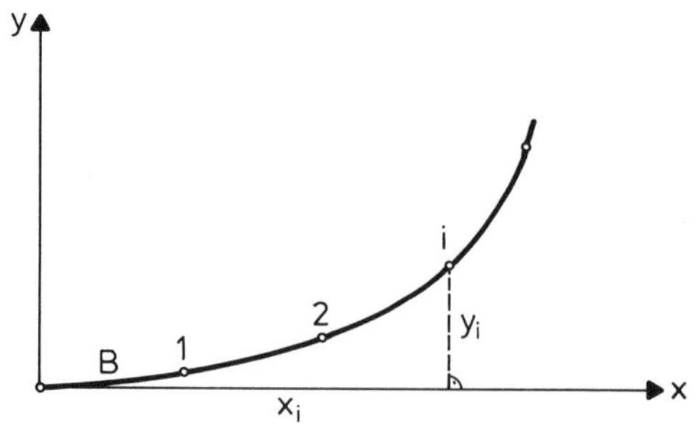

Fig. 13.7.13 Equal arc lengths starting at spiral origin

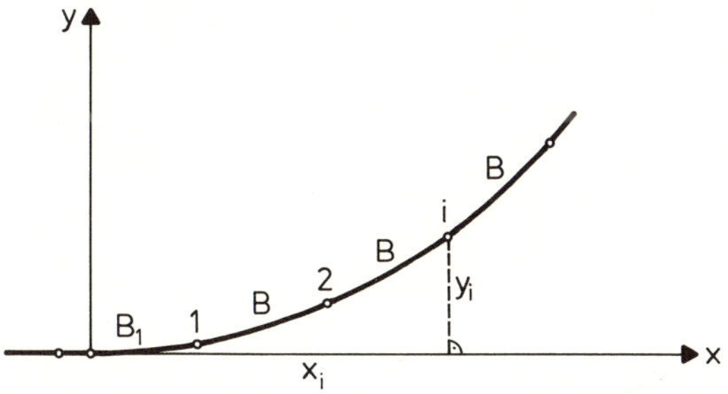

Fig. 13.7.14 Equal arc lengths, based on continuous stationing

As a rule, equal distances along the arc are being staked. If the origin of the spiral coincides with the origin of the stationing (Fig. 13.7.13), then the individual arc lengths l of the unit spiral are:

$$\begin{aligned} l_1 &= \frac{B}{A} = b \\ l_2 &= 2b \\ l_3 &= 3b \\ &\vdots \end{aligned} \tag{13.7.20}$$

For continuous stationing, it is very unlikely that an equal station coincides with the spiral origin. Then the following applies (Fig. 13.7.14)

$$\begin{aligned} l_1 &= \frac{B_1}{A} \\ l_2 &= \frac{B_1 + B}{A} \\ l_3 &= \frac{B_1 + 2B}{A} \\ &\vdots \\ l_{n-1} &= \frac{B_1 + (n-2)B}{A} \\ l_n &= \frac{B_1 + (n-2)B + B_n}{A} \end{aligned} \tag{13.7.21}$$

a) *Staking from main tangent or chord*

First the arc lengths on the unit spiral are computed according to equations (13.7.20) or (13.7.21). Since A is known, the orthogonal staking values with respect to the main tangent are (Fig. 13.7.15):

$$X_i = \left(l_i - \frac{l_i^5}{40} + \frac{l_i^9}{3456} - + \ldots\right) A$$

$$Y_i = \left(\frac{l_i^3}{6} - \frac{l_i^7}{336} + \frac{l_i^{11}}{42240} - + \ldots\right) A. \tag{13.7.22}$$

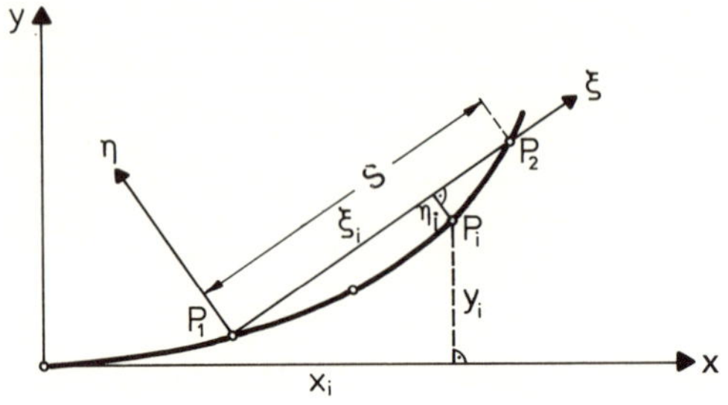

Fig. 13.7.15 Staking of a spiral from the main tangent or chord

Long ordinates can be avoided by staking from a chord. The staking values are obtained by coordinate transformation. Solving equation (5.2.9a) for ξ and η, we get:

$$\begin{aligned} \xi &= (X_i - X_1)a - (Y_i - Y_1)o \\ \eta &= (Y_i - Y_1)a + (X_i - X_1)o \end{aligned} \tag{13.7.23}$$

where

$$o = \sin\varphi; \quad a = \cos\varphi$$

$$\varphi = \arcsin\frac{X_2 - X_1}{S} = \arccos\frac{Y_2 - Y_1}{S}$$

$$S = \sqrt{(X_2 - X_1)^2 + (Y_2 - Y)^2}$$

Example:

Staking from chord

Given:

	l	X [m]	Y [m]
P_1	0.2	70.00	0.47
P_2	0.9	309.00	42.03
P_i	0.3		
⋮			

A = 350

Wanted: ξ_i, η_i.

Computations:

$Y_i = 1.57$ m
$X_i = 104.98$ m
$\cos\varphi = 0.1713$
$\sin\varphi = 0.9852$
$\xi_i = 4.91$ m

$\eta_i = 34.65$ m

b) *Staking from freely selected survey lines*

The orthogonal staking of spiral points from the main tangent is only economical for short ordinates, e.g., flat spirals. Furthermore, the principal points of the main tangent are often obstructed by construction equipment, etc. These difficulties can be overcome by staking from a freely selected survey line. This method has gained considerable popularity with the advent of small pocket computers which can be taken to the field. The staking procedure is as follows:

- Computation of rectangular coordinates of the route points in the regional coordinate system as defined by survey control points. This is done in the office.
- Selection of a survey line in the field for optimal working conditions.
- Field computation of orthogonal staking values in respect to the selected survey line
- Staking of the points.

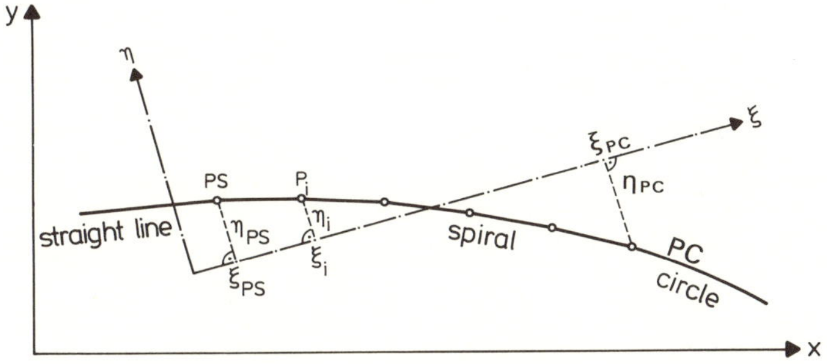

Fig. 13.7.16 Staking from freely selected survey line

The survey line corresponds to the ξ-axis of a local coordinate system (Fig. 13.7.16). By measuring 2 points (e.g., PS and PC) orthogonally (section 8.3.1.1) with respect to the survey line, they are coordinated in the ξ, η system. Knowing two points in both systems, all other x, y coordinates can be transformed into ξ, η coordinates (orthogonal staking values) using a similarity transformation:

Solving equation (5.2.9a) for ξ_i and η_i we get:

$$\xi_i = (x_i - x_0)\frac{1}{q}\cos\varphi - (y_i - y_0)\frac{1}{q}\sin\varphi \qquad (13.7.24)$$

$$\eta_i = (x_i - x_0)\frac{1}{q}\sin\varphi + (y_i - y_0)\frac{1}{q}\cos\varphi$$

with

$$x_0 = x_{PS} - a\xi_{PS} - o\eta_{PS}$$

$$y_0 = y_{PS} + o\xi_{PS} - a\eta_{PS}$$

$$q = \frac{\sqrt{(x_{PC} - x_{PS})^2 + (y_{PC} - y_{PS})^2}}{\sqrt{(\xi_{PC} - \xi_{PS})^2 + (\eta_{PC} - \eta_{PS})^2}} = \frac{s}{\sigma} \qquad (13.7.25)$$

$$\varphi = \arccos \frac{y_{PC} - y_{PS}}{s} - \arccos \frac{\eta_{PC} - \eta_{PS}}{\sigma}. \qquad (13.7.26)$$

Example:

Staking from freely selected survey line

Given:

	x [m]	y [m]
PS	0724.65	5433.32
P_i	0764.36	5336.13
PC	0807.46	5131.78

	ξ [m]	η [m]
PS	2.84	22.63
PC	305.63	−54.95

Wanted: ξ_i, η_i

Computations:

$q = 1.000427$
$\varphi = 66.96976$ gon $= 60.27279°$
$o = 0.868767$
$a = +0.496083$
$x_0 = 703.581$ m
$y_0 = 5424.561$ m
$\underline{\underline{\xi_i = 106.89 \text{ m}}}$
$\underline{\underline{\eta_i = -8.93 \text{ m}}}$

c) *Staking from a traverse near the centre line*

Often it is advantageous to stake from a traverse, located near the centre line (Fig. 13.7.17).

Either polar or orthogonal coordinates with respect to the traverse legs are computed for the spiral points.

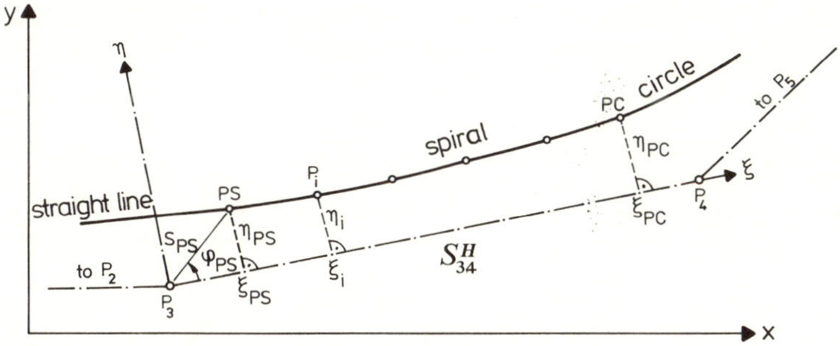

Fig. 13.7.17 Staking from traverse

Given are the x, y coordinates of the PS and PC as well as the points P_i on the spiral plus the traverse points. Wanted are

i) orthogonal staking values for the spiral points P_i in the ξ, η system, with the traverse leg $\overline{P_3 P_4} = S^H_{34}$ being the ξ-axis

ii) polar staking values with respect to the traverse leg.

In this case $\xi_3 = \eta_3 = 0; \quad \xi_4 = s_3; \quad \eta_4 = 0$

then

$$\xi_i = (x_i - x_3)\frac{1}{q}\cos\varphi - (y_i - y_3)\frac{1}{q}\sin\varphi$$
$$\eta_i = (x_i - x_3)\frac{1}{q}\sin\varphi + (y_i - y_3)\frac{1}{q}\cos\varphi \qquad (13.7.27)$$
$$\varphi = \arccos\frac{y_4 - y_3}{s_{34}} - \arccos\frac{\eta_4 - \eta_3}{S^H_{34}}; \; q = \frac{s_{34}}{S^H_{34}}$$

$$s_{34} = \sqrt{(x_4 - x_3)^2 + (y_4 - y_3)^2}$$

The polar staking values for PS and P_i are:

$$\varphi_{PS} = \arctan\frac{\eta_{PS}}{\xi_{PS}}; \quad S_{PS} = \frac{1}{q}\sqrt{\xi_{PS}^2 + \eta_{PS}^2}$$
$$\varphi_i = \arctan\frac{\eta_i}{\xi_i}; \quad S_i = \frac{1}{q}\sqrt{\xi_i^2 + \eta_i^2}\,. \qquad (13.7.28)$$

Example:

Stacking from traverse

Given:

	x [m]	y [m]
PS	820.63	438.27
P_i	852.25	461.11
PC	901.76	494.99
P_3	795.45	452.33
P_4	908.32	518.19

Measured $S_{34}^H = 130.71$ m

Wanted: $\xi_i, \eta_i, \varphi_i, S_i^H$

Computations:

$s_{34} = 130.680$ m

$\varphi = -33.6263 \text{ gon} = -30.2637° = 366.3737 \text{ gon} = 329.7363°$

$q = -0.999770$

	ξ [m]	η [m]	φ [°]	[gon]	S [m]
PS	14.67	−24.84	300.5582	333.9536	28.84
P_i	53.50	−21.04	338.5234	376.1371	57.48
PC	113.35	−16.74	351.6009	390.6677	114.55

d) *Staking with deflection angles and chord lengths*

With the growing popularity of electronic tacheometers, polar staking becomes more important. The deflection angle plus chord length method (Fig. 13.7.18) is computationally simple.

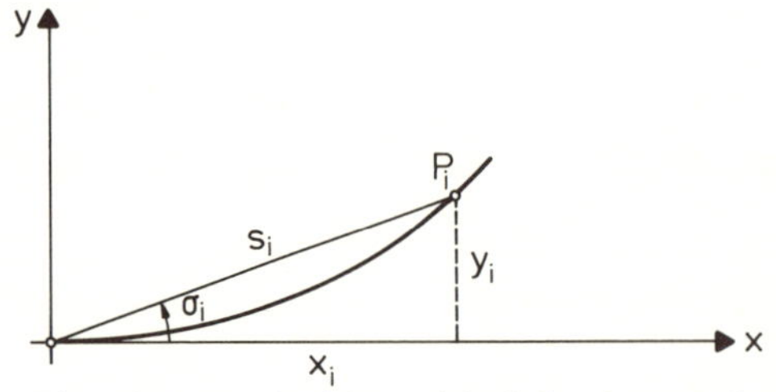

Fig. 13.7.18 Staking with deflection angles and chord lengths

The polar coordinates are:

$$s = \sqrt{x_i^2 + y_i^2} \qquad (13.7.29)$$
$$\sigma_i = \arctan(y/x).$$

The disadvantage of this method is the length of the staking distances, especially for long spiral branches. By selecting a free station, higher accuracies and better efficiency is usually obtained.

c) *Staking from a freely selected station*

When staking a number of coordinated points from a freely selected station (Fig. 13.7.19), first, an instrument station is selected close to the points to be staked, such that

- directions and distances can be measured to at least two fixed control points
- there is clear sight to as many points to be staked as possible
- the staking distances are as short as possible.

Previously staked route point can be utilized in place of control monuments.

The next step consists of computing the following: see section 6.6.

- coordinates (x_s, y_s) of the instrument station S
- the angle φ between the positive y-axis and the zero direction of the theodolite circle
- scale factor q, which represents the difference between the measuring unit of the instrument and the coordinate network after the measured distances have been reduced to the reference horizon. If, however, distances S_i^H, reduced to the measuring horizon (at elevation H) are introduced, the scale factor also contains a component for the reduction from measuring horizon to reference horizon.

The second solution is presented here, as the necessary staking distances can be reduced to the measuring horizon by simply multiplying with the inverse of the scale factor.

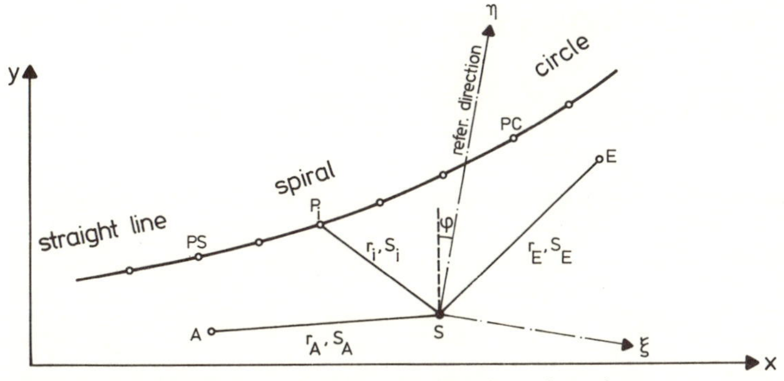

Fig. 13.7.19 Polar staking from freely selected station

Finally, the polar staking values are computed:

$$S_i = \frac{1}{q}\sqrt{(x_i - x_s)^2 + (y_i - y_s)^2} \qquad (13.7.30)$$

$$r_i = \alpha_i + \varphi$$

with

$$\alpha_i = \arctan \frac{x_i - x_s}{y_i - y_s} \tag{13.7.31}$$

Sometimes it is advantageous for the working progress to select an approximate point P_n and to stake the spiral point P_i with short orthogonal values. With the direction r_n and the distance S_n^H, the staking elements become (Fig. 13.7.20).

$$\begin{aligned} \mathrm{d}x' &= S_i^H \cos(r_i - r_n) - S_n^H \\ \mathrm{d}y' &= S_i^H \sin(r_i - r_n)\,. \end{aligned} \tag{13.7.32}$$

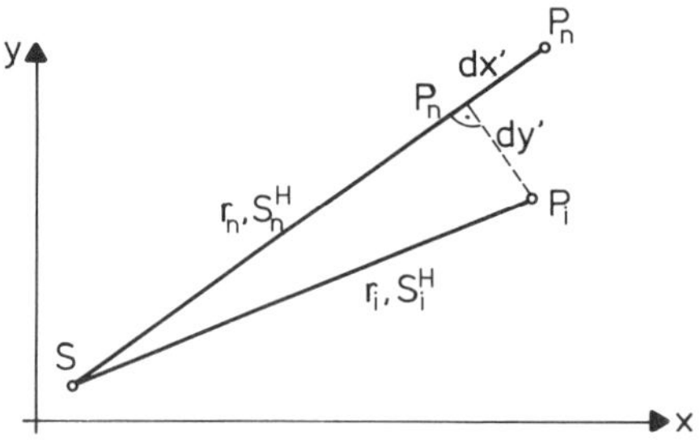

Fig. 13.7.20 Orthogonal elements for P_i with respect to an approximate point P_n

13.7.4 Approximations for Flat Spirals

If L_1 and L_2 are the arc lengths for points P_1 and P_2 on the spiral, then the deflection angle of the chord $P_1 P_2$ is approximately (Fig. 13.7.21):

$$\varphi_{12}(\text{gon}) \approx \frac{1}{6A^2}(L_1^2 + L_1 L_2 + L_2^2). \tag{13.7.33}$$

If the mid-ordinate remains less than 0.02 A, then this formula provides values which are correct to within 2 mgon (6″).

Based on equation (13.7.33) the following staking methods are derived:

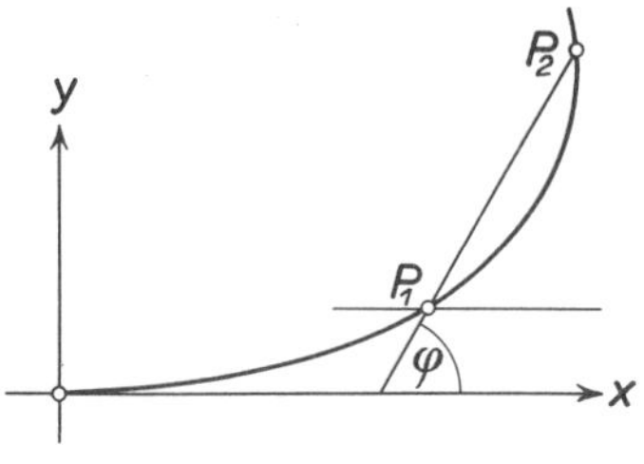

Fig. 13.7.21 Azimuth of spiral chord

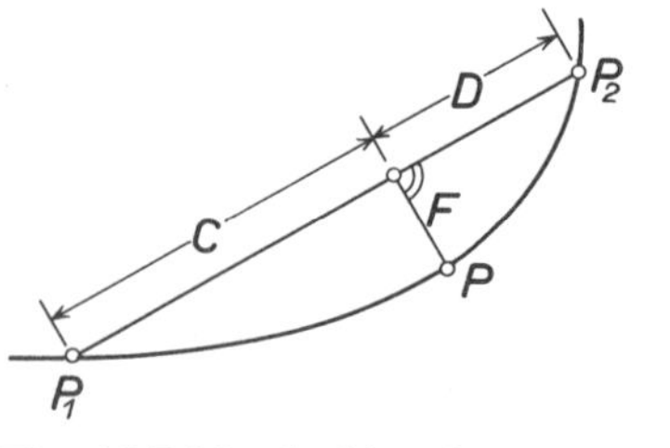

Fig. 13.7.22 Staking from chord

a) *Staking from the chord* (Fig. 13.7.22):

$$C = L - L_1; \quad D = L_2 - L; \quad F \approx \frac{CD}{6A^2}(L_1 + L_2 + L). \tag{13.7.34}$$

With $B = L - L_1 = L_2 - L$ and $C = D$ the mid-ordinate becomes

$$F_0 \approx \frac{B^2 L}{2A^2}. \tag{13.7.35}$$

b) *Staking from the secant* (Fig. 13.7.23):

If P is in the direction of an increasing curvature, then

$$U = L - L_2; \quad V \approx \frac{(L - L_1)(L - L_2)(L_1 + L_2 + L)}{6A^2}, \tag{13.7.36}$$

while P_1 and P_2 have to be interchanged for P in direction of decreasing curvature.

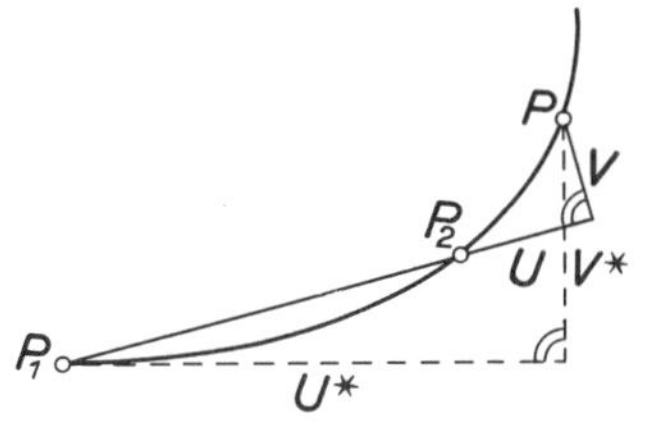

Fig. 13.7.23 Staking from secant

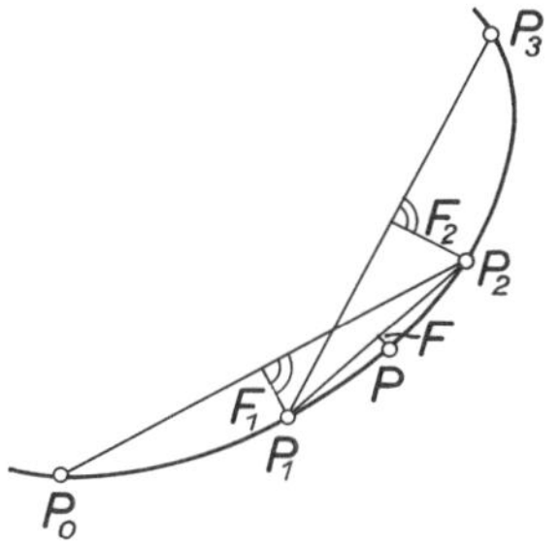

Fig. 13.7.24 Two-eighth's method

c) *Staking from the tangent* (dashed lines in Fig. 13.7.8):

$$U^* = L - L_1; \quad V^* \approx \frac{(L - L_1)^2 (2L_1 + L)}{6A^2}. \tag{13.7.37}$$

d) *Two-eighths method* (Fig. 13.7.24):

This method corresponds to the quarter method for circular arcs, and requires four spiral points, staked at equal distances. After measuring the mid-ordinates F_1 and F_2, the mid-ordinate F at P is

$$F \approx \frac{F_1}{8} + \frac{F_2}{8}. \tag{13.7.38}$$

e) *Staking of a chord traverse:*

Chords and traverse angles for selected chord lengths are best computed from rectangular coordinates. The traverse angles for short chords (Fig. 13.7.25) are computed by applying equation (13.7.33) twice as

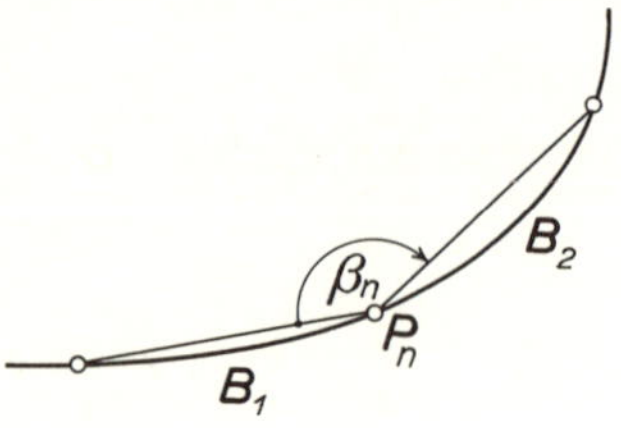

Fig. 13.7.25 Chord traverse

$$\beta_n(\text{gon}) \approx 180° - \frac{1}{6A^2}\,\frac{180°}{\pi}(B_1 + B_2)(3L_n + B_2 - B_1) \tag{13.7.39}$$

where B_1 and B_2 are the short arc sections, and L_n the arc length at point P_n.

14. Engineering Surveys

14.1 Tasks of and Special Problems in Engineering Surveys

14.1.1 General Considerations

Surveys in connection with planning, staking, and monitoring of technical objects often require that the surveyor cooperates and is in direct contact with engineers of other disciplines. Such projects require not only solid knowledge and experience in all aspects of practical and theoretical surveying, geodesy and associated subjects, but also sufficient understanding of structural methods and of machinery as well as knowledge of fundamental physics.

The field of engineering surveys has been growing significantly in recent years. The surveying engineer is nowadays not only involved in railway, highway and hydrological projects, but also when locating energy distribution systems, erecting large buildings, setting up and aligning machines, mounting structures from prefabricated parts, building towers, adjusting antennas, etc. Accuracies in position and height of ± 0.1 mm for distances of up to 1 km may be required, e.g., when staking an electron accelerator.

A surveying engineer faces three different tasks with each project.

1) Mapping of the local conditions as the basis for preliminary and final design.

2) Transfer of the final design into the field and checking of completed sections of the project.

3) Monitoring of the structure and its effect on surrounding areas for horizontal and vertical movements during construction and afterwards.

Generally, the civil engineer expects from the surveying engineer reliable numerical data, which can be directly utilized in further design considerations. The surveying engineer has to keep up with the construction progress and the tempo dictated by the use of modern machines. Furthermore, he must be versed in highly precise measuring methods, e.g., during bridge construction, tunnelling, etc.

These demands force the surveying engineer to use photogrammetric methods, EDM instruments, recording theodolites, as well as special instrumentation, such as plummets, gyro theodolites, lasers and, of course, electronic computers. The latter used to take over repetitive computation, but have long become vital for actual design work, which means that the surveying data have been integrated into the design. The surveying engineer, therefore, has to utilize measuring methods which are economical.

With the aid of error theory and mathematical statistics, he has to determine which measuring accuracy is appropriate for a given construction phase. For this, he must know specifications and tolerances which he often obtains on site only when discussing it with the engineer of the respective discipline. He also has to forcefully insist on measuring efforts which he considers necessary to meet these specifications.

Of the three types of jobs, namely topographic survey, staking, and monitoring, the survey does not generally differ too much from the methods of detail surveys and topographic mapping. Instrumental errors and projection distortions rarely need to be considered because most of them are eliminated by the observation and computation methods.

When staking, however, the surveying engineer not only utilizes the already mentioned new approaches, but occasionally has to resort to totally unconventional methods. Furthermore, he has to always consider the effects of instrumental errors and of projection distortions as well as, in some cases (e.g., tunnelling), the influences of refraction and of the deflection of the vertical.

Thus, the surveying engineer needs to be well versed in practical surveying as well as in the geodetic sciences. Thus, engineering surveying is not a specific area like photogrammetry but rather is the application of the whole field of surveying engineering under difficult circumstances.

It is thus only possible to present a limited selection of examples for the rapidly growing area of engineering surveying. Sections 14.2 and 14.3 deal with earthwork computations; in section 14.4 some of the surveying tasks in highway construction are mentioned; section 14.5 contains examples for measurements on engineering constructions, such as bridges and tunnels; and section 14.6 deals with monitoring of large dams by surveying means.

Recently, a number of special instruments have been developed to simplify and accelerate engineering surveys. The most important of these are optical precision plummets (section 14.1.2) and alignment devices with lasers (section 14.1.3).

14.1.2 Optical Precision Plummets

Practically all modern instruments are now equipped with optical plummets for centring above an observation station. Their accuracy, however, is insufficient for plumbing as required during construction of high-rise buildings, cooling towers, smoke stacks, piers, television antennas, etc., for mounting elevators, or for the determination of deformations. For projects like these, separate precision plummets either for zenith or nadir plumbing are being built, equipped with more accurate levelling devices, larger magnifications, and telescopes with longer ranges.

The collimation axis can be set vertical with the aid of spirit levels, compensators, or a mercury horizon. The telescopes of these optical plummets are either fixed vertically, or else attached to a horizontal axis. For the latter instruments, plumbing has to be arranged in two vertical planes normal to each other.

Optical precision plummets are usually levelled with the aid of cross bubbles. The manufacturers also offer auxiliary devices to adapt the plummets for the different tasks mentioned (e.g., base plates for setting up on building parts rather than using tripods, targets with checker board fields for estimation, and electrical illumination of the instrument for night operation). The plummets themselves are usually manufactured by incorporating parts of other instruments. Breithaupt, for instance uses the telescope of the precision level Nabon for the "Telim" (Fig. 14.1.1); and Zeiss, Jena the Ni 007 telescope for the PZL 100.

Fig. 14.1.1 Telim optical plummet by *Breithaupt*

Fig. 14.1.2 *Kern* precision plummet

A precision plummet can also be obtained by attaching an objective lens prism to the front of a quality engineer's level, e.g., the Ni 2 by Zeiss, Oberkochen.

With proper arrangement, the alignment lasers discussed in section 14.1.3.2 can also be utilized for precision plumbing (Fig. 14.1.2). Geofeinmechanik in Essen-Kettwig, an alignment laser manufacturer, supplies the LL 12 laser plummet which is listed at the bottom of Table 14.1.1.

The most important condition for the adjustment of all optical plummets is perfect parallelism between collimation axis ZZ and vertical axis VV. This is achieved by means of the bubble axis LL with the aid of the conditions $LL \perp VV$ and $ZZ // VV$ as discussed with the levels (section 9.1.2). Instruments with compensator levelling have to be adjusted in two mutually perpendicular planes.

Table 14.1.1 Precision plummets

Manufacturer	Model	Telescope			Dimensions			Weight		Maximum range	Accuracy of one plumbing	Special remarks
		Magnification	Lens diameter	Shortest Object dist.	Bubble sensitivity	Length	Height	Instrument	Container			
		×	mm	m	×	mm	mm	kg	kg	m		
Breithaupt, Kassel	Telim	42	50	2.0	6	270	350	7.1	6.5	600	1 : 100000	Cross bubble
Jenoptik Jena	PZL 100	31.5	40	2.2	–	–	335	4.8	2.8	200	1 : 100000	8′-bull's-eye bubble
Kern & Co. Aarau (Switzerland)	OL	22.5	30	0.8	6	135	160	3.7	1.8	100	1 : 50000	diff. telescope for up and down
Theis (previously known as *Fennel*)	Ferlo	30	40	1.6	6	125	210	3.2	2.8	150	1 : 100000	4′-bull's-eye bubble
Wild-Heerbrugg (Switzerland)	ZL	24	36	0.9	6	260	170	3.1	2.2	200	1 : 200000	also with nadir plummet
Geofeinmechanik Essen-Kettwig	LL 12	Laser with 2 mW			9	555	210	8.0		(100)	2.5 mm –100 m	see sect. 14.1.3.2

Note: Alignment instruments described in section 14.1.3.2 can be utilized for plumbing in conjunction with detecting prisms.

14.1.3 Alignment Instruments with Lasers

14.1.3.1 Gas Lasers

Using the method described in section 13.2, only short distances can be staked and even then with limited accuracy. Today lasers are used for aligning of long highway sections, rails or channels, for staking of axes and foundations of machines, and for observing deformations on large structures (e.g., power dams) under varying loads and temperature conditions.

The word "laser" stands for "light amplification by simulated emission of radiation". While the rays of the sun or a light bulb consist of a mixture of different wavelengths, i.e., they are incoherent, lasers consist of monochromatic and coherent light rays which are narrowly bundled. There are several types of lasers. Gas lasers are especially suited for surveying purposes, for they have good density and excellent directional stability coupled with a small angle of divergence, which can be reduced to about 1″ by attaching an optical system.

The gas laser (Fig. 14.1.3) contains a 9 : 1 mixture of helium and neon (He-Ne laser). If this mixture is excited by a direct current, alternating effects between the He and Ne atoms occur. The photon radiation caused by this is reflected by mirrors onto the quartz plates whose tilt angle (Brewster angle) is calculated such that a light ray which is parallel to the axis of the tube is not reflected on their surface but passes through the closed tube without loss of energy.

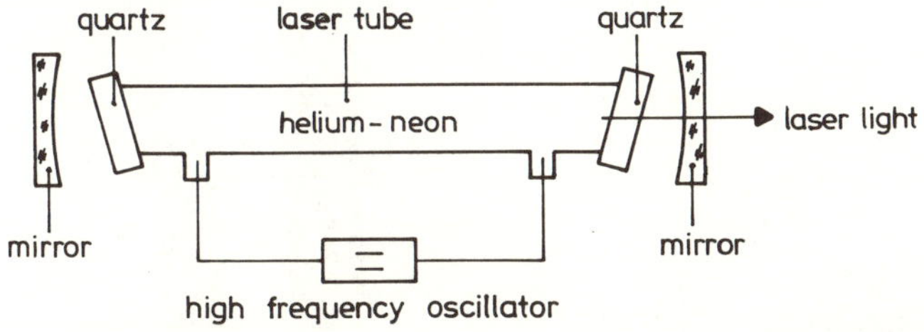

Fig. 14.1.3 Principle of gas laser

These emitted rays are received for up to 60 metres directly on the rod using a template or a milk glass for manually determining the location of the light spot. Beyond 60 m, photo diodes are used as laser detectors which provide a range of 250 to 300 m. The alignment or height accuracy is quoted as 2 mm per 60 m. Since the survey lasers have only 0.001 to 0.005 W, they are not dangerous for humans, as long as the rays do not directly hit the eye.

As illustrated in Fig. 14.1.3, a surveying laser consists of a 12 V battery to generate the voltage, the laser tube, and the optics for bundling of the rays. Furthermore, a support with a micrometer device for tilting of the tube by about $\pm 30°$, as well as 1 or 2 spirit bubbles for levelling and sighting a telescope, whose axis is parallel to the laser tube, make up the instrument. Occasionally a horizontal and vertical circle is added. One or two laser rods complete the equipment.

If a laser tube is attached to the telescope of a theodolite or level such that their axes are parallel, a survey laser is obtained. Both types of instruments are classified as self-contained laser instruments, while lasers as attachments are also used in connection with surveying equipment.

Most survey lasers have been designed for construction and tunnelling and are available with varying auxiliary devices. For surveying work, either survey lasers or simple construction lasers are used. The following sections provide a short overview of these instruments.

14.1.3.2 Self-Contained Laser Instruments

Spectra Physics, U.S.A., offers a large variety of surveying lasers, construction lasers, tunnelling lasers, etc., and is steadily expanding its offerings.

The electronic automatic laser levels (Models 944 SL and 945 GS, Fig. 14.1.4) combined with the model 970 automatic laser rod (Fig. 14.1.5) form a very useful system for the surveyor, especially for construction work. It operates as follows. After the laser level has been roughly levelled with the aid of a spirit level, the electronic system

Fig. 14.1.4 945 GS laser level

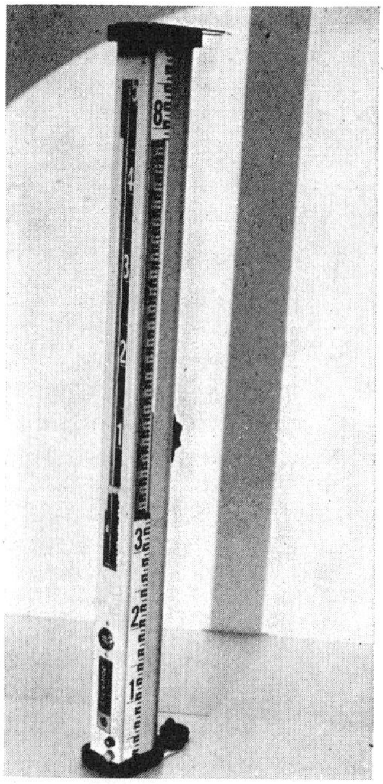

Fig. 14.1.5 Laser rod, model 970

scans the edges of the level bubble. It then corrects the levelling and projects a horizontal light plane (or vertical if desired) which is detected at the rod by a laser detector moving up and down. If the levelling of the instrument is disturbed by vibrations, etc., for longer than 3 seconds, then the laser is switched off by a warning system. It is only switched on after the automatic electronic levelling has readjusted the instrument. The tilt range of the automatic level is ± 8 degrees.

After the laser rod has been positioned above the point to be surveyed, the operating knob is depressed. The detector then starts moving up and down. As soon as it catches the rotating laser beam it stops and orients itself into the axis of the laser beam.

Model 945 GS can also be used for vertical planes (plumbing and alignments), again with a ± 8 degree range for automatic levelling.

The Model 842 Rodolite construction laser used for aligning and construction control corresponds to the laser level. This laser produces a pencil-thin beam which rotates automatically at different velocities in a horizontal and a vertical plane. A second beam is emitted exactly normal to the light plane and can be used for plumbing.

Other Spectra Physics instruments which are of interest for surveying engineers are the LT 3 alignment laser and the 611 laser theodolite where the optical axis of the telescope coincides with the axis of the laser beam.

Visomat-Laier A.G. in Neuendorf (Switzerland) produces surveying instruments specifially designed for work on construction sites. Among others, the company developed the "fully automatic VAK-1 construction level" which operates with visible light and which is levelled automatically with the aid of a mechanical pendulum supported in ball bearings and dampened magnetically. Because of a special axis

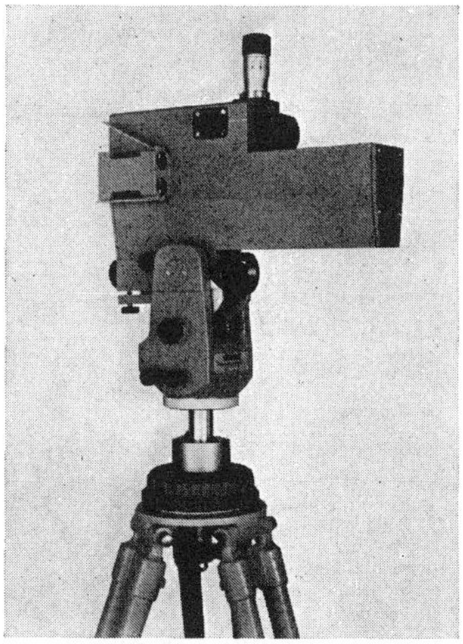

Fig. 14.1.6 *Visomat* multipurpose laser

system, which is also used with the LS 5 construction laser (Fig. 14.1.6), it can be shifted vertically by 120 mm and thus can permit the often preferred "levelling without computing".

The company recognized early the advantages of the laser beam which led to the design of the multi-purpose LS 5 construction laser (Fig. 14.1.6), which has an efficient surveying component and permits horizontal and vertical alignments in one working sequence. The "fully automatic LS 7 canal laser" automatically stabilizes slope and alignment setting. A "pilot" which is connected to the laser by cable is inserted into the line of sight at about 10 metres from the instrument in order to stabilize the vertical plane. If deviations of 0.01% or more occur in either elevation or alignment, the laser automatically indicates this by blinking at the laser beam. This blinking continues until the deviations are corrected. The laser beam also blinks when the instrument system is incorrectly set up.

14.1.3.3 Laser Attachments

Laser attachments to surveying instruments are produced by Spectra Physics (U.S.A.), AGA (Sweden), as well as Kern and Wild from Switzerland.

In addition to the various alignment lasers, Spectra Physics offers the laser distance meter 3 G geodolite which uses an infrared gas laser ($\lambda = 0.6328$ μm) as a light source. This provides a range of 40 miles (65 km) at day, and 60 miles (95 km) at night. Since the late 1960 s, AGA has produced the Laser Geodimeter 8 which utilizes a 5 mW He-Ne laser. This has now been replaced by Model 600 which only requires a 1 mW He-Ne laser. The range is 40 km at day or night.

The Kern DKM-2-AL laser theodolite (Fig. 14.1.7) utilizes a Spectra Physics 120 T

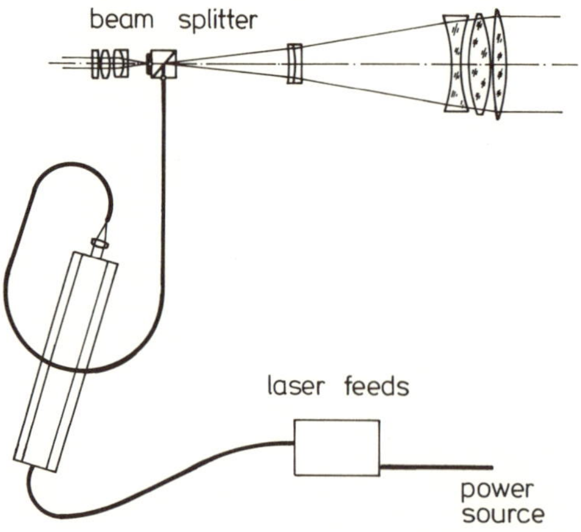

Fig. 14.1.7 Feeding laser light into the telescope on the *Kern* DKM 2-AL

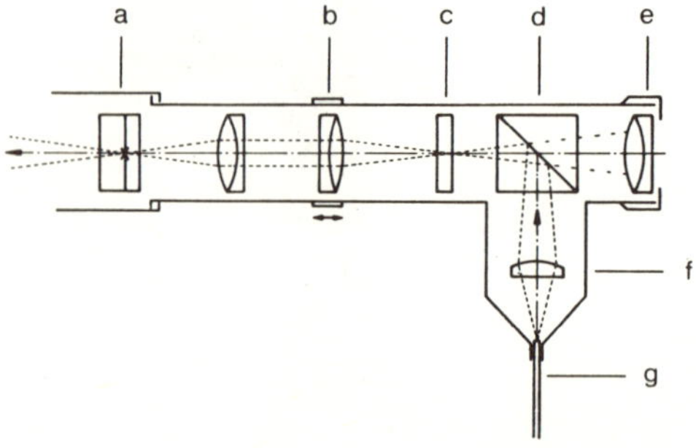

Fig. 14.1.8 Section through the GLO 2 laser eyepiece by *Wild*

gas laser whose beam is concentrated with a lens to a light ray of 0.08 mm diameter. Through fibre optics it illuminates the cross hair which is located inside the beam splitter. This beam splitter directs the laser beam into the optical axis of the theodolite telescope which causes the cross hair to be projected through the objective lens onto the target. The instrument has a range of about 60 km at day and night.

Wild Heerbrugg developed the GL 02 laser eyepiece (Fig. 14.1.8) which can be exchanged via the bayonet fitting against the standard eyepieces of the T1, T2, T16 theodolites and of the NA2 and NAK2 levels.

In Fig. 14.1.8, *a* represents the reticule with the cross hair, *g* the fiber optics carrying the laser light, *f* the collecting lens, and *d* the beam splitter. *b* and *c* serve to adjust the laser beam onto the reticule plane, while *e* is the actual eyepiece.

14.2 Earthwork Computation

Virtually all large engineering structures require earthworks. Thus, in hilly and mountainous terrain, it is desirable to design the projects such that the earthworks remain low or at least that there is a balance between cut and fill. The mathematical foundations for earthwork computation are the well-known formulae for calculating the volume of prisms and pyramids, as well as Simpson's and Guldin's rules. Generally, earthwork computation is based on the method used for surveying the earth's surface, such as profiles and cross sections, area levelling, and topographic surveys.

Due to irregularities of the earth's surface, the earthwork volumes will only be approximated using the formulae mentioned. Systematic errors have to be expected. Therefore, the effect of errors is included with the treatment of the different methods.

14.2.1 Earthwork Computation Using Cross Sections

This type of problem occurs primarily for cut and fill in road construction. The most important methods are Simpson's rule for the mass computation along straight road sections, and Guldin's rule for curve sections.

14.2.1.1 Simpson's Rule

This formula is used for the computation of volumes of dams or cuts in smoothly inclined terrain.

Using the neighbouring cross sectional areas F_1 and F_2 (Fig. 14.2.1), the cross section area F_m in the middle, and the length l between F_1 and F_2, the volume is

$$V = \frac{1}{6}(F_1 + 4F_m + F_2)\,l\,. \tag{14.2.1}$$

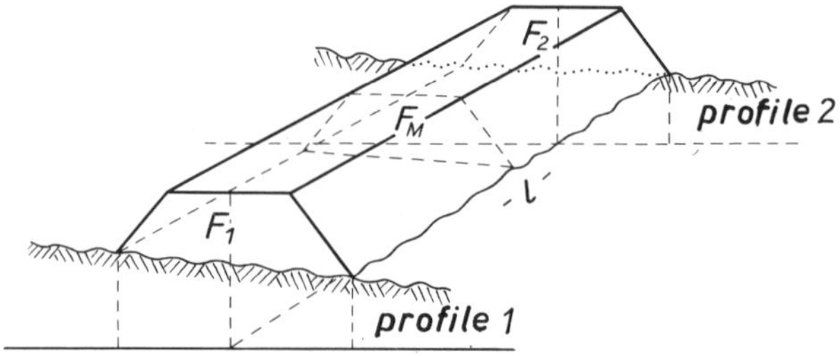

Fig. 14.2.1 Simpson's rule

For practical purposes, often $^1/_2\,(F_1 + F_2)$ is used for F_m which leads to the approximate formula

$$\bar{V} = \frac{1}{2}(F_1 + F_2)\,l \approx F_m l\,. \tag{14.2.2}$$

These formulae hold for fill, cut, and partial cut. In the latter case, the fill is considered positive, and the cut negative, which results in the difference of the two partial volumes (Fig. 14.2.2).

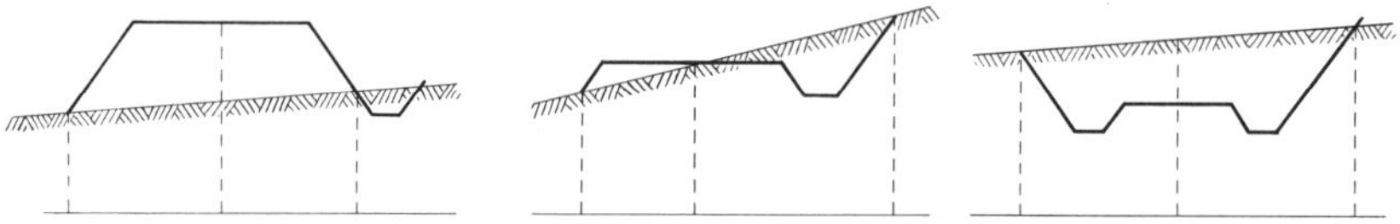

Fig. 14.2.2 Regular cross sections

The cross sectional areas F_1 and F_2 are best obtained by either using a planmeter or numerically from coordinates. There are some special equations for frequently occurring cases (*Heimerle* 1929), for instance, provides a formula for ramps with width b, height h, inclination $1:m$, and slope ratio $1:n$ as

$$V = \frac{h^2}{6}(m-n)\left(3b + 2hn\left(1 - \frac{n}{m}\right)\right). \tag{14.2.3}$$

14.2.1.2 Guldin's Rule

This formula provides the volume of a rotational body generated by rotating a plane area F around a fixed axis and bounded by the cross sections F_1 and F_2. The rule simple states (Fig. 14.2.3):

V = cross sectional area · path length of its centre of gravity. (14.2.4)

If the planes of the cross sections are selected such that they are vertical and pass through the centre of the circle, then Guldin's rule can be considered as a general case of Simpson's rule. However, the separation of the cross section has to be measured along the arc, not the chord. Furthermore, one has to consider that the centre of gravity of the cross sectional areas does not usually fall onto the axis of the road.

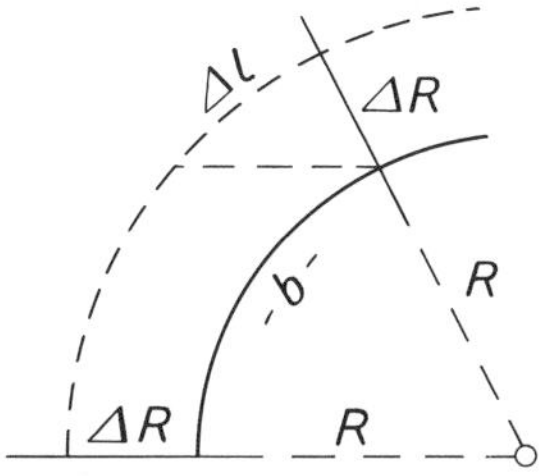

Fig. 14.2.3 Guldin's rule

14.2.1.3 Accuracy Considerations

a) Errors occurring when applying Simpson's rule according to equation (14.2.2) are estimated from the two extreme cases:

Case 1: Straight or screw prisms:

$$F_1 = F_2 = F_m, \quad \text{and} \quad V = \bar{V}.$$

Case 2: Straight or screw pyramids:

$$F_1 > 0; \quad F_2 \to 0; \quad F_m = {}^1/_4 \, F_1 .$$

Therefore, using equation (14.2.1):

$$V = \frac{1}{3} F_1 \, l$$

while according to equation (14.2.2) we get

$$\bar{V} = \frac{1}{2} F_1 \, l .$$

All practical cases fall between cases 1 and 2.

Thus

$$V \leqq \bar{V} .$$

When using the approximate formula, the resulting volume is always too large. The result is favouring the contractor who charges for the earthwork.

If the separation of the cross sections is reduced to one half, then the error is reduced to one quarter. Experience shows that the relative error of the volume is about

$$\frac{\Delta V}{V} \approx 0.01 \text{ to } 0.05\,.$$

b) Guldin's rule has errors if the axis of the road does not coincide with the centre of gravity of the profile.

Let us assume that

$$F_1 = F_2 = F\,.$$

According to equation (14.2.4) we get

$$V = F(l + \Delta l)$$

which is approximated to

$$V' = Fl.$$

As evident from Fig. 14.2.3

$$\Delta l \approx \Delta R \cdot \frac{1}{R}$$

and therefore

$$V = Fl\left(1 + \frac{\Delta R}{R}\right).$$

The difference is then

$$\Delta V = V' - V = -Fl \cdot \frac{\Delta R}{R}\,.$$

This leads to the relative error in volume

$$\frac{\Delta V}{V'} = -\frac{\Delta R}{R}\,. \tag{14.2.5}$$

Thus, the relative volume error increases with increasing ΔR, but decreases for larger radii. Especially for partial cuts, where the centre of gravity deviates significantly from the axis, the ΔR influence is critical.

Example:

For $R = 150$ m and $\Delta R = 3$ m, the ratio $\Delta V/V'$ becomes 0.02. This is a maximum value which only rarely reaches this magnitude.

c) Error caused by insufficient approximation of the terrain within cross sections as well as between them have to be expected. Experience shows that they range from 1 % to 4 %.

For earthwork computation using Simpson's or Guldin's rules, a height error of about 1 % to 5 % can be expected. Considering the ratio given at the end of section 14.2.1.3 a) it is desirable to keep the systematic errors small by selecting shorter spacing between cross sections. The approximation errors mentioned in section 14.2.1.3 c) are of a random nature, but are also reduced when the cross section spacing is decreased, although not as rapidly.

14.2.2 Earthwork Computation Using Area Levelling

In moderately rolling terrain, an area is designated for a sports field, factory yard, parking lot, etc., and thus has to be transformed to a horizontal plane. For such a project, usually a square grid is superimposed onto the terrain as the base for area levelling. If such an arrangement is not suitable, a continuous series of triangles (triangular grid) can be used instead. All corner points are levelled or else measured by stadia.

Such a survey provides the base for a height plan onto which the project is plotted. If the project has a predetermined ground elevation, it is used as the reference plane for cut and fill. The above mentioned squares or triangles are projected, and the volume is equal to the sum of all square or triangular prisms above and below this plane. If, as illustrated in Figs. 14.2.4 and 14.2.5, F is the area of the basic figure (e.g., triangle or square), and h_i are the vertical differences between the reference plane and terrain surface, measured along the prism edges, then the prism volumes are obtained as

$$V_3 = F_3 \frac{h_1 + h_2 + h_3}{3}; \quad V_4 = F_4 \frac{h_1 + h_2 + h_3 + h_4}{4}.$$

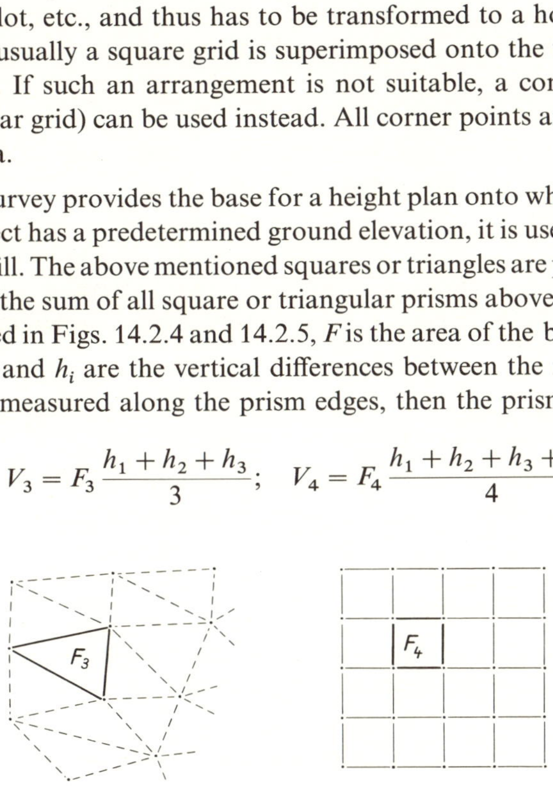

Fig. 14.2.4 Triangular and square grid

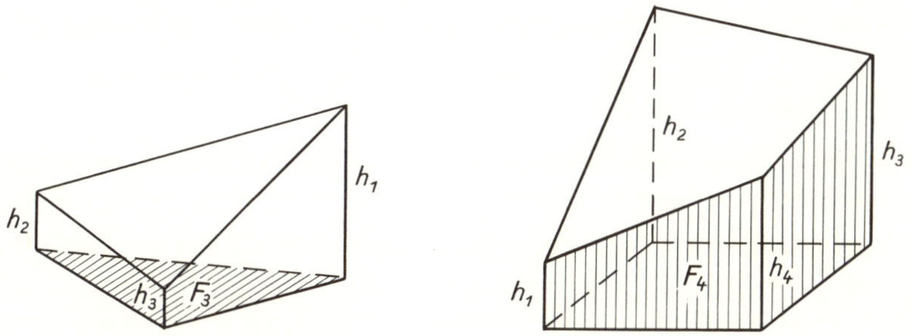

Fig. 14.2.5 Volumes of prisms

Triangular grids generally provide a better approximation to the terrain than square grids. For road planning, it is possible to have one triangle side coincide with a cross section. Position and elevation of the grid points are best determined by tacheometric means, preferably with a recording tacheometer (section 12.1.4). Thus the directions (α), distances (s), and zenith angles (z) from the instrument station P_0 to the vertices of the triangle F_3 are to be observed. The elevations of the vertices are obtained as

$$h_j = h_0 + s_j \cot z_j + i - t \,, \tag{14.2.7}$$

where h_0 is the MSL height of the instrument station, i the instrument height above terrain, and t the target height or the rod reading. The area F_s is computed by either

$$2F_3 = s_1 s_2 \sin(\alpha_2 - \alpha_1) + s_2 s_3 \sin(\alpha_3 - \alpha_2) + s_1 s_3 \sin(\alpha_1 - \alpha_3) \tag{14.2.8}$$

or from the general formula suitable for all types of polygons:

$$2F = \sum_{j=1}^{j=n} s_{j-1} \cdot s_j \sin(\alpha_j - \alpha_{j-1}) \,. \tag{14.2.9}$$

Since the observer has freedom of choice when selecting the points, resurveys usually lead to somewhat different results. In spite of it, the relative error of the volume can be kept within

$$\frac{\Delta V}{V} = 0.5\% \,.$$

When using square grids, no special area computation is required as the area F_4 of the grid squares is known. The method, however, is rather inflexible. Even in reasonably varied terrain, the sides of the grid squares have to be reduced in such a manner that the method becomes uneconomical. This lack of flexibility, however, has an advantage: The total volume of the body covered by the square grid (Fig. 14.2.4) can be computed in one shot by using the corner elevations of the total area once each, the border elevations twice, and the interior elevation four times each, and then multiplying the sum by $^1/_4\, F_4$.

14.2.3 Earthwork Computation Using Contour Lines

When looking at Fig. 14.2.6, immediately a number of practical applications come to mind, such as hydro dams, gravel pits, open pit mines, or with the opposite sign, stock piles, land fills, etc.

Here again, the project is first plotted onto an existing or specially prepared contour plan. There are several ways of carrying out the earthwork computation. If the objects are small, profiles and cross sections can be obtained from the contours, and the computation follows the approach discussed in section 14.2.1. The object can also be divided into prisms and treated as in section 14.2.2. In both cases, however, the terrain heights have to be interpolated.

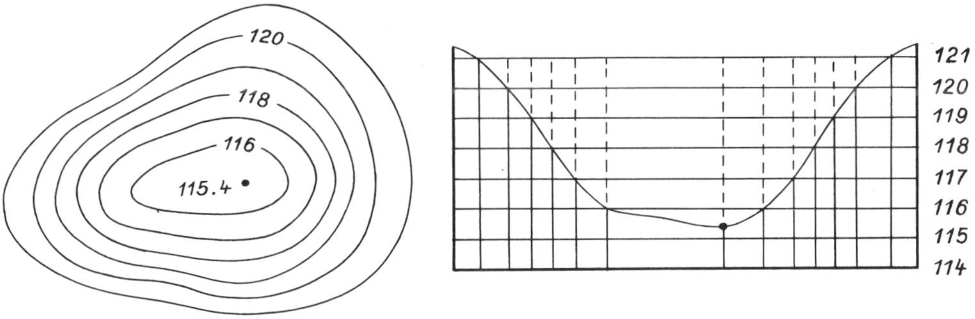

Fig. 14.2.6 Contour plan and vertical section

This leads to a loss in accuracy which means that the result is inferior to the methods discussed in sections 14.2.1 and 14.2.2.

For larger objects, therefore, the computation is directly based on the contour plan. The volume is computed as follows: If F_1, $F_2 \ldots F_n$ are the areas enclosed by one particular contour line, and if h is their vertical separation in all cases, then the volume of the body becomes

$$V = \frac{F_1 + F_2}{2} h + \frac{F_2 + F_3}{2} h + \ldots \frac{F_{n-1} + F_n}{2} h$$

or $$V = (^1/_2 F_1 + F_2 + F_3 + \ldots + F_{n-1} + {}^1/_2 F_n) h\,. \qquad (14.2.10)$$

The easiest and safest way of obtaining these areas within contour lines is by using a planimeter.

The same result as in equation (14.2.10) is obtained when the mean between F_i and F_{i+1} is used in equation (14.2.1), which is a good approximation considering the small separation of the contour lines. Thus equation (14.2.10) provides a satisfactory accuracy. The uncertainty of the results obtained when using equation (14.2.10) is primarily a function of the course of the contour lines and amounts to about 1%.

In special cases, the volumes of bodies have to be determined which can be expressed by specific formulae, e.g., cylinder, sphere, pyramid, cone or prism. In these cases, the specific mathematical formula should be utilized.

14.2.4 Earthwork Computation Using Profile Diagram and Mass Profiles

When designing roadways, it is possible to utilize longitudinal profiles for earthwork computation, especially if the vertical scale is exaggerated and the gradient of the designed project is plotted. The profile diagram and the mass profile provide some technical means for this. There are several solutions for each. The method by Goering

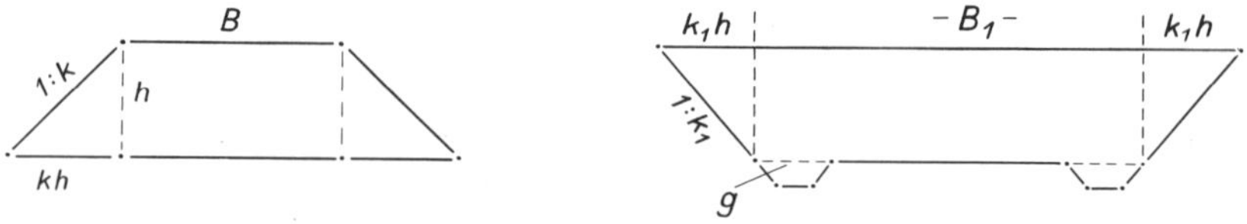

Fig. 14.2.7 Cross section areas

is the easiest. It is, however, only valid if the terrain is not sloped normal to the axis of the road, and if both dams and cuts have the same slope ratio.

As evident from Fig. 14.2.7, we get

$$F_{\text{Dam}} = Bh + kh^2; \quad F_{\text{Cut}} = B_1 h + 2g + k_1 h^2. \tag{14.2.11}$$

Goering proposes a graphical evaluation of these equations using the "profile diagram" (Fig. 14.2.8) with which the products in equation (14.2.11) can easily be read and, after addition, result in the value for the cross section area in m^2. For the profile diagram, the same height ratio is selected as in the longitudinal profile of the road, shown in Goering's illustration (Fig. 14.2.9 top). This figure is a reduction of a drawing where 1 cm represented 2.5 m in height, and 50 m along the road. Thus the values for h, without being read numerically, can be directly transferred with a compass from longitudinal profile to profile diagram. Then for each height h, the components of the cross section area as given in equation (14.2.11) can be measured in the profile diagram.

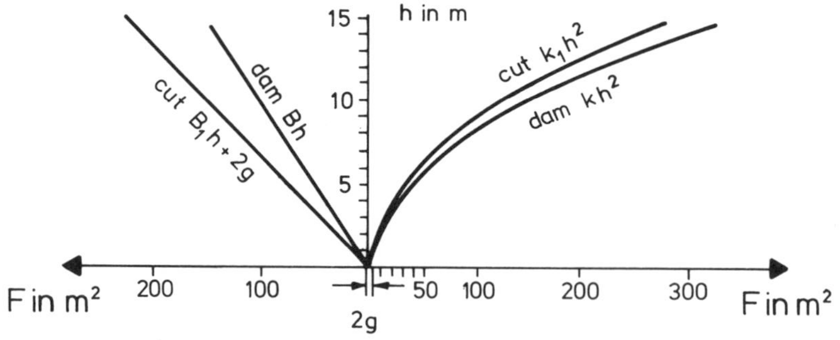

Fig. 14.2.8 Profile diagram

The area values for the cross sections can be plotted below the longitudinal profile (see centre part of Fig. 14.2.9) based on the zero line as the abscissa axis. Cuts are plotted above the reference line, and dams below. By connecting the plotted points, an area profile is obtained, whose ordinates represent cross section areas. The original diagram for Fig. 14.2.9 is based on a scale where 1 cm represents 40 m^2.

By graphically integrating the area profile, the mass profile as shown on the bottom of Fig. 14.2.9 is obtained. The centre line of each trapezium formed by neighbouring

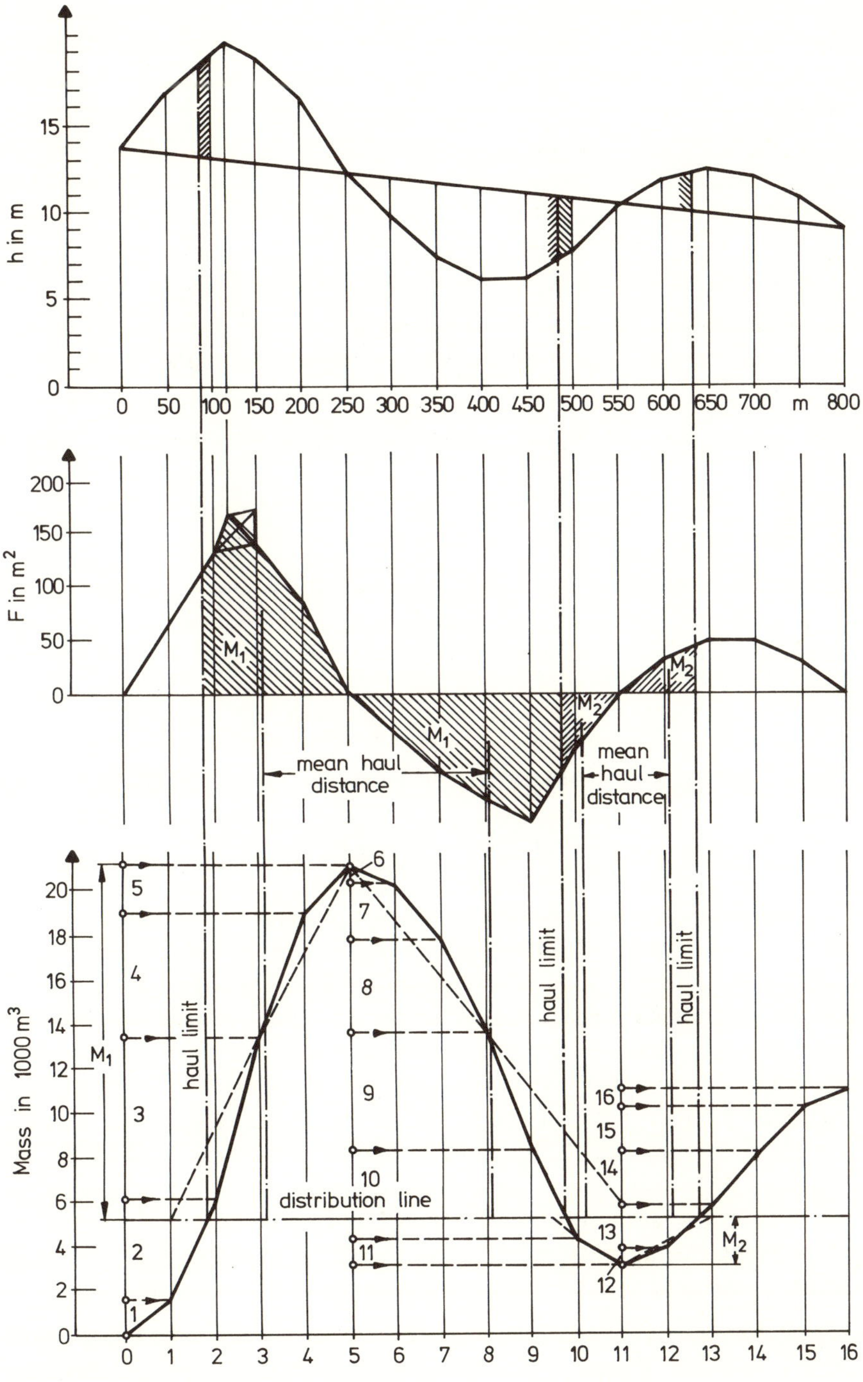

Fig. 14.2.9 Mass profile

ordinates, the reference line, and the area profile, are to be extracted. After summation the values are to be plotted on the right-hand ordinate for the mass profile. Goering uses a somewhat different approach. Again, he extracts the centre lines of each trapezium. However, he transfers them, separately for cut and fill consecutively, onto a "mass ordinate" at the beginning or end of the respective line. Cut masses are plotted positively, fills negatively. The terminal points of the distance segments of the mass ordinate (numbered in Fig. 14.2.9), are projected onto the corresponding ordinates based on the number and arrow and thus represent these terminal points there. Connecting these latter points generates the mass profile.

Since the mean heights of the area profile are plotted without change as masses, and since the separation of the cross sections is 40 m on the ground, the mass represented by 1 cm is $50\ m^2 \cdot 40\ m = 2000\ m^3$.

The mass profile is a technical tool for balancing masses as much as possible while keeping the haul distances short. This is accomplished with the aid of a vertical "distribution line" (Fig. 14.2.9 bottom) which is drawn into the mass profile and which correlates ordinates for cut with ordinates for fill.

For practical design work, several gradients and distribution lines are studied in order to balance the masses with short haul distances. The points of intersection between distribution lines and the mass profile provide haul limits as plotted in Fig. 14.2.9.

14.3 Earthwork Computation Using Digital Terrain Models

14.3.1 Creation of a Digital Terrain Model (DTM)

When planning large traffic projects, the main purpose, i.e., the construction of a highway between *A* and *B*, is known, while its exact location remains open. Usually, a preanalysis has been conducted to select a most suitable route using available maps. This route was then transferred onto the terrain, and profiles and cross sections were measured to provide data for the final design. One of the most important criteria is the amount of necessary structures, such as bridges and tunnels, together with the amount of earth to be moved which usually can be determined only by specific earthwork computations. If the evaluation indicated that the designed route was not very favourable, a new one was designed and staked. If transformation of the original profiles onto the new route was not feasible, new profiles and cross sections were measured for further evaluations. Occasionally, a third or even fourth route was studied in the same manner until the optimum solution was found.

Such an approach is very unsatisfactory because of the continued requirement of new field surveys. It is, therefore, desirable to compare the different alternatives on the basis of a single extended numerical (digital) terrain survey and to do this with the aid of a computer in the office.

This method, therefore, does not consider a specific route but rather a band of interest which is defined with the aid of available maps in the scales 1 : 20000 to 1 : 50000. This band contains all alternatives under study and has a width of 200 to 1000 m. Utilizing optimization computations, the different routes are compared considering technical and structural aspects as well as traffic flow and so on.

In order to conduct this comparative evaluation, the band of interest has to be measured with an accuracy suitable for 1 : 2000 to 1 : 5000 maps. This can be done by differential levelling, stadia, or photogrammetry and provides the data for the digital terrain model. It is arranged such that the characteristic terrain points are defined numerically in position and height. Self recording electronic tacheometers which record direction, distance and zenith angle (e.g., Zeiss Reg Elta) are most suitable for this purpose. When using an analogue plotter for photogrammetric evaluation, it is necessary to also register points along the contours or else measure all significant terrain points numerically in the plotter. After transforming these data from ground and aerial surveys into a rectangular coordinate system, the digital terrain model is obtained. A mathematical interpolation surface is fitted onto the terrain points – often this is rather cumbersome – and the various alternatives are compared with the aid of the computer.

14.3.2 Mathematical Representation of the Terrain Model

The problem is reasonably simple, if one assumes that the terrain model already exists. Then the important heights for the route, i.e. break points on profiles and cross sections (Fig. 14.3.1), have to be interpolated.

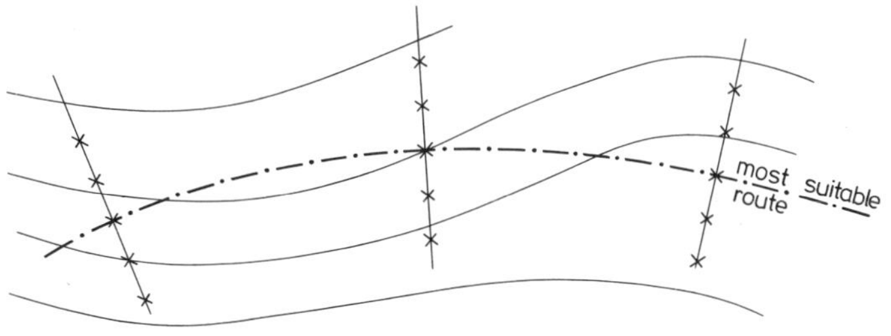

Fig. 14.3.1 Contours and cross sections

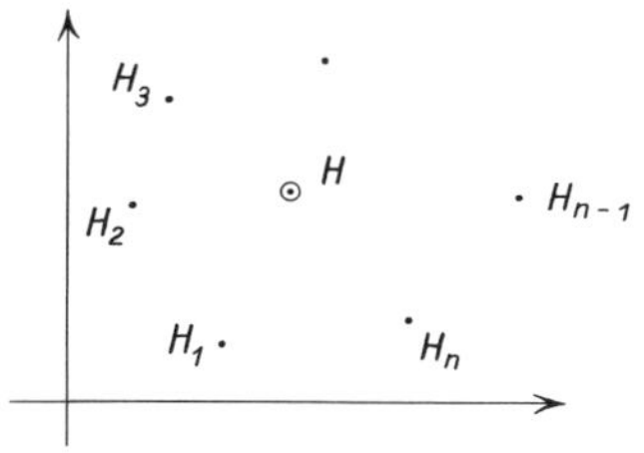

Fig. 14.3.2 Interpolation of elevations from planimetric coordinates

When looking at the terrain differentially, it is a smooth surface which is described as a function of the planimetric coordinates. In Fig. 14.3.2, the points $H_1, H_2 \ldots H_n$ represent digitally determined terrain points transformed into a spherical coordinate system.

The point H is a new point whose elevation is to be interpolated:

Given: $H_1, H_2 \ldots H_4$;

Wanted: $H = fct\,(H_1 \ldots H_4)\,.$ (14.3.1)

A general surface is described by the following equation:

$$H(x, y) = a_{00} + a_{10}x + a_{01}y + a_{11}xy + a_{20}x^2 + a_{02}y^2 + \ldots\,. \tag{14.3.2}$$

By extending such a polynomial expression, any surface can be described. Thus for the given points, n equations for the determination of the coefficients can be written as:

$$\left.\begin{array}{l} H_1 = a_{00} + a_{10}x_1 + a_{01}y_1 + a_{11}x_1y_1 + a_{20}x_1^2 + a_{02}y_1^2 + \ldots \\ H_2 = a_{00} + a_{10}x_2 + a_{01}y_2 + a_{11}x_2y_2 + a_{20}x_2^2 + a_{02}y_2^2 + \ldots \\ H_3 = a_{00} + a_{10}x_3 + a_{01}y_3 + a_{11}x_3y_3 + a_{20}x_3^2 + a_{02}y_3^2 + \ldots \\ \vdots \\ H_n = a_{00} + a_{10}x_n + a_{01}y_n + a_{11}x_ny_n + a_{20}x_n^2 + a_{02}y_n^2 + \ldots \end{array}\right\} \tag{14.3.3}$$

However, to solve such a system does not make sense. It is much more realistic to select a low order polynomial and permit residuals v_i. Thus, a least-squares adjustment with the following observation equations is obtained:

$$\left.\begin{array}{l} v_1 = a_{00} + a_{10}x_1 + a_{01}y_1 + a_{11}x_1y_1 + a_{22}x_1^2 + a_{02}y_1^2 + \ldots - H_1 \\ v_2 = a_{00} + a_{10}x_2 + a_{01}y_2 + a_{11}x_2y_2 + a_{22}x_2^2 + a_{02}y_2^2 + \ldots - H_2 \\ v_3 = a_{00} + a_{10}x_3 + a_{01}y_3 + a_{11}x_3y_3 + a_{22}x_3^2 + a_{02}y_3^2 + \ldots - H_3 \\ \vdots \\ v_n = a_{00} + a_{10}x_n + a_{01}y_n + a_{11}x_ny_n + a_{22}x_n^2 + a_{02}y_n^2 + \ldots - H_n \end{array}\right\} \tag{14.3.4}$$

With n coefficients, the standard deviation

$$\sigma_0 = \pm\sqrt{\frac{[vv]}{n-u}} = r$$

provides a limit of "roughness" which amounts to about 5 cm for grassland, and 10 to 15 cm for ploughed fields. r also provides an aid when judging whether the degree of the chosen polynomial is sufficient.

Special types of approximations, such as orthogonal Tschebyscheff polynomials, have been mentioned in the literature, as they provide computational advantages. Furthermore, refinements have been proposed which are based on the theory of stochastic processes, e.g., least-squares prediction, collocation, etc.

14.3.3 Mass Balancing and Optimal Route Location

With the aid of digital terrain models, it is now possible to utilize the computer for solving the mass balancing problem, which was dealt with conventionally in section 14.2.4. The demands for an optimum route location, however, go further than that. Among others, the following is required.

1) Minimum cut and fill masses.

2) Minimum sum of masses and minimum difference between cut and fill.

3) Minimum driving time and minimum fuel consumption for a representative trailer truck.

These requirements are further constraint by

1) given height of existing points, e.g., intersections, bridges, tunnels

2) maximum grades

3) curvature which prohibits small radii.

Some of the above mentioned conditions are contradictory. It is therefore necessary to optimize several models based on various conditions, and then to select one.

Thus the task of investigating various alternatives for route location depends heavily on the mathematical modelling of the problem suitable for computer processing. This requires a solid mathematical background in addition to the necessary engineering knowledge.

While in the past deterministic models were developed, today the design is the centre of the modelling and subjected simultaneously to virtually all possible technical design considerations. The alternatives, which are possible within certain imposed limitations, are expressed as a series of linear elements, realized on the computer in a random fashion. It is thus a stochastic process which requires a stochastic model for simulation.

This connection to random events led to the name of the best known method for optimizing road adjustments, the Monte Carlo method. The reader is advised to consult special literature on this subject for a more detailed treatment.

14.4 Engineering Surveys in Connection with Transportation Routes

14.4.1 Production of Maps for Design Purposes

The decision on whether and where a highway or railway is to be built is largely a political one. As already mentioned in section 14.3.1, it is the task of the project engineer to define the band of interest during a preliminary analysis using maps of 1 : 20000 to 1 : 50000 scale. Furthermore, he has to compare different alternatives under technical and economical considerations, and then select the most suitable route. He will likely select an alternative which requires the least expected earth movement and the fewest major structures while being adaptable to other economical aspects, such as grades, traffic safety, etc.

In the second phase (preliminary design), maps of 1 : 2000 to 1 : 5000 have been used to graphically design the route using circle and spiral templates. The maps were produced by photogrammetry (occasionally by stadia), so that local constraints could be considered in the design.

The project engineer defined the tangents and their points of intersection during this phase, thus providing the basis of the tangent polygon from which the curves are staked. Then the profiles and cross sections necessary for earthwork computation were measured locally. This graphical solution was computed numerically in a third phase for the generation of staking plans. Thus, the final design was obtained and staked in the field just prior to the commencement of the construction.

When proceeding in this manner, it could happen that the axis has to be changed because of the earthwork computation results, which means that a new series of cross sections and profiles are needed. In order to avoid this, the first phase is retained, just like in the recently developed method of comparative study of alternatives (section 14.3.3). However, all project computations are carried out in coordinates using the general provincial or state system. Then the whole band of interest is carefully surveyed by stadia or photogrammetry and represented as a digital terrain model (section 14.3.1). This model, which basically transfers the terrain to the office, is used for a comparative investigation of various alternatives using general or specific criteria (section 14.3.3). Then only the final design is transferred onto construction and lay-out plans.

In this case, it is no longer necessary to stake a tangent polygon. Instead certain points of the centre line are selected and graphically coordinated. These coordinates, together with the parameters of the geometric elements, form the input to a computer program which fits the axis into the graphically determined coordinates while holding the parameters of the geometric elements.

This type of design requires a careful densification of the horizontal network, especially near buildings, where a high neighbouring accuracy is essential. Whether this network is densified by trigonometric means (triangulation and trilateration), or with the aid of precision traverses, depends largely on the local conditions. Today distances are measured with EDM equipment. All control points are to be carefully monumented. Since many of them get damaged or lost during construction, they are to be witnessed by short local linear measurements and also place markers on buildings or bedrock which permit a fast recovery.

In a similar manner, a network of vertical control points has to be established. Again, it is desirable to tie it into the local datum. Usually groups of bench marks are established every 1 to 2 km along the axis of the road, again with referencing in the surrounding area. The local datum should be used in all design phases so that the same elevations can be used for profiling and staking. Usually the planimetric control points are also tied into the vertical network.

14.4.2 Surveys for Staking and Monitoring

Since the final design provides horizontal coordinates and sea level elevations for all points, several possible methods are available for staking. The most important tool is an efficient mini-computer with which the point coordinates can be transformed onto reference lines selected on site or else decomposed to direction and distance for polar staking.

It is not necessary to know the principal points at the time of the terrain survey. Thus, topographic survey, design, and staking can be handled independently. The common link for all three phases is the common coordinate and height system.

During construction, the surveyor has to perform a number of tasks other than staking the designed route. Especially for large structures, like bridges and tunnels, a number of movements may occur, such as settling, shifts in position and height, deformation of the structure and of the surrounding terrain. It is very important to register these in time for corrective measures. Thus, during and after construction, these structures have to be monitored (sections 14.5.1 and 14.6).

14.4.3 Measuring and Numerical Accuracy for Traffic Routes

There are no general accuracy specifications. When investigating such a project, map scale and accuracy depend on the planning phase for which the map is needed.

In the first phase of a route design project, contours are only required if the vertical alignment is critical. For all other cases, orthophotos with a few spot heights are sufficient. The choice of an adequate accuracy is economically important, as increasing accuracy requirements mean sharply increasing surveying costs.

Different considerations apply to the staking phase. In this phase the standard deviation is not usually a suitable criterion, as outliers of three times the amount have to be expected. Thus the civil engineer uses tolerances which indicate the maximum permissible errors, the absolute limit for the surveyor. It is, therefore, very important to know how much of a tolerance can be accepted by the surveyor. Often the civil engineer likes to claim the whole tolerance for his purposes (see also section 14.5.4, which deals with structures, where this is even more critical). As a rule, the relative accuracy between staked project points is of prime importance. The highest requirements in highway design are placed on the edges of the road, which are highly visible and should not deviate by more than 2 cm from the design.

A high relative accuracy is best achieved by a high absolute accuracy. Thus, the coordinates of the staked principal points should have standard deviations of 1 to 2 cm with respect to the overlaying coordinate system. This accuracy is also needed when staking structures like bridges, which often are built prior to the actual highway.

Again a high relative accuracy is required for elevations. The requirements are even higher than for the positions. A good absolute accuracy is again the simplest way to achieve this. Especially, in view of possible settling, a standard deviation of $< \pm 3$ mm per km should be obtained for double levelling.

These points apply with respective modifications to other routes, such as railroads, where they have to be more stringent at times, or for water courses, where the planimetric accuracy can be relaxed.

14.5 Staking of Structures

14.5.1 General Considerations

The involvement of a surveyor during planning of a structure (e.g., bridge) is usually not as extensive as for traffic routes. However, its transfer to the field, i.e. staking, is much more critical. In addition, monitoring of the structure is required during and after construction.

The monitoring deals primarily with settlements and deformations within the structure as well as with its influence on the surrounding area. Since the stability of the materials is utilized to its fullest extent, continuous checks under loading have to be taken to ensure that critical limits are not exceeded. Such observations are not only necessary for the safety of the structure but also provide important input for physicists, and structural and soils engineers to refine their theories.

Monitoring often continues after completion of the construction work. This is especially true for hydro dams. An example for this is given in section 14.6, while in this section the emphasis is placed on problems associated with layout and staking.

The basis for staking and monitoring of structures is a good geodetic network. Especially for smaller projects, a modern horizontal network together with a levelling net is usually sufficient. For higher demands, e.g., the construction of a longer tunnel, this is generally not suitable. A horizontal network has often been established over a longer period of time with subsequent densifications, which provide sufficient neighbouring accuracy but leave discrepancies which are significant over large distances. In such cases, a special local construction network is established using triangulation, trilateration, and precision traversing. This network is simultaneously and rigorously adjusted and subsequently fitted into the provincial or state network with the aid of a Helmert transformation (section 5.2.1.2). Even if the distances are measured with modern short range EDM equipment (section 4.4), a test line is to be established. It is used to compare the various EDM instruments with each other as well as with other distance measuring tools.

If several companies are involved in the construction, their instruments have to be carefully calibrated, because unfavourable circumstances may cause a summation of small errors which then might exceed the given tolerances. Different firms may base their values on different reference temperatures which has to be taken into account.

Although a modern levelling net is usually sufficient, it is advisable to observe zenith angles at the horizontal points in mountainous terrain in order to detect possible short-comings of the vertical network.

Important control points have to be properly monumented and referenced beyond the expected area of influence. Bench marks have to extend below the frost line if possible.

Secondary control points can be marked with wooden pegs. The centre is usually marked with a nail and protected by brightly painted lathes.

Finally, all control points have to be coordinated. If the structure is part of a larger project, e.g., a bridge or a tunnel as part of a route, the provincial or state system is most suitable.

To have an overview of the various surveying tasks, the surveying work should be properly designed, planned, and executed.

14.5.2 Bridge Staking

The surveying work for bridge construction of medium length (a few hundred metres to 1 km) is usually divided into 3 main segments:

1) Basic surveys, i.e. establishing a horizontal and vertical network, and staking of the bridge axis.
2) Construction surveys:
   a) Alignment of the piers into the bridge axis and measuring of tilt and alignment of sunken foundations
   b) Staking and marking of the bridge axis, of pier axes, and of the elevation of the piers. For suspension bridges this includes:
      - measurements for the construction and erection of pylons;
      - determination of cable lengths.

3) Structural monitoring:

a) Monitoring of the piers for settlement and tilting.

b) Measurement of creeping and similar horizontal movements.

A few remarks should illustrate these points:

1) Basic surveys: Fig. 14.5.1 schematically shows a horizontal geodetic network established symmetrical to the bridge axis, which coincides with one of its sides. Since during construction the bridge axis is often obstructed, parallel side axes at 50 to 100 m distance are established (not shown in the figure).

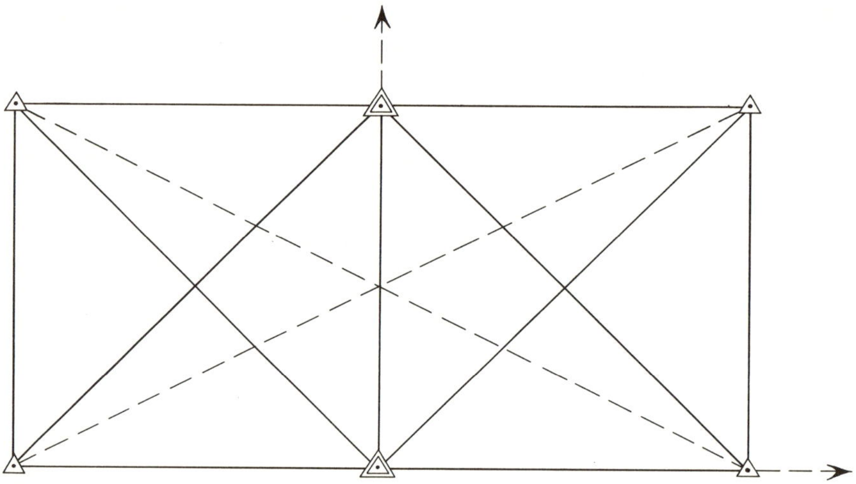

Fig. 14.5.1 Schematic network for bridge staking

The following observation methods are applied: Angular measurements with a precision theodolite; distance measurements with a short range EDM instrument (section 4.4.7.1); determination of elevation difference between abutments by precision levelling utilizing a nearby bridge, or by river crossing techniques (section 9.2.5.1), or by hydrostatic levelling.

For the basic surveys for a bridge, accuracies of $\pm 1''$ for angles, ± 1 to 2 mm for distances, and ± 0.1 mm for elevations should be achieved. If all measurements and computations are independently checked, the residuals may be so small for small bridges that a least-squares adjustment is not really necessary. Even then it is recommended, as it provides the best protection from hidden errors.

For large bridges, a significantly higher effort is required, namely a combined angle and distances net with proper consideration of refraction conditions during observation. For the elevation transfer, it is advisable to select two different methods, e.g., geometric and trigonometric levelling or hydrostatic levelling.

Finally, the observations have to be subjected to a rigid least-squares adjustment as well as a thorough statistical evaluation. Without it, it is impossible to guarantee the required accuracy.

2) Construction surveys: The piers can be aligned using distances; it is, however, more economical to use the classical intersection approach. For height monitoring, plugs are placed at every 2.5 m elevation. With the aid of the marks defining bridge axis and pier axes, as well as with the bench marks on the piers, the superstructure can be controlled in planimetry and height. The position of pylons can be determined by theodolite plumbing from shore pillars or with trigonometric means for the piers.

3) Structural monitoring: Observations to detect settlements or tilting are regularly carried out during construction. The results are usually plotted. Tests under loads are also performed. Measurements to detect creeping and other lateral movements are designed in cooperation with the civil engineer, and are also graphically evaluated.

Completed bridges may suffer from movements of the foundation caused by terrain movements, which lead to deformations, even for small bridges. Directional measurements to characteristic bridge points using forced centring are well suited to monitor this aspect.

14.5.3 Tunnel Staking

Surveying work in connection with the construction of a tunnel can be arranged systematically in a manner similar to that stated in section 14.5.2 for bridge construction. A formal surveying project is to be set up to cover such an extensive undertaking. To avoid repetitions, only a summary of the main points pertaining to tunnel work is given.

The surveying work for tunnels in mountainous areas are of special interest, because the work proceedes nearly always from both ends.

Unlike other engineering surveys, there is no question that in spite of very careful measurements, the necessary accuracy in position and height can only be guaranteed after a rigid least-squares adjustment. Ever since the staking of the St. Gotthardt Tunnel by C. Kappe in the 1870s, this requirement has been widely accepted, especially for long or curved tunnels.

Variances and covariances are computed for the critical points in order to judge the accuracy of the tunnel direction as staked at both ends. The surveyor is then in a positon to estimate the discrepancy which can be expected in the middle. This difference, however, is the prime indicator of the reliability of the surveying work.

A three-dimensional control net is necessary in order to connect the axis points at each portal. The design of this net depends largely on the terrain conditions. A combination of triangulation and traversing is most suitable, with the scale transfer being based primarily on the traverse measured with short range EDM, while the triangulation facilitates the directional transfer.

A rigorous preanalysis is used to decide whether a closed traverse is sufficient for a short tunnel.

After an optimal net configuration has been selected for the given circumstances, the measurements are carried out. In addition to the geometry, physical conditions (refraction, deflection of the vertical, etc.) have to be considered.

After determining the axis points near both portals, they are monumented, usually with concrete pillars. As with elevation, reference targets are placed at proper distances, which permit precision orientation of the theodolite set up at the observation pillars. The actual staking is usually based on an open traverse with forced centring measured with precision theodolite and EDM equipment. Based on the traverse (for curves, usually a chord traverse is selected), staking plans are prepared which contain distance and direction for the points to be staked. For this purpose it is necessary to apply the proper corrections, as the design is based on the geoid or a computational surface as its approximation, while the actual work takes place at a different level. For staking values, the usual reductions have to be applied with opposite sign.

Since modern tunnel construction progresses by several decametres per day, directional surveys are usually carried out by construction engineers, often utilizing laser equipment. It is, however, not uncommon for larger discrepancies to occur when proceeding in this manner. It is therefore necessary for an experienced surveying engineer to perform a principal control survey after every few hundred metres of progress. This commences with a check of the position of the axis points at both portals using the control net. From these, a precision traverse with forced centring is observed along the staked axis. This is done at slack times during the construction. The traverse legs should be long and not closer than a half metre from the tunnel walls because of disturbing side refraction.

In addition to the problems common to surveys at construction sites, these observations are subjected to some unique difficulties. The transport of equipment is cumbersome. Voice communication between observer and assistants inside the tunnel is difficult if not impossible because of the specific acoustic circumstances inside a tunnel, and radio instruments fail, just as do electro-magnetic distance meters. Electro-optical distance meters do function, but not as smoothly as in open terrain. Direction measurements are strongly influenced by refraction within the tunnel. Finally, one has to expect changes in the point marks as well. All these difficulties have to be overcome, because possible deviations of the tunnel from the design can only be detected from the results of the principal control survey. If it was carried out in time, these deviations can be corrected for without much trouble. The culmination of the tunnel staking work comes when both crews meet. After that, a complete control survey of the entire tunnel is carried out to obtain a final connection to the exterior net. Such a complete control survey is of no interest to the construction personnel, and is often considered an unnecessary disturbance. For a surveying engineer, however, it is essential. All assumptions on which the survey work was based contained an undetermined influence of deflection of the vertical, as curvature of the plumb line and of refraction. All of these were modelled according to theories using unknown or estimated characteristics of the surrounding terrain. A comparison between an estimated model and actually achieved values is therefore an indispensible tool for improving these assumptions and for obtaining an even higher accuracy for future projects.

14.5.4 Staking Accuracy for Engineering Projects

There is no unique answer as far as staking accuracy for larger engineering structures is concerned. Accuracies and tolerances are presently being investigated by various international bodies. Starting values for any such studies are obviously the tolerances acceptable to the civil engineer.

Discussions between civil and surveying engineers are often hampered when both sides use different measures for accuracy. While the surveying engineer normally utilizes the standard deviation, general measuring techniques consider deviations as errors. Often these are ignored, if they fall between the maximum and minimum permissible values. Some measures used in engineering are as follows:

Nominal value: value representing size and position of a building element as entered in the site plans.

Actual value: value obtained by measuring the particular building elements.

Limit value: allowable extreme value for the actual value.

Maximum value: the upper limit value.

Minimum value: the lower limit value.

Tolerance: difference between maximum and minimum values.

Discrepancy value: difference between actual and nominal values.

Based on this, the following accuracies are required:

	for 4 m	for 100 m
Limit value	±3 mm	±40 mm
Standard deviation	±0.6 mm	±8 mm

These values represent a high measuring effort, especially when the often turbulent conditions of a construction site are considered.

Some actual surveying accuracies were given for the 2.65 km long tunnel under the Elbe River near Hamburg. In this case, a geodetic determination was only possible once, and differences in staking cannot be repaired or eliminated without significant additional costs.

The following tolerances were provided by the civil engineers for the tunnel elements of 132 m length, 41.7 m width, and 8.4 m height which had to be sunk:

1) – Separation of air vents (total distance: 1057.05 m)	±5 cm
– Direction of tunnel axis	±2 cm
2) Local (secondary) control net	
– position	±0.8 cm/100 m
– elevation	±0.2 cm
3) Cross sectional tolerance	
– wall	+4 cm
– ceiling	+3 cm
– floor	+7 cm
4) Fittings at the ends	
– tilt, plane, perpendicularity	<0.4 cm

In spite of the fact that Hamburg has a modern urban control net in Gauss-Krueger coordinates (section 5.3.4), a special precision control network had to be established and measured with precision theodolites and EDM equipment.

A local secondary control network for the tunnel elements was derived from this primary precision net. The urban bench mark net consisting of 330 points in the area was sufficient for vertical control.

Using these as reference systems, the required building tolerances were checked after completion of the tunnel. There were no deviations from the staking values larger than 4 mm per 100 m.

When comparing these accuracy requirements obtained by rather different means, a reasonably good agreement is obtained. Thus, accuracy requirements for surveying as derived from building tolerances can be used for guidance, especially for large structures.

14.6 Monitoring of Power Dams

All engineering structures, especially large ones, have to be checked after completion for years or decades by the civil engineer as well as the surveying engineer. The latter checks for positional or height changes relative to the surrounding terrain, which may be cause for corrective measures. Such monitoring is done initially at short intervals, and later at more extended periods. In response to a number of catastrophies, power dams have to be carefully monitored. In fact, in some countries this has been legislated. Thus the monitoring of a power dam is given here as an example to represent the surveying monitoring of engineering structures.

14.6.1 Physical and Geodetic Methods

Both the number and the size of power dams have significantly increased with the growing need to harness water for consumption as electrical energy. Several failures in recent years have drawn public attention to the monitoring of these dams. This is especially true in the Alpine countries, where a large number of immense power dams are concentrated in a relatively small area. Switzerland has the highest density and thus has contributed much to the methodology of power dam monitoring.

Due to the change in water levels behind the dam as well as temperature changes, all dams continuously are subject to horizontal and vertical movements. For most dams, the influence of the water level is larger than the temperature influence. Usually down stream movements in the summer time due to water accumulation are of the same amount as upstream movements in the winter when the water level drops. These movements amount to a few centimetres only. Swiss experiences, however, show that a dam of 150 m width and 120 m height suffers elastic deformation of 6 to 10 cm in the centre with smaller amounts towards the edges. Usually even the surrounding bedrock is subjected to elastic deformations due to loading. Thus only points beyond the influence region of the pressure zone (50 m or more distant) can be considered as fixed.

All these experience values are average values which provide a certain guidance but may be different for different projects. Abnormal changes on dams are much more dangerous. In order to recognize these as early as possible, it is necessary to monitor them at regular time intervals. The movements of a dam determined by precise measurements are divided into relative movement, i.e. between neighbouring building elements, and absolute ones, i.e. with respect to a previously established and coordinated initial position.

Physical and geodetic methods can be used to study deformations. The physical methods utilize observations of free hanging or horizontal pendula, frequency shifts of an oscillating string, changes in electric resistance due to extension of a conductor, and others. These methods are primarily used to detect relative movements.

Geodetic methods are primarily used to detect absolute movements. Among these are precision levelling and alignment. In the 1920s, trigonometric methods became popular, and more recently precision traverses have shown great potential.

14.6.2 Geodetic Monitoring Method

Precision levelling is performed along the crest as well as in corridors inside the dam. Automatic levels with micrometer and invar rods are used in order to obtain a standard deviation of < 1 mm/km. Furthermore, levelling along the foot of the wall prior to pouring the foundation and at regular intervals during the construction are very useful in determining the behaviour of the terrain under the weight of the dam. When the head pond is being filled, such levelling will also provide values on the effect of water pressure on the dam and surrounding area.

Alignment methods are used to determine deformations of the dam when raising the water level. Two pillars are placed to the side of the dam such that they are outside of the influence sphere and that a line connecting the two will be tangential to the top of the dam. One pillar carries a tribrack for the alignment telescope; the other a target. One or more movable targets on top of the dam permit direct reading of the deformation. Some instrument manufacturers offer "power dam equipment", consisting of a combination of a high powered telescope with specially designed targets. The method is simple enough to be applied by on-site personnel.

The *trigonometric method* is based on two to three observation stations which should be as close as possible to the dam for accuracy and geometric reasons, yet must be located beyond the influence sphere of the head pond. Since both requirements cannot always be met, geological conditions have to be considered because of possible movements. It is thus essential to tie these points into a trigonometric network which is farther removed. Orientation directions to fixed points in bedrock are used on both sides of the dam. All pillars have to be suitable for forces centring and should be protected from direct sun light. A network of 6 to 10 points is used to reference the observation stations, with all possible directions being observed. Fig. 14.6.1 shows such a network for a small power dam. All observations are rigorously adjusted by least squares.

A series of targets on the dam are measured by intersection from the observation stations with the aid of a precision theodolite. Additional targets are placed in bed-

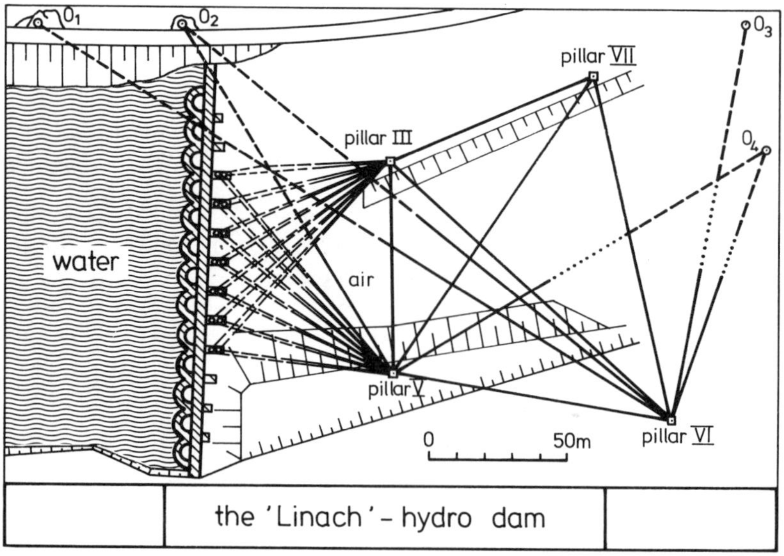

Fig. 14.6.1 Monitoring network for the Linach hydro dam

rock to aid in preorientation of subsequently observed sets of directions and for checking the reliability of the stations. If there are any doubts, the positions of the observation stations have to be checked or even redetermined from the reference network. The sights to distant targets have to be used for orientation purposes.

The elevations are determined by trigonometric means as well, using precison theodolites with automatic index. In order to determine the elevation of a target at 50 to 100 m distance to about 0.1 mm, an angular accuracy of 0.1 mgon (3″) is necessary. Because of the rather disturbing refraction conditions in narrow valleys, this cannot usually be reached. The elevations are also adjusted rigorously by least squares.

Precision traversing requires the least effort. Traverses are extended beyond the dam onto bedrock. They are usually measured in inside corridors which have pillars with forced centring tribracks. The traverses have a low curvature with nearly equal sides (20–40 m). The observations are not influenced by the weather, thus ideal conditions are available. Precision theodolites and invar wires or tapes, or the Mekometer III, provide a high neighbouring accuracy.

The absolute accuracy is lower, because the traverses have to be tied into trigonometric points on bedrock. This reduces the originally favourable circumstances. An accuracy loss by a factor 4 to 6 may occur. Therefore precision traversing is usually utilized for relative monitoring only.

The use of *floating plumb bobs* may provide a solution for this dilemma. They are rigid and located in shafts which extend for several metres into the bedrock below the dam. The bottom element of such a plumb bob can be tilted, while the upper third consists of a floating device which can move freely in a water basin. When properly manufactured, they will move directly into the physical plumb direction.

It is apparent that in this manner the foot of the floating plumb, which is anchored in the bedrock, can be plumbed into the control tunnels with a very high accuracy. These floating plumbs serve as reference points for the mentioned precision traverses which are then fitted in between two plumb points similar to mining traverses. Positional changes of the traverse pillars determined in this way will now practically represent absolute shifts. A connection to the reference network to secure the observation stations below the dam – usually by trigonometric means – is still recommended in all cases.

14.6.3 Computations and Presentation of the Results

Based on the investigations carried out at the Institute for Geodesy and Photogrammetry at the ETH in Zurich, a combination of precision trigonometric and traversing observations with subsequent least-squares adjustment has been accepted as the most suitable approach for monitoring power dams. By comparing the coordinates of various epochs, the most probable values of the deformation vectors are obtained. This method which has been fully accepted for larger projects, should, however, be used for shorter dams as well. In the latter case, horizontal and vertical adjustments can be separated thereby somewhat reducing the effort. Although a three-dimensional adjustment provides higher reliability, experience shows that nearly identical values for horizontal and vertical shifts are obtained during separate adjustments, thus making a combined adjustment uneconomical.

The same applies basically to including traverse observations in the adjustment. Because of the small redundancy in traverses, the expectations will not justify the inclusion of traverse observations into a combined adjustment.

In order to determine the accuracy of the results, it is best to compute the variance matrices of the adjusted unknowns from which standard deviations and confidence ellipses can be described if needed. All results, however, should be subjected to statistical tests in order to gain more insight into the deformations of power dams, which will be useful for future projects.

Finally, the results of dam monitoring surveys have to be presented numerically and/or graphically. Usually the movements of pillars and wall monuments on the dam are given separately in tabular form. Graphical displays usually are based on a large scale plan into which the horizontal movements are plotted in an exaggerated scale. Relative elevation changes can be presented in a similar manner.

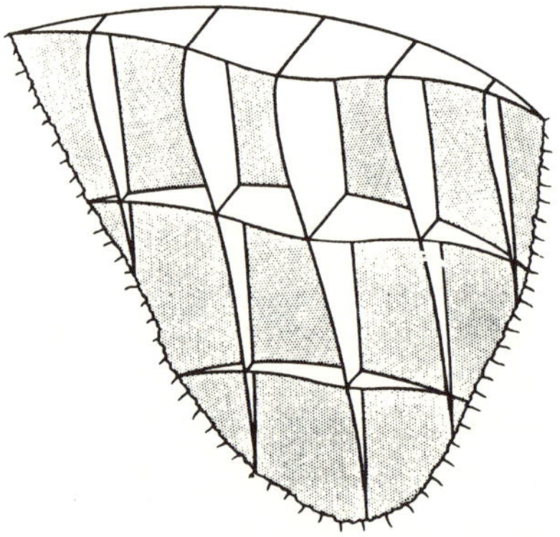

Fig. 14.6.2 Axonometric presentation of dam deformation

Relative deformations are then presented in horizontal or vertical sections. The most popular representation in an axonometric illustration combined with numerical tables, as illustrated in Fig. 14.6.2.

References

Aeschlimann, H. (1975). Eine verbesserte Zwangszentrierung. Vermessungswesen, Photogrammetrie, Kulturtechnik, Schweiz.

Aeschlimann, H. (1979). Messen der Kippachsneigung mit dem Kern DKM 2–AM, Vermessungstechnik, Schweiz.

Angus-Leppan, P. V. (1979). Refraction in levelling – its variation with ground slope and meteorological conditions. Australian Journal for Geodesy, Photogrammetry & Surveying, 31, pp. 27–41.

Angus-Leppan, P. V. (1983). Preliminary study for a new levelling system. Australian Journal for Geodesy, Photogrammetry & Surveying, 39, pp. 69–81.

Anschütz u. Co, Kiel (1915). Kreiselkompaß für Vermessungszwecke. Deutsches Reichs-Patent (DRP) Nr. 322144.

Augath, W. (1976). Untersuchungen zum Aufbau geodätischer Lagenetze. Wissenschaftliche Arbeiten der Lehrstühle für Geodäsie, Photogrammetrie und Kartographie an der Universität Hannover, 72.

Babbage, G. (1977). Operation of the inertial survey system: the manager's view. Proceedings, 1 st ISS-Symposium, Ottawa.

Barr and Carriere (1981). Accuracy criteria for inertial surveys. Proceedings, FIG XVI. Int. Congress, Montreaux 512.2.

Barrel, H. and J. E. Sears (1939). The refraction and dispersion of air for the visible spectrum. Phil. Trans. Royal Soc., London, A.

Becker, J. M. (1985). The Swedish experience with motorized levelling – New techniques and tests. Third International Symposium on the North American Vertical Datum, Rockville, Maryland, U.S.A.

Bergstrand, E. (1943). Measurement of distances by high frequency light signalling. Arkiv for Mathematik, Astronomy och Fysik, 29 A.

Beutler, G., D. A. Davidson, R. B. Langley, R. Santerre, P. Vaníček and D. E. Wells (1984). Some theoretical and practical aspects of geodetic positioning using carrier phase difference observations of GPS Satellites. Department of Surveying Engineering Technical Report 109, University of New Brunswick, Frederiction, Canada.

Black, H. D. (1978). An easily implemented algorithm for the tropospheric range correction. Journ. Geophys. Res., 83 (B 4).

Bullock, M. C. and R. E. Warren (1976). Electronic Total Station speeds survey operations. Hewlett-Packard-Journal, 27.

Chrzanowski, A., T. Greening, W. Kornacki, J. Secord, S. Vamosi and Y. Q. Chen (1985). Applications and limitations of precise trigonometric height traversing. Third International Symposium on the North American Vertical Datum, Rockville, Maryland, U.S.A.

Cooper, M. A. R. (1982). Modern theodolites and levels. Granada Publishing Limited, London, Toronto, Sydney, New York.

Deumlich, F. (1982). Surveying instruments. Walter de Gruyter, Berlin, New York.

Drodofsky, M. (1960). Stromübergangsnivellement mit dem Zeiss-Nivellier Ni 2. Zeitschrift für Vermessungswesen, 85.

Drodofsky, M. (1964). Stromübergangsnivellement mit dem Zeiss-Nivellier Ni 2. Zeitschrift für Vermessungswesen, 89.

Essen, L. and K. D. Froome (1951). The refraction indices and dielectric constants of air and its principal constituents at 24000 Mc/s. Proc. Phys. Soc., Series B, 64.

Grabe, M. and F. Lebowsky (1979): Prozeßrechnergesteuerte Prüfung von Nivellierlatten. Physikalische-Technische Bundesanstalt, Braunschweig, Bericht 25.

Grafarend, E. (1974). Fehlertheoretisch günstigstes Vorwärtseinschneiden. Zeitschrift für Vermessungswesen, 93.

Grafarend, E., R. Heister, H. Kelm, H. Kropff and B. Schaffrin (1979). Optimierung Geodätischer Meßoperationen. Herbert Wichmann Verlag, Karlsruhe.

Gründig, L. and R. Joeckel (1983). Vergleich der hierarchischen Polygonnetzberechnung mit der strengen Ausgleichung. Allgemeine Vermessungs Nachrichten 90.

Heer, R. (1984). Application of different refraction models on measuring results of the "Leveling Test Loop Koblenz". In: Pelzer, H. and W. Niemeier (eds.) "Precise Levelling". Dümmler Verlag, Bonn.

Heimerle, A. (1929). Erdbau. Im Taschenbuch für Landvermessung und Kulturtechnik, Stuttgart.

Holdahl, S. R. (1981). A model for temperature stratification for correction of levelling refraction. Bull. Geod., 55.

Holdahl, S. R. (1983). The correction for levelling refraction and its impact on definition of the North American Vertical Datum. Surveying and Mapping, 43, No 2.

Hopfield, H. (1979). Improvements in the tropospheric refraction correction for range measurements. Proceedings of the Symposium on Satellite Doppler Tracking and its Geodetic Applications, The Royal Society of London.

Joos, D. K. (1975). Fehlermodell eines Inertialnavigationssystems für integrierte Navigation durch Ausgleichung nach der Methode der kleinsten Quadrate, Dissertation, Universität Stuttgart.

Jordan, W., O. Eggert and M. Kneißl (1956). "Höhenmessung und Tachymetrie". Band III. Metzlersche Verlagsbuchhandlung, Stuttgart.

Kahmen, H. (1978). Elektronische Meßverfahren in der Geodäsie. 2nd Edit. Herbert Wichmann Verlag, Karlsruhe.

Kakkuri, J. (1966). Versuche mit dem automatischen Doppelinstrument Zeiss Ni 2 beim Stromübergangsnivellement. Zeitschrift für Vermessungswesen, 91.

Kakkuri, J. (1969). Über den Einfluß von Temperaturveränderungen auf das bei Stromübergängen benutzte automatische Doppelinstrument Zeiss Ni 2. Zeitschrift für Vermessungswesen, 94.

Kasper, H., W. Schürba and H. Lorenz (1968). Die Klothoide als Trassierungselement. Dümmler Verlag, Bonn.

Koch, K. R. (1980). Parameterschätzung und Hypothesentests in linearen Modellen. Dümmler Verlag, Bonn.

Krack, K. (1977). Die Blockausgleichung zur Berechnung umfangreicher tachymetrischer Aufnahmen. Deutsche Geodätische Kommission, Reihe C, 231, München.

Kukkamäki, T. J. 1938. Über die nivellitische Refraktion. Veröffentlichungen des Finnischen Geodätischen Instituts, No. 25, Helsinki.

Langley, R. B., J. D. McLaughlin and D. E. Wells (1982). The potential engineering and land survey market for GPS. Conference on "Engineering Applications of Space Surveying Technology". Nashville, Tennessee.

Leitz, H. (1984). The influence of the geomagnetic field on the Zeiss Ni 1 Precision Level. In: Pelzer, H. and W. Niemeier (eds.) "Precise Levelling". Dümmler Verlag, Bonn.

Moeller, D. (1962). Beiträge zur barometrischen Höhenmessung. DGK, Reihe C.
Pelzer, H. (1984). Systematic instrumental errors in precise levelling. In: Pelzer, H. and W. Niemeier (eds.) "Precise levelling". Dümmler Verlag, Bonn.
Remondi, B. W. (1984). Using the Global Positioning System (GPS) phase observable for relative geodesy: modelling, processing, and results. Ph. D. Dissertation, Center of Space Research, The University of Texas at Austin, Austin, Texas.
Rueger, J. M. and F. K. Brunner (1981). Practical results of EDM-height traversing. The Australian Surveyor, 30, No 6.
Rueger, J. M. and F. K. Brunner (1982). EDM-height traversing versus geodetic leveling. The Canadian Surveyor, 36, No 1.
Rumpf, W. E. and H. Meurisch (1981). Systematische Änderungen der Ziellinie eines Präzisionskompensator – Nivelliers (insbesondere des Zeiss Ni 1) durch Magnetische Gleich- und Wechselfelder. International Congress of Surveyors (FIG) Commission 5, Montreux, Switzerland.
Ruopp, M. (1971). Genauigkeitsbetrachtungen zur polaren Punktbestimmung mit elektronischen Tachymetern bei freier Standpunktswahl. Allgemeine Vermessungsnachrichten 78.
Schlemmer, H. (1984). A new technic to produce precise graduations on invar tape. In: Pelzer, H. and W. Niemeier (eds.) "Precise Levelling". Dümmler Verlag, Bonn.
Schmitt, G. (1979). Zur Numerik der Gewichtsoptimierung in geodätischen Netzen. Habilitationsschrift, Techn. Universität Karlsruhe.
Schuler, M. (1922). Die theoretischen Grundlagen des Vermessungskreisels. Mitteil. aus dem Markscheidewesen 28.
Schwarz, K. P. 1980. Inertial surveying systems-experience and prognosis. The Canadian Surveyor, 34.
Schwarz, K. P. and M. Gonthier (1981). Adjustment problems in inertial positioning. Proceedings, Int. Symposium on Geodetic Network and Comp., München.
Spilker, J. J. (1978). GPS signal structure and performance characteristics. Navigation 25.
Tille, R. (1984). Einige moderne Verfahren zur Kalibrierung von Nivellierlatten. In: Schriftenreihe Wissenschaftlicher Studiengang Vermessungswesen, Hochschule der Bundeswehr München, Heft 10.
Torge, W. (1980). Geodesy. Walter de Gruyter, Berlin, New York.
Van Dierendonk, A. J., S. S. Russel, E. R. Kopiteke and M. Birnbaum (1978). The GPS navigation message. Navigation 25.
Vaníček, P. and E. Krakiwsky (1986). Geodesy: the concepts. North-Holland, Amsterdam, New York, Oxford, Tokyo.
Wadley, T. L. (1956). The Tellurometer system of distance measurement. Emp. Surv. Rev., pp. 100–111 and 146–160.
Wells, D. E., W. Lindlohr, B. Schaffrin and E. Grafarend (1985). GPS design using undifferenced carrier beat phase observations. Department of Surveying Engineering Technical Report 116, University of New Brunswick, Fredericton, Canada.
Whalen, C. T. (1985). Trigonometric motorized levelling at the National Geodetic Survey. Third International Symposium on the North American Vertical Datum, Rochville, Maryland, U.S.A.
Whalen, C. T. and W. E. Strange (1983). The 1981 Sangus to Palmdale, California, levelling refraction test. NOAA Technical Report NOS 98 NGS 27.
Wickham, J. H. (1977). Inertial surveying technology in private practice. Proceedings, 1 st ISS-Symposium, Ottawa.

Wolf, H. (1975). Ausgleichungsrechnung nach der Methode der kleinsten Quadrate. Dümmler Verlag, Bonn.

Zippelt, K. (1983). Measurement of level rod's temperature and effects on precise levelling. In: Pelzer, H. and W. Niemeier (eds.) "Precise Levelling". Dümmler Verlag, Bonn.

Author Index

Subject Index